MICROBIAL REACTION TO ENVIRONMENT

Other Publications of the
Society for General Microbiology

THE JOURNAL OF GENERAL MICROBIOLOGY
(*Cambridge University Press*)

MICROBIAL GENETICS
TENTH SYMPOSIUM OF THE SOCIETY
(*Cambridge University Press*)

VIRUS GROWTH AND VARIATION
NINTH SYMPOSIUM OF THE SOCIETY
(*Cambridge University Press*)

THE STRATEGY OF CHEMOTHERAPY
EIGHTH SYMPOSIUM OF THE SOCIETY
(*Cambridge University Press*)

MICROBIAL ECOLOGY
SEVENTH SYMPOSIUM OF THE SOCIETY
(*Cambridge University Press*)

BACTERIAL ANATOMY
SIXTH SYMPOSIUM OF THE SOCIETY
(*Cambridge University Press*)

MECHANISMS OF MICROBIAL PATHOGENICITY
FIFTH SYMPOSIUM OF THE SOCIETY
(*Cambridge University Press*)

AUTOTROPHIC MICRO-ORGANISMS
FOURTH SYMPOSIUM OF THE SOCIETY
(*Cambridge University Press*)

ADAPTATION IN MICRO-ORGANISMS
THIRD SYMPOSIUM OF THE SOCIETY
(*Cambridge University Press*)

THE NATURE OF VIRUS MULTIPLICATION
SECOND SYMPOSIUM OF THE SOCIETY
(*Cambridge University Press*)

THE NATURE OF THE BACTERIAL SURFACE
FIRST SYMPOSIUM OF THE SOCIETY
(*Blackwell's Scientific Publications Limited*)

MICROBIAL REACTION TO ENVIRONMENT

ELEVENTH SYMPOSIUM OF THE SOCIETY FOR GENERAL MICROBIOLOGY HELD AT THE ROYAL INSTITUTION, LONDON APRIL 1961

CAMBRIDGE
Published for the Society for General Microbiology
AT THE UNIVERSITY PRESS
1961

PUBLISHED BY
THE SYNDICS OF THE CAMBRIDGE UNIVERSITY PRESS

Bentley House, 200 Euston Road, London, N.W.1
American Branch: 32 East 57th Street, New York, 22, N.Y.
West African Office: P.O. Box 33, Ibadan, Nigeria

Printed in Great Britain at the University Press, Cambridge
(Brooke Crutchley, University Printer)

CONTRIBUTORS

BAWDEN, F. C., Rothamsted Experimental Station, Harpenden, Hertfordshire, England.

CANTINO, E. C., Department of Botany and Plant Pathology, Michigan State University, East Lansing, Michigan, U.S.A.

DUGUID, J. P., Bacteriology Department, University of Edinburgh, Teviot Place, Edinburgh, 8.

FISHER, K. W., Medical Research Council, Microbial Genetics Research Unit, Hammersmith Hospital, Ducane Road, London, W. 12.

GOODER, H., Streptococcus Reference Laboratory, Central Public Health Laboratory, Colindale, London, N.W. 9.

HERBERT, D., Microbiological Research Establishment, Porton, Wiltshire, England.

HUTNER, S. H., Haskins Laboratories, 305 East 43rd Street, New York 17, New York, U.S.A.

KERRIDGE, D., Medical Research Council, Chemical Microbiology Research Unit, Department of Biochemistry, Tennis Court Road, Cambridge.

LACEY, B. W., Department of Bacteriology, Westminster Medical School, London, S.W.1.

MAXTED, W. R., Streptococcus Reference Laboratory, Central Public Health Laboratory, Colindale, London, N.W. 9.

MEYNELL, G. G., Lister Institute of Preventive Medicine, Chelsea Bridge Road, London, S.W. 1.

MURRELL, W. G., C.S.I.R.O. Division of Food Preservation, Homebush, N.S.W., Australia.

PARDEE, A. B., Department of Biochemistry and Virology, University of California, Berkeley, California, U.S.A.

SCOPES, A. W., Brewing Industry Research Foundation, Nutfeld, Surrey, England.

SERRES, F. J. de, Biology Division, Oak Ridge National Laboratory, Oak Ridge, Tennessee, U.S.A.

CONTRIBUTORS

WILKINSON, J. F., Bacteriology Department, University of Edinburgh, Teviot Place, Edinburgh, 8.

WILLIAMSON, D. H., Brewing Industry Research Foundation, Nutfield, Surrey, England.

WOODRUFF, H. B., Merck, Sharp & Dohme Research Laboratories, Division of Merck & Co., Inc., Rahway, New Jersey, U.S.A.

CONTENTS

CONTENTS

EDITORS' PREFACE

The environment can alter the properties of a population of micro-organisms in two ways: either through genetic variation, following mutation or selection, or through non-genetic variation, that is, by producing changes in behaviour unaccompanied by alterations in genetic structure. The second effect forms the chief subject of this volume which contains the papers to be discussed at the Eleventh Symposium of the Society for General Microbiology.

G. G. MEYNELL
H. GOODER

Lister Institute of Preventive Medicine, S.W. 1,
and The Central Public Health Laboratory, Colindale

THE ENVIRONMENT AND GROWTH: PROTOZOAN ORIGINS OF METAZOAN RESPONSITIVITIES

S. H. HUTNER

Haskins Laboratories, 305 *East* 43*rd Street*, *New York* 17, *New York*, *U.S.A.*

Let dogma eat dogma

DON MARQUIS: *The Almost Perfect State* (1927)

As we uncover the rich microbiological stew that is this Symposium, I propose to add a dash of anthropocentric seasoning.

With the Darwin centennial in mind, we may ponder on the microbial origins of the devices by which man responds to the environment. That this responsitivity somehow rests on animality—more narrowly, metazoan animality—with like obtuseness is neglected by biologists and deplored by theologians. Worse, man talks like an *arriviste*: in Western culture 'animal' is a taunt, and the distinction of animals, phagotrophy, has a pejorative connotation, as in our opening quotation (a demonstration that phagotrophy so inescapably colours our thinking that scientific method itself is expressible in terms of phagotrophy).

Darwin, no snob, was 'ardently theriophilous', delighting in 'the excellence of every living thing' (Fitch, 1959). Thanks to this spirit surviving in protozoologists, several different kinds of protozoa are in pure culture: material is at hand for a decently comprehensive biochemistry of animals.

MEMBRANE-FORMATION AND PREY-CATCHING

Huxley (1876) defined animals as eaters. More elaborately, Haldane (1954*a*) described animals as wanderers seeking fuel and ready-made spare parts; if one adds 'characteristically ingested in chunks' one has a workable primary definition. That certain phytoflagellates which, like *Euglena*, are non-ingesters, are treated in texts on protozoa and in zoology courses has practical justification: some euglenoids are skilful animals; seemingly, protozoologists, in laying out their academic domain, have indulged in a predation rivalling that of some of their pets.

My immediate object is to consider the question: Wherein do the growth-responses of protozoa offer homologies to human responses?

Conversely, how do metazoan responses illuminate protozoan animality? Since, for predators, growth and behaviour centre on capture, digestion, and assimilation of prey, behaviour and growth in them are intimately joined, and so one may find a hormone for one organism a conventional growth factor for another. Some examples will be mentioned later.

For a unicellular predator to tell self from prey—a function of the apparatus of immunity—it must compartmentalize itself, if but to guard its deoxyribonucleic acid from insidious usurpers such as the temperate bacteriophages (Jacob, Schaeffer & Wollman, 1960). (One wonders whether the absence of phage-like viruses in protozoa, related algae, and fungi denotes a phylogenetic immunity conferred by envelopment of the nucleus in a membrane, or merely inadequate technical methods.) The quintessential specialization of Protozoa, and of phagocytic cells generally, may be synthesis, on cue, of a special membrane—the food-vacuole membrane—along with digestive enzymes and expanded excretory organelles.

Membrane-making takes work, as inferred from the heightened O_2 consumption and diminution in intracellular glycogen of leucocytes ingesting bacteria or inert particles (Becker, Munder & Fischer, 1958; Elsbach, 1959; Karnovsky & Sbarra, 1960). The latter workers noted also increased uptake of acetate and phosphate, interpreted as reflecting the rich lipid content of the membranes. In keeping, furthermore, glycolysis inhibitors stop phagocytosis in polymorphonuclear leucocytes, and use of exogenous glucose increases during phagocytosis (Cohn & Morse, 1960).

Membrane-formation is spectacular in the slime mould *Physarum polycephalum* (Stewart & Stewart, 1960). On cutting a vigorously streaming plasmodium, a globule wells out and sets firm, soon becoming a viable independent plasmodium. Also on slow desiccation 'the non-cellular, continuous sheet of fluid protoplasm divides into a brittle mass of small cell-like structures...increasing tremendously the total area of surface membrane'.

Many ciliates regenerate well. Thus fragments of *Tetrahymena pyriformis* as small as thirds may regenerate (Albach & Corliss, 1959). Since ciliates are commonly under high internal pressure, a healing membrane must form fast or the animal bursts. *T. pyriformis* is easily grown in chemically defined media, which opens the way to close analysis of conditions for regeneration.

Indiscriminate, i.e. cue-less, gluttony seems allied with prodigies of membrane-formation. The monstrous multinucleate amoeba *Pelomyxa palustris* eats everything in its path (Kudo, 1960). Feeding is by posterior viscous villi; when body movement is reversed, the new posterior end develops villi. Multiplication by other than simple division has not been observed nor have the complicated reproductive structures met in Acrasieae and slime moulds; it may be that in the large amoebae obvious structural complexity has been sacrificed to membrane-formation.

Roth (1960) traced formation of many small vesicles (0·2–0·5μ dia.) at the cytoplasmic side of the food-vacuole membrane. He concluded that micropinocytosis (ingestion of fine droplets) mediated transfer of the contents of food vacuoles into the cytoplasm and calculated that, starting with $20 \times 10^3 \mu^2$ of surface area of food vacuole, over the 720 min. required for digestion each μ^2 of original vacuole membrane duplicated its surface area every minute. To account for pinocytosis allowing passage into the cytoplasm of substances which cannot diffuse across the plasmalemma, he postulated progressive changes in the membrane. One such change might be increased acid phosphatases—a change especially striking in rat macrophages ingesting erythrocytes in Kupffer cells of liver, and in *Amoeba proteus* (Novikoff, 1960). (*A. proteus* should not be called, according to Kudo, 1959, '*Chaos chaos*'.) Kudo notes that in no large free-living amoeba has morphological change been correlated with environmental or internal conditions.

The cues for pinocytosis in *Amoeba proteus* are rather non-specific: simple cations, rabbit gamma-globulin, gelatin and glutamate (reviewed by Holter, 1959*a*, *b*). Marshall, Schumaker & Brandt (1959), generalizing from the binding of lysozyme by homogenates of *A. proteus*, suggest that pinocytosis begins with binding of proteins or other inducers to the cell surface, the essential groups of the inducer carrying a positive charge—a theory supported by the strong attachment of toluidine blue to the surface of *A. proteus* and its ability to induce pinocytosis (Rustad, 1959).

Another specialization for phagocytosis is a sticky surface or at least one binding protein. It is as disconcerting to a naïve sense of the phyletic fitness of things to see an amoeba capturing ciliates as to see a turtle catching ducklings or a spider snaring humming birds. Stickiness is in some way associated with fine fibrous extensions of the plasmalemma in *Amoeba proteus* and *Pelomyxa carolinensis* (Pappas, 1959); how the posterior villi of *Pelomyxa*, previously mentioned, help stickiness is not clear.

The relation of 'lysosomes', widely distributed bodies having single membranes and high in acid phosphatases (Novikoff, 1960) to phagocytosis and pinocytosis generally is a new subject (de Duve, 1960).

Amoeba proteus is so seductive a research object that the non-microbiologist is apt to be shocked to learn it is not in pure culture. It can be reared fairly dependably on *Tetrahymena* supplemented with bacteria and yeasts. It is easily freed of other organisms; the problem is finding a non-toxic soluble medium (Prescott, 1959). Some small soil amoeba of the *Hartmannella* type do grow in autoclaved peptone media; the minimal nutritional requirements of one include thiamine, vitamin B_{12}, and some still-unsorted amino acids (Adam, 1959); dependence on carbohydrate and utilization of acetate has been demonstrated in a similar form (Band, 1959). What induces the stage in the 'life cycle' of '*A. proteus*' marked by sporulation and the hatching out of small amoebae (reviewed by Jepps, 1956), is unknown.

GROWTH FACTORS AND MICROBIAL ECOLOGY

Cue-less gluttony may prevail in the tiny (down to 1 μ) photosynthetic chrysomonads such as those beautifully figured by Parke, Manton & Clarke (1955, 1959) who fed them colloidal graphite to observe phagocytosis. Knight-Jones (1957) supposes that speed is one of their nutritional specializations; μ-flagellates, especially abundant in warmer seas, '...dash about...squandering the energy which they can get from sunshine, their movements and relatively large surface area presumably helping them to find and mop up the sparse nutrients which they need just for growth'. This is an exceedingly successful way of life: Knight-Jones (1951), citing results by H. W. Harvey and his own, notes that 90% of the plankton passed through a 40–50 μ mesh net consisted of flagellates. Indeed, μ-flagellates may be the most abundant organisms on earth: the waters around the British Isles harbour $\sim$1000/ml.

A litre of sea water may hold a hundred different flagellates and diatoms. Predilections for different growth factors, joined with differences in efficiency of absorption of minerals and use of light, may underlie the succession of species in natural waters and the occasional wealth of forms (Provasoli, 1958*a*). Fine subdivisions of ecological niches are set up by differences in utilization of congeners of vitamin B_{12} by various micro-organisms (Droop, McLaughlin, Pintner & Provasoli, 1959). In marine littoral diatoms, variations in respect of obligate photoautotrophy and heterotrophy, superimposed on variations in auxotrophy (thiamine and cobalamins), reveal more modes of subdivisions of environments (Lewin & Lewin, 1960). Superposition of variations in ability to digest microbial bodies upon graduations in efficiency in trapping light define still more niches.

Work with non-protozoan micro-organisms and metazoa provide

some working hypotheses for sorting out the specialties of protozoa. The soil microflora is a consortium of chemical specialists. Bacteria sorted out as to growth requirements (Lochhead, 1958) are gradually being matched up with those isolated from enrichment cultures with rare substrates. That the growth of some seaweeds may depend on environmental factors also operating on protist and other groups which include prototrophic and auxotrophic members is suggested by the cobalamin requirement of the red alga *Goniotrichum elegans* (Fries, 1960). As in plant tissue cultures where hormones often serve as exogenous growth factors, the lines between hormones and growth factors are blurred in the green marine alga *Ulva lactuca*: for normal development of the thallus, a sea-water culture medum needs supplementation with indoleacetic acid, kinetin, and gibberellin (Provasoli, 1958*b*). The *Ulva* situation is complicated: sea water cannot yet be replaced by the usual synthetic substitutes, and some sea waters are inert (Provasoli, personal communication). A start has been made in identifying auxin-like substances in sea water (Bentley, 1959). If to these newly recognized specializations one adds variations in ability to assimilate microbial prey, scope for chemical specialization in protozoa is wide indeed. Man, on this scale of nutritional fastidiousness, viewed in the light of the Knight-Lwoff canon (see Snell, 1951) of loss of function in evolution, has boxed himself into an ecological corner: he needs one growth factor (vitamin B_{12}) primarily of bacterial origin (a reminder of a filter-feeding past? Berrill, 1955); the auto-oxidizable ascorbic acid, apparently absent from most microbes (except perhaps for photosynthetic bacteria); vitamin D, for which he needs light or other metazoa; and vitamin E, quite unstable, which, with ascorbic acid, restricts him to fresh foods. Inability to digest cellulose and most polysaccharides eliminates higher plants as a source of fuel. And, finally, he requires sundry other lipid, poorly diffusible growth factors. Were man a protozoon, he would be described as a virtually obligate phagotroph: demanding a great variety of food (and so, obligately omnivorous), preferably in the form of fresh bodies, not carrion, still with their antioxidants. This nutritional complexity would strike the investigator as improbable if not downright irrational—the outcome of a tortuous gastronomic history. The difficulties in defining the growth requirements of carnivorous protozoa, e.g. *Amoeba proteus*, assume in this perspective a more understandable cast. The origin of the thyroid in the mucus-secreting particle-trapping endostyle of early chordates presents a striking demonstration that a mucoid food-catching membrane may have remarkable biological potentialities.

SYMBIOSIS

Ruminants have largely escaped the biochemical and ecological trap described for man by evolving into fermentation tanks—thereby helping man to survive in *his* trap. Likewise many protozoa and invertebrate metazoa have symbionts. Cultivation of the symbiotic zooxanthellae of coelenterates led to the proof (foreshadowed by a few purely observational data) that their zooxanthellae are morphologically primitive dinoflagellates (McLaughlin & Zahl, 1959), which adds to the likelihood that this symbiosis was anciently established. The symbionts of radiolarians and other marine non-coelenterate invertebrates may include such 'primitive' flagellates as cryptomonads (McLaughlin & Zahl, 1961). Quite likely invertebrates, notably corals, have sheltered, as in a museum, an array of primitive flagellates. Work on zooxanthellid life-cycles and the environmental control of host-symbiont relations, especially as influenced by light, has barely begun (Freudenthal, 1959).

METAZOAN ORIGINS, PREDATION, AND ENZYMIC VERSATILITY

The varied compositions of bacterial cell walls (Salton, 1960) implies that even the most versatile predator cannot utilize all microbial prey. Nanney (1960) supposes that an organism carries, for economy, not all useful enzymes, but the blueprints for the enzymes in its genetic library, and has evolved switches for turning on and off enzyme production; and that persistence of environmentally induced differences in micro-organisms of the same genotype (exemplified by temperature-induced antigenic differences in *Paramecium* and induction of permeases in *Escherichia coli*) foreshadows the embryological differentiations of genotypically identical cells.

Naegleria gruberi, an amoeba, can grow a uroid (tuft of villi) at its rear when put in a 'more than usually watery' medium, and also anterior flagella which enable it to swim off (Willmer, 1960). Impressed by this apposition of amoeboid feeding and flagellate dispersal stages, and the prevalence of these cell types in the embryos of many invertebrates, Willmer thinks that the primitive metazoon might have been a symbiosis of 'saprobiotic' flagellate cells and phagocytic amoeboid cells. This theory is opposed to the theory of J. Hadži (see Hanson, 1958) deriving the metazoa from the cellularization of a large multinucleate ciliate, such as the ultra-voracious *Dileptus*.

PROTISTAN HORMONES AS PHYLOGENETIC MARKERS: STRESS IN *OCHROMONAS*

With patent disagreement about how the metazoa arose, the search for phylogenetic markers continues. A conservative behaviour pattern bespeaks a conservative metabolic pattern, and conversely (Lorenz, 1960). Resemblances in mediators of behaviour—hormones—in metazoa and protozoa might reveal uniquely metazoan-like patterns, assuming that nicely archaic protozoa are extant. Hormones controlling intake and interconversion of foodstuffs ought to be especially deep-seated in animals; in metazoa these processes are conspicuously under hormonal control. Zuckerman's (1952) idea that the adaptive reactions of an animal in response to changes in its environment are mediated through the nervous and endocrine systems, may be juxtaposed with Haldane's (1954*b*) idea that the hormones of multicellular animals originated in the intracellular signals of protozoa and later in the signals of neighbouring cells. The problem comes down to the cellular targets of hormones in metazoa and the origins of neurohumoral systems in invertebrates and protozoa. Developments in microbiology hint that matters may not be as bleak as Stettin (1959) makes out for metazoan hormones or Galston & Purves (1960)—to take a look at the metaphyta—for auxin. Thus acrasin, the aggregation hormone of *Dictyostelium*, is replaceable by the weakly active Δ^{22}-stigmasten-3β-ol, isolated from *D. discoideum* (Heftmann, Wright & Liddel, 1959), a finding not yet reconciled with the high activity of female sex hormones, e.g. oestrone sulphate is active at 0·05 μg./ml. (Wright & Anderson, 1958). This particular stigmastenol is probably active as a vitamin for *Paramecium* and the guinea-pig (Conner & van Wagtendonk, 1955).

A certain similarity in steroid relations attends the non-plasmodial *Labyrinthula* which streams, cell by cell, toward food, by an unidentified cell-to-cell signal, effective even among widely scattered cells (Aschner & Kogan, 1959). Some labyrinthulas have a steroid growth requirement. Vishniac (1957), seeking transformation products, fed labelled cholesterol (which satisfies the requirement) to *L. minuta* var. *atlantica*. The only steroid recovered was cholesterol itself.

Metazoan and metaphytan hormones presumably evolved from a common protistan source; the resemblances between indoleacetic acid and serotonin speaks in favour of this theory: zoological interest attends tracing the origins of metaphytan hormones. As tissue cultures of brown or red seaweeds—probably nearer to the metazoa than are the chlorophytes—are not yet available, we are thrown back on the growth processes and behaviour of protozoa for evidences of the protistan origins of metazoan hormones, and perhaps of metaphytan hormones also. The contributions in the present Symposium on adaptive enzymes (see contributions by Pardee, Kerridge, etc.), bearing on cellular memory, might with equal justice be considered to shed light on the primordial molecular adjustments of behaviour to the environment. It is unfortunate that knowledge of adaptive behaviour in protozoa is so scanty as compared with bacteria and viruses, except for drug resistance—but here genetic analysis is lacking because of the lack of demonstrable sexuality in the investigated pathogens.

TEMPERATURE STRESS IN FLAGELLATES AS AN APPROACH TO ENDOCRINOLOGY

As one views the many failures to demonstrate interpretable responses in micro-organisms of metazoan hormones which regulate the main metabolic pathways, one wonders how best to arrange conditions for possible demonstration of such responses—if demonstrable at all. In metazoa, hormones regulate the interconversion of fat, protein, and carbohydrate: e.g. insulin affects utilization of glucose and synthesis of long-chain fatty acids; the corticosteroids gluconeogenesis from protein; and anterior pituitary hormones can induce ketosis (Krebs, 1960). The line we are following starts with the assumption that an omnivorous predator is adept at interconverting the building blocks for fat, protein, and carbohydrate, and that these interconversions are under control of the intracellular equivalent of hormones. Then, by applying metabolic stress, such as that induced by raising the incubation temperature and so speeding metabolism, it was hoped that the need for intracellular hormones would outrun synthesis, and that these hormones or their precursors could be supplied exogenously. In essence, this is a crude attempt to study the 'adaptation energy' invoked by Selye (1959).

The initial experiments were on *Ochromonas malhamensis*, a choice dictated by its celebrated metazoan-like requirement for vitamin B_{12}, its phagotrophy, and considerations pointing to the primitiveness of chrysomonad animals. The early work showed that, as in thyroid-treated birds and mammals, the vitamin B_{12} requirement went up steeply with temperature; amino-acid antagonisms, and stimulations by purines, pyrimidines, and folic acid also appeared (Hutner *et al.* 1957). Enhanced metal requirements with heightened temperature were then recognized (Hutner *et al.* 1958*a*).

Poisoning by dinitrophenol, studied as a possible gauge of how damage to oxidative phosphorylation figures in thermal injury or in the inhibition by various drugs being considered as alternative stress-inducing agents, proved different from thermal injury, as it was annulled by phosphate + glutamate (or compounds convertible to glutamate) (Hutner *et al.* 1958*b*).

We then decided to shift to *Ochromonas danica* (Pringsheim, 1955), a voracious, deep olive-green, exceptionally large, hardy chrysomonad, not requiring vitamin B_{12} (Heinrich, 1955). Identification of its 'temperature factors' has been speeded by the remarkable promotion of growth exerted by ketone-body acids (see Krebs, 1960). As this work is in progress, newer results will be outlined as they bear on the themes of this Introduction. We had long been disturbed by the seeming indispensability of carbohydrate for *Tetrahymena* (Holz, Erwin & Davis, 1959) and for *O. malhamensis*—a dietary restriction seemingly at variance with their omnivorousness, and which implied that the basal media for temperature experiments were unphysiological. The answers came from three directions:

(1) *Microbiological*: a polymer of β-hydroxybutyric acid was a food reserve of bacilli (Macrae & Wilkinson, 1958) and Athiorhodaceae (Stanier, Doudoroff, Kunisawa & Contopoulu, 1959); also *Ochromonas malhamensis* can utilize acetone (Vishniac & Reazin, 1958).

(2) *Metazoan*: as chief substrates, ruminants use fatty acids, notably acetic, propionic, butyric, and branched-chain acids, rather than carbohydrate (Annison & Lewis, 1959); also, many individual acids of the ketone-body cycle are readily oxidized by birds and mammals (see Krebs, 1960).

(3) *Enzymological*: β-hydroxybutyrate is readily oxidized by preparations of mitochondria from many sources.

Inclusion of β-hydroxybutyrate, butyrate, or amino acids yielding ketone bodies, in media containing acetate and a 'glycogenic' compound such as lactate or glycerol, allowed a full replacement of carbohydrate for *Tetrahymena* (D. Cox, personal communication) as well as *Ochromonas danica*; as expected, monobutyrin proved favourable, especially for *O. danica*. With media amply fortified with these compounds, the limit for *O. danica* could be raised from 36° to *c.* 39·2° and of *Tetrahymena pyriformis* var. 1 from *c.* 39·5° to 41·8°; high concentrations of amino acids, e.g. glycine, serine, leucine, isoleucine, and valine and methionine were required—a suggestive parallel, we think, to the long-known intense catabolism of protein and fatty acid in mammalian stress (Wilson, Moore & Jepson, 1958; Cuthbertson, 1960).

Apparently, as the temperature is raised, one mode of substrate utilization after another comes into play, some to capacity, until, at the present upper temperature limit, an exaggerated catabolism of amino acids seems to be the source of the extra energy for growth (Frank *et al.* 1960). Dissimilation of propionic acid (a vitamin B_{12}-requiring reaction —Smith & Monty, 1959) coming in quantity from dissimilation of certain amino acids—e.g. valine, isoleucine, and methionine (see Greenberg, 1954), would account for the heightened B_{12} requirement seen in thyroid-treated higher animals and in *Ochromonas malhamensis* grown at elevated temperatures (Hutner *et al.* 1957).

Not unexpectedly, complex interactions of B vitamins, amino acids and substrates have been encountered and, at the upper limit, lipid requirements.

A comparison with the factors restoring the growth of rats treated with thyro-active materials might be instructive. Increasing the casein (Overby, Fredrickson & Frost, 1959) or lipids in the rations (Dryden, Riedel & Hartman, 1960) increased growth, but the active components in these and other supplements have not been identified. As the liver temperature of a normal rat exposed to 3° may rise abruptly 1° (Stoner, 1960), cold-stress may provide leads; nutritional information on cold-exposed rats seems wholly lacking.

Whether our results with *Ochromonas danica* mean anything for metazoa is problematical; they of course point to the molecular events that underlie environmental temperature barriers; as yet the only detailed information along these lines comes from an analysis of 'temperature' mutants. We have not yet sought seriously to push the temperature limits higher by feeding live food or using temperature-adapted inocula.

Oxyrrhis marine, a colourless marine dinoflagellate, seems to flout the idea that aerobic phagotrophs readily interconvert exogenous substrates: although able to feed on intact yeast cells, in pure culture no good substitute for acetate was found: many carbohydrates, Krebs-cycle acids, and butyric and propionic acids, also whole milk, egg yolk, and plant oils (Droop, 1959). This adds to the interest in working out the substrates for *Peranema*, which should reflect the characteristically 'acetate flagellate' habits of some of their *Euglena–Astasia* relatives.

INTERPRETATIONS AND A PROPHECY

What's the use of trying to educate ourselves?—we'll never know the ultimate secrets of life anyhow.
True, true—but let us joyously confirm our eternal ignorance through a million suggestive and stimulating contacts.

DON MARQUIS: *The Almost Perfect State* (1927).

Photosynthesis, sex, and animal origins

With protozoa so diversely made and variously adept, ancestor-hunting seems more of a needle-in-a-haystack enterprise. The hunt widens as, unexpectedly, the electron microscope confers patents of animality: the malaria organism is a phagotroph (Rudzinska & Trager, 1960), and *Trypanosoma mega* has a proper mouth (Steinert & Novikoff, 1960). The choice of experimental protozoa is narrowed by the desirability of the organism not being so highly adapted to life in nutritional deserts—fresh water or marine—that it copes poorly with culture media fortified for rapid growth, nor rendered so fastidious by the lush pickings of intracellular life as to demand almost prohibitively unstable enzymic cofactors. Provenance, however, is not much of a guide: *Ochromonas malhamensis*, for instance, which flourishes at American room temperatures in rich broths, came from cold, nutrient-poor Malham Tarn in Yorkshire (Pringsheim, 1952).

Resourcefulness is increasing in analysing the requirements of fastidious protozoa. Thus the carnivorous ciliate *Euplotes patella* can be reared on *Tetrahymena* if a guanine-antagonist such as 8-azaguanine is supplied (Lilly & Henry, 1956). (One wonders whether a high-ciliate diet would induce gout in a pig—pigs are susceptible to guanine gout: Bendich, 1955).

With the elucidation of most basic metabolic pathways, we can strive to discern common features in the responses of man and microbe to environmental change, e.g. it is now operationally useful to approach the eye-spot + flagellum + interposed transducer as 'the most elementary nervous system' (Wolken, 1956).

Desire to discern the connexions of environment, behaviour, and growth grows as old preoccupations lose urgency. The evolutionary success and venerability of voracious photosynthesizers—and laboratory domestication of some of them—makes pointless worry about the propriety of animals having photosynthesis; the problem in such protozoa is, in the formulation of Myers & Graham (1956), knowing how photosynthesis stretches time between bites. The ease with which *Euglena gracilis* can be 'cured' of its chloroplasts with streptomycin, or with heat or anti-histamines (see Greenblatt & Schiff, 1959) or ultra-violet-irradiation (Lyman, Epstein & Schiff, 1959) without hurt to its heterotrophic capacities, and the speed with which many dark-grown protists regrow chloroplasts in light, imply that the difference between photosynthetic and heterotrophic organisms need go no deeper than photosynthesis itself. Nevertheless, bits of the photosynthetic apparatus remain imbedded in us—archaic, but with new functions: e.g. the phytyl side-chains of vitamin K and ubiquinone, and the carotenoid precursors of vitamin A. The role of light in forming vitamin D seems as difficult to put into a phyletic context as the morphogenetic effects of light on fungi, which will be mentioned in this Symposium by Dr Cantino.

The origin of phagotrophy seems less mysterious thanks to news from microbial genetics. In genetic transformation in bacteria, DNA is taken into a bacterium, and in copulation in *Escherichia coli*, a chromosome from a 'male' passes into a 'female': the bacterial wall has gates. Conceivably, certain bacteria learned to keep the gate open wide and long enough to take in materials besides genetic blueprints; eventually, we may speculate, they regulated the intake of genetic and non-genetic materials by evolving recognition signals for each material along with different depots for each. They even began hunting non-genetic particles to engulf—they were animals. Engulfment of the sperm in metazoan fertilization can thus be regarded as evidence of an exceedingly conservative gate-mechanism. The theory of Bennett (1956) that membrane flow and vesiculation assist active transport in cells generally, may not be incompatible with the signal-and-compartmentation theory.

Photoreduction may occur in *Ochromonas malhamensis*; if shown to support growth it would mean that *O. malhamensis* had to some extent a bacterial-type photosynthesis and thus helps bridge the gap between protozoa and bacteria. *O. malhamensis* did grow in light under initially anaerobic conditions (its ability to evolve O_2 is low), and reduced CO_2 in the presence of isopropanol with the production of acetone (Vishniac & Reazin, 1958; for discussion of photoreduction, see Woods & Lascelles, 1954).

Microbial signals and hormones

The prophecy ventured is: as the main metabolic pathways are charted and as microbial genetics is consolidated, interest will mount in the mediation of morphogenesis and metabolism. In microbial genetics, for example, the environmental conditions for genetic interchange—a neglected problem—will be examined. We see gropings toward a coherent science of signals (*hormionics*?), comprising endocrinology (chemical communication amongst cells and tissues of a metazoon), signals (phoromones) (Karlson & Lüscher, 1959) between individuals of the same species, and intracellular signals.

Microbiology provides endless fascinating habits awaiting analysis. Only a few can be cited to put alongside those mentioned earlier. Friend-or-foe recognitions come up in great variety in protozoa: why, Pringsheim (1959) asks, do peranemas not attack each other in cultures? And, as Roth (1959) and others note, they invariably engulf euglenas front end first. In tetrahymenid ciliates, cannibalism is common and often leads to giantism, sometimes with drastic changes in the organization of the mouth (see, for example, Williams, 1960); the chemical determinants are unknown. In short, the perceptions of protozoon predators are a vast subject for research. Only in passing can it be noted that the hormones for sexual behaviour in algae and phycomycetes are just beginning to be explored, and that micro-organisms with complex life-cycles such as slime moulds are in pure culture (Rusch, 1959) and hence accessible, as are Acrasieae, to experimental analysis. The flagellates of the wood-roach enter a sexual phase when ecdysone, the moulting hormone, is administered to the host (Cleveland, 1959), as does *Opalina* after injection of testosterone in host male frogs (el Mofty & Smyth, 1960). Unfortunately, cultivation methods for these anaerobic protozoa are undeveloped. Lack of culture methods likewise impedes analysis of the action of hormones on cellular targets; pure cultures would be highly desirable in following up such observations as that diethylstilboestrol induces a lasting phagocytosis of carbon by mouse leucocytes (Ware & Nicol, 1960), and that leucocyte aggregation in mice after injection of *Escherichia coli* or lipopolysaccharide preparations is inhibited by corticoids (Meier & Ecklin, 1960). Similar situations, if found in protozoa, might be easier to analyse. Thus phagocytosis and digestion of *Thiobacillus thioparus* by *Ochromonas danica*, a biotin-requirer, can be followed by differential assay with *O. danica* and by a bacterium which responds to free, not bound, biotin (Aaronson & Baker, 1959). One envisions use of protozoa to carry out opsonic-index tests of host resistance: there ought to be instances where a bacterium can be ingested by a protozoan only if coated with antibody, and the use of leucocytes for the purpose might be looked back upon as an historical accident.

Attempts to throw chemical bridges across the gulf separating signal systems of protozoa and metazoa reveal difficulties. Working down from the vertebrates, Berg, Gorbman & Kobayaski (1959) state that no invertebrate has been shown to respond to a thyroid hormone, and Dodd (1959) could not detect posterior pituitary hormones, melanophore-stimulating hormone, or gonadotrophin in extracts of the neural complex of an ascidian. In *Tetrahymena* and *Ochromonas malhamensis*, inhibition of growth and motility was annulled by histidine, not by histamine (Sanders & Nathan, 1959): working upwards presents difficulties too. The action of other neuroeffector drugs on micro-organisms is outside the scope of this Introduction other than to say that the data are scant or uninterpretable. Quite likely recent progress in identifying insect and crustacean hormones will help bridge the endocrinological protozoan-metazoan gap.

Concealed in the welter of microbial intra- and extracellular signals are, one hopes, clues to the beginnings of the metazoan signals; at this stage it seems wise to explore as many 'systems' as possible.

This Symposium may, then, speed progress towards a future one on some such topic as *Microbial Morphogenesis: Endogenous and Exogenous Regulators*, or, more remotely, *Microbial Hormones*.

For a certain discretion in exposing these notions I acknowledge the cautionary influence of Pirie's 1st and 2nd Laws:

'Just as a cynic can assess roughly the eminence of a scientist by the length of time for which his theories are able to hold up the development of science after his death, so the value of concepts, in their own field, is measured by the amount of harm they do when it is assumed that they apply in others' (Pirie, 1951).

I am indebted to colleagues cited in text and references for magnanimity in letting me ransack their brains and notebooks.

REFERENCES

AARONSON, S. & BAKER, H. (1959). A comparative biochemical study of two species of *Ochromonas*. *J. Protozool.* **6**, 282.

ADAM, K. M. G. (1959). The growth of *Acanthamoeba* sp. in a chemically defined medium. *J. gen. Microbiol.* **21**, 519.

ALBACH, R. A. & CORLISS, J. O. (1959). Regeneration in *Tetrahymena pyriformis*. *Trans. Amer. micr. Soc.* **78**, 276.

ANNISON, E. G. & LEWIS, D. (1959). *Metabolism in the Rumen*. London: Methuen; New York: John Wiley and Son.

ASCHNER, M. & KOGAN, S. (1959). Observations on the growth of *Labyrinthula macrocystis*. *Bull. Res. Coun. Israel*, 8D, 15.

BAND, R. N. (1959). Nutritional and related biological studies on the free-living soil amoeba, *Hartmannella rhysodes*. *J. gen. Microbiol.* **21**, 80.

BECKER, H., MUNDER, G. & FISCHER, H. (1958). Über den Leukocytenstoffwechsel bei der Phagocytose. *Hoppe-Seyl. Z.* **313**, 266.

BENDICH, A. (1955). Chemistry of purines and pyrimidines. In *The Nucleic Acids*, vol. 1, p. 81. Edited by E. Chargaff and J. N. Davidson. New York: Academic Press.

BENNETT, H. S. (1956). The concepts of membrane flow and membrane vesiculation as mechanisms for active transport and ion pumping. *J. biophys. biochem. Cytol.* (Suppl.), **2**, 99.

BENTLEY, J. A. (1959). Plant hormones in marine phytoplankton, zooplankton and sea water. *Preprints 1st Int. Congr. Oceanogr., New York*, p. 910. Edited by M. Sears. Washington: Amer. Ass. Advanc. Sci.

BERG, O., GORBMAN, A. & KOBAYASHI, H. (1959). The thyroid hormones in invertebrates and lower vertebrates. In *Comparative Endocrinology*, p. 302. Edited by A. Gorbman. New York: John Wiley; London: Chapman and Hall.

BERRILL, N. J. (1955). *The Origin of Vertebrates.* Oxford: Clarendon Press.

CLEVELAND, L. R. (1959). Sex induced with ecdysone. *Proc. nat. Acad. Sci., Wash.* **45**, 747.

COHN, Z. A. & MORSE, S. I. (1960). Functional and metabolic properties of polymorphonuclear leucocytes. I. Observations on the requirements and consequences of particle ingestion. *J. exp. Med.* **111**, 667.

CONNER, R. L. & WAGTENDONK, W. J. VAN (1955). Steroid requirements of *Paramecium aurelia. J. gen. Microbiol.* **12**, 31.

CUTHBERTSON, D. P. (1960). The disturbance of protein metabolism following physical injury. In *The Biochemical Response to Injury*, p. 193. Edited by H. B. Stoner and C. J. Threlfall. Oxford: Blackwell Scientific Publications; Springfield, Illinois: C. C. Thomas.

DODD, J. M. (1959). Discussion in *Comparative Endocrinology*, p. 262. Edited by A. Gorbman. New York: John Wiley; London: Chapman and Hall.

DROOP, M. R. (1959). A note on some physical conditions for cultivating *Oxyrrhis marina. J. Mar. biol. Ass. U.K.* **38**, 599.

DROOP, M. R., MCLAUGHLIN, J. J. A., PINTNER, I. J. & PROVASOLI, L. (1959). Specificity of some protophytes toward vitamin B_{12}-like compounds. *Preprints 1st Int. Congr. Oceanogr., New York*, p. 916. Edited by M. Sears. Washington: Amer. Ass. Advanc. Sci.

DRYDEN, L. P., RIEDEL, G. H. & HARTMAN, A. M. (1960). Unidentified nutrients required by the hyperthyroid rat. *J. Nutr.* **70**, 547.

DE DUVE, C. (1960). La localisation des enzymes dans les éléments figurés des cellules. *Bull. Soc. Chim. biol., Paris*, **42**, 11.

ELSBACH, P. (1959). Composition and synthesis of lipides in resting and phagocytizing leukocytes. *J. exp. Med.* **110**, 969.

FITCH, R. E. (1959). Charles Darwin: science and the saintly sentiments. *Columbia Univ. Forum*, **2**, 7.

FRANK, O., HUTNER, S. H., BAKER, H., COX, D., PACKER, E., SIEGEL, S., AARONSON, S. & AMSTERDAM, D. (1960). Sugar- and glycerol-free media for *Ochromonas danica. J. Protozool.* **7** (Suppl.), 13.

FREUDENTHAL, H. (1959). Doctoral dissertation, New York University; submitted for publication.

FRIES, L. (1960). The influence of different B_{12} analogues on the growth of *Goniotrichum elegans* (Chauv.). *Physiol. Plant.* **13**, 264.

GALSTON, A. W. & PURVES, W. K. (1960). The mechanism of action of auxin. *Annu. Rev. Pl. Physiol.* **11**, 239.

GREENBERG, D. M. (1954). Carbon catabolism of amino acids. In *Chemical Pathways of Metabolism*, vol. II, p. 47. Edited by D. M. Greenberg. New York: Academic Press.

GREENBLATT, C. L. & SCHIFF, J. A. (1959). A pheophytin-like pigment in dark-adapted *Euglena gracilis. J. Protozool.* **6**, 23.

HALDANE, J. B. S. (1954*a*). The origins of life. *New Biol.* **16**, 12.

HALDANE, J. B. S. (1954*b*). La signalisation animale. *Ann. biol., Paris*, **30**, 89.

HANSON, E. D. (1958). On the origin of the Eumetazoa. *Syst. Zool.* **7**, 16.
HEFTMANN, E., WRIGHT, B. E. & LIDDEL, G. U. (1959). Identification of a sterol with acrasin activity in a slime mold. *J. Amer. chem. Soc.* **81**, 6525.
HEINRICH, H. C. (1955). Der B-Vitamin-Bedarf der Chrysophyceen *Ochromonas danica* nom. provis. Pringsheim und '*Ochromonas malhamensis*' Pringsheim. *Naturwissenschaften*, **14**, 1.
HOLTER, H. (1959*a*). Pinocytosis. *Int. Rev. Cytol.* **8**, 481.
HOLTER, H. (1959*b*). Problems of pinocytosis, with special regard to amoebae. *Ann. N.Y. Acad. Sci.* **78**, 524.
HOLZ, G. G., Jr., ERWIN, J. A. & DAVIS, R. J. (1959). Some physiological characteristics of the mating types and varieties of *Tetrahymena pyriformis*. *J. Protozool.* **6**, 149.
HUTNER, S. H., AARONSON, S., NATHAN, H. A., BAKER, H., SCHER, S. & CURY, A. (1958*a*). Trace elements in microorganisms: the temperature factor approach. In *Trace Elements*, p. 47. Edited by C. A. Lamb, O. G. Bentley and J. M. Beattie. New York: Academic Press.
HUTNER, S. H., BAKER, H., AARONSON, S., NATHAN, H. A., RODRIGUEZ, E., LOCKWOOD, S., SANDERS, M. & PETERSEN, R. A. (1957). Growing *Ochromonas malhamensis* above 35° C. *J. Protozool.* **4**, 259.
HUTNER, S. H., NATHAN, H. A., AARONSON, S., BAKER, H. & SCHER, S. (1958*b*). General considerations in the use of microorganisms in screening antitumor agents. *Ann. N.Y. Acad. Sci.* **76**, 457.
HUXLEY, T. H. (1876). On the border territory between animal and vegetable kingdoms. In *Collected Essays* (1908), vol. 8, p. 162. London: Macmillan.
JACOB, F., SCHAEFFER, P. & WOLLMAN, E. L. (1960). Episomic elements in bacteria. In *Microbial Genetics. Symp. Soc. gen. Microbiol.* **10**, 67.
JEPPS, M. W. (1956). *The Protozoa, Sarcodina.* Edinburgh: Oliver and Boyd.
KARLSON, P. & LÜSCHER, M. (1959). 'Pheromones': a new term for a class of active substrates. *Nature, Lond.* **183**, 55.
KARNOVSKY, M. L. & SBARRA, A. J. (1960). Metabolic changes accompanying the ingestion of particulate matter by cells. *Amer. J. clin. Nutr.* **8**, 147.
KNIGHT-JONES, E. W. (1951). Preliminary studies of nanoplankton and ultraplankton systematics and abundance by a quantitative culture method. *J. Cons. int. Explor. Mer*, **20**, 140.
KNIGHT-JONES, E. W. (1957). Marine biology in Wales. *Inaug. Lecture, Prof. Zool., Univ. College, Swansea*, 26 pp. Oxford University Press.
KREBS, H. (1960). Biochemical aspects of ketosis. *Proc. roy. Soc. Med.* **53**, 71.
KUDO, R. R. (1959). *Pelomyxa* and related organisms. *Ann. N.Y. Acad. Sci.* **78**, 474.
KUDO, R. R. (1960). *Pelomyxa palustris* Greeff. I. Cultivation and general observations. *J. Protozool.* **4**, 154.
LEWIN, J. C. & LEWIN, R. A. (1960). Auxotrophy and heterotrophy in marine littoral diatoms. *Canad. J. Microbiol.* **6**, 127.
LILLY, D. M. & HENRY, S. M. (1956). Supplementary factors in the nutrition of *Euplotes*. *J. Protozool.* **3**, 200.
LOCHHEAD, A. G. (1958). Soil bacteria and growth-promoting substances. *Bact. Rev.* **22**, 145.
LORENZ, K. (1960). Methods of approach to the problems of behavior. *Harvey Lect.* **54**, 60.
LYMAN, H., EPSTEIN, H. T. & SCHIFF, J. (1959). Ultraviolet inactivation and photoreactivation of chloroplast development in *Euglena* without cell death. *J. Protozool.* **6**, 264.
MCLAUGHLIN, J. J. A. & ZAHL, P. A. (1959). Axenic zooxanthellae from various invertebrate hosts. *Ann. N.Y. Acad. Sci.* **77**, 55.

McLaughlin, J. J. A. & Zahl, P. A. (1961). Associations in animals. In *Biochemistry and Physiology of Algae*. Edited by R. A. Lewin. New York and London: Academic Press (in the Press).

Macrae, R. M. & Wilkinson, J. F. (1958). Poly-β-hydroxybutyrate metabolism in washed suspensions of *Bacillus cereus* and *Bacillus megaterium*. *J. gen. Microbiol.* **19**, 210.

Marshall, J. M., Jr., Schumaker, V. N. & Brandt, P. W. (1959). Pinocytosis in amoebae. *Ann. N.Y. Acad. Sci.* **78**, 515.

Meier, R. & Ecklin, B. (1960). Die Wirkung des Hydrocortisons auf die infektionsbedingte lokale Leukozytenansammlung. *Experientia*, **16**, 204.

Mofty, M. el & Smyth, J. D. (1960). Endocrine control of sexual reproduction in *Opalina ranarum* in *Rana temporaria*. *Nature, Lond.* **186**, 559.

Myers, J. & Graham, J. (1956). The role of photosynthesis in the physiology of *Ochromonas*. *J. cell. comp. Physiol.* **47**, 397.

Nanney, D. L. (1960). Microbiology, developmental genetics and evolution. *Amer. Nat.* **94**, 167.

Novikoff, A. B. (1960). Biochemical and staining reactions of cytoplasmic constituents. In *Developing Cell Systems and Their Control*. Edited by D. Rudnick. New York: Ronald Press (in the Press).

Overby, L. R., Fredrickson, R. L. & Frost, D. V. (1959). The antithyrotoxic factor in liver. III. Comparative activity of liver residue and other proteins. *J. Nutr.* **69**, 412.

Pappas, G. D. (1959). Electron microscope studies on amoebae. *Ann. N.Y. Acad. Sci.* **78**, 448.

Parke, M., Manton, I. & Clarke, B. (1955). Studies on marine flagellates. II. Three new species of *Chrysochromulina*. *J. Mar. biol. Ass. U.K.* **34**, 579.

Parke, M., Manton, I. & Clarke, B. (1959). Studies on marine flagellates. V. Morphology and microanatomy of *Chrysochromulina strobilus* sp.nov. *J. Mar. biol. Ass. U.K.* **38**, 169.

Pirie, N. W. (1951). Concepts out of context: The pied pipers of science. *Brit. J. Phil. Sci.* **2**, 269.

Prescott, D. M. (1959). Microtechniques in amoebae studies. *Ann. N.Y. Acad. Sci.* **78**, 655.

Pringsheim, E. G. (1952). On the nutrition of *Ochromonas*. *Quart. J. micr. Sci.* **93**, 71.

Pringsheim, E. (1955). Über *Ochromonas danica* n.sp. und andere Arten der Gattung. *Arch. Mikrobiol.* **23**, 181.

Pringsheim, E. G. (1959). Phagotrophic. *Handb. Pflphysiol.* **11**, 179. Berlin: Springer Verlag.

Provasoli, L. (1958*a*). Nutrition and ecology of protozoa and algae. *Annu. Rev. Microbiol.* **12**, 279.

Provasoli, L. (1958*b*). Effect of plant hormones on *Ulva*. *Biol. Bull., Wood's Hole*, **114**, 375.

Roth, L. E. (1959). An electron-microscope study of the cytology of the protozoan *Peranema trichophorum*. *J. Protozool.* **6**, 107.

Roth, L. E. (1960). Electron microscopy of pinocytosis and food vacuoles in *Pelomyxa*. *J. Protozool.* **7**, 176.

Rudzinska, M. A. & Trager, W. (1960). Phagotrophy and two new structures in the malaria parasite. *J. biophys. biochem. Cytol.* **6**, 103.

Rusch, H. P. (1959). The organization of growth processes. In *Biological Organization: Cellular and Subcellular*, p. 263. Edited by C. W. Waddington. London and New York: Pergamon Press.

Rustad, R. C. (1959). Molecular orientation at the surface of *Amoebae* during pinocytosis. *Nature, Lond.* **183**, 1058.

SALTON, M. R. J. (1960). Surface layers of the bacterial cell. In *The Bacteria*, Vol. 1, p. 97. Edited by I. C. Gunsalus and R. Y. Stanier. New York: Academic Press.

SANDERS, M. & NATHAN, H. A. (1959). Protozoa as pharmacological tools: the antihistamines. *J. gen. Microbiol.* **21**, 264.

SBARRA, A. J. & KARNOVSKY, M. L. (1959). The biochemical basis of phagocytosis. I. Metabolic changes during the ingestion of particles by polymorphonuclear leukocytes. *J. biol. Chem.* **234**, 1355.

SELYE, H. (1959). Perspectives in stress research. *Perspect. Biol. Med.* **2**, 403.

SMITH, R. M. & MONTY, K. J. (1959). Vitamins B_{12} and propionate metabolism. *Biochem. biophys. res. Commun.* **1**, 105.

SNELL, E. E. (1951). Bacterial nutrition—chemical factors. In *Bacterial Physiology*, p. 214. Edited by C. H. Werkman and P. W. Wilson. New York: Academic Press.

STANIER, R. Y., DOUDOROFF, M., KUNISAWA, R. & CONTOPOULU, R. (1959). The role of organic substrates in bacterial photosynthesis. *Proc. nat. Acad. Sci., Wash.* **45**, 1246.

STEINERT, M. & NOVIKOFF, A. B. (1960). The existence of a cytostome and the occurrence of pinocytosis in the trypanosome, *T. mega*. *J. biophys. biochem. Cytol.* (in the Press).

STETTIN, D. (1959). Hormone regulation. In *Biophysical Science—a Study Program*, p. 563. Edited by J. L. Oncley. New York: John Wiley and Sons, Inc. Also: *Rev. mod. Phys.* (1959), **31**, 563.

STEWART, B. T. & STEWART, P. A. (1960). Electron microscopical studies of plasma membrane formation in slime molds. *Norelco Reporter*, **7**, 21.

STONER, H. B. (1960). In Discussion, *The Biochemical Response to Injury*, p. 193. Edited by H. B. Stoner and C. J. Threlfall. Oxford: Blackwell Scientific Publications; Springfield, Illinois: C. C. Thomas.

VISHNIAC, H. S. (1957). The occurrence of cholesterol in *Labyrinthula*. *Biochim. biophys. Acta*, **26**, 430.

VISHNIAC, W. & REAZIN, G. H., Jr. (1958). Photoreduction in *Ochromonas malhamensis*. In *Research in Photosynthesis*, p. 239. Edited by H. Gaffron *et al.* New York and London: Interscience.

WARE, C. C. & NICOL, T. (1960). Duration of stimulation of phagocytic activity of the reticulo-endothelial system after cessation of treatment with diethylstilboestrol. *Nature, Lond.* **186**, 974.

WILLIAMS, N. E. (1960). The polymorphic life history of *Tetrahymena patula*. *J. Protozool.* **7**, 10.

WILLMER, E. N. (1960). *Cytology and Evolution*. New York and London: Academic Press.

WILSON, G. M., MOORE, F. D. & JEPSON, R. P. (1958). The metabolic disturbances following injury. In *Metabolic Disturbances in Clinical Medicine*, p. 72. Edited by G. A. Smart. London: J. and A. Churchill.

WOLKEN, J. J. (1956). A molecular morphology of *Euglena gracilis* var. *bacillaris*. *J. Protozool.* **3**, 211.

WOODS, D. D. & LASCELLES, J. (1954). The no man's land between the autotrophic and heterotrophic ways of life. In *Autotrophic Micro-organisms. Symp. Soc. gen. Microbiol.* **4**, 1.

WRIGHT, B. E. & ANDERSON, M. L. (1958). Enzyme patterns during differentiation in the slime mold. In *The Chemical Basis of Development*, p. 296. Edited by W. D. McElroy and B. Glass. Baltimore: The Johns Hopkins Press.

ZUCKERMAN, S. (1952). The influence of environmental changes on the pituitary. *Ciba Found. Colloq. Endocr.* **4**, 213.

RESPONSE OF ENZYME SYNTHESIS AND ACTIVITY TO ENVIRONMENT

A. B. PARDEE
Departments of Biochemistry and Virology, University of California, Berkeley, California, U.S.A.

THE PROBLEM OF GROWTH IN UNFAVOURABLE ENVIRONMENTS

Bacteria appear to have evolved so as to multiply as rapidly as possible. However, only a few rich environments permit the most rapid possible multiplication of which the bacteria are capable. Lacking the homeostatic mechanisms that higher organisms have developed to compensate for unfavourable aspects of their environments, bacteria have evolved metabolic tricks. These are responses of enzyme synthesis and activity which tend to bring the intracellular environment into a condition suitable for rapid growth. Responses of enzyme synthesis and activity of this type are the subject of the present article.

The fundamental postulate of the present hypothesis relating enzyme activities to environment—that enzyme activities adjust themselves to permit as rapid growth as possible—is not intended to imply a purposeful arrangement by the bacterium, but rather, a situation that exists as a result of evolution. A teleological approach is not suggested. Bacteria have developed for billions of generations in environments as competitive as those faced by any other form of life; rapid reproduction must be an important element in their survival. It is not a matter for special wonder that through evolution the bacteria have gained abilities to grow rapidly and to adjust their enzyme activities to achieve this end.

Bacteria are exposed at different times to different environments in nature. Enzymes vital for the decomposition of a nutrient or the production of a metabolite under one set of conditions are useless, or can even be detrimental for growth, under another. Therefore, as is well known, compromises are made: the bacteria produce different enzymes and the activities of these enzymes differ in each environment. Natural environments are too numerous and complex to be described, and the relative importance to any given bacterium of ability to grow in any given environment cannot be stated quantitatively. Therefore one cannot predict what enzymes will be made under each set of conditions. We will have to restrict ourselves to discussing the effects of various chemical and physical forces in well-defined environments on the enzymic make-up of the bacteria. A few examples of these effects will be presented—such as will, it is hoped, help the reader to appreciate the kind of data that lead to the stated conclusions; but no attempt will be made to give exhaustive references to the subject. Recent reviews on control of enzyme activity (Pardee, 1959) and on enzyme formation in bacteria (Pardee, 1961) summarize the author's knowledge of these areas.

The present ideas regarding the chemical mechanisms by which enzyme synthesis is regulated quantitatively and qualitatively will not be detailed here. Instead, the less fashionable approach will be used of simply describing some of the general and also some of the less usual instances of environmental effects on enzyme composition. This means of presentation lacks the over-all coherence of a mechanistic interpretation. However, if one restricts oneself to what appears to be interpretable in terms of a mechanism, one loses certain data of interest in terms of functioning of the whole organism.

SPECIFIC ENVIRONMENTAL EFFECTS ON ENZYME FORMATION

Early studies

Since the beginning of the present century, striking differences in the enzymic activities of bacteria grown in different environments were noted, and the significance of these observations for growth were well recognized. The data were reviewed by Karström (1938). Karström's own work serves as an example of these researches. He studied the ability of bacteria grown on a given sugar to ferment a variety of sugars. Fermentation of a sugar often depended on prior growth on that sugar. For example, *Escherichia coli* grown on maltose was able to ferment maltose but not lactose, and vice versa. Karström postulated that growth on a given carbon source could cause formation of a special enzyme capable of fermenting that carbon source. These enzymes were named 'adaptive enzymes'; but now they are usually called 'inducible enzymes'.

Karström found that some carbon sources such as glucose were fermented no matter what medium was used for growth of the bacteria. Enzymes involved in this sort of fermentation were named 'constitutive' to differentiate them from the inducible ones. However, even glucose fermentation is not a completely fixed property of the bacteria but varies at least fivefold depending on prior growth conditions (Stephenson &

Gale, 1937). Constitutivity appeared to be an idealized extreme response of enzyme formation to nutritional conditions.

Although these earlier observations did not permit a precise definition of constitutivity, a distinction between constitutivity and inducibility does exist. Wild-type *Escherichia coli* are inducible for amylomaltase and β-galactosidase; from these bacteria one can obtain mutants constitutive for these enzymes (Cohen-Bazire & Jolit, 1953). Mutants of *E. coli* constitutive for formation of enzymes of tryptophan synthesis (Cohen & Jacob, 1959) and aspartate transcarbamylase (Shepherdson & Pardee, 1960), and of *Bacillus cereus* for penicillinase (Pollock, 1959) have also been obtained.

The early observed striking changes in enzyme formation and the ease of investigation of the phenomenon led to numerous works and hypotheses designed to provide an explanation for induction. As early as 1938, a mass action hypothesis involving enzyme-substrate combination was proposed (Yudkin, 1938). The ingenious facts and experiments on mechanisms of enzyme induction have been reviewed extraordinarily often. Recent comprehensive reviews are those of Pollock (1959) and Halvorson (1960).

For some time it was debated whether the enzyme composition of each bacterium could change or whether selection of mutants was responsible for the changed activity of the culture. The ultimate capacity of an organism to synthesize enzymes lies in its genetic composition. Since in many of the earlier experiments on enzyme induction, selection of genetically different organisms was possible, the role of selection had to be defined. Now it is known that both the appearance of enzyme in every cell (Benzer, 1953) and also genetic selection can account for changes in activities of populations, in different cases. We will restrict ourselves to the former phenomenon.

Gale (1943) and his associates carried out numerous experiments on effects of environment on enzyme formation. Inductions by substrates were described for a dozen enzymes of *Escherichia coli*. The effect of pH on the production of these enzymes was also determined. Certain amino acid decarboxylases were formed only in acid media; their reactions tend to increase the pH, bringing it toward the optimal value for growth. Conversely, amino acid deaminases were produced only in alkaline media; these enzymes tend to decrease the pH.

Certain other enzymes that destroy toxic products of metabolism are formed in very different amounts at different pH. For instance, alcohol and formic dehydrogenases are produced in the smallest amounts at approximately neutral pH where their specific activities are greatest.

The resulting total enzyme activity was approximately constant over a wide pH range (Gale, 1943). Thus, approximately 2·5 times as much alcohol dehydrogenase is formed at pH 5 as at pH 8, but the specific activity of the enzyme is only about one-fourth as great at the former pH as at the latter. Gale (1943) suggests that by this means the bacteria maintain an enzyme activity adequate to remove the toxic metabolites. The mechanism, possibly a stimulation of enzyme formation by the toxic metabolites, remains undetermined.

Gale's work (1943) also included studies of the effects of oxygen and glucose on enzyme formation. These topics will be discussed at greater length in a later section of this article.

The effect of temperature on enzyme formation has not been extensively studied. The subject was reviewed in 1953 (Knox, 1953). The most interesting observation is that *Salmonella paratyphi* B formed adaptive tetrathionate reductase below 37°, but not at 44°. This temperature effect, unlike most, was specific because the activity of the preformed enzyme and the growth of the bacteria adapted to tetrathionate were not abolished at 44°. This effect is usually attributed to a temperature-sensitive enzyme-forming system—another way of describing the data.

Enzyme induction

Concepts and problems of adaptive enzyme formation were made more precise in a review by Monod (1947). Many instances of induced enzyme formation were listed, the role of the gene was discussed, and the importance of determining the mechanism of induction was stressed. Emphasis was placed on 'diauxie' which is defined as the preferential utilization of one carbon compound when several are available. Monod showed, for example, that when glucose and lactose are both available to a culture of *Escherichia coli*, the bacteria first grow on glucose and then, after a lag, grow on lactose. This is because glucose prevents the formation of β-galactosidase, an enzyme essential for lactose catabolism. With this article of Monod's, the mere description of adaptive enzyme phenomena and the interpretation of enzyme changes in terms of advantage to the bacteria essentially ended. Emphasis shifted away from the entire organism and toward the determination of the chemical mechanism of enzyme induction. Obviously, though, induction is also significant in adjusting growth to environment.

Foremost in the investigations of enzyme induction have been the numerous studies on β-galactosidase. One knows now that this enzyme and its induction are under the control of specific genes; that it is induced by some compounds with almost no affinity for the enzyme and is not induced by other compounds which have an affinity for the enzyme; that it is created completely from amino acids almost at once after the inducer is added and stops being made almost as soon as the inducer is removed. Conditions, defined as 'gratuitous', have been devised under which the induction of the enzyme can be studied without influencing the over-all metabolism of the bacteria. This work has been summarized in recent reviews (Cohn, 1957; Monod, 1958).

So much is known about induction of the β-galactosidase of *Escherichia coli* that it has become a sort of yardstick against which to compare data obtained with other inducible enzymes. Whether all or even most inducible enzymes behave in this way is open to question. Several other enzymes, including D-serine deaminase (Pardee & Prestidge, 1955), β-glucuronidase (Stoeber, 1957), and tryptophanase (Pardee & Prestidge, unpublished), are induced similarly. However, other enzymes, of which the penicillinase of *Bacillus cereus* is best known, show a pattern of induction that is at least superficially dissimilar (Pollock, 1959). After the inducer, penicillin, is removed, penicillinase continues to be made for many generations. The induced penicillinase-forming system seems to be stable, unlike that for β-galactosidase synthesis which appears to be unstable.

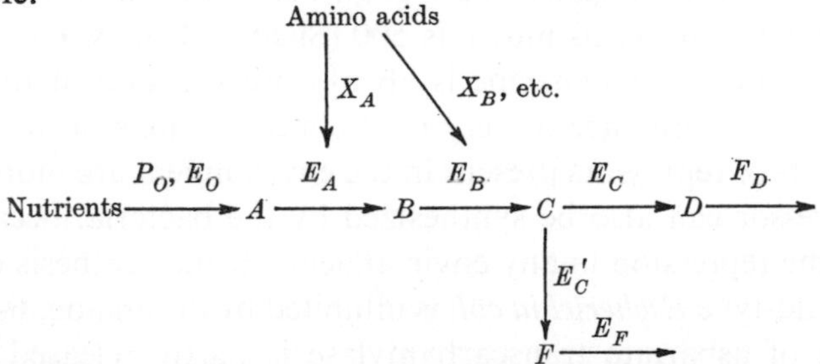

Fig. 1. Schematic representation of feedback inhibition in a metabolic pathway. P_O is a permease; $A, \ldots, F$ are metabolites; $E_O, E_A, \ldots, E_F$ are enzymes; and X_A, X_B are the 'systems' for synthesis of E_A, E_B, etc. Compound D could repress formation of enzyme E_A by interfering with X_A. It could also inhibit the activity of E_A.

Enzyme repression

Repression is a reduction of the rate of synthesis of a specific enzyme, or set of enzymes, by a nutrient or metabolite called the repressor. In numerous cases, the repressor is not the product of the enzyme-catalysed reaction but is a metabolite at the end of the metabolic pathway in which the enzyme catalyses the first step. This situation leads to a 'feedback' type of phenomenon in which an end product prevents overproduction of enzymes involved in its own synthesis (see Fig. 1).

Several enzymes are often repressed by one repressor. For example, arginine represses at least three enzymes in its pathway—acetylornithinase (Vogel, 1957), ornithine transcarbamylase and argininosuccinic synthetase (Gorini & Maas, 1958). All of a set of enzymes can be repressed in proportion, as is the case with four enzymes of histidine synthesis (Ames & Garry, 1959). Repression of three enzymes of pyrimidine synthesis is not proportional (Yates & Pardee, 1957).

Induction and repression are not different phenomena. On the basis of experiments with β-galactosidase (Pardee, Jacob & Monod, 1959) and tryptophan synthetase (Cohen & Jacob, 1959) of *Escherichia coli*, inducers appear to act by competing with repressors; the rate of enzyme synthesis depends on the fraction of some site in the bacteria, perhaps on the genetic material (Jacob, Perrin, Sanchez & Monod, 1960), not combined with repressor. Recently, direct evidence has been presented for repression by arginine and induction by ornithine of ornithine transcarbamylase (Gorini, 1960).

When the repressor can only be made from a nutrient fed to the bacteria, repression can vary over wide limits. For example, pyrimidine-requiring *Escherichia coli* mutants grown on excess uracil contain aspartate transcarbamylase at a specific activity of about 0·5 units/mg. protein. If the same organisms are grown slowly on limiting uracil, the specific activity rises to as much as 500 (Shepherdson & Pardee, 1960). Curiously, some of this enzyme is always present, even if the bacteria are grown in the presence of very high concentrations of uracil.

The effects of repressors present in the environment are more complex if the repressor can also be synthesized by the bacteria, because there will be some repression in any environment. If the synthesis of pyrimidines by wild-type *Escherichia coli* is inhibited by the analog, 6-azauracil, repression of aspartate transcarbamylase is partly released (Yates & Pardee, 1957). Or, if the bacteria are made leaky with penicillin, arginine no longer represses ornithine transcarbamylase (Rogers & Novelli, 1959; Maas, 1959). The repression of ornithine transcarbamylase is constant and only moderately strong in wild-type *E. coli* when arginine is provided to a culture growing in the chemostat, provided the arginine is supplied no faster than it can be built into proteins. With excess arginine the repression becomes much stronger (Gorini, 1958). Repression here conserves the synthetic powers of the bacteria in response to metabolites made by the cells as well as to nutrients in the environment.

Permeability and permeases

The effects on enzyme formation exerted by substances in the environment is greatly influenced by the cells' permeability. The intracellular environment is directly involved in enzyme formation; it differs from the extracellular environment because of the barriers of the bacterial wall and membrane. These exclude some compounds from the cell; for example, citrate is not metabolized because it cannot penetrate *Escherichia coli* (Stern & Ochoa, 1951; Lara & Stokes, 1952). Riboflavin cannot readily enter *E. coli* (Wilson & Pardee, unpublished). Some compounds appear to enter the bacteria by simple diffusion (Roberts *et al.* 1957); thymine probably enters *E. coli* in this way (Pardee, 1957). Still other compounds are brought into the bacteria by active transport mechanisms which require energy and which can concentrate the nutrient in the cell. A thoroughly studied instance of this sort is the accumulation of glutamate in *Staphylococcus aureus* (Gale, 1953). The amino acid can make up as much as 1% of the cells' dry weight.

These active transport mechanisms have been named 'permeases'. The galactoside permease has been studied very extensively (Cohen & Monod, 1957). Permease formation is under the control of special genes, because some mutants of *Escherichia coli* cannot concentrate galactosides but can still form β-galactosidase. This permease, in the wild-type bacteria, is induced by a variety of α- and β-galactosides and thiogalactosides. Usually the same compounds are effective for induction of the permease and the galactosidase. The induction of permeases is very much like the induction of enzymes.

Bacteria that lack a permease behave differently than do bacteria which possess it. *Escherichia coli* mutants that lack galactoside-permease cannot use lactose and are not induced to form β-galactosidase unless very high concentrations of the sugar are present. The wild-type bacteria grown on most carbon sources do not possess the permease. The latter has to be induced before it concentrates the inducer sufficiently to permit induction of β-galactosidase. For this reason, there can be a considerable lag before first the permease and then the enzymes are induced (Pardee, 1957). Once the permease is formed, the bacteria can induce β-galactosidase and grow at much lower concentrations of lactose than they could initially use (Novick & Weiner, 1957). This is because the sugar is at a much higher concentration inside these bacteria than outside them. As another example of the effects of permeases on bacterial growth, arginine represses the formation of an acetyl-ornithine permease; for this reason, among others, arginine is used preferentially to acetylornithine when both are available to *E. coli* (Vogel, 1960).

Competition between nutrients

Instances in which one nutrient interferes with the utilization of another have already been mentioned: one compound represses enzyme or permease formation and thereby blocks the uptake or conversion of a nutrient. Since the repressor is often used more economically than the compound whose use it prevents, fewer metabolic processes result. Diauxie provides an excellent example of competition between nutrients.

Diauxie by glucose has been studied most. Glucose prevents the formation of a great many enzymes, and hence the utilization of numerous compounds. The mechanism is not understood. Glucose itself does not appear to be the inhibitory material; it must be metabolized. According to one hypothesis, glucose is converted into repressors which prevent the formation of numerous enzymes (Neidhardt & Magasanik, 1957). Evidence for this view is that glucose is used more rapidly than most other carbon sources and produces appreciably larger pools of metabolic intermediates. Furthermore, a sufficient source of nitrogen is required for glucose to produce its inhibitory effect on histidase formation. Presumably the true repressor of histidase synthesis contains nitrogen and is not formed under nitrogen-deficient conditions.

Mutant organisms that utilize glucose slowly do not show this diauxie (Englesberg, 1959; Magasanik & Bojarska, 1960). Possibly such mutants do not accumulate the repressors. However, certain of these mutants, derived from *Salmonella typhimurium*, possess three times as much acid phosphatase as the wild-type. Perhaps, as an alternative to the previous hypothesis, phosphorylation of glucose in the wild-type bacteria creates a shortage of phosphate which prevents the induction of the other enzymes; diauxie is interpreted as a competition for phosphate (Englesberg, 1959). The hypothesis that a limiting nutrient is responsible for diauxie is opposed by the work of Neidhardt & Magasanik (1957) and of Cohn & Horibata (1959). In this area, much work and many hypotheses have led to little agreement.

Preferential syntheses

Under conditions where growth of an organism is blocked by lack of a nutrient, one enzyme—that which normally alleviates the deficiency—is sometimes preferentially synthesized. As examples, when *Escherichia coli* is put into a medium in which lactose must be hydrolysed to provide energy, β-galactosidase is made in large amounts; at the same time there is virtually no protein synthesis (Pardee, 1955; Rickenberg & Lester, 1955). If *E. coli* is deprived of pyrimidines, aspartate transcarbamylase is synthesized to the extent of perhaps 15% of the total protein (Yates & Pardee, 1957). Under these conditions β-galactosidase cannot be formed, but ornithine transcarbamylase can be produced (Rogers & Novelli, 1959). *E. coli* deprived of ammonia cannot produce β-galactosidase or synthesize net protein, but forms nitrate reductase and tetrathionase (Wainwright & Nevill, 1956).

The explanation for these preferential syntheses is obscure. Pre-existing proteins break down at an accelerated rate when bacteria are starved. This provides an amino acid pool, which may explain how specific enzymes are formed when there is no net protein synthesis (Mandelstam & Halvorson, 1960). However, this does not account for the preferential synthesis of one enzyme relative to others. A working hypothesis is that when the bacteria are starved for a particular nutrient, certain repressors are absent, and so the corresponding enzyme is synthesized, from the small amino acid pool produced by protein turnover, at a greater rate than the repressed enzymes (Magasanik, 1957).

Coenzymes and enzyme synthesis

Coenzymes are more necessary for synthesis of certain enzymes than for other enzymes. Evidence on this point is not extensive, but production of enzymes of *Lactobacillus* 30*a* requiring vitamin B_6 as cofactor depends to different extents on the coenzyme supply: only 1/100 the amount of vitamin B_6 is required for L-histidine decarboxylase as for L-ornithine decarboxylase (Guirard & Snell, 1954). Synthesis of the protein part of an enzyme in the absence of the coenzyme has been reported; the tyrosine decarboxylase of *Streptococcus faecalis* is produced in the absence of vitamin B_6 (Bellamy & Gunsalus, 1945). Also the apoenzymes for glucose dehydrogenation are made by *Hemophilus parainfluenzae* lacking diphosphopyridine nucleotide (Lwoff & Lwoff, 1937).

However, the apoenzyme of tryptophanase does not appear to be produced by vitamin B_6-deficient *Escherichia coli* (Pardee & Prestidge, unpublished). Pyridoxal-requiring mutants of this organism deprived of vitamin B_6 for 2·5 hr. were still capable of β-galactosidase synthesis, yet after induction by tryptophan showed only a weak tryptophanase activity when the assay medium was supplemented with pyridoxal phosphate. In any event, no marked responses such as induction or repression can be related to the coenzyme supply.

Virtually nothing is known about the effect of environment on the rates of formation of enzymes required for coenzyme synthesis. The ability of *Escherichia coli* to synthesize flavins is scarcely affected by variations in the chemical composition of the growth medium (Wilson & Pardee, unpublished).

Special nutritional requirements for enzyme synthesis

Media quite adequate for growth are not always sufficiently rich to permit synthesis of certain enzymes. The nutrients which must be added to permit enzyme synthesis do not seem to function as inducers. For example, complex media are required for production of arginine decarboxylase by *Staphylococcus aureus* (Gale, 1940) or lysine decarboxylase of *Escherichia coli* (Sher & Mallette, 1954). The requirements for arginine decarboxylase have been studied in more detail recently (Melnykovych & Snell, 1958). Five amino acids were necessary for the production of this enzyme by aerated cultures of *E. coli*. Ferric ion was also required at 3 μM/l.—three times the amount required for maximum growth in 24 hr. Iron is not required to activate an apoenzyme. Possibly these data are explained (for amino acid requirements, at least) on the basis of different avidities of enzyme-forming systems for nutrients, certain systems having such low affinity for the nutrients that the medium must be specially supplemented (Spiegelman, Halvorson & Ben-Ishai, 1955).

Environment and enzyme sequences

Entire pathways of metabolism are brought into action by environmental changes. An apparently straightforward case is known as 'sequential induction' (Stanier, 1951): a nutrient induces an enzyme and is converted by it to a metabolite which in turn induces a second enzyme. The latter enzyme changes the metabolite into another compound, which induces another enzyme and so on, until a whole sequence of enzymes are induced and an entire new pathway comes into existence.

The entire sequence of enzymes in a pathway need not be induced in order to set the pathway into action. Formation of a key enzyme can suffice for the change if the other enzymes are present. Conversely, if a key enzyme is not formed, an entire pathway may become inactive. A remarkable example of the selective utilization of three pathways by *Micrococcus denitrificans* has been examined by Kornberg, Collins & Bigley (1960). When this organism is grown in nitrate medium it contains carboxydismutase (a key enzyme of the CO_2-fixing pathway) and lacks isocitritase, an enzyme of the glyoxylate shunt. When grown on acetate, the bacteria possess the latter enzyme and lack the former. In a medium containing succinate, neither of these enzymes is formed and the bacteria obtain their carbon compounds from the Krebs cycle.

Environmental changes bring about rapid responses in the synthesis of some of these enzymes, and these shift the entire carbon-source metabolism of the bacteria.

As other examples of altered pathways brought about by environmental changes, one can cite the formation of the photosynthetic apparatus, including pigments, upon conversion from heterotrophic to photosynthetic growth of *Rhodopseudomonas spheroides* (Griffeths, Sistrom, Cohen-Bazire & Stanier, 1955), and the increases of the Krebs cycle enzymes on conversion of *Pasteurella pestis* from anaerobic to aerobic growth (Englesberg & Levy, 1955).

Biosynthetic pathways are affected similarly to catabolic ones by environmental conditions. This was first shown by Cohen & Fowler (1948) in studies on the nutritional requirements of *Escherichia coli* infected by bacteriophage T2. Since infection with this phage prevents the formation of many bacterial enzymes, the synthetic capacities of the infected cells are those imposed by the medium in which the bacteria were grown before infection. If, on the one hand, the bacteria are grown on a synthetic medium or a rich medium and infected in the same medium, each infected bacterium yields *c.* 100 phage particles. If, on the other hand, growth is in a rich medium and the bacteria are infected in a synthetic medium, the yield of viruses is much smaller. These results suggested that bacteria grown in a rich medium lack enzymes required for synthesis of some of the nutrients found in such a medium. This supposition soon was directly demonstrated by enzyme assays (see, for example, Cohn & Monod, 1953). These are early examples of repression, discussed above. Numerous demonstrations of shutting off biosynthetic pathways by repression are now known.

ENVIRONMENT AND ENZYME ACTIVITY

Stimulation of activity

Enzymic activities are affected by physical and chemical properties of the environment in well-known ways. These alterations in enzyme activity are found inside bacteria as well as with purified enzymes, and must be compensated for to provide the balanced metabolism essential for rapid growth. The environment of importance here is of course the intracellular one; the extracellular environment is quite different because of permeability factors which exclude some nutrients and concentrate others, and because biosynthetic mechanisms produce metabolites inside the cell which are virtually absent outside it.

The most obvious way by which environment affects enzyme activity is by the degree to which it saturates the enzymes with substrates and cofactors. Extremely low concentrations of substrates in the medium can saturate intracellular enzymes because permeases can create high concentrations of these substrates inside the bacteria. Probably this is why amino acids, which are concentrated inside the cell (Britten, Roberts & French, 1955; Cohen & Rickenberg, 1956), saturate the protein-synthesizing apparatus at extracellular concentrations as low as 10^{-8} M (Novick & Szilard, 1954). The intracellular concentration of most nutrients must be in a steady-state balance between the rate of entry and the rate of utilization. Depending on the circumstances, this concentration can be ample, or the rate of entry can be a limiting factor in growth.

The initial reactions of sequences are most evidently dependent on nutrient concentration. Subsequent steps in pathways depend upon the previous reactions for their substrates. Under steady-state conditions, the latter substrate concentrations are fixed to correspond to the total flow of metabolites, and the environment affects these reaction rates only indirectly, to the extent that it determines the slowest enzyme of the sequence (Hearon, 1952). This leads one to the pacemaker concept (Krebs & Kornberg, 1957): a key enzyme of a sequence is considered to set the pace for the sequence and the others merely keep up with it.

Inhibition of activity

Numerous early observations suggested that the rates of metabolic sequences are somehow determined by the concentrations of their products. One example is the inhibition by preformed aspartate of the biosynthesis of aspartate by *Lactobacillus arabinosus* (Lardy, Potter & Burris, 1949). The preferential utilization of numerous preformed nutrients was established in the course of applications of the isotope competition technique (Roberts *et al.* 1957). *Escherichia coli* was provided with both a preformed, non-radioactive amino acid (or nucleic acid base) and ^{14}C-glucose. Then the added compound was isolated from the bacterial macromolecules and its specific radioactivity was measured in order to determine what fraction of it was derived from glucose. In most cases, the preformed compound was used almost exclusively. Furthermore, a corresponding amount of radioactive compound was not found elsewhere in the culture: the added compound had prevented synthesis of more of itself.

Certain compounds inhibit an early step in their own synthetic pathway. As evidence, the production by a mutant organism of an intermediate in the blocked metabolic pathway is prevented by the end product. The purine precursor, 5-NH_2-4-imidazole ribotide, is produced by purine-requiring mutants in large amounts (Gots, 1957), but only when no adenine or guanine is present in the medium. This permits the conclusion that purines inhibit an early enzyme in their pathway. Such an inhibition has recently been demonstrated with a purified enzyme preparation (from pigeon liver): the reaction that forms the first specific intermediate of purine synthesis is inhibited competitively by purine nucleoside phosphates (Wyngaarden & Ashton, 1959).

An elegant and early demonstration of the same phenomenon, this time in the tryptophan pathway, was made by Novick & Szilard (1954). The production of an intermediate of tryptophan synthesis by a tryptophan-requiring mutant of *Escherichia coli* was studied in the chemostat. The intermediate was not produced when the organism was grown on excess tryptophan. However, growth on limiting amounts of tryptophan permitted production of the intermediate at about 4 times the rate at which the wild-type bacteria produced tryptophan. The change in the rate of intermediate production on alteration of tryptophan concentration was rapid, occurring in 10 min. or less. Therefore, tryptophan must inhibit the activity of an early enzyme in its own synthetic pathway.

The inhibition of an early step in the arginine synthetic pathway was ingeniously demonstrated by the experiment, cited earlier, on repression of ornithine transcarbamylase by arginine (Gorini, 1958). The enzyme was strongly repressed only when more arginine was made available to the bacteria, growing in the chemostat, than could be incorporated into protein. Therefore, a surplus of arginine could not have been synthesized by the bacteria under any circumstances (although enough arginine was always available to permit the bacteria to grow at the same rate). Just enough arginine must have been synthesized to make up the difference between the supply of preformed arginine and the amount required for protein synthesis.

Feedback inhibition

Inhibition by a compound of an early step in its own synthetic pathway is known as 'feedback inhibition'. Several examples have been cited above, and many others are known. Direct demonstrations of feedback inhibition in cell-free extracts or with purified enzymes are becoming fairly common. An example found in the purine pathway was cited above. Others include the inhibition of threonine deaminase by isoleucine (Umbarger, 1956; Umbarger & Brown, 1958) and inhibition of aspartate transcarbamylase by cytidylic acid (Yates & Pardee, 1956).

Feedback inhibitions serve admirably to prevent overproduction of metabolites. Since an excess of the end product of a pathway inhibits the entire pathway, the rate of the pathway is automatically adjusted to the rate of end product removal. The effectiveness of feedback inhibition in preventing wasteful overproduction of metabolites is indicated by the following experiment (Yates & Pardee, 1956). Normally, in salts-glycerol medium, about 4 % of the nitrogen and 2·2 % of the carbon utilized by *Escherichia coli* are built into pyrimidines. However, when feedback inhibition (and repression) in the pyrimidine pathway is released by placing a pyrimidine-requiring mutant into pyrimidine-deficient medium, approximately 50 % of the nitrogen and 16 % of the carbon is built into pyrimidine precursors. Therefore, the pathway of pyrimidine biosynthesis must be up to 90 % inhibited during normal growth.

SELECTIVE ADVANTAGES OF REGULATORY MECHANISMS

Growth rate as a guiding principle

This article has so far presented little more than a compilation of the factors that determine enzyme concentrations and activities, together with some indications of how these factors may aid in the economical growth of bacteria. Also, rapid growth was stated to be the fundamental function supported by these metabolic adjustments. We will now turn to some general problems regarding the above matters.

Maximal growth rate has been suggested as the guiding principle of bacterial metabolism. Why is this choice preferable to others such as maximal efficiency of carbon or nitrogen utilization, maximal efficiency of energy production, or even creation of maximum entropy (disorder)? If one accepts an evolutionary explanation for development of metabolic control mechanisms, as proposed below, it is much easier to see how the most rapidly growing bacteria would be selected, rather than those which fulfil best the requirements mentioned above. In many circumstances, maximum efficiency and maximum growth rate are not exclusive. However, under other circumstances, efficiency is sacrificed in favour of growth rate. One such instance is when *Escherichia coli* is grown aerobically on glucose. The glucose at first is partly converted to acetate (Roberts *et al.* 1957). This is not the most efficient utilization of the glucose, because if acetate were converted to CO_2 and water (as it is after the glucose is all gone), more energy and carbon would be available. But these reactions do permit rapid growth because *E. coli* grows much more rapidly on glucose than on acetate, and probably at least as rapidly on glucose alone as on glucose + acetate. Growth rate here appears a better criterion than efficiency.

Economical metabolism and rapid growth

What is the relation between the economical responses described earlier and rapid growth? We will consider growth simply as mass increase and will set aside the equally important but poorly understood problem of cell division, even though the latter is another aspect of growth. We must inquire how the growth rate of bacteria is limited. Under unusually favourable conditions, in very rich media, the growth rate reaches an upper limit: bacteria can double their mass in about 15 min. The protein-synthesizing systems are probably saturated and function at their maximal rates. In less nutritionally rich environments, the bacteria grow more slowly; nutrients, supplied and synthesized, are not adequate to saturate the growth machinery of the cell. Apparently the ability of the bacterial enzymes to synthesize nutrients is inadequate to provide an excess of all nutrients.

A limited number of enzyme molecules are present in a bacterium (about two million in an *Escherichia coli* cell). There are probably a few thousand different kinds. Each theoretically possible combination of enzymes will synthesize the metabolites needed by the cell at different rates. For a given combination of enzymes, we imagine that some enzymes can supply the needs of the cell; others are in suboptimal amounts (for that environment). These latter enzymes would make amounts of metabolites that limit the growth rate and the rate of total enzyme synthesis. If the bacterium formed more than the optimal amount of one enzyme, these limiting metabolites would not be available for synthesis of other enzymes. Over-production of one enzyme would rob the other enzyme-forming mechanisms, and the bacterial enzyme composition would deviate from the conditions optimal for rapid growth. Therefore, prevention of overproduction of enzymes would permit more rapid growth.

One example consistent with this model is that constitutive producers of β-galactosidase grow considerably more slowly than the wild-type bacteria (Novick, 1961). Another example is provided by comparing the growth rate of wild-type cells with that of mutants which cannot produce a given enzyme. In media where the action of the missing enzyme is not required, several such mutants were reported to grow a few per cent more rapidly than the wild-type bacteria (Roepke, Libby & Small, 1944). The difference in growth rate might be attributed to more efficient utilization by the mutants of limiting nutrients which, in the wild-type strain, would be used for production of an enzyme which was not utilized, and had no selective advantage in such media.

Slower growth rate also could result if the action of some enzyme was too rapid, i.e. if it overproduced a metabolite. Overproduction of one metabolite would rob other enzymes of the substrates and energy they require to form their products. These products, being produced too slowly, would diminish the growth rate.

Regulatory mechanisms are vital to the rapid growth of bacteria in a second way. They can prevent production of excesses of destructive substances. Enzymes in excess, such as alkaline phosphatase (Torriani, 1960), can damage the cell. Toxic metabolites, for example, surpluses of many amino acids, inhibit bacterial growth (Rowley, 1953). Control mechanisms can prevent their formation.

The evolutionary basis of metabolic regulation

Induction-repression and feedback-inhibition aid bacteria to grow rapidly. These means by which bacteria are flexibly adjusted to their environments are truly a cause for wonder. They are not present by chance or owing to a wilful act of the bacteria, but have developed as special properties of the enzyme-forming systems and enzymes through natural selection (Stanier, 1953). They are the counterparts of the horns, poison glands, rumens and protective colorations so familiar, yet so remarkable, in higher organisms.

Natural selection is expected to be extraordinarily effective for selection of bacteria, owing to the large numbers of bacteria, their appreciable mutation rate, their rapid growth, and the competitive environments in which they are often found. Indeed, the action of natural selection has been investigated in the laboratory under the controlled conditions of the chemostat (Novick & Szilard, 1950). Under competitive conditions (tryptophan limiting) a more rapidly-growing variant of *Escherichia coli* appeared in 10 days. (A greater efficiency in concentrating the limiting component of the medium has been suggested as a cause of the more rapid growth: Novick, 1961.) Natural selection of rapidly growing mutants is therefore an established phenomenon. The wild-type bacteria grew faster than certain mutants obtained from them; therefore, selection had probably approached a limit.

Alternatively, regulatory mechanisms are not to be thought of in a teleological way—as if the bacteria were able to arrange matters for their own convenience. As stressed by Monod (personal communication), induction mechanisms, and others, are completely mechanical; they can readily be turned to the disadvantage of the bacteria. For example, enzymes are induced by compounds that are not metabolized: thiogalactosides are powerful inducers of β-galactosidase, yet they are not substrates. The enzyme, which can reach 5% of the total bacterial protein, is useless to the bacteria under these conditions. As another example of the production of a useless enzyme, the huge amount of aspartate transcarbamylase formed by uracil-starved *Escherichia coli* is not useful in mutants which cannot convert carbamyl aspartate to pyrimidines.

Feedback inhibition, also, can work to the detriment of bacteria. Some analogs of metabolites exhibit feedback inhibition in the normal pathway. By doing so, they prevent synthesis of the end product. This 'false' feedback inhibition must be a common feature of inhibition by metabolic analogs; otherwise the normal compound, synthesized by the bacteria, would overcome the inhibition. As examples, the purine pathway is inhibited at an early step by 6-thioguanine, 6-mercaptopurine, or 2, 6-diaminopurine (Gots & Gollub, 1959). Tryptophan synthesis is blocked by several tryptophan analogs, of which the action of 5-methyltryptophan has been particularly carefully studied (Moyed, 1960). Bacteria appear to lack the discrimination to decide their metabolic affairs.

A selective mechanism for permitting rapid growth must be more useful in one environment than in another. Since each regulatory mechanism must itself require the expenditure of energy and metabolites for its formation and action, under which circumstances is such a mechanism advantageous and under which is it costly? Do the advantageous situations outweigh the others? Nature has apparently provided a variety of solutions to the problem of meeting diverse environments. Presumably organisms such as lactobacilli, which cannot synthesize numerous amino acids, usually have media available containing a plentiful supply of amino acids. The saving in not having to form the enzymes must be more advantageous than the disadvantage of not occasionally finding the proper medium. Similarly, obligate anaerobes presumably possess a growth advantage in not forming the respiratory enzymes and pigments in their usual ecological niche.

With other organisms, flexible metabolic systems must provide a great advantage as the bacteria pass from one environment to another, or when they convert their environment to a new one as a consequence of metabolic activity. Here inducibility, whereby energy for enzyme synthesis is only expended in case of availability of the substrate, would seem to furnish an excellent compromise between constitutivity and total inability to synthesize the enzyme. One would expect, then, a correlation between the ecological niches in which an organism is found (Williams & Spicer, 1957) and the kinds and variability of its enzymes.

Perhaps bacteria normally found in very difficult environments, methane bacteria for example, must divert a great part of their enzymes to eking out an energy supply, and therefore have only a limited complement of enzymes for biosynthesis. It is not surprising that bacteria grow at different rates even after the innumerable generations available for their development and specialization.

There is no reason to believe that control mechanisms for all enzymes are similar. The reactions in a bacterium proceed at vastly different rates, and enzymes are presumably formed in correspondingly different amounts (McIlwain, 1946). As a rough approximation, we can assume that there are three kinds of enzymes, quantitatively speaking. The first kind catabolize a primary carbon source, the second are found in pathways of biosynthesis of major cell metabolites such as amino acids, and the third are required for biosynthesis of minor constituents such as coenzymes. In an organism such as *Escherichia coli*, the three kinds of enzymes must convert about 10^8, 10^6, and 10^3 molecules of substrate/min., respectively, in each cell. Assuming a reasonable turnover number, an enzyme of the first sort could comprise several per cent of the total cell protein, and the last could be present to the extent of only a few molecules per cell (McIlwain, 1946). Preventing the overproduction of the first kind of enzyme could be vital to the growth rate, but one wonders if the effort expended in setting up a delicate regulatory system for the last kind would be repaid. A moderate overproduction of a coenzyme would not constitute a serious drain on the synthetic capacities of the cell. Data on the production of riboflavin coenzymes by *E. coli* suggest that feedback mechanisms are not important in flavin synthesis (Wilson & Pardee, unpublished). Possibly, reactions of this sort are limited by a mechanism which holds the enzyme synthesizing system to a low, relatively constant rate of enzyme production.

CONCLUSIONS

Bacteria are thought to respond to environmental stresses by altering their metabolic patterns so as to grow as rapidly as possible under the circumstances. Two main types of response are illustrated; one is a response of enzyme synthesis and the other a response of enzyme activity. Both can lead to more efficient functioning of the bacteria, and this in turn can permit more rapid growth. These response mechanisms are suggested to originate by mutation and to be chosen by natural selection.

The responses of enzyme synthesis and activity to environment would appear to be basic to rapid growth rate and also to the complex responses in product excretion, physiology and morphology. An understanding of the interactions of enzyme formation and activity with the environment will guide us in interpreting the complex processes of mass increase, cell division and differentiation.

REFERENCES

AMES, B. N. & GARRY, B. (1959). Coordinate repression of the synthesis of four histidine enzymes by histidine. *Proc. nat. Acad. Sci., Wash.* **45**, 1453.

BELLAMY, W. D. & GUNSALUS, I. C. (1945). Tyrosine decarboxylase. II. Pyridoxin deficient medium for apoenzyme production. *J. Bact.* **50**, 95.

BENZER, S. (1953). Induced synthesis of enzymes in bacteria analysed at the cellular level. *Biochim. biophys. Acta,* **11**, 383.

BRITTEN, R. J., ROBERTS, R. B. & FRENCH, E. F. (1955). Amino acid absorption and protein synthesis in *Escherichia coli. Proc. nat. Acad. Sci., Wash.* **41**, 863.

COHEN, G. N. & JACOB, F. (1959). Sur la répression de la synthèse des enzymes intervenant dans la formation du tryptophane chez *E. coli. C.R. Acad. Sci., Paris,* **248**, 3490.

COHEN, G. N. & MONOD, J. (1957). Bacterial permeases. *Bact. Rev.* **21**, 169.

COHEN, G. N. & RICKENBERG, H. V. (1956). Concentration spécifique réversible des aminoacides chez *Escherichia coli. Ann. Inst. Pasteur,* **91**, 693.

COHEN, S. S. & FOWLER, C. B. (1948). Chemical studies in host-virus interactions. V. Some additional methods of determining nutritional requirements for virus multiplication. *J. exp. Med.* **87**, 275.

COHEN-BAZIRE, G. & JOLIT, M. (1953). Isolement par sélection de mutants d'*Escherichia coli* synthétisant spontanément l'amylomaltase et la β-galactosidase. *Ann. Inst. Pasteur,* **84**, 937.

COHN, M. (1957). Contributions of studies on the β-galactosidase of *Escherichia coli* to our understanding of enzyme synthesis. *Bact. Rev.* **21**, 140.

COHN, M. & HORIBATA, K. (1959). Physiology of the inhibition by glucose of the induced synthesis of the β-galactosidase-enzyme system of *Escherichia coli. J. Bact.* **78**, 624.

COHN, M. & MONOD, J. (1953). Specific inhibition and induction of enzyme biosynthesis. In *Adaptation in micro-organisms. Symp. Soc. gen. Microbiol.* **3**, 132.

ENGLESBERG, E. (1959). Glucose inhibition and the diauxie phenomenon. *Proc. nat. Acad. Sci., Wash.* **45**, 1494.

ENGLESBERG, E. & LEVY, J. B. (1955). Induced synthesis of tricarboxylic acid cycle enzymes as correlated with the oxidation of acetate and glucose by *Pasteurella pestis. J. Bact.* **69**, 418.

GALE, E. F. (1940). The production of amines by bacteria. I. The decarboxylation of amino acids by strains of *Bacterium coli. Biochem. J.* **34**, 392.

GALE, E. F. (1943). Factors influencing the enzymic activities of bacteria. *Bact. Rev.* **7**, 139.

GALE, E. F. (1953). Assimilation of amino acids by gram-positive bacteria and some actions of antibiotics thereon. *Advanc. Protein Chem.* **8**, 285.

GORINI, L. (1958). Regulation en retour (feedback control) de la synthèse de l'arginine chez *Escherichia coli. Bull. Soc. Chim. biol., Paris,* **40**, 1939.

GORINI, L. (1960). Antagonism between substrate and repressor in controlling the formation of a biosynthetic enzyme. *Proc. nat. Acad. Sci., Wash.* **46**, 682.

GORINI, L. & MAAS, W. K. (1958). Feedback control of the formation of biosynthetic enzymes. In *The Chemical Basis of Development*, p. 469. Edited by W. D. McElroy and B. Glass. Baltimore: Johns Hopkins Press.

GOTS, J. S. (1957). Purine metabolism in bacteria. V. Feed-back inhibition. *J. biol. Chem.* **228**, 57.

GOTS, J. S. & GOLLUB, E. G. (1959). Purine analogs as feedback inhibitors. *Proc. Soc. exp. Biol., N.Y.* **101**, 641.

GRIFFETHS, M., SISTROM, W. R., COHEN-BAZIRE, G. & STANIER, R. Y. (1955). Function of carotenoids in photosynthesis. *Nature, Lond.* **176**, 1211.

GUIRARD, B. M. & SNELL, E. E. (1954). Pyridoxal phosphate and metal ions as cofactors for histidine decarboxylase. *J. Amer. chem. Soc.* **76**, 4745.

HALVORSON, H. O. (1960). The induced synthesis of proteins. *Advanc. Enzymol.* **22**, 99.

HEARON, J. Z. (1952). Rate behaviour of metabolic systems. *Physiol. Rev.* **32**, 499.

JACOB, F., PERRIN, D., SANCHEZ, C. & MONOD, J. (1960). L'opéron: groupe de gènes à expression coordonnée par un opérateur. *C.R. Acad. Sci., Paris*, **250**, 1727.

KARSTRÖM, H. (1938). Enzymatische Adaptation bei Mikroorganismen. *Ergebn. Enzymforsch.* **7**, 350.

KNOX, R. (1953). The effect of temperature on enzymic adaptation, growth and drug resistance. In *Adaptation in Micro-organisms. Symp. Soc. gen. Microbiol.* **3**, 184.

KORNBERG, H. L., COLLINS, J. F. & BIGLEY, D. (1960). The influence of growth substrates on metabolic pathways in *Micrococcus denitrificans*. *Biochim. biophys. Acta*, **39**, 9.

KREBS, H. A. & KORNBERG, H. L. (1957). *Energy Transformations in Living Matter*. Berlin: Springer-Verlag.

LARA, F. J. S. & STOKES, J. L. (1952). Oxidation of citrate by *Escherichia coli*. *J. Bact.* **63**, 415.

LARDY, H. A., POTTER, R. L. & BURRIS, R. H. (1949). Metabolic functions of biotin. I. The role of biotin in bicarbonate utilization by *Lactobacillus arabinosus* studied with C^{14}. *J. biol. Chem.* **179**, 721.

LWOFF, A. & LWOFF, M. (1937). Studies on codehydrogenases. II. Physiological function of growth factor 'V'. *Proc. Roy. Soc.* B, **122**, 360.

MAAS, W. K. (1959). Effect of penicillin on the uptake of amino acids in bacteria. *Biochem. biophys. Res. Comm.* **1**, 13.

MCILWAIN, H. (1946). The magnitude of microbial reactions involving vitamin-like compounds. *Nature, Lond.* **158**, 898.

MAGASANIK, A. K. & BOJARSKA, A. (1960). Enzyme induction and repression by glucose in *Aerobacter aerogenes*. *Biochem. biophys. Res. Comm.* **2**, 77.

MAGASANIK, B. (1957). Nutrition of bacteria and fungi. *Annu. Rev. Microbiol.* **11**, 221.

MANDELSTAM, J. & HALVORSON, H. (1960). Turnover of protein and nucleic acid in soluble and ribosome fractions of non-growing *Escherichia coli*. *Biochim. biophys. Acta*, **40**, 43.

MELNYKOVYCH, G. & SNELL, E. E. (1958). Nutritional requirements for the formation of arginine decarboxylase in *Escherichia coli*. *J. Bact.* **76**, 518.

MONOD, J. (1947). The phenomenon of enzymatic adaptation and its bearings on problems of genetics and cellular differentiation. *Growth*, **11**, 223.

MONOD, J. (1958). An outline of enzyme induction. *Rec. Trav. chim. Pays-Bas*, **77**, 569.

MOYED, H. S. (1960). False feedback inhibition: inhibition of tryptophan biosynthesis by 5-methyltryptophan. *J. biol. Chem.* **235**, 1098.

NEIDHARDT, F. C. & MAGASANIK, B. (1957). Reversal of the glucose inhibition of histidase biosynthesis in *Aerobacter aerogenes*. *J. Bact.* **73**, 253.

NOVICK, A. (1961). Bacteria with high specific enzyme levels. In *Growth: Molecule, Cell and Organism* (in the Press). New York: Basic Books.

NOVICK, A. & SZILARD, L. (1950). Experiments with the chemostat on spontaneous mutations of bacteria. *Proc. nat. Acad. Sci., Wash.* **36**, 708.

NOVICK, A. & SZILARD, L. (1954). Experiments with the chemostat on the rates of amino acid synthesis in bacteria. In *Dynamics of Growth Processes*, p. 21. Edited by E. J. Boell. Princeton: Princeton University Press.

NOVICK, A. & WEINER, M. (1957). Enzyme induction as an all-or-none phenomenon. *Proc. nat. Acad. Sci., Wash.* **43**, 553.

PARDEE, A. B. (1955). Effect of energy supply on enzyme induction by pyrimidine-requiring mutants of *Escherichia coli*. *J. Bact.* **69**, 233.

PARDEE, A. B. (1957). An inducible mechanism for accumulation of melibiose in *Escherichia coli*. *J. Bact.* **73**, 376.

PARDEE, A. B. (1959). The control of enzyme activity. In *The Enzymes*, 2nd ed., p. 681. Edited by P. D. Boyer, H. A. Lardy and K. Myrback. New York: Academic Press.

PARDEE, A. B. (1961). The synthesis of enzymes. In *The Bacteria*. Edited by I. C. Gunsalus & R. Y. Stanier (in the Press). New York: Academic Press.

PARDEE, A. B., JACOB, F. & MONOD, J. (1959). The genetic control and cytoplasmic expression of 'inducibility' in the synthesis of β-galactosidase by *Escherichia coli*. *J. molec. Biol.* **1**, 165.

PARDEE, A. B. & PRESTIDGE, L. S. (1955). Induced formation of serine and threonine deaminases by *Escherichia coli*. *J. Bact.* **70**, 667.

POLLOCK, M. R. (1959). Induced formation of enzymes. In *The Enzymes*, 2nd ed., p. 619. Edited by P. D. Boyer, H. A. Lardy and K. Myrback. New York: Academic Press.

RICKENBERG, H. V. & LESTER, G. (1955). The preferential synthesis of β-galactosidase in *Escherichia coli*. *J. gen. Microbiol.* **13**, 279.

ROBERTS, R. B., ABELSON, P. H., COWIE, D. B., BOLTON, E. T. & BRITTEN, R. J. (1957). Studies of biosynthesis in *Escherichia coli*. *Publ. Carneg. Instn*, no. 607.

ROEPKE, R. R., LIBBY, R. L. & SMALL, M. H. (1944). Mutation or variation of *Escherichia coli* with respect to growth requirements. *J. Bact.* **48**, 401.

ROGERS, P. & NOVELLI, G. D. (1959). Formation of ornithine transcarbamylase in cells and protoplasts of *Escherichia coli*. *Biochim. biophys. Acta*, **33**, 423.

ROWLEY, D. (1953). Interrelationships between amino acids in the growth of coliform organisms. *J. gen. Microbiol.* **9**, 37.

SHEPHERDSON, M. & PARDEE, A. B. (1960). Production and crystallization of aspartate transcarbamylase. *J. biol. Chem.* **235**, 3233.

SHER, I. H. & MALLETTE, M. F. (1954). Purification and study of L-lysine decarboxylase from *Escherichia coli* B. *Arch. Biochem. Biophys.* **53**, 354.

SPIEGELMAN, S., HALVORSON, H. O. & BEN-ISHAI, R. (1955). Free amino acids and the enzyme-forming mechanism. In *Amino Acid Metabolism*, p. 124. Edited by W. D. McElroy and B. Glass. Baltimore: Johns Hopkins Press.

STANIER, R. Y. (1951). Enzymatic adaptation in bacteria. *Annu. Rev. Microbiol.* **5**, 35.

STANIER, R. Y. (1953). Adaptation, evolutionary and physiological: or Darwinism among the micro-organisms. In *Adaptation in Microorganisms*. *Sym. Soc. gen. Microbiol.* **3**, 1.

STEPHENSON, M. & GALE, E. F. (1937). The adaptability of glucozymase and galactozymase in *Bacterium coli*. *Biochem. J.* **31**, 13.

STERN, J. R. & OCHOA, S. (1951). Enzymatic synthesis of citric acid. I. Synthesis with soluble enzymes. *J. biol. Chem.* **191**, 161.

STOEBER, F. (1957). Sur la biosynthèse induite de la β-glucuronidase chez *Escherichia coli*. *C.R. Acad. Sci., Paris*, **244**, 950.

TORRIANI, A. (1960). Influence of inorganic phosphate in the formation of phosphatases by *E. coli*. *Biochim. biophys. Acta*, **38**, 460.

UMBARGER, H. E. (1956). Evidence for a negative feedback mechanism in the biosynthesis of isoleucine. *Science*, **123**, 848.

UMBARGER, H. E. & BROWN, B. (1958). Isoleucine and valine metabolism in *Escherichia coli*. VII. A negative feedback mechanism controlling isoleucine biosynthesis. *J. biol. Chem.* **233**, 415.

VOGEL, H. J. (1957). Repression and induction as control mechanisms of enzyme biogenesis: The 'adaptive' formation of acetylornithinase. In *The Chemical Basis of Heredity*, p. 276. Edited by W. D. McElroy and B. Glass. Baltimore: Johns Hopkins Press.

VOGEL, H. J. (1960). Repression of an acetylornithine permeation system. *Proc. nat. Acad. Sci., Wash.* **46**, 488.

WAINWRIGHT, S. D. & NEVILL, A. (1956). The influence of depletion of nitrogenous reserves upon the phenomenon of induced enzyme biosynthesis in cells of *Escherichia coli*. *J. gen. Microbiol.* **14**, 47.

WILLIAMS, R. E. O. & SPICER, C. C. (1957). *Microbial Ecology*. *7th Symp. Soc. gen. Microbiol.*

WYNGAARDEN, J. B. & ASHTON, D. M. (1959). The relations of activity of phosphoribosylpyrophosphate amidotransferase by purine ribonucleotides: a potential feedback control of purine biosynthesis. *J. biol. Chem.* **234**, 1492.

YATES, R. A. & PARDEE, A. B. (1956). Control of pyrimidine synthesis in *Escherichia coli* by a feedback mechanism. *J. biol. Chem.* **221**, 757.

YATES, R. A. & PARDEE, A. B. (1957). Control by uracil of formation of enzymes required for orotate synthesis. *J. biol. Chem.* **227**, 677.

YUDKIN, J. (1938). Enzyme variation in micro-organisms. *Biol. Rev.* **13**, 93.

THE EFFECT OF ENVIRONMENT ON THE FORMATION OF BACTERIAL FLAGELLA

D. KERRIDGE

Medical Research Council, Chemical Microbiology Research Unit, Department of Biochemistry, Tennis Court Road, Cambridge

Since the formation of the Society for General Microbiology, bacterial flagella have been the subject of contributions at two of its symposia. In the first symposium on the Nature of the Bacterial Surface, Pijper (1949) discussed the function of bacterial flagella, and in the symposium on Bacterial Anatomy, Stocker (1956*a*) covered their structure and morphology and the genetic control of motility. It is appropriate, therefore, that the third contribution on bacterial flagella should be concerned with their formation. It is intended that this article will be complementary to the other two, but it is inevitable that there will be some overlap in the reviewing of the literature.

STRUCTURE OF BACTERIAL FLAGELLA

It has been known for some time that flagella can be detached from the bacterial cell by shaking (Orcutt, 1924; Yokota, 1925; Balteanu, 1926; Craigie, 1931), but it was not until Gard (1944) showed that purified preparations of flagella could be obtained in high yield by differential high-speed centrifugation, that chemical analysis was possible. The experimental findings of Weibull and collaborators confirmed the idea put forward by Migula (1897) and Boivin & Mesrobeanu (1938) that flagella are protein in nature. Much information on the nature of bacterial flagella has been provided by the studies of Weibull (Weibull & Tiselius, 1945; Weibull, 1948, 1949*a*, *b*, 1950*a–c*, 1951*a*, *b*, 1953). The flagella from *Proteus vulgaris* and *Bacillus subtilis* consist of at least 98% protein. Exposure of flagella to acid, alkali, or heat results in their irreversible disintegration; the soluble material obtained after the treatment of flagella with acid is protein in nature, having an approximate molecular weight of 40,000, and is homogeneous in the ultracentrifuge. The generic name 'flagellin' has been proposed by Astbury, Beighton & Weibull (1955) for the constituent protein of the bacterial flagella. Flagellins from *P. vulgaris* and *B. subtilis* are unusual proteins in that they contain no detectable histidine, tryptophan, hydroxyproline or cysteine/cystine.

Table 1. *Amino acid composition of bacterial flagellins*

	g. amino acid residues/10^5 g. protein		
		Salmonella typhimurium	
	Proteus vulgaris	(a)	(b)
Glycine	86	79	81
Alanine	110	147	150
Valine	64	60	60
Leucine	86	74	76
*iso*Leucine	56	49	46
Glutamic acid	97	96	96
Aspartic acid	166	155	153
Serine	68	59	57
Threonine	85	101	99
Tyrosine	13	20	27
Phenylalanine	24	13	14
Tryptophan	0	0	0
Proline	0	13	9
Cystine/2	6	0	0
Methionine	7	3	3
Lysine	49	33	31
ϵ-*N*-methyl lysine	0	28	24
Arginine	40	24	29
Histidine	0	4	0
Total residues	957	958	955
Amide groups	144	156	152

(a) and (b) are values obtained in two separate determinations. Data from Koffler (1957) and Ambler & Rees (1959).

More detailed chemical analyses have been performed by Koffler and his collaborators on flagella isolated from *Proteus vulgaris*, *Bacillus subtilis* and *Serratia marcescens* (a detailed account of their work is given in Kobayashi, Rinker & Koffler, 1959) and by Ambler & Rees (1959) on flagella from *Salmonella typhimurium*. Both groups analysed the soluble protein obtained after acid treatment of flagella. The data are very interesting in that not only are there quantitative differences in the amino acid composition of flagellins from different sources, but also qualitative differences (Table 1). The most striking of these is the finding that flagellin from *Salmonella typhimurium*, strain SW 1061, contains the amino acid, ϵ-*N*-methyl lysine, which had not previously been found in biological material. Flagella from *S. typhimurium* also differ from flagella of *P. vulgaris* in that proline and histidine are present in, and cysteine absent from, flagellin from the former organism.

SYNTHESIS OF BACTERIAL FLAGELLA

Consisting as they do of up to 2% of the bacterial dry weight, the formation of flagella is a suitable system for studying the synthesis of a single protein. It must always be remembered, however, that flagella are organized structures with a definite function in the cell (Weibull, 1951*c*, 1960; van Iterson, 1953; Kvittingen, 1955; Pijper, 1957) and any studies on their synthesis must be extended to include not only the formation of the individual flagellin molecules, but also the arrangement of these molecules within the flagellum and their relationship to its functioning. The absence of any detectable enzyme activity in the flagellum (Barlow & Blum, 1952) renders the quantitative assay of flagellin tedious. In this respect, bacterial flagella differ from those of certain phytoflagellates which have been shown to possess apyrase activity (Tibbs, 1957; Brokaw, 1960). Qualitatively, the synthesis of functional flagella can be assayed readily by estimating the motility of the bacteria, while the total number of flagella can be obtained by examination of stained preparations.

During the germination of spores of *Bacillus vulgatus*, *B. cereus* and *B. flavus*, the *de novo* appearance of flagella occurs in a random manner on the bacterial surface before the completion of germination. Leifson (1931) estimated the growth rate of the flagella as 1 μ every 2–3 min., and the organisms became motile after the formation of 2 or 3 short flagella. Bisset & Hale (1951), in a study of the growth of microcysts (a resting stage produced during the stationary phase of growth) of *Salmonella typhi*, *Proteus vulgaris*, *Escherichia coli* and *Pseudomonas fluorescens*, found that flagella appeared after about 2 hr. incubation in a nutrient medium, the formation of flagella being complete after 4 hr. Unlike the germinating spore, where there is an apparently random appearance of flagella on the bacterial surface, the germinating microcyst first produces a polar flagellum, although the mature organisms may have either polar or peritrichate flagellation.

The finding that flagella could be removed from a logarithmically growing culture of *Salmonella typhimurium* without affecting the growth rate, and that on continued incubation in a nutrient medium the bacteria rapidly regenerated their flagella, provided an additional impetus to the study of the formation of bacterial flagella (Stocker, 1956*a*, 1957). The regeneration of flagella, after mechanical removal by treatment in a Waring Blendor, is completely inhibited by growth-inhibitory concentrations of chloramphenicol. In a later paper, Stocker & Campbell (1959) surveyed the methods for the removal of flagella and the effects of these treatments on bacterial viability. Removal of the flagella from a culture of *S. typhimurium*, growing logarithmically in a nutrient medium, does not affect the growth rate of the bacteria which recovered their motility within 1 generation time. On the other hand, regeneration of flagella occurred only very slowly, or not at all, if the bacteria were in the stationary phase of growth when the flagella were detached from them. By measuring the lengths of the flagella and determining the no. of flagella/bacterium, Stocker & Campbell obtained an estimate of the rate of growth of the flagella. This was found to be 0·1 μ/min. at 37° and 0·085 μ/min. at 25°, values somewhat lower than those reported by Leifson (1931) for the development of flagella in germinating spores. The mechanical removal of flagella does not affect the rate of synthesis of flagella by the bacteria. But the average value for flagella/bacterium in samples taken less than 1 generation time after deflagellation was less than in the controls, and this deficit can be explained by the assumption that the flagellum increases in length for only a limited period after its inception. Confirmation of these data concerning the formation of flagella by *S. typhimurium* was obtained by studying the uptake of ^{14}C labelled leucine into logarithmically growing cultures of the bacteria (Kerridge, unpublished results). Samples were taken at intervals after the addition of the ^{14}C labelled amino acid; the bacteria were harvested by centrifugation and washed twice with cold saline containing 20 μg. chloramphenicol/ml. The flagella were mechanically detached from the bacteria and after the removal of the bacteria by centrifugation, the flagella were precipitated by the addition of specific antiserum. The initial rates of incorporation of ^{14}C labelled leucine into the general cell protein and into the flagella were identical, but after 30 min. incubation, there was some fall off in the rate of incorporation into the general cell protein (Fig. 1). A similar result was obtained using ^{14}C labelled glutamic acid as the marker. Mechanical removal of the flagella, before the addition of the ^{14}C labelled amino acid, did not have any effect on the rates of incorporation of the tracer into either the general cell protein, or the flagella. In studying the formation of flagella, one can therefore regard the flagellin as a constitutive protein.

DISTRIBUTION OF FLAGELLA AT CELL DIVISION

As experimental findings on the effects of inhibitors on the formation of bacterial flagella have shed some light on this controversial topic, it is relevant at this point to include a section on the distribution of cellular constituents at cell division. With the monoflagellate organisms, for example, the *Caulobacter* (Houwink, 1951; Bowers, Weaver, Grula & Edwards, 1954) and the *Pseudomonas* spp., it is obvious that unless the flagellum divides by a longitudinal split, it must be retained by only one of the two daughter bacteria and the other will produce a flagellum *de novo*. The problem in this case is at what time the new flagellum appears relative to the separation of the daughter bacteria. In a study of synchronously dividing *P. aeruginosa*, Jacherts (1960) found that the

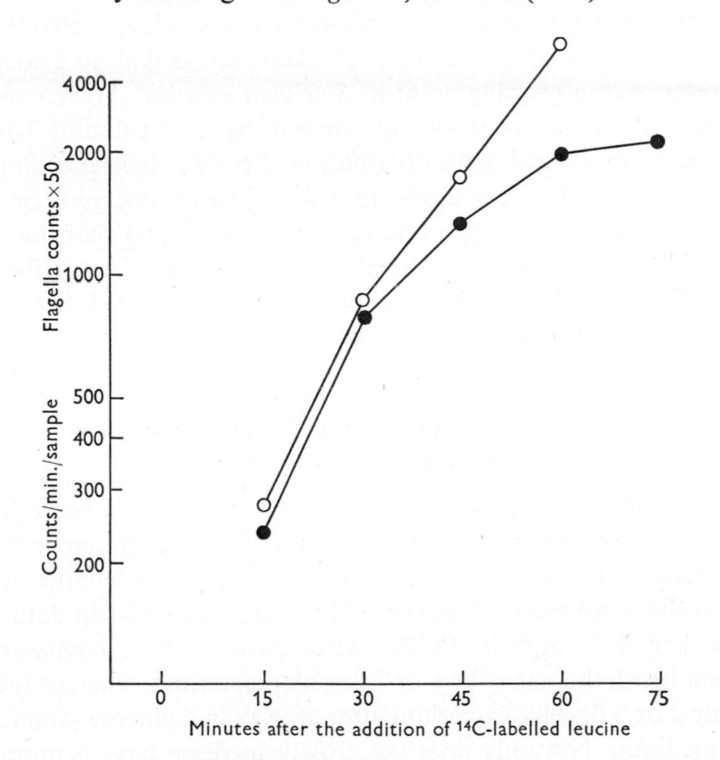

Fig. 1. The incorporation of ^{14}C-labelled leucine into cellular protein and flagella by *Salmonella typhimurium* growing logarithmically at 37°. —●—●—, Cellular protein; —○—○—, flagella.

new flagellum appears at the time of division of the cells, and is fully formed within 30 sec. of the separation of the daughter bacteria. This rate of synthesis is much faster than that found by Stocker & Campbell (1959) for the synthesis of flagella by logarithmically growing cultures of *Salmonella typhimurium*.

There is much conflicting evidence for the distribution of flagella in peritrichously flagellated bacteria. Bisset (1951), Bisset & Pease (1957) and Bisset & Hale (1960) have presented evidence that the bacterial cell grows by means of a 'growing point' and, as a result, the flagella of the parent bacterium are not distributed between the daughter bacteria, but are all retained by one of them. In contrast to these results, Quadling & Stocker (1956) and Quadling (1958) find that there is a distribution of flagella from the parent bacterium between the daughter bacteria in the case of *Salmonella typhimurium*, suggesting that there is no evidence for a growing point, but that growth occurs by intercalation. Data supporting this latter view were obtained from a study of the effects of inhibitors on the growth of *S. typhimurium* (Kerridge, 1960). Williams (1959), using cellular inclusions as markers, could find no evidence for the presence of a growing point in *Spirillum annulus*. More conclusive evidence in favour of bacterial growth by intercalation has been obtained by investigating the distribution of tritium-labelled compounds using a microradioautographic technique (van Tubergen, Setlow & Sherk, 1960). van Tubergen and his colleagues found that the ribonucleic acid (RNA), protein and cell wall, but not the deoxyribonucleic acid (DNA) of *Escherichia coli* were equally divided between the daughter bacteria after cell division.

NUTRIENTS AND THE FORMATION OF BACTERIAL FLAGELLA

In *Salmonella typhimurium* the mean number of flagella/bacterium can be varied over rather wide limits by alteration of the environmental conditions, although the distribution of the flagellar lengths does not vary to the same extent (Stocker & Quadling, unpublished data quoted in Stocker & Campbell, 1959). After growth of *S. typhimurium* in nutrient broth there may be 6 or 7 flagella/bacterium, whereas there will be only 2 or 3 flagella/bacterium after growth in a glucose–ammonium–salts medium. Not only does the growth medium have a pronounced effect on the number of flagella/bacterium, but it also affects the ability of the organisms to regenerate their flagella after mechanical removal. *S. typhimurium* strain LT-2 grows well in a glucose–ammonium–salts medium, but the bacteria grown under these conditions do not readily regenerate flagella after their mechanical detachment. Of a number of nitrogen sources tested for their ability to promote regeneration of flagella in a defined medium, only a complete mixture of amino acids gives consistently reproducible results comparable with those obtained with organisms incubated in a nutrient broth (Kerridge, 1959*b*).

Synthesis of flagella by nutritionally-exacting mutants of Salmonella typhimurium

Flagellin is similar to the M protein of group A haemolytic streptococci in being an incomplete protein lacking certain of the naturally occurring amino acids. Lancefield (1943) found that it is possible to remove the M protein from streptococci by digestion with trypsin without affecting their viability. The trypsin-treated organisms required amino acids, polypeptides and glucose for synthesis of the M protein. Tryptophan and phenylalanine are essential for growth of streptococci, but not for synthesis of the M protein in a chemically defined medium (Fox & Krampitz, 1956). The formation of specific enzymes has been followed in the majority of studies on the formation of single proteins, and although the synthesis of only one protein is followed, other proteins are also being synthesized. It was with the idea of overcoming this limitation that the study of the formation of flagella by amino acid-requiring mutants of *Salmonella typhimurium* was begun (Kerridge, 1959*b*).

All the amino acid-requiring mutants of *Salmonella typhimurium* examined are able to grow and regenerate flagella after mechanical removal when incubated in the presence of glucose and a complete mixture of the amino acids. In the absence of the amino acid required for growth, there is little or no net synthesis of RNA or protein, but formation of DNA continues. Pardee & Prestridge (1956) obtained similar results with auxotrophic mutants of *Escherichia coli*. Although the synthesis of RNA and protein is prevented when auxotrophic strains are incubated in the absence of their growth requirement, certain strains regenerate flagella when incubated under these conditions.

The amino acid-requiring strains of *Salmonella typhimurium* can be conveniently divided into two major and one intermediate group on the basis of their ability to regenerate flagella in the absence of their growth requirement (Table 2). The amino acids leucine, tyrosine and glutamic acid are all present in flagellin and strains exacting for these amino acids are unable to regenerate flagella when incubated in the presence of a nitrogen and energy source but in the absence of their growth requirement. Strains SL 282 (tryptophan-requiring) and SL 281 (cysteine-requiring) are exacting for amino acids absent from flagellin and when incubated in the absence of the amino acid required for growth are nevertheless able to regenerate flagella, although there is little or no net synthesis of protein under these conditions.

The behaviour of the bacteria in the intermediate group was rather difficult to reconcile with that of the two major groups, since it is

Table 2. *The effect of essential amino acid nutrients on growth and regeneration of flagella by amino acid-requiring strains of* Salmonella typhimurium

Strain	Growth requirement	Starvation	Complete regeneration medium: Suspension optical density (% increase)	Complete regeneration medium: Motility (%)	Regeneration medium lacking growth requirement: Suspension optical density (% increase)	Regeneration medium lacking growth requirement: Motility (%)	Control (phosphate buffer + glucose): Suspension optical density (% increase)	Control (phosphate buffer + glucose): Motility (%)
Group a								
SL186	Leucine	−	35	40 (50)	7	< 1 (1)	0	< 1
SL186	Leucine	+	65	10 (60)	13	< 1 (1)	−1	< 1
SL185	Tyrosine	−	78	50 (40)	−3	< 1 (1)	0	< 1
SL443	Glutamic acid	−	90	30	−13	< 1	−15	< 1
Group b								
SL291	Cysteine	−	60	40 (39)	16	50 (36)	−7	< 1
SL282	Tryptophan	−	45	70 (90)	2	65 (85)	−12	< 1
SL282	Tryptophan	+	78	80	−1	50	−1	< 1
Group c								
SL286	Histidine	−	66	70	−1	30	−17	10
SL286	Histidine	+	106	80	−12	10	−13	< 2
SL287	Histidine	−	100	70	6	40	−15	< 2
SL287	Histidine	+	120	70	−6	40	2	< 2
SL262	Methionine	−	76	60	13	10	−10	< 1

The deflagellated bacteria were incubated aerobically for 2 hr. at 37° in a medium consisting of 0·01 M phosphate buffer pH 7·0, a complete mixture of amino acids (0·1 mg. of each/ml.) and glucose (0·2 %, w/v). Starved preparations were obtained by incubating the bacteria in a glucose–ammonium–salts medium for 30 min. at 37° prior to the mechanical removal of flagella. Growth of the bacteria was measured by determining the optical density of the suspensions at the beginning and end of the incubation.

The motility of the bacteria at the end of the incubation period was estimated microscopically after dilution with nutrient broth containing 10 μg. chloramphenicol/ml. The figures in parentheses are % flagellated bacteria in stained preparations.

plausible to assume that in the absence of the growth requirement, strains requiring an amino acid for growth would synthesize proteins which lacked that amino acid but not proteins containing it. Histidine and methionine occur to a limited extent in flagellin, and although the synthesis of flagella might be accounted for by the presence of these amino acids in an amino acid pool, the ability of starved preparations of bacteria to regenerate flagella makes this unlikely. The demonstration by Mandelstam (1957, 1958) and Mandelstam & Halvorson (1960) of protein turnover by resting suspensions of *Escherichia coli* suggests a possible explanation. If a similar turnover occurred in *Salmonella typhimurium*, then a limited amount of histidine and methionine would be made available for the synthesis of protein, and the relative amount of a particular protein synthesized would depend on the number of histidine or methionine residues required per molecule. Flagellin would be at an advantage as it contains only one histidine or methionine residue per molecule of approximate molecular weight 30,000. Consequently the synthesis of flagellin by the auxotrophic strains of *S. typhimurium* in the absence of histidine or methionine would be more probable than that of other proteins containing more histidine or methionine residues/molecule. The failure of strains SL 186 (leucine-requiring), SL 185 (tyrosine-requiring) and SL 443 (glutamic acid-requiring) to regenerate flagella in the absence of their growth requirement, although protein turnover may be occurring during incubation, can be explained by assuming that as the amino acids, leucine, tyrosine and glutamic acid are relatively abundant in flagellin, this protein will be at a disadvantage in competing for the limited quantity of these amino acids released during protein turnover.

Effects of inhibitors on the formation of bacterial flagella

Amino acid analogues. Certain amino acid analogues are incorporated into bacterial proteins, resulting in many cases in the formation of protein with either reduced or no biological activity (Munier & Cohen, 1956, 1959; Baker, Johnson & Fox, 1958; Yoshida, 1960). Bacterial flagella, consisting as they do of at least 98% of a single well characterized protein, provide a suitable system for the study of the effects of these compounds on protein formation.

A number of these compounds inhibit the growth of *Salmonella typhimurium* in a glucose–ammonium–salts medium, but apart from *p*-fluorophenylalanine (*p*-FP), none inhibit the regeneration of functional flagella by deflagellated *S. typhimurium* (Table 3; Kerridge, 1959*a*, 1960). After incubation in the presence of *p*-FP (0·005 M) the bacteria have little or no translational motility, although flagella can be demonstrated both serologically and by staining. The flagella synthesized in the presence of *p*-FP are still sinusoidal, but are abnormal in having a mean wavelength only half that of the control culture (1·04 μ, standard deviation = 0·15 μ compared to 2·04 μ, standard deviation = 0·1 μ). Removal of the analogue after the formation of the aberrant flagella does not result in a return to normal morphology. If the bacteria are deflagellated after incubation in the presence of *p*-FP and incubated in a nutrient medium, there is a considerable lag before the recovery of motility.

The occurrence of two distinct wavelengths for bacterial flagella, one usually half the other, has been noted for a considerable time (see

Table 3. *The effect of amino acid analogues on growth and regeneration of flagella by* Salmonella typhimurium

Analogue	Inhibition of: (i) Growth (measured by the time taken in min. for the optical density to double after addition of analogue)	(ii) Synthesis of flagella: Regeneration	Function	Morphology
Control	60	+	Motile	Normal
p-FP (0·005 M)	102	+	Non-motile	Aberrant
β-2-thienylalanine (0·002 M)	195	+	Motile	Normal
Ethionine (0·005 M)	72	+	Motile	Normal
Norvaline (0·002 M)	90	+	Motile	Normal
Norleucine (0·002 M)	78	+	Motile	Normal
5-Methyl-tryptophan (0·005 M)	·	+	Motile	Normal

Growth inhibition was determined by turbidity measurements after the addition of the analogue to *Salmonella typhimurium*, strain SW 1061, growing logarithmically in a minimal medium. Synthesis of flagella was followed in the regeneration medium, the homologous amino acid being omitted when the effect of the analogue was being investigated. Data from Kerridge (1959*a*).

Pijper, 1957), and the term 'biplicity' has been proposed for this phenomenon (Pijper, 1955). In the majority of cases where biplicity has been recorded, there has been no attempt to correlate the morphology of the flagella with either environmental conditions or the motility of the bacteria. Leifson, Carhart & Fulton (1954) reported that for certain *Proteus* spp., the change from one form to the other is dependent on the pH value of the suspending medium and not on the growth conditions *per se*. Addition of either acid or alkali to the bacterial suspension results in the predominance of one type of flagella. The change in morphology occurs within a very narrow range of pH values; in an acid medium, the culture possesses predominantly 'curly', short wavelength, flagella and in normal or alkaline media the normal form predominates. Intermediate forms can be found where the bacteria possess flagella having both long and short wavelengths. These findings were not confirmed by Pijper, Neser & Abraham (1956) who, in a survey of flagella wavelengths, were unable to show similar pH effects. They were, however, able to demonstrate some effect of the pH value of the medium in that, with strains of *Salmonella*, *Sarcina* and *Proteus*, it was found that the wavelength of the flagellum is at a minimum at pH 7, and increasing alkalinity or acidity results in an increase in the wavelength. The motility of *Bacillus brevis* was found by Shoesmith (1960) to be dependent on the pH value of the suspending medium. At pH 7 all the bacteria were motile, having an average speed of 21 μ/sec.; lowering the pH value to 5 resulted in a reduction in the average speed to 10 μ/sec. and a fall in the number of motile bacteria. The temperature of the suspending medium was also found to have an effect on the wavelength of the flagellum; the wavelength increasing with rise in temperature. Changes in wavelength resulting from variations in the temperature or the pH value of the medium were freely reversible.

Purine and pyrimidine analogues. Since the finding by Matthews (1953) that the purine analogue 8-azaguanine is incorporated into the RNA of tobacco mosaic virus, numerous workers have studied the effects of purine and pyrimidine analogues on nucleic acid and protein synthesis in bacteria. In bacteria, the general effect of incubation in the presence of 8-azaguanine is to inhibit protein synthesis, but not RNA, DNA, or cell-wall synthesis (Richmond, 1959; Chantrenne & Devreux, 1960). Creaser (1956) produced evidence that in *Staphylococcus aureus* the synthesis of inducible enzymes is more sensitive to inhibition by 8-azaguanine than the synthesis of constitutive enzymes. Richmond (1959) was unable to show a similar differentiation in *Bacillus subtilis* R. Incubation of *B. megaterium* in the presence of 2-thiouracil results in linear growth and the incorporation of the analogue into the RNA replacing uracil (Hamers, 1956). Jeener, Hamers-Casterman & Mairesse (1959) showed that after induction of a lysogenic strain of *B. megaterium*, 2-thiouracil exerts a greater inhibitory effect on the synthesis of bacteriophage nucleic acid and protein than on the synthesis of bacterial proteins. Although 6-azauracil inhibits the synthesis of DNA by *Escherichia coli* B, it was suggested that the effects of the analogue are not necessarily due to incorporation of the inhibitor into the nucleic acid, but could be due to an inhibition of cell-wall synthesis (Otsugi & Takagi, 1959). 5-Fluorouracil is incorporated into the RNA of *E. coli* (Cohen *et al.* 1958) and is also converted into fluorodeoxyuridine (FDUR) which interferes with the functioning of the thymidylate synthetase, so inhibiting the formation of DNA.

A number of these analogues have a marked inhibitory effect on the growth of *Salmonella typhimurium*, strain SW 1061, but none prevent the regeneration of functional flagella after the flagella have been mechanically detached from the bacteria (Kerridge, 1960). Microscopic examination of stained preparations of the bacteria after 2 hr. incubation in the presence of the analogue, showed that for bacteria incubated in the presence of either 5-fluoro-uracil or FDUR there is a marked increase in the number of flagella/bacterium compared with the control culture (Table 4); for example, after regeneration in the presence of

Table 4. *The effect of purine and pyrimidine analogues on nucleic acid and protein synthesis and the regeneration of flagella by* Salmonella typhimurium, *strain SW* 1061

Analogue	Conc. $\times 10^{-4}$ M	Regeneration of flagella after 2 hr. incubation: Flagellated bacteria (%)	Mean no. flagella/ bacterium	Percentage inhibition of synthesis after 2 hr. incubation: Total nucleic acid	DNA	Protein
Control	0	96	3·1	—	—	—
8-Azaguanine	6·5	95	2·6	50	16	51
2-Thiouracil	30	80	2·4	86	73	88
6-Azauracil	4·3	65	1·9	92	58	100
5-Fluorouracil	7·4	95	7·9	—	—	—
Fluorodeoxyuridine	1·0	48	5·2	38	90	36

The organisms were incubated aerobically at 37° for 2 hr. in a buffered salts medium containing Difco vitamin-free casamino acids (0·2%, w/v), glucose (0·2%, w/v) and the analogues as required. Total nucleic acid was determined by measuring the extinction at 260 mμ of the 0·5 N perchloric acid extract; DNA was determined by the method of Burton (1956); protein was measured using the method of Lowry, Rosebrough, Farr & Randall (1951). The regeneration of flagella was determined by examining stained preparations of the bacteria.

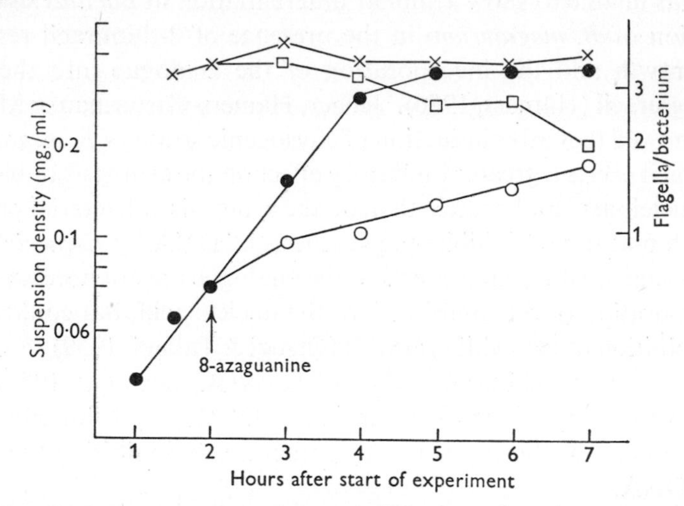

Fig. 2. The effect of 8-azaguanine on the optical density and on the mean no. of flagella/ bacterium of *Salmonella typhimurium* growing logarithmically in a casamino acids–glucose medium. —●—●—, suspension density of control; —○—○—, suspension density + 0·001 M 8-azaguanine; —×—×—, flagella/bacterium of control. —□—□—, Flagella/ bacterium, + 0·001 M 8-azaguanine.

5-fluorouracil (0·0007 M), there were 7·9 flagella/bacterium compared with 3·7 in the control, while with FDUR (0·0001 M) there were 5·2 flagella/bacterium. This effect was not observed with the other analogues tested and may have resulted from an inhibition of cell division in the absence of a comparable effect on the formation of flagella.

During logarithmic growth of *Salmonella typhimurium* in minimal medium at 37° the mean number of flagella/bacterium remains approximately constant. Addition of either 2-thiouracil or 8-azaguanine (0·001 M) results in immediate reduction in the growth rate of the bacteria (Fig. 2). There is little change in the number of flagella/bacterium for the first 2 hr. after the addition of the analogue and then the number of flagella/bacterium decreases until, 5 hr. after the addition of the inhibitor, it has fallen to approximately half that of the control culture. The bacteria are still motile at the end of the incubation, 2-thiouracil and 8-azaguanine apparently having no effect on the functioning of preformed flagella. By analogy with the results obtained for regeneration in the presence of 5-fluorouracil and FDUR, it would appear that both 2-thiouracil and 8-azaguanine inhibit the formation of new flagella without completely inhibiting bacterial growth and cell division, so that the number of flagella/bacterium is reduced.

THE EFFECT OF TEMPERATURE ON BACTERIAL MOTILITY AND THE FORMATION OF FLAGELLA

In general, studies on the effects of temperature on biological systems have concentrated on the thermal denaturation of proteins and the thermal death-point of bacteria; there have been few investigations of the effects of temperature on bacterial morphology and physiology (see Johnson, 1957). I exclude from this generalization the studies on thermophilic and cryophilic bacteria, and limit my remarks to the effects of extremes of temperature on mesophilic bacteria.

In the majority of cases, raising the growth temperature of the bacteria results in the loss of certain cellular functions. This loss of function can result in the production of filamentous forms of *Acetobacter* spp. (Hansen, 1894); additional nutritional requirements (Ware, 1951; Hills & Spurr, 1952; Maas & Davis, 1952); or the failure to produce a particular inducible enzyme (Pollock, 1945; Knox, 1950, 1951, 1953; Knox & Collard, 1952). Various explanations have been given for these effects. Hills & Spurr (1952) suggested that the effects of temperature on nutritional requirements for growth could be due to the enzymic processes leading to the synthesis of essential nutrients being better coordinated at the optimum temperature. At the higher temperature, catabolic processes leading to the death of the organisms are accelerated relative to synthetic processes and the organisms become dependent on a wider range of preformed nutrients. The additional requirement for pantothenate for growth above 30° by a temperature-sensitive mutant of *Escherichia coli* was found by Maas & Davis (1952) to result from the production of an excessively heat-labile pantothenate-synthesizing enzyme.

Similarly, certain strains of bacteria become non-motile on incubation at high growth temperatures. One exception is a motile variant isolated from an O strain of *Salmonella paratyphi* C, which is non-motile after growth at 20° but motile after growth at 37° (Quadling, 1958). Early observations stressed the importance of this loss of motility in relationship to bacterial classification (Arkwright, 1927; Jordan, Caldwell & Reiter, 1934; Boquet, 1937) and to the distribution of the flagella on the bacterial surface (Paterson, 1939; Griffin & Robbins, 1944). Although these workers had shown that growth at high temperatures results in the loss of motility, it was not until the studies of Preston & Maitland (1952) that any attempt was made to correlate the growth temperature with the ability of the bacteria to produce flagella. *Pasteurella pseudotuberculosis* is flagellated and motile after growth at 22° but not after growth at 37°. Preston & Maitland (1952) found that transferring a non-motile stationary phase culture of this organism to 22° after 4 days growth at 37° resulted in 30% of the bacteria becoming motile within 24 hr. The authors claimed that the formation of flagella occurred in the absence of cell division but, as only 30% of the bacteria were viable at the time of transfer, this interpretation is open to criticism. Raising the incubation temperature after growth at 22° resulted in a fall in the number of motile bacteria, less than 5% remaining motile after 2 days at 37°; here again interpretation is difficult as the viable count fell to 12% by the end of the incubation period. It is implicit here that the loss of motility was due to the death of the bacteria, rather than failure to produce flagellated cells.

More definite data on the effect of incubation temperature on the formation of bacterial flagella were obtained by Quadling & Stocker (1956) and Quadling (1958). The loss of motility of *Salmonella typhimurium* strain LT2 associated with growth of the bacteria at 44° was found to be due to a failure to produce new flagella, rather than to impairment of function of pre-existing flagella. Similarly, Bisset & Pease (1957) found that growth of both *S. typhimurium* and *Proteus vulgaris* at 44° resulted in the production of non-motile bacteria. In contrast to Quadling & Stocker (1956), who incubated the bacteria in a liquid medium, Bisset & Pease examined microcolonies produced from single flagellate bacteria after incubation at 44° and found that pre-existing flagella were retained by the parent bacterium.

Results similar to those of Quadling & Stocker (1956) were obtained by Kerridge (1960) and a typical experimental result is shown in Fig. 3. *Salmonella typhimurium*, strain SW 1061, continues to grow logarithmically after raising the temperature from 37° to 44°. The number of flagella/bacterium remains approximately constant for one generation time after the transfer, and then shows a progressive fall. Eventually the majority of bacteria, although still motile, possess only one flagellum; on continued growth at 44° these motile bacteria are diluted out and the

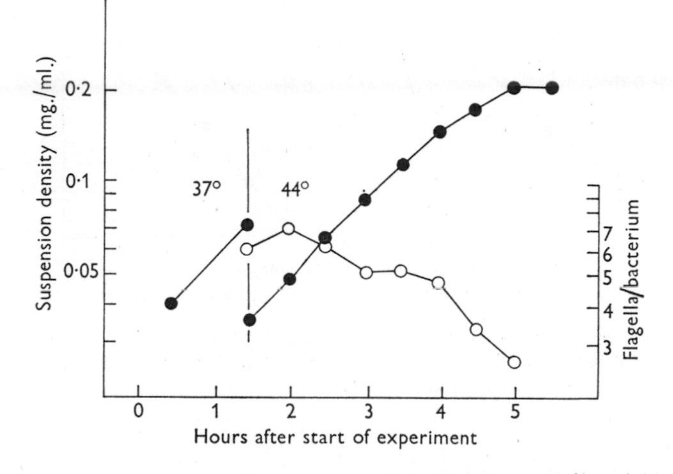

Fig. 3. The effect of raising the temperature of incubation from 37° to 44° on the optical density and on the mean number of flagella/bacterium of *Salmonella typhimurium* growing in a casamino acids–glucose medium. —●—●—, Suspension density, —○—○—, flagella/bacterium.

culture becomes non-motile. The change to a non-motile state during growth at 44° is a phenotypic effect resulting from the increased temperature, and on transferring a logarithmically growing non-motile culture from 44° to 37° the bacteria recover their motility after a short lag. Fig. 4 gives a comparison between the regeneration of flagella after their mechanical removal and the recovery of motility after growth at 44°. The rates of growth and recovery of motility are similar in both cases, but there is a considerable difference in the time before the appearance of flagella. The rates of incorporation of ^{14}C labelled leucine into the general cell protein and the flagella again support the view that flagellin is a constitutive protein and growth at 44° does not affect the subsequent rate of synthesis of flagella during incubation at 37°.

Regeneration of flagella after mechanical removal can occur at 44°, so it is unlikely that the loss of motility during growth at 44° is due to an inhibition of synthesis of flagella *per se*. One possible explanation for the failure to produce flagella at 44° is that the system involved in the formation of a flagellum is not produced at this temperature. The lag before the appearance of flagella on lowering the growth temperature from 44° to 37° would then represent the time required by the bacteria to form these systems.

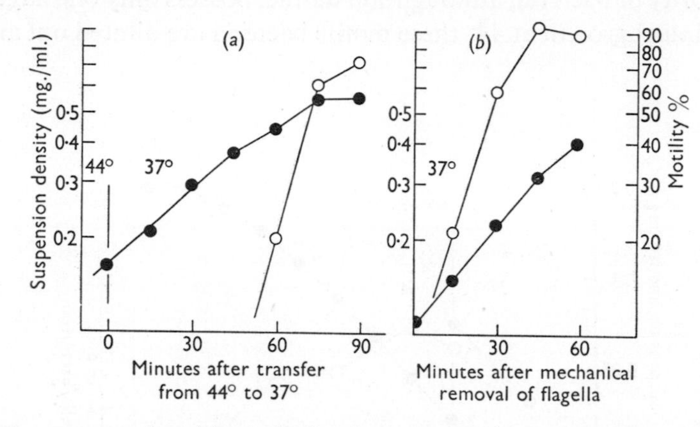

Fig. 4. The recovery of motility during incubation at 37° either after growth at 44° (*a*) or after mechanical removal of flagella in the Waring blendor (*b*). —●—●—, Suspension density; —○—○—, percentage motile bacteria.

SITE OF FLAGELLUM SYNTHESIS IN BACTERIA

There is very good evidence that the microsomal particles of mammalian tissues are involved in the synthesis of proteins. In *in vivo* studies, Littlefield, Keller, Gross & Zamecnik (1955) found immediately after administering a single ^{14}C labelled amino acid, that the amino acid recovered from the ribonucleo-protein particles of guinea-pig liver was more heavily labelled than that recovered from other cell fractions. By *in vitro* studies, Kirsch, Seikevitz & Palade (1960) have shown that these ribonucleo-protein particles can incorporate ^{14}C labelled amino acids into protein.

There is rather conflicting evidence concerning the site of protein synthesis in bacteria. Butler and his collaborators (Hunter, Crathorn & Butler, 1957; Butler, Crathorn & Hunter, 1958; Hunter, Brookes, Crathorn & Butler, 1959), and also Beljanski & Ochoa (1958*a*, *b*), and Spiegelman (1958) have presented evidence for the involvement of the bacterial cytoplasmic membrane in protein synthesis, and have shown that under certain conditions their membrane preparations can incorporate ^{14}C labelled amino acids into proteins. In contrast, McQuillen, Roberts & Britten (1959), studying the incorporation of $^{35}SO_4^{++}$

and ^{14}C-labelled amino acids into *Escherichia coli*, found that the radioactivity rapidly appeared in protein bound to ribosomes having sedimentation constants of 70 and 80 Svedbergs. The rates of incorporation were such that it was possible to account for all the protein synthesis of the cell by this pathway. The discrepancy between these findings on the site of bacterial protein synthesis may be accounted for by the failure to ensure a complete separation of ribosomes from the bacterial membranes.

Although synthesis of flagellin molecules may be mediated by individual ribosomes, aggregation of the molecules into functional flagella will require more complex systems. It is probable that the synthesis of individual flagellin molecules and their co-ordination to form a flagellum occurs at closely linked sites in the cell. However, a recent paper by Weinstein, Koffler & Moskowitz (1960) suggests that this may not necessarily be the case, since they were able to detect flagellin in the supernatant after high-speed centrifugation of the lysate obtained from penicillin-induced spheroplasts of *Proteus vulgaris*. The internal flagellin was of the order of 1–3 % of the extracellular flagellin. No other method of cell breakage was reported and it is possible that the presence of flagellin in the soluble fraction may be a direct result of the method of breakage. Penicillin does not itself inhibit the formation of flagella by *Salmonella typhimurium*, but it has been impossible to detect the synthesis of flagella (Kerridge, 1960) after the formation of penicillin-induced spheroplasts. It may be that, although penicillin does not prevent the formation of individual flagellin molecules, the disorganization of the cell walls produced by the antibiotic in some way prevents the formation of flagella.

INTRACELLULAR STRUCTURES ASSOCIATED WITH THE FLAGELLA IN BACTERIA

Cytological studies on certain motile bacteria have shown that in many cases the flagella originate from basal granules within the cell. These granules (blepharoplasts) have been demonstrated in autolysed preparations from *Vibrio metchnikovii*, *Proteus vulgaris*, *Chromobacterium violaceum* (van Iterson, 1953); *V. chlorae*, *Rhodospirillum rubrum* (Grace, 1954) and *V. comma* (Tawara, 1957). With the exception of *R. rubrum*, where several flagella are attached to one basal granule, each flagellum arises from a single basal granule. The electron micrographs are very convincing, but it is a little unfortunate that in all these cases the preparations were from old or autolysed preparations, and can be criticized on the grounds that the granules could be artefacts, arising from an aggregation of cellular material around the base of the flagellum during ageing or autolysis. In this connexion, it is interesting to note that Weibull (1960) reports that he was unable to demonstrate granules at the bases of flagella released by osmotic lysis of protoplasts of *Bacillus megaterium.* In their papers on the unilinear transmission of motility in *Salmonella typhimurium,* Stocker & Quadling (Stocker, 1956*b*; Quadling & Stocker, 1957; Quadling, 1958) refer to a 'motility conferring (MC) particle'. In the first paper, Stocker states that the MC particle could be either a basal granule or a flagellum but, in a later paper, Quadling & Stocker (1957) identify the MC particle with the flagellum.

THE EFFECT OF INHIBITORS ON THE RECOVERY OF MOTILITY BY *SALMONELLA TYPHIMURIUM* AFTER GROWTH AT 44°

It has been shown above that *Salmonella typhimurium* grown at 44° and then transferred to 37° proceed to produce flagella after a lag period. If this lag is due to the time required by the cells to synthesize a system involved in the formation of the flagellum, then investigation of the factors affecting this lag should provide information on the nature of this system. It is, of course, necessary to distinguish between inhibition of formation of the synthesizing system, and inhibition of synthesis of the flagellum itself. As none of the purine, pyrimidine or amino acid analogues tested prevented regeneration of flagella by *Salmonella typhimurium,* strain SW 1061, it is possible to test directly the effect of these compounds on the formation of the generating system (Kerridge, 1960).

The addition of either 2-thiouracil or 8-azaguanine (0·001 M) to a logarithmically growing non-motile culture of *Salmonella typhimurium* immediately after changing the incubation temperature from 44° to 37°, results in a progressive inhibition of growth (Fig. 5). After 4 hr. incubation at 37° less than 1% of the bacteria are motile, although in the control culture 80% of the bacteria are motile after 2 hr. incubation at 37°. As neither analogue prevents regeneration of functional flagella by deflagellated *S. typhimurium,* it would appear probable that they are affecting the formation of the synthesizing system. Data supporting this view have been obtained by delaying the time of addition of 8-azaguanine (0·001 M) after the change in the incubation temperature. The results are shown in Fig. 6. Addition of 8-azaguanine either immediately after reducing the temperature or 30 min. later has a marked inhibitory effect on the recovery of motility; if however the analogue is added 60 min. after lowering the temperature, the bacteria recover their motility, even though at the time of addition less than 5 % are motile.

FDUR is a specific inhibitor of DNA synthesis in *Escherichia coli* and addition of this compound to a culture of the bacteria results in 'thymineless death' (Cohen *et al.* 1958). Unlike thiouracil and 8-azaguanine,

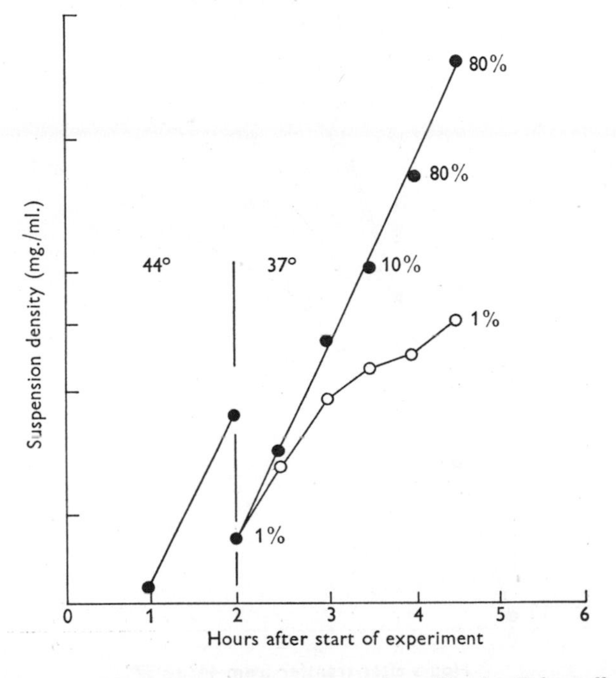

Fig. 5. The effect of 2-thiouracil on the recovery of motility by *Salmonella typhimurium* during incubation at 37° after growth at 44°. —●—●—, Suspension density, control. —○—○—, Suspension density, +0·001 M 2-thiouracil. The figures next to the curves give the percentage motile bacteria at the corresponding times.

FDUR does not completely prevent the formation of flagella by *Salmonella typhimurium* after growth at 44° (Fig. 7). At a concentration of 10^{-4} M, FDUR caused a reduction in the optical density of the bacterial suspension, probably due to cell lysis, but after 3 hr. incubation, microscopic examination of the suspension showed that approximately 15 % of the bacteria were motile and flagellated.

None of the amino acid analogues tested prevented the regeneration of flagella, but both *p*-FP and *β*-2-thienylalanine (0·005 M) inhibited the recovery of motility by non-motile cultures of *Salmonella typhimurium* after alteration of the growth temperature. The results for *p*-FP are shown in Fig. 8; *β*-2-thienylalanine gave essentially similar results.

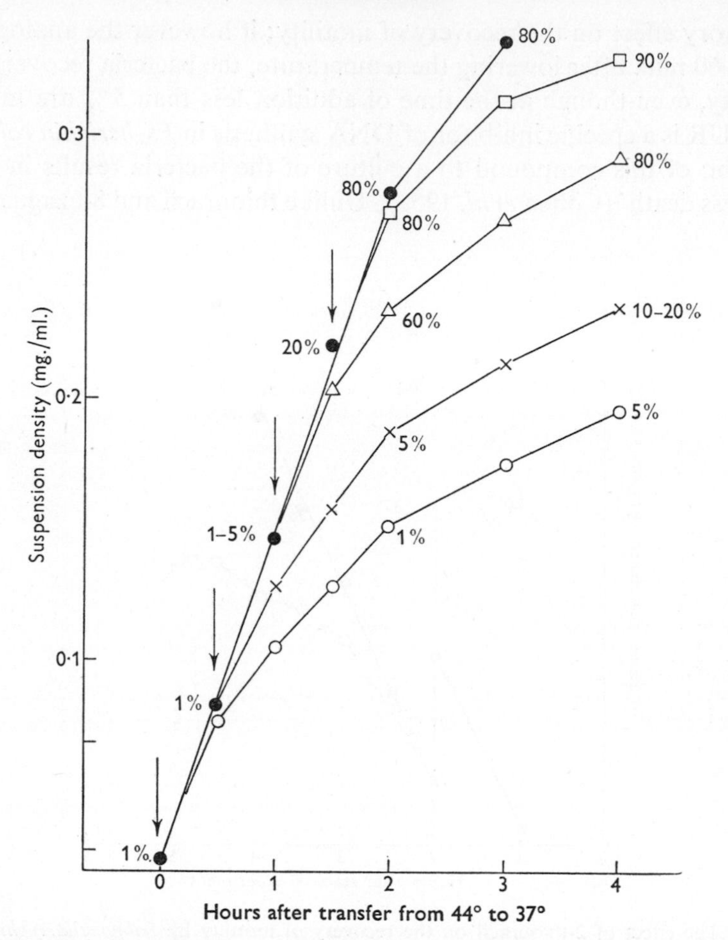

Fig. 6. The effect of 8-azaguanine on the recovery of motility by *Salmonella typhimurium* when added at different times during growth at 37° after growth at 44°. 8-azaguanine (0·001 M) added at the following times after transfer to 37°: 0 min., —○—○—; 30 min., —×—×—; 60 min., —△—△—; 90 min., —□—□—; control, —●—●—. The figures next to the curves give the percentage motile bacteria at the corresponding times.

CONCLUSION AND SPECULATION

Many of the studies of the effect of environment on bacterial flagella have been concerned primarily with the functioning of the flagella and not with their formation. Pijper (1947) and Shoesmith (1960) studied the effect of viscosity on bacterial motility and found that it was possible to increase the viscosity considerably without affecting motility, but that motility was reduced in media of high viscosity; neither of them attempted to study the effects of viscosity on the formation of flagella by growing cultures of the bacteria. De Robertis & Peluffo (1951)

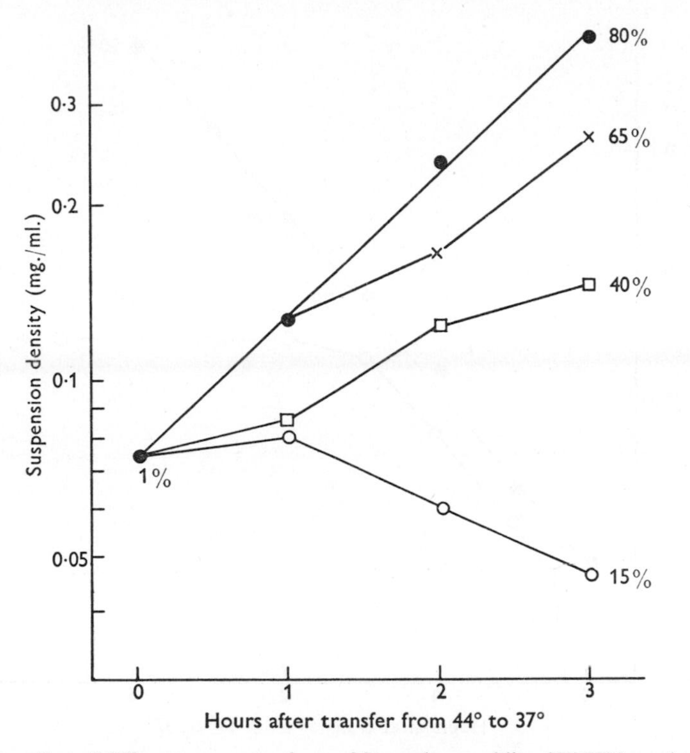

Fig. 7. The effect of different concentrations of fluorodeoxyuridine (FDUR) on the recovery of motility during incubation at 37° by *Salmonella typhimurium* previously grown at 44°. —●—●—, Control; —×—×—, $+10^{-6}$ M FDUR; —□—□—, $+10^{-5}$ M FDUR; —○—○—, $+10^{-4}$ M FDUR. The figures next to the curves give the percentage motile bacteria at the corresponding times.

studied chemical stimulation and inhibition of motility of *Proteus vulgaris* but did not test the effects of any of the compounds on the formation of the flagella. Studies of this type should not be limited to bacterial motility, but should be extended to include both the formation and morphology of the bacterial flagellum.

The data on the effects of the purine and pyrimidine analogues on the formation of flagella have a marked similarity to those obtained by Jeener *et al.* (1959) in their studies on inhibition of bacteriophage synthesis in an induced lysogenic strain of *Bacillus megaterium*. They interpreted their findings by postulating that, in contrast to the bacterial RNA, the RNA involved in the synthesis of phage protein was entirely formed after the addition of the analogue. The fraction of RNA which determined phage protein specificity would then undergo structural modifications on a much larger scale than the RNA determining the specificity of the bacterial proteins. The selective action of thiouracil

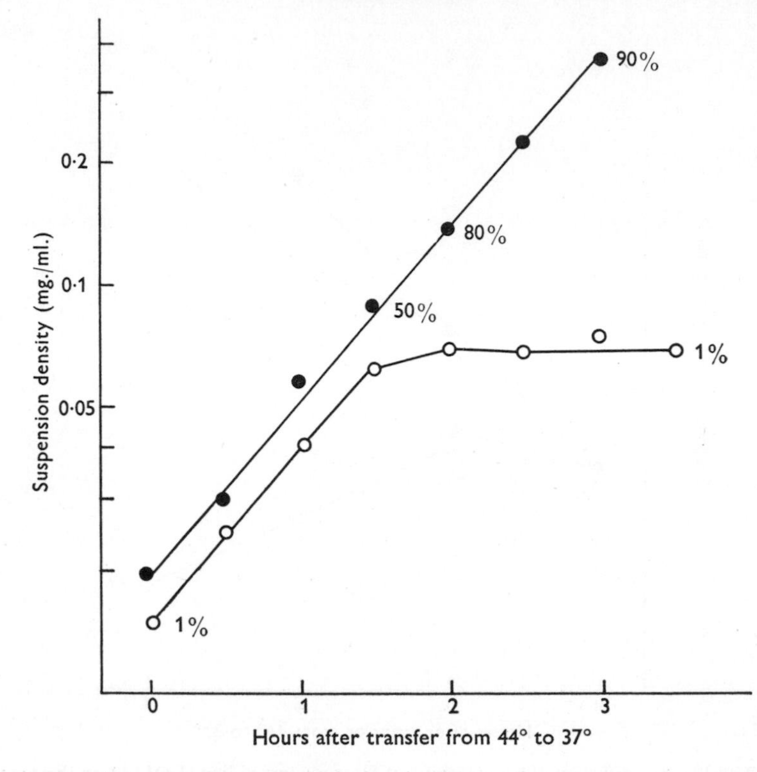

Fig. 8. The effect of *p*-fluorophenylalanine (*p*-FP) on the recovery of motility during incubation at 37° by *Salmonella typhimurium* previously grown at 44°. —●—●—, Control, —○—○— +0·005 M *p*-FP. The figures next to the curves give the percentage motile bacteria at the corresponding times.

and azaguanine on phage protein synthesis would thus be easily explained.

The most probable explanation of the lack of motility of *Salmonella typhimurium* grown at 44° is that the increased temperature exerts a differential effect on the growth of the bacteria and on the formation of the functional systems involved in the synthesis of the flagella. The lag before the recovery of motility after lowering the incubation temperature to 37° (Fig. 4*a*) would then be explained by assuming that this is the time required by the bacteria for the development of the synthesizing systems. As flagella consist entirely of protein it is a logical assumption that this synthesizing system will contain RNA, and by analogy with the results of Jeener *et al.* (1959), the presence of thiouracil or azaguanine would probably result in the formation of a non-functional system.

So far it has been impossible to locate the system involved in the synthesis of flagella by *Salmonella typhimurium* by either cytological or biochemical techniques (Kerridge, unpublished data). It is, however, possible to assign certain properties to this synthesizing system.

1. The synthesizing system is not self-reproducing (cf. van Iterson, 1953), unless, of course, the high growth temperatures result in the formation of systems incapable of the synthesis of flagella, but capable of replication.
2. The system has a limited life. For example, one does not normally find bacteria with immensely long flagella, and in experiments involving the regeneration of flagella, there are always fewer flagella/bacterium after regeneration than in the control culture. This latter effect could, however, be due to damage to the synthesizing system as a result of ripping-out the flagellum.
3. If the synthesizing system is involved in both the synthesis and functioning of the flagellum, then the functional life greatly exceeds the synthetic life.
4. The results obtained with the analogues, thiouracil and azaguanine, suggest that synthesis of a biologically active RNA is necessary for the formation of a functional synthesizing system, but that the concomitant synthesis of a functional RNA is not necessary for the synthesis of a flagellum by an existent synthesizing system.
5. The increase in the number of flagella/bacterium after regeneration in the presence of 5-fluoro-uracil and FDUR, and the ability of the bacteria to recover their motility when incubated at 37° in the presence of FDUR, after growth at 44°, implies that the synthesis of DNA is not necessary for the formation of the synthesizing system.
6. The effects of the amino acid analogues, *p*-FP and β-2-thienylalanine, suggest that synthesis of protein is also involved in the formation of the synthesizing system.
7. Incubation of *Salmonella typhimurium*, in the presence of 8-azaguanine at 44°, has no inhibitory effect on the subsequent formation of flagella at 37° after the removal of the inhibitor. This would suggest that high molecular weight precursors of the synthesizing system are absent from bacteria grown at 44°, but are synthesized *de novo* after lowering the incubation temperature.

Although the data suggest that a ribonucleo-protein complex is involved in the synthesis of the flagellum, it is impossible to confirm this until the synthesizing system has been isolated and characterized. If the basal granules that have been demonstrated in other bacteria are equivalent to the synthesizing system in *Salmonella typhimurium*, it should be possible, using normal fractionation techniques, to isolate the units responsible for the formation of flagella. One other very important problem that so far has not been mentioned in this article is the mechanism of energy supply, and the conversion of chemical to mechanical energy. If the energy-supplying mechanism is closely associated with the synthetic mechanism, the isolation of an intracellular structure associated with the flagellum might then provide information on the energetics of motility.

REFERENCES

AMBLER, R. P. & REES, M. W. (1959). ϵ-*N*-Methyl lysine in bacterial flagellar protein. *Nature, Lond.* **186**, 56.

ARKWRIGHT, J. A. (1927). The importance of motility of bacteria in classification and diagnosis with special reference to *B. pseudotuberculosis rodentium*. *Lancet*, i, 13.

ASTBURY, W. T., BEIGHTON, E. & WEIBULL, C. (1955). The structure of bacterial flagella. In *Fibrous Proteins and their Biological Significance*. *Symp. Soc. exp. Biol.* **9**, 282.

BAKER, R. S., JOHNSON, J. E. & FOX, S. W. (1958). Incorporation of *p*-fluorophenylalanine into proteins of *Lactobacillus arabinosus*. *Biochim. biophys. Acta*, **28**, 318.

BALTEANU, I. (1926). The receptor structure of *Vibrio cholerae* (*V. comma*) with observations in cholera and cholera-like organisms. *J. Path. Bact.* **29**, 251.

BARLOW, G. H. & BLUM, J. J. (1952). On the 'contractibility' of bacterial flagella. *Science*, **116**, 572.

BELJANSKI, M. & OCHOA, S. (1958*a*). Protein synthesis by a cell-free bacterial system. *Proc. nat. Acad. Sci., Wash.* **44**, 494.

BELJANSKI, M. & OCHOA, S. (1958*b*). Protein synthesis by a cell-free bacterial system. II. Further studies on the amino acid incorporation enzyme. *Proc. nat. Acad. Sci., Wash.* **44**, 1157.

BISSET, K. A. (1951). The development of the surface structure in dividing bacteria. *J. gen. Microbiol.* **5**, 155.

BISSET, K. A. & HALE, C. M. F. (1951). The development of bacterial flagella in the germinating microcyst. *J. gen. Microbiol.* **5**, 150.

BISSET, K. A. & HALE, C. M. F. (1960). Flagellar pattern and growth of *Bacillus* spp. *J. gen. Microbiol.* **22**, 536.

BISSET, K. A. & PEASE, P. (1957). The distribution of flagella in dividing bacteria. *J. gen. Microbiol.* **16**, 382.

BOIVIN, A. & MESROBEANU, L. (1938). Sur la résistance à l'acide trichloroacétique de l'antigène flagellaire (antigène H) du bacille typhique et sur la nature chimique possible de cet antigène. *C.R. Soc. Biol., Paris*, **129**, 136.

BOQUET, P. (1937). Recherches expérimentales sur la pseudo-tuberculose des rongeurs. *Ann. Inst. Pasteur*, **59**, 341.

BOWERS, L. E., WEAVER, R. H., GRULA, E. A. & EDWARDS, O. F. (1954). Studies on a strain of *Caulobacter* from water. I. Isolation and identification as *Caulobacter vibroides* Henrici & Johnson, with emended description. *J. Bact.* **68**, 194.

BROKAW, C. J. (1960). Decreased adenosine triphosphatase activity of flagella from a paralysed mutant of *Chlamydomonas moewusii*. *Exp. Cell Res.* **19**, 430.

BURTON, K. (1956). A study of the conditions and mechanism of the diphenylamine reaction for the colorimetric estimation of desoxyribonucleic acid. *Biochem. J.* **62**, 315.

BUTLER, J. A. V., CRATHORN, A. R. & HUNTER, G. D. (1958). The site of protein synthesis in *Bacillus megaterium*. *Biochem. J.* **69**, 544.

CHANTRENNE, H. & DEVREUX, S. (1960). Action de l'8-azaguanine sur la synthèse des protéines et des acides nucléïques chez *Bacillus cereus*. *Biochim. biophys. Acta*, **39**, 486.

COHEN, S. S., FLAKS, J. G., BARNER, H. D., LOEB, H. R. & LICHTENSTEIN, J. (1958). The mode of action of 5-fluorouracil and its derivatives. *Proc. nat. Acad. Sci., Wash.* **44**, 1004.

CRAIGIE, J. (1931). Studies on the serological reactions of the flagella of *B. typhosus*. *J. Immunol.* **21**, 417.

CREASER, E. H. (1956). The assimilation of amino acids by bacteria. 22. The effect of 8-azaguanine upon enzyme formation in *Staphylococcus aureus*. *Biochem. J.* **64**, 539.

FOX, E. N. & KRAMPITZ, L. O. (1956). Studies on the biosynthesis of the M protein of group A haemolytic streptococci. *J. Bact.* **71**, 454.

GARD, S. (1944). Preparation of bacterial flagella. *Ark. Kemi Min. Geol.* **19** A, no. 21.

GRACE, J. B. (1954). Some observations on the flagella and blepharoplasts of *Spirillum* and *Vibrio* spp. *J. gen. Microbiol.* **10**, 325.

GRIFFIN, A. M. & ROBBINS, H. L. (1944). The flagellation of *Listeria monocytogenes*. *J. Bact.* **48**, 114.

HAMERS, R. (1956). Incorporation of ^{35}S thiouracil in *Bacillus megaterium*. *Biochim. biophys. Acta*, **21**, 170.

HANSEN, E. C. (1894). Recherches sur les bactéries acétifiantes. *C.R. Lab. Carlsberg*, **3**, 182.

HILLS, G. M. & SPURR, E. D. (1952). The effect of temperature on the nutritional requirements of *Pasteurella pestis*. *J. gen. Microbiol.* **6**, 64.

HOUWINK, A. L. (1951). *Caulobacter* versus *Bacillus* spec.div. *Nature, Lond.* **168**, 654.

HUNTER, G. D., BROOKES, P., CRATHORN, A. R. & BUTLER, J. A. V. (1959). Intermediate reactions in protein synthesis by isolated cytoplasmic membrane fraction of *Bacillus megaterium*. *Biochem. J.* **73**, 369.

HUNTER, D. G., CRATHORN, A. R. & BUTLER, J. A. V. (1957). Sites of the incorporation of an amino acid into proteins of *Bacillus megaterium*. *Nature, Lond.* **180**, 383.

JACHERTS, D. (1960). Untersuchungen über die Bildung von Geisseln von *Pseudomonas aeruginosa* in synchronisierten Kulturen. *Zbl. Bakt.* (II. Abt), **113**, 111.

JEENER, R., HAMERS-CASTERMAN, C. & MAIRESSE, N. (1959). On the inhibition of phage production by 2-thiouracil and 8-azaguanine in an induced lysogenic *Bacillus megaterium*. *Biochim. biophys. Acta*, **35**, 166.

JOHNSON, F. H. (1957). The action of pressure and temperature. In *Microbial Ecology*. *Symp. Soc. gen. Microbiol.* **7**, 134.

JORDAN, E. O., CALDWELL, M. E. & REITER, D. (1934). Bacterial motility. *J. Bact.* **27**, 165.

KERRIDGE, D. (1959*a*). The effect of amino acid analogues on the synthesis of bacterial flagella. *Biochim. biophys. Acta*, **31**, 579.

KERRIDGE, D. (1959*b*). Synthesis of flagella by amino acid-requiring mutants of *Salmonella typhimurium*. *J. gen. Microbiol.* **21**, 168.

KERRIDGE, D. (1960). The effects of inhibitors on the formation of flagella by *Salmonella typhimurium*. *J. gen. Microbiol.* **23**, 519.

KIRSCH, J. F., SIEKEVITZ, P. & PALADE, G. E. (1960). Amino acid incorporation *in vitro* by ribonucleoprotein particles detached from guinea pig liver microsomes. *J. biol. Chem.* **235**, 1419.

KNOX, R. (1950). Tetrathionase: the differential effect of temperature on growth and adaptation. *J. gen. Microbiol.* **4**, 388.

KNOX, R. (1951). The formation of bacterial urease. *J. gen. Microbiol.* **5**, xx.

KNOX, R. (1953). The effect of temperature on enzymic adaptation, growth and drug resistance. In *Adaptation in Micro-organisms*. *Symp. Soc. gen. Microbiol.* **3**, 184.

KNOX, R. & COLLARD, P. (1952). The effect of temperature on the sensitivity of *Bacillus cereus* to penicillin. *J. gen. Microbiol.* **6**, 369.

KOBAYASHI, T., RINKER, J. N. & KOFFLER, H. (1959). Purification and chemical properties of flagellin. *Arch. Biochem. Biophys.* **84**, 342.

KOFFLER, H. (1957). Protoplasmic differences between mesophiles and thermophiles. *Bact. Rev.* **21**, 227.

KVITTINGEN, J. (1955). Some observations on the nature and significance of bacterial flagella. *Acta path. microbiol. scand.* **37**, 89.

LANCEFIELD, R. C. (1943). Studies on the antigenic composition of Group A haemolytic streptococci. I. Effects of proteolytic enzymes on streptococcal cells. *J. exp. Med.* **78**, 465.

LEIFSON, E. (1931). Development of flagella on germinating spores. *J. Bact.* **21**, 357.

LEIFSON, E., CARHART, S. R. & FULTON, M. (1954). Morphological characteristics of flagella of *Proteus* and related species. *J. Bact.* **69**, 73.

LITTLEFIELD, J. W., KELLER, E. B., GROSS, J. & ZAMECNIK, P. C. (1955). Studies on cytoplasmic ribonucleoprotein particles from the liver of the rat. *J. biol. Chem.* **217**, 111.

LOWRY, O. H., ROSEBROUGH, N. J., FARR, A. L. & RANDALL, R. J. (1951). Protein measurements with the Folin phenol reagent. *J. biol. Chem.* **193**, 265.

MAAS, W. K. & DAVIS, B. D. (1952). Production of an altered pantothenate-synthesising enzyme by a temperature-sensitive mutant of *Escherichia coli*. *Proc. nat. Acad. Sci., Wash.* **38**, 785.

MANDELSTAM, J. (1957). Turnover of protein in starved bacteria and its relationship to the induced synthesis of enzyme. *Nature, Lond.* **179**, 1181.

MANDELSTAM, J. (1958). Turnover of protein in growing and non-growing populations of *Escherichia coli*. *Biochem. J.* **69**, 110.

MANDELSTAM, J. & HALVORSON, H. (1960). Turnover of protein and nucleic acid in soluble and ribosome fractions of non-growing *Escherichia coli*. *Biochim. biophys. Acta*, **46**, 43.

MATTHEWS, R. E. F. (1953). Incorporation of 8-azaguanine into nucleic acid of T.M.V. *Nature, Lond.* **171**, 1065.

MCQUILLEN, K., ROBERTS, R. B. & BRITTEN, R. J. (1959). Synthesis of nascent protein by ribosomes in *Escherichia coli*. *Proc. nat. Acad. Sci., Wash.* **45**, 1437.

MIGULA, W. (1897). *System der Bakterien*. Jena: Fischer.

MUNIER, R. & COHEN, G. N. (1956). Incorporation d'analogues structuraux d'amino acides dans les protéines bactériennes. *Biochim. biophys. Acta*, **21**, 593.

MUNIER, R. & COHEN, G. N. (1959). Incorporation d'analogues structuraux dans les protéines bactériennes au cours de leur synthèses *in vivo*. *Biochim. biophys. Acta*, **31**, 378.

ORCUTT, M. L. (1924). Flagellar agglutinins. *J. exp. Med.* **40**, 43.

OTSUGI, N. & TAKAGI, Y. (1959). Effect of 6-azauracil on cells and sub-cellular preparations of *Escherichia coli*. *J. Biochem.* (*Tokyo*), **46**, 791.

PARDEE, A. B. & PRESTRIDGE, L. S. (1956). The dependence of nucleic acid synthesis on the presence of amino acids in *Escherichia coli*. *J. Bact.* **71**, 677.

PATERSON, J. S. (1939). Flagellar antigens of the genus *Listerella*. *J. Path. Bact.* **48**, 25.

PIJPER, A. (1947). Methylcellulose and bacterial motility. *J. Bact.* **53**, 257.

PIJPER, A. (1949). Bacterial surface, flagella and motility. In *The Nature of the Bacterial Surface*. *Symp. Soc. gen. Microbiol.* **1**, 144.

PIJPER, A. (1955). Shape of bacterial flagella. *Nature, Lond.* **175**, 214.

PIJPER, A. (1957). Bacterial flagella and motility. *Ergebn. Hyg. Bakt.* **30**, 37.

PIJPER, A., NESER, M. L. & ABRAHAM, G. (1956). The wavelengths of bacterial flagella. *J. gen. Microbiol.* **14**, 371.

POLLOCK, M. R. (1945). The influence of temperature on the adaptation of tetrathionase in washed suspensions of *Bact. paratyphosum* B. *Brit. J. exp. Path.* **26**, 410.

PRESTON, N. W. & MAITLAND, H. B. (1952). The influence of temperature on the motility of *Pasteurella pseudotuberculosis*. *J. gen. Microbiol.* **7**, 117.

QUADLING, C. (1958). The unilinear transmission of motility and its material basis in *Salmonella*. *J. gen. Microbiol.* **18**, 227.

QUADLING, C. & STOCKER, B. A. D. (1956). An experimentally induced transition from the flagellated to the non-flagellated state in *Salmonella*: the fate of parental flagella at cell division. *J. gen. Microbiol.* **15**, i.

QUADLING, C. & STOCKER, B. A. D. (1957). The occurrence of rare motile bacteria in some non-motile *Salmonella* strains. *J. gen. Microbiol.* **17**, 424.

RICHMOND, M. H. (1959). Effect of inhibitors on lytic enzyme synthesis in *Bacillus subtilis* R. *Biochim. biophys. Acta*, **34**, 325.

ROBERTIS, E. DE, & PELUFFO, C. A. (1951). Chemical stimulation and inhibition of bacterial motility studied with a new method. *Proc. Soc. exp. Biol., N.Y.* **78**, 584.

SHOESMITH, J. G. (1960). The measurement of bacterial motility. *J. gen. Microbiol.* **22**, 528.

SPIEGELMAN, S. (1958). Protein and nucleic acid synthesis in subcellular fractions of bacterial cells. *VIIth International Congress Microbiol.* p. 81.

STOCKER, B. A. D. (1956*a*). Bacterial flagella: morphology, constitution and inheritance. In *Bacterial Anatomy. Symp. Soc. gen. Microbiol.* **6**, 19.

STOCKER, B. A. D. (1956*b*). Abortive transduction of motility in *Salmonella*; a non-replicated gene transmitted through many generations to a single descendant. *J. gen. Microbiol.* **15**, 575.

STOCKER, B. A. D. (1957). Methods of removing flagella from live bacteria; effects on motility. *J. Path. Bact.* **73**, 314.

STOCKER, B. A. D. & CAMPBELL, J. C. (1959). The effect of non-lethal deflagellation on bacterial motility and observations on flagellar regeneration. *J. gen. Microbiol.* **20**, 670.

TAWARA, J. (1957). Electron-microscopic study of the flagella of *Vibrio comma*. *J. Bact.* **73**, 89.

TIBBS, J. (1957). The nature of algal and related flagella. *Biochim. biophys. Acta*, **23**, 275.

VAN ITERSON, W. (1953). Some remarks on the present state of our knowledge of bacterial flagellation. In *Bacterial Cytology, Symp. 6th Congr. Int. Microbiol. Rome*, p. 24.

VAN TUBERGEN, R. P., SETLOW, R. B. & SHERK, T. (1960). Quantitative radio-autographic studies on log phase cultures of *E. coli*: The distribution of parental H^3-DNA, H^3-RNA, H^3-protein and H^3-cell wall in progeny cells. Paper read before Soc. gen. Microbiol., April, 1960.

WARE, G. C. (1951). Nutritional requirements of *Bacterium coli* at 44°. *J. gen. Microbiol.* **5**, 880.

WEIBULL, C. (1948). Some chemical and physico-chemical properties of the flagella of *Proteus vulgaris*. *Biochim. biophys. Acta*, **2**, 351.

WEIBULL, C. (1949*a*). Chemical and physicochemical properties of the flagella of *Proteus vulgaris* and *Bacillus subtilis*. *Biochim. biophys. Acta*, **3**, 378.

WEIBULL, C. (1949*b*). Morphological studies on salt precipitated bacterial flagella. *Ark. Kemi*, **1**, 21.

WEIBULL, C. (1950*a*). Ordered aggregation of salted out and dried bacterial flagella. *Ark. Kemi Min. Geol.* **1**, 573.

WEIBULL, C. (1950*b*). Electrophoretic and titrimetric measurements on bacterial flagella. *Acta chem. scand.* **4**, 260.

WEIBULL, C. (1950*c*). Investigations on bacterial flagella. *Acta chem. scand.* **4**, 258.

Weibull, C. (1951 *a*). Bacterial flagella as fibrous macromolecules. *Disc. Faraday Soc.* **11**, 195.

Weibull, C. (1951 *b*). Some analytical evidence for the purity of *Proteus* flagellar protein. *Acta chem. scand.* **5**, 529.

Weibull, C. (1951 *c*). Movement of bacterial flagella. *Nature, Lond.* **167**, 511.

Weibull, C. (1953). The free amino acids of the *Proteus* flagellar protein. Quantitative determination of the DNP amino acids using paper chromatography. *Acta chem. scand.* **7**, 335.

Weibull, C. (1960). Bacterial motility. In *The Bacteria*, vol. 1. Edited by I. C. Gunsalus and R. Y. Stanier. New York and London: Academic Press.

Weibull, C. & Tiselius, A. (1945). Note on the acid hydrolysis of bacterial flagella. *Ark. Kemi Min. Geol.* **20** B, no. 3.

Weinstein, D., Koffler, H. & Moskowitz, M. (1960). Evidence for the intracellular occurrence of flagellar proteins. *Bact. Proc.* p. 63.

Williams, M. A. (1959). Cell division and elongation in *Spirillum analus*. *J. Bact.* **78**, 374.

Yokota, K. (1925). Neue Untersuchungen zur Kenntnis der Bakteriengeisseln. 2. Mitteilung. *Zbl. Bakt.*, 1. *Abt. Orig.*, **95**, 261.

Yoshida, A. (1960). Studies on the mechanisms of protein synthesis; incorporation of *p*-fluorophenylalanine into α-amylase of *Bacillus subtilis*. *Biochim. biophys. Acta*, **41**, 98.

ENVIRONMENTALLY INDUCED CHANGES IN BACTERIAL MORPHOLOGY

J. P. DUGUID AND J. F. WILKINSON

Bacteriology Department, University of Edinburgh, Teviot Place, Edinburgh 8

From early in the era of pure-culture bacteriology it was recognized that bacterial strains can undergo marked morphological changes which are determined by the age and other conditions of culture. Interest has centred primarily on the vexed question of whether these changes are physiological adaptations or pathological failures of homeostasis in growth: whether, for instance, an abnormally shaped cell is a life-cycle stage specially adapted for growth, survival or dissemination in particular environmental conditions, or an involution form which is weakened in viability by its enforced deviation from normal composition.

The study of these changes can yield valuable information on the composition, synthesis and function of the cell structures involved. Thus, Wiame's (1946*a*, *b*; 1947*a*, *b*) observation that the volutin content of *Saccharomyces cerevisiae* varied with the supply of phosphate, led to his identification of polymetaphosphate as the metachromatic component of volutin granules and to the suggestion that volutin has an energy storage function. These conclusions were later extended to bacteria as a result of similar findings with *Klebsiella aerogenes* (Smith, Wilkinson & Duguid, 1954) and other species.

Most of the morphological changes are produced in growing or ageing cultures and seem to result from the environmental conditions influencing cell metabolism so as to increase or decrease the synthesis of a particular component relative to general growth. Thus, the synthesis of one component may be inhibited while protoplasmic growth is otherwise allowed to proceed, e.g. the inhibition of cell wall synthesis by penicillin (Duguid, 1946) and of flagellar synthesis by a raised temperature of incubation (Quadling & Stocker, 1956). Conversely, protoplasmic growth may be inhibited while a particular component continues to accumulate, e.g. capsular substance (Duguid, 1948) and volutin (Smith *et al.* 1954). The change to an abnormal cell form is thus the result of unbalanced growth. It is probably best to regard as 'normal' the cell form exhibited during free exponential growth in the most favourable medium, since the familiar stationary-phase cells of a 24 hr. culture are arrested in growth by starvation or intoxication which may cause a one-sided inhibition of the synthetic processes.

Some morphological changes can be induced independently of growth, for instance, the rapid plasmolysis or plasmoptysis caused by transfer to a medium of higher or lower osmotic pressure (Fischer, 1900). In certain cases, moreover, washed bacilli suspended in incomplete nutrient media which do not allow increase of cell count or nitrogen content may nevertheless synthesize and accumulate capsular substance (Bernheimer, 1953), volutin granules (Sall, Mudd & Davis, 1956) or lipid granules (Macrae & Wilkinson, 1958*a*), or dissimilate and lose these granules.

Apart from directly impressing non-genetic variations, the environmental conditions may also change morphology by selectively favouring the outgrowth of genetic variants. Thus, cultivation of pneumococci on agar favours replacement of the short diplococcal forms by filamentous, chain-form mutants whose mode of growth enables them to extend outwards from the parent colony to unexhausted medium, while cultivation in liquid medium favours reversion to short forms able to grow diffusely instead of as a concentrated sediment (Dawson, 1934; Austrian, 1953). The genetic variations are usually distinguishable because they initially affect only a minority of cells in the culture, and the variant may be isolated and subcultivated in absence of the selective conditions without immediately reverting. In contrast, the directly impressed changes affect the majority of cells from the time of their first exposure to the inducing conditions, are completed usually within a few generations' growth, and are rapidly reversed on subcultivation in absence of the inducing conditions. However, genetic variations may occur which are reversible with a high frequency in both directions, e.g. the H-antigen phase variation in Salmonella, and a reversible morphologic change might be due to rapid environmental selection of such mutants. This is possibly the mechanism of the environmentally determined fimbrial phase variation in enterobacteria (Duguid & Gillies, 1957).

VARIATIONS IN CELL SHAPE AND SIZE

Gross changes in cell shape and size were the earliest observed morphological variations and their interpretation has been the subject of much controversy. The monomorphistic view, now generally accepted, supposes that eubacteria reproduce exclusively by fission in a cell form which remains fairly constant under physiological conditions, and that abnormally shaped cells are degenerate involution forms. The pleomorphistic view regards the abnormal forms as physiologically determinate stages in a complex bacterial life cycle, each adapted for a special vegetative or reproductive function like the various structures of a higher fungus. Thus, eubacteria have been thought capable of assuming a filamentous and branching mode of growth, of budding, growing as an amorphous plasmodium or 'symplasm', undergoing cell fusion, and forming 'microcysts', 'conidia', 'arthrospores', 'chlamydospores' and 'zygospores'; inclusion granules have been interpreted as fertile 'gonidia' capable of developing into new cells. Notable proponents of these views were Hort (1920), Löhnis (1921), Almquist (1922) and Mellon (1942). More recently, Bisset (1949) has figured a reproductive cycle for Gram-negative bacilli which culminates in the development of microcysts, either directly from the vegetative cells in ageing cultures or as a consequence of cell fusion.

Hort's work was specially important because he made continuous warm-stage observations which proved that some of the abnormal forms were fertile and capable of growing into normal cells, and also that in some conditions (acid) they showed growth when the accompanying normal forms did not. Like other pleomorphists, he applied the term 'involution form' only to cells which were obviously dead and autolysing, and did not entertain the now accepted view that cells may be grossly deformed by sublethal degrees of a pathological change, but still be capable of recovery when transferred to favourable conditions.

Types of morphological variants

In his classic review, Löhnis (1921) reproduced 297 illustrations of morphological variants reported in the earlier literature. These, and the similar forms later produced by penicillin, glycine and other methods, are of four main kinds (Fig. 1). Most observations have been made on Gram-negative bacilli, but comparable variants have been found in many families of bacteria.

(1) *Long forms* are bacilli which have become filamentous by the continuance of axial growth without corresponding cell division. They may be septate, with complete transverse cell walls, or non-septate, and may have their nuclear bodies normally distributed, widely separated, concentrated medially or fused into an axial filament. Some show branching, localized swelling or plasmoptysis.

(2) *Branching forms* are short or long forms which grow with three-point or multiple-point branching. They have been seen to yield normal bacilli by segmentation (Gardner, 1925).

(3) *Large and globular forms* (Pl. 1, figs. 1, 2) are cells which have

undergone three-dimensional enlargement, apparently through the yielding of defective cell walls to osmotic and growth pressures. Free protoplasts constitute a special class of spherical forms which are produced by complete removal of the cell wall and can be maintained without lysis only in special osmotically protective media (Weibull, 1956).

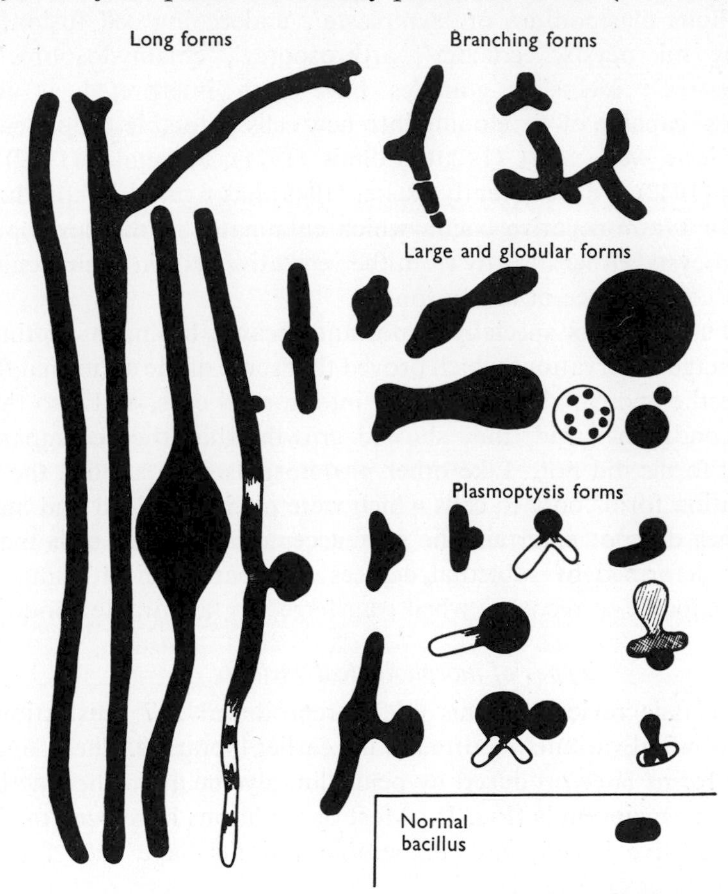

Fig. 1. Abnormal cell forms induced in enterobacteria by many kinds of antibacterial agents (see Text). × *c.* 4000.

Spheroplasts are similar osmotically sensitive globules which are covered by a weakened cell wall (McQuillen, 1960). L-Forms consist of plastic bodies ranging from minute filterable granules of $0 \cdot 2\mu$ diam. to large globules of 10μ or more (Dienes & Weinberger, 1951) and are distinguished from other globular forms by their power of continued growth in suitable media, apparently by a process of 'sprouting' (Liebermeister, 1960).

(4) *Plasmoptysis forms* were originally described by Fischer (1900)

and later observations have been reviewed by Weibull (1956). They consist of short, long or large forms which extrude a globule of protoplasm from the side or end. In some plasmoptysed cells the globule appears to be enclosed by an intact covering of expanded cell wall (Pl. 1, figs. 3, 4), since these cells remain for a time unlysed, retaining normal density and stainability. In other cases, the ejected protoplasm seems to be naked since it rapidly disperses and the cell becomes lysed. Plasmoptysis may be produced directly by transfer of cells from saline solution to distilled water, dilute acid or glycerol solution, or it may occur as an end-result of growth under conditions interfering with cell-wall synthesis. Plasmoptysis globules have geen interpreted variously as buds, branches, conidia, plasmodia, zygospores and microcysts.

Inducing agents

Life-cycle interpretations have been generally abandoned because the abnormal forms were never proved to serve a useful reproductive purpose and because they were found to develop most commonly in conditions obviously unfavourable to growth and likely to cause involution. Aberrant forms were first observed mainly in ageing cultures, where the cells are liable to damage from starvation, deficiency of oxygen, the accumulation of toxic metabolites and the concentration of solutes by drying (see Gillespie & Rettger, 1939). It was later found that they develop more rapidly and abundantly on media containing certain antibacterial substances in amounts just insufficient to prevent growth. All the inoculated cells may become abnormal during the first few hours of growth on such media. Hughes (1956) has listed many of the inducing agents and the most commonly used have been the following.

(1) *Inorganic salts.* After Hankin & Leumann (cited by Matzuschita, 1900) had discovered the effect of 3 % NaCl in promoting abnormal growth of *Pasteurella pestis*, Matzuschita (1900) showed that higher concentrations of NaCl (5–10 %) had a similar effect on a wide range of bacteria. Other salts have proved even more active. Thus, with *Vibrio* El Tor, Eisler (1909) found 0·5–1·0 % LiCl, $Ca(NO_3)_2$ or Mg $(NO_3)_2$ as effective as 3–4 % NaCl. Ionic imbalance appeared to be operative, since the effect of a monovalent cation could be prevented by a balancing addition of a bivalent cation, and that of a bivalent cation by addition of a monovalent or different bivalent cation. More recently, a deprivation of Mg^{++}, K^+ or Na^+ has been shown to induce filament or chain formation in *Clostridium welchii* (Webb, 1949; Shankar & Bard, 1952), and an insufficient NaCl concentration, the formation of spheroplasts in a halophilic bacterium (Takahashi & Gibbons, 1959).

(2) *Acidity and alkalinity*. Involution forms develop at an early stage in cultures on media initially poised at a low or high pH, but at a later stage on initially neutral media which contain a fermentable sugar and so become acid in the course of growth (Hort, 1920; Reed & Orr, 1923; Duguid, 1948).

(3) *Amino acids*. Single amino acids added to the culture medium in excessive concentration, e.g. glycine, DL-methionine or L-phenylalanine at 0·5–2·0 %, are inhibitory to growth and productive of grotesque involution forms (Gordon & Gordon, 1943; Dienes & Zamecnik, 1952; Welsch & Osterreith, 1958). The unnatural D-isomers appear to be more effective than the L-isomers. Thus, Tuttle & Gest (1960) found that *Rhodospirillum rubrum* gave nearly normal growth on media containing any one of 10 L-amino acids as the sole carbon and nitrogen source, but developed elongated, swollen and distorted forms when supplied instead with the corresponding D-amino acid.

(4) *Antibacterial agents* of many kinds can induce abnormal development, e.g. methyl violet, proflavine, phenol, *m*-cresol, bile salts and detergents. Among the antibiotics penicillin is particularly effective (Gardner, 1945).

(5) *Starvation of growth factors* which are required preferentially for cell wall, cytoplasmic or nuclear synthesis may lead to unbalanced growth and abnormal development: e.g. deprivation of thymidine (Jeener & Jeener, 1952), thymine (Cohen & Barner, 1954), diaminopimelic acid (Meadow, Hoare & Work, 1957), lysine or valine (Shockman, Kolb & Toennies, 1958), and hexosamine derivatives (Park, 1958; Glick, Sall, Zilliken & Mudd, 1960).

Interference with cell wall synthesis

It is noteworthy that these very varied agents produce generally similar effects. A wide range of long, large, branching and plasmoptysis forms can be induced by penicillin, and a similar range of forms by glycine, LiCl, acid or an excessive incubation temperature (Pl. 1, figs. 1–5). Since the cell wall is responsible for maintaining the shape of bacteria and transverse walls are essential for fission, it is reasonable to accept Park's (1958) suggestion that many of the inducing agents have a common mode of action in interfering with cell wall synthesis. The enzymes responsible for wall synthesis may be located at the surface of the cytoplasmic membrane and thus be specially exposed to the influence of external inhibitors.

The diversity of changes produced by different concentrations of penicillin are well seen in Gram-negative bacilli such as *Escherichia coli*, but comparable reactions are shown by vibrios, cocci and Gram-positive bacilli (Duguid, 1946). Formation of the transverse cell walls is most readily inhibited. On media containing about the minimum inhibiting concentration of penicillin, the inoculated bacilli may first undergo 1–4 apparently normal divisions; they then grow without further formation of cross walls and elongate into filaments of 10–100μ or more. The filaments contain regularly distributed nuclear bodies and for a time remain viable, motile and capable of segmentation into normal bacilli if transferred to penicillin-free medium. After some hours on the penicillin medium a defective formation of the lateral walls becomes apparent in the development of one or more localized swellings, or 'bulbs', and then in an abrupt or gradual lysis by plasmoptysis; globules of protoplasm are ejected from the bulbs, or elsewhere, and soon disperse. With higher concentrations of penicillin there is less preliminary growth and earlier lysis. At very high concentrations the cells do not elongate at all, but swell into spindles and spheres, or extrude globules, or merely fade away without change of shape. This lysis without preliminary growth may be due to an autolytic decomposition of the cell wall which is not counteracted by reparative synthesis (Park, 1958). Light microscopic findings have been amplified by electron microscopy. Murray, Francombe & Mayall (1959) demonstrated a thinning of the peripheral cell walls in sections of *Staphylococcus aureus* grown in penicillin, but found the main lesions to be in the developing cross-walls, which showed loss of density and gross irregularity of shape.

The specific inhibition of cell wall synthesis has been confirmed by the production with penicillin of stabilized spheroplasts (Liebermeister & Kellenberger, 1956; Lederberg, 1956) and of L-forms which continue to grow in special media at otherwise inhibitory concentrations (Dienes & Weinberger, 1951; Lederberg & St Clair, 1958). Park and Strominger obtained evidence that in *Staphylococcus aureus* the inhibition is of the polymerization of the cell wall mucopeptide from a uridine pyrophosphate-muramic acid-peptide component (Park, 1958). McQuillen (1960) has summarized work showing that penicillin, bacitracin, cycloserine and glycine similarly inhibit cell wall synthesis in *S. aureus* causing the accumulation of cell wall precursors, induce spheroplast formation in *Escherichia coli*, and are not inhibitory to protein synthesis.

The cell walls of most Gram-positive and Gram-negative bacteria appear to be alike in having as their basic rigid component a 'mucocomplex', or 'mucopeptide', which consists mainly of the hexosamines, *N*-acetyl-glucosamine and *N*-acetyl-muramic acid, with sometimes additional sugars, and several amino acids including D-alanine, D-glutamic acid and either lysine or diaminopimelic acid (DAP) (reviewed by Salton, 1960). DAP is not a component of the structures in the protoplast and is thus unnecessary for protoplasmic growth. Meadow *et al.* (1957) found that an *Escherichia coli* strain which required an external supply of DAP underwent globular swelling and lysis when grown on a DAP-limited medium; this was probably due to protoplasmic growth continuing in the absence of proper cell wall formation. In confirmation, McQuillen (1958) and Lederberg & St Clair (1958) obtained stable spheroplasts by preparing such DAP-deficient growths in an osmotically protective medium.

Similar effects can result from deprivation of hexosamine derivatives. Park & György (see Park, 1958) found that *Lactobacillus bifidus* var. *pennsylvanicus* produced large forms when grown on a medium deficient in a glucosamine derivative required as a growth factor for cell wall synthesis. Glick *et al.* (1960) showed that on media supplemented with sufficient α, β-methyl-*N*-acetyl-D-glucosaminide, a precursor of muramic acid, the lactobacilli grew as slender unbranched rods; on slightly deficient media they produced branching forms in the later stages of culture when the glucosaminide was exhausted, and in unsupplemented medium produced branching and plasmoptysis forms at an early stage.

Interference with cell division

The long forms induced by certain conditions are septate, with apparently complete transverse cell walls, and show no undue tendency to lysis. Their formation therefore does not seem to involve an inhibition of cell wall synthesis, but probably results from a failure in the mechanism for splitting of the cross-walls and cell separation. *Clostridium welchii* and other Gram-positive bacilli were found by Webb (1949) to form septate filaments ('chains') when grown in peptone media in which the magnesium ion concentration was slightly deficient (0·0006 %) or excessive (0·05 %), and mainly non-septate filaments when the Mg^{++} concentration was grossly deficient (0·00003 %) or excessive (0·1–0·5 %). Shankar & Bard (1952) obtained filamentous growth when either Mg^{++} or K^{+} were deficient, and chain formation when Na^{+} was deficient. Webb (1951) found that the septate filaments could be broken into bacilli by digestion with lysozyme or with a lysozyme-like enzyme extracted from a normally grown culture. This suggests that a lysozyme-like enzyme is involved in the normal mechanism of cell separation. Lominski, Cameron & Wyllie (1958) advanced this hypothesis on the basis of comparable findings with *Streptococcus faecalis*. The streptococcus was induced to grow in abnormally long chains by addition of suramin, crystal violet, bile or a detergent to the culture medium. The inducing agent was thought to inhibit the formation or activity of a cell-separation enzyme, since the long chains could be broken up by brief incubation with the cell free supernatant fluid from a normal short-form culture presumed to contain this enzyme. The extra-cytoplasmic location of the site of action of a cell wall splitting enzyme would account for its special susceptibility to external inhibitors.

Interference with nuclear synthesis

Yet another mechanism giving long forms is the specific inhibition of synthesis of deoxyribonucleic acid (DNA). Thymine and thymidine are nucleic acid components found in DNA but not in ribonucleic acid (RNA) or other non-nuclear components. A deficiency of either may inhibit nuclear growth while permitting continued cytoplasmic growth and cell wall formation. Jeener & Jeener (1952) cultivated a strain of *Thermobacterium acidophilus* which required thymidine as a growth factor; this was supplied by addition of DNA to the medium. On a medium deficient in DNA, the bacillus grew into very long filaments whose staining showed them to contain abundant RNA but only a few widely spaced DNA-containing nuclear bodies. When later supplied with DNA the filaments showed rapid nuclear multiplication and then segmented into normal bacilli. Zamenhof, de Giovanni & Rich (1956) induced development of similar viable, poorly nucleated filaments by cultivating *Escherichia coli* in presence of the thymine analogue, 5-bromouracil. Since there is no reason to believe that thymine or thymidine is required directly for septum formation or fission, these findings suggest that nuclear multiplication supplies the normal stimulus for fission.

Somewhat different behaviour was displayed in comparable conditions by the thymine-requiring *Escherichia coli* strain of Cohen & Barner (1954). When placed in a thymine-free medium the cells underwent only a limited enlargement, doubling in protein and RNA content without increase of DNA, and then rapidly died. In this case the unbalanced growth produced a change which was irreversible on restoration of thymine.

Abnormal development in apparently favourable conditions

Life-cycle explanations of aberrant cell forms have gained support from the finding that these forms sometimes develop in conditions which appear perfectly favourable for normal growth, for instance, amid a majority of normally growing cells in a young culture on rich medium. It may seem on first consideration that an involutionary mechanism is hardly possible in such cases, but this conclusion overlooks the differences in physiological state and genotype that may exist between the cells in a pure culture.

Cultures are usually inoculated from old growths in which many of the cells are dead and many of the survivors either damaged or mutated. Some of the latter may be affected in such a way that they will undergo a period of abnormal development on the fresh medium before either recovering by segmentation into normal forms or dying. It is well known from warm-stage observations that cells previously damaged by an antibacterial agent, e.g. LiCl (Wyckoff, 1933; Richter, 1934) or antibiotics (Pulvertaft, 1952), are liable to develop abnormally for a time after transfer to a favourable medium. Damage sustained in the previous culture probably accounts for the branching forms of enterobacteria which Gardner (1925) observed to develop from about 0·1 % of cells taken from 2 to 10 month-old cultures, and for the scanty long forms, described by Robinow (1944) and Bisset (1949), which commonly develop in the early hours of subculture on nutrient agar. Bisset interpreted these long forms as stages in a life cycle involving nuclear fusion and reorganization. Certainly they often remain viable and later segment into normal bacilli. Hughes (1953) demonstrated this for single long forms which he isolated with a micromanipulator and transferred to fresh medium. Nevertheless, there is some evidence that these forms are essentially involutionary. We have watched the long forms of *Klebsiella aerogenes* growing among the normal cells in a warm-stage preparation; about half of them ultimately segmented into normal bacilli, but the remainder underwent lysis.

Certain bacteria, e.g. *Streptobacillus moniliformis*, are especially liable to filament formation, swelling and plasmoptysis, even when grown on the best available culture medium. This may merely reflect an unexplained deficiency of the medium in supplying the organism's growth requirements. The stable, or rarely reverting, type of L-form growth in *S. moniliformis* and other bacteria (e.g. the 3A type of Dienes & Weinberger, 1951) presumably constitutes a heritably distinct phase of the organism which may lack a gene specifically required for cell wall synthesis; it seems unlikely that this phase has any significance as part of a natural life cycle. The unstable type of L-form growth (e.g. the 3B type of Dienes & Weinberger), which reverts *en masse* to the normal form within a few hours of withdrawal from the inducing agent (e.g.

penicillin), is probably best regarded as an involutionary growth wherein the pathogenic effect of the inducer is not so great as to render the cells non-viable in the special cultural conditions used.

CAPSULES, INTRACELLULAR POLYSACCHARIDE, LIPID AND VOLUTIN GRANULES

If a bacterium is grown in a variety of media so that different constituents act as the growth-limiting nutrient, very great changes occur in the amount of certain of the morphological components, particularly the polysaccharide capsule, intracellular polysaccharide, and lipid and volutin granules (reviewed in more detail by Wilkinson & Duguid, 1960). These can be estimated either microscopically, with or without prior staining, or by chemical estimation of the characteristic structure (see Table 1). It is now known that each of these structures contains a

Table 1. *Characterization of polysaccharide capsules, intracellular polysaccharide, volutin granules and lipid granules*

(References in Wilkinson & Duguid, 1960)

	Polysaccharide capsules	Intracellular polysaccharide	Volutin granules	Lipid granules
Microscopic demonstration	Visible as clear zones with light microscope using wet-film India ink method	Stained by periodic acid-Schiff method;* sometimes visible in ultrathin sections as clear areas by electron microscope	Refractile by light microscope; stain metachromatically with toluidine blue; opaque by electron microscope	Highly refractile by light microscope; stain with fat-soluble dyes; clear areas in sections by electron microscope
Characteristic component	Heteropolysaccharide	Homopolysaccharide (?usually glycogen in enterobacteria)	Polymetaphosphate	Poly-β-hydroxybutyrate (e.g. in *Bacillus* sp.)
Other components	Minor ones may be present	.	Uncertain but probably include RNA and lipid	Minor ones such as lipid and protein may occur

* Cell wall and microcapsular polysaccharide may also be stained.

characteristic polymeric substance as its principal component, i.e. polysaccharide, polymetaphosphate or poly-β-hydroxybutyrate. It is to be expected that the synthesis of a single such substance will have simpler requirements than those of protoplasmic growth generally, so that cultures may be halted in general growth by conditions allowing the continued synthesis and accumulation of this substance during the stationary phase. The morphology is thus altered by a form of unbalanced growth. Conversely, when exhaustion of the carbon and energy source is responsible for the onset of the stationary phase, the cells may be expected to economize in the production of these partly dispensable components in the final stages of growth.

Variation in stationary phase cells

(1) *Capsular and intracellular polysaccharide.* Table 2 summarizes some of the results obtained in our laboratory from experiments using *Klebsiella aerogenes*, strain A3. This organism was grown in a synthetic

Table 2. *Capsule size, polysaccharide, volutin and polymetaphosphate in* Klebsiella aerogenes, *strain A3, grown on chemically defined media with different nutrient deficiencies**

	Exponential phase cells†	Stationary phase cells from cultures in which growth was limited by: Deficiency of source of					
		Carbon, energy	Phosphorus	Nitrogen	Sulphur	Potassium	Developing acidity
1. Capsule diameter (μ)	2·5	1·6	4·3	4·3	2·9	1·8	2·0
2. Intracellular polysaccharide staining	±	±	++	++	++	+	+
3. Polysaccharide in cell and capsule	1·9	1·5	15·9	15·5	6·5	3·2	1·0
In loose slime	0·3	0·3	23·8	13·3	4·0	0·7	0·7
4. Volutin granules	−	−	−	+	++	−	+++
5. Cell polymetaphosphate	0·3	0·1	0·1	8·5	27·3	0·1	28·0

* Inoculated from energy-limited cultures. † Similar results on all media.

1–3: Representative results from Duguid & Wilkinson (1953, 1954) for cultures grown at 35° on lactose agar media. 1 = average transverse diameter of capsule and cell. 3 = polysaccharide mg./mg. total non-dialysable N in culture.

4–5: From Duguid, Smith & Wilkinson (1954) and Smith *et al.* (1954). Acid cultures of strain A3 were grown at 35° on a glucose agar medium. Other results are for the non-capsulate mutant A3 (0) grown at 37° in aerated liquid medium; this mutant behaved like the parent strain as regards volutin production. 5 = trichloracetic acid insoluble polymetaphosphate as μg. P/mg. total cell N.

medium containing varying amounts of glucose or lactose (carbon and energy source), ammonium salt (nitrogen source), phosphate (phosphorus source) and potassium ions as the principal nutrients. It is evident that there were wide differences between stationary phase cells limited by different deficiencies. The capsule size and amount of polysaccharide per cell were minimal in carbon- and energy-limited cultures. As the concentration of the nitrogen source in the medium was decreased until it became the growth-limiting nutrient, the components all rose to the maximal level given in the fourth column of Table 2. In a similar way, the polysaccharide components increased to a characteristic maximum when growth was limited by deficiencies of other nutrients, though not in poorly buffered cultures whose growth was halted by increasing acidity. Thus, the volume of the capsule per standard-sized cell in the nitrogen-deficient and phosphorus-deficient cultures was about 20 times that in the carbon and energy deficient cultures.

Similar observations have been made by Holme & Palmstierna (1956*a*, *b*) for glycogen synthesis in *Escherichia coli*, strain B. For example, the glycogen content of nitrogen-deficient cells was as high as 25 % of the total cell dry weight as against about 5 % in carbon- and energy-starved cells. This could be correlated with the number of glycogen granules present in ultrathin sections examined under the electron microscope (Cedergren & Holme, 1959).

(2) *Volutin granules*. Although some micro-organisms (e.g. yeasts, corynebacteria and mycobacteria) commonly form volutin during the later stages of growth, bacteria in general only form visible granules under very specialized growth conditions. Thus *Klebsiella aerogenes* is completely devoid of volutin when growth is limited by a deficiency in the carbon and energy source, as occurs in conventional defined and protein-digest media. It accumulates volutin in nitrogen- or sulphur-starved cultures and, to a much greater extent, in poorly buffered cultures wherein growth is halted by the development of a low pH (*c*. 4·5) before the sugar supply is exhausted (Table 2). If the polymetaphosphate content of the stationary phase cultures is measured, there is a close correspondence to the degree of volutin stainability. This correspondence is also shown during the striking changes which occur when volutin-free phosphate-deficient cells are transferred into a medium containing a carbon and energy source, together with phosphate, potassium and magnesium ions. Volutin granules appear within a few minutes, increase rapidly in size (until up to 10–20 % of the cell volume is occupied) and then, after about 60 min., gradually disappear; the level of polymetaphosphate shows a correspondingly large increase (e.g. × 250) and decrease (Smith *et al.* 1954).

(3) *Lipid granules*. The influence of cultural conditions on the formation of lipid granules has been studied most closely in species of *Bacillus*, which show wide variations in their number and size in stationary phase cultures. Although the literature is conflicting with regard to the conditions most suitable for production, carbohydrate-rich media generally give the greatest yields. Grelet (1952) has shown that nitrogen, sulphur or potassium deficiency is favourable, while carbon and energy, or manganese deficiency is unfavourable. If the amount of poly-β-hydroxybutyrate per cell is measured, the level increases up to sevenfold in sugar-rich media as compared with carbon- and energy-deficient media

(Macrae & Wilkinson, 1958*b*). The nature of the carbon source is also important, a further doubling of poly-β-hydroxybutyrate production up to about 50% of the cell dry weight being obtained by the addition of acetate to the medium.

Whether lipid or polysaccharide is accumulated in the stationary phase cells depends on the bacterium studied, but both substances are stored in some species. Dagley & Johnson (1953) studied stationary phase cells of *Escherichia coli* grown in media containing varying amounts of glucose and acetate as the carbon sources. While the presence of acetate stimulated lipid production and depressed polysaccharide production, the presence of glucose had the opposite effect.

Variation in exponential phase cells

Exponential phase cells in batch culture seem to give a similar morphological picture regardless of the nature of the final exhausting nutrient. As shown in Table 2, all exponential phase cells of *Klebsiella aerogenes* bear moderately sized capsules, have a level of polysaccharide little above that of carbon- and energy-deficient cells and contain no volutin granules visible in the light microscope. Similarly, the amount of lipid granules and poly-β-hydroxybutyrate is relatively small in all exponential phase cells of *Bacillus megaterium* (e.g. Norris & Greenstreet, 1958). It may be presumed that during the exponential phase, the level of all nutrients is sufficiently high to prevent an alteration in the level of intermediates within the cell, or that the control mechanisms governing enzyme activity are then sufficiently versatile to prevent any appreciable change in the amount of the characteristic polymers.

Whereas, in batch culture, the limiting nutrient may reach a sufficiently low level to alter the character of growing cells only in the short period between the end of the exponential phase and the beginning of the stationary phase, it is possible to obtain this stage indefinitely using continuous culture methods. By altering the level of nutrients entering the culture vessel and the flow rate, it should be possible to make the utilization of any nutrient the factor limiting the growth rate. Further, the effect of any level of a nutrient such as the carbon and energy source can be studied at will under a variety of other environmental conditions. However, this type of study is still in its infancy, particularly with regard to the compounds and structures under consideration. Holme (1957) grew *Escherichia coli* in continuous culture using the nitrogen source as the limiting nutrient and showed that the glycogen formed per cell increased considerably with decreasing dilution rates. This indicated that the synthesis of glycogen increased with a reduction in the rate of synthesis of nitrogen-containing compounds. Further, glycogen synthesis occurred more readily with glucose as the carbon source than with lactate. Unpublished experiments in this laboratory using *Klebsiella aerogenes* grown under continuous culture with the level of different nutrients limiting the rate of growth, showed differences in the capsule size and amount of polysaccharide per cell which were intermediate between those of the corresponding exponential and stationary phases in batch culture.

Production in 'non-growing' cells

It may be imagined that conditions in a batch culture just prior to and during the stationary phase are analogous to those in a washed suspension. The cells will utilize, as far as possible, the remaining nutrients, thus undergoing a form of unbalanced growth. Further metabolism results in the synthesis of those cell components which do not require the growth-limiting nutrient either as a building block or as cofactor for synthesis. For example, in nitrogen-limited cultures a sufficiency of the carbon source and a supply of various inorganic ions will remain at the onset of the stationary phase. The carbon source will be catabolized, part of it being assimilated into non-nitrogenous polymers. Consequently the same phenomenon should occur if, say, glucose and suitable inorganic salts are added to a washed bacterial suspension. This is often so. For instance, when washed, carbon- and energy-starved cells of *Klebsiella aerogenes*, strain A3, were suspended in an aerated glucose–salts solution at 35°, they increased their average capsule diameter from 1·5μ to 3·3μ in 18 hr. and 4·5μ in 72 hr., and their polysaccharide content (mg./mg. N) from 0·73 to 6·0 in 18 hr. and 15·6 in 72 hr., without showing any increase of cell count or nitrogen content (unpublished experiment; see also Wilkinson & Stark, 1956). Potassium ions are essential, explaining the low level of polysaccharide in potassium-limited stationary phase cells (Table 1). Similarly, a washed suspension of *Bacillus megaterium* will synthesize poly-β-hydroxybutyrate in presence of a suitable carbon and energy source (Macrae & Wilkinson, 1958*a*); wide differences were found in the suitability of different substrates for this synthesis and there was an optimum low oxygen tension, perhaps explaining some of the variability of results obtained by previous observers using batch cultures.

It might be expected that volutin and polymetaphosphate would also be formed in washed suspensions if a source of energy, phosphate and other essential ions were provided. This occurs in corynebacteria (Sall *et al.* 1956), but we have been unable to demonstrate it in enterobacteria. The latter organisms may possess a much more effective control system regulating polymetaphosphate synthesis, so that they accumulate volutin only under exceptional circumstances, such as in the period between the end of the exponential phase and the beginning of the stationary phase, or when the enzymes metabolizing phosphate are altered in level after growth in phosphate-deficient media.

Breakdown in 'non-growing' cultures

A gradual breakdown and disappearance of volutin, lipid and intracellular polysaccharide can occur in the absence of an exogenous carbon and energy source, both in the stationary phase of growth and in washed suspension. Intracellular polysaccharides and lipids probably act as major substrates for endogenous respiration, although other substances such as amino acids may also be responsible, at any rate under certain conditions (Dawes & Holms, 1958). Polysaccharides will be catabolized

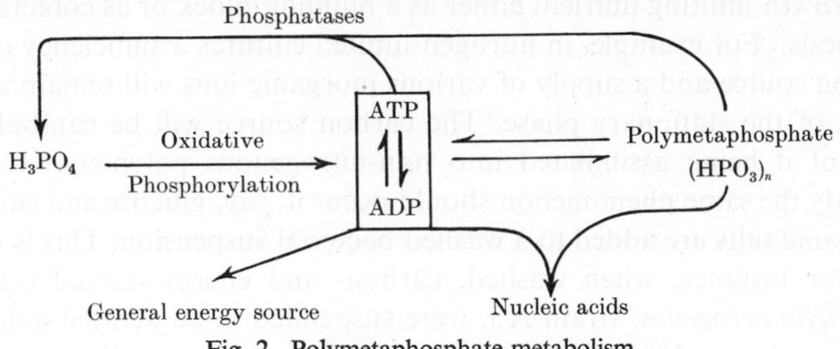

Fig. 2. Polymetaphosphate metabolism.

via monosaccharides or their derivatives, and lipids via β-hydroxybutyrate, acetate or their derivatives. Energy and carbon intermediates will result. Polymetaphosphate contains phosphate units linked by 'energy-rich' bonds and an enzyme has been purified from *Escherichia coli* catalysing a reversible transfer of terminal phosphate units from ATP to polymetaphosphate (Kornberg, 1957). Further, a close relationship between polymetaphosphate and nucleic acid metabolism has been disclosed (e.g. Mudd, Yoshida & Koike, 1958; Sall, Mudd & Takagi, 1958), which may be bound up with the presence of RNA as an important component of volutin granules (Widra, 1959). As a result of this and other work, we may represent polymetaphosphate metabolism as shown in Fig. 2.

The position with regard to the breakdown of extracellular polysaccharide is variable and most species seem to be unable to catabolize polysaccharides produced outside the cytoplasmic membrane; presumably no extracellular polysaccharidases are formed, although there are exceptions to this rule (e.g. the hyaluronic acid-hyaluronidase system of *Streptococcus pyogenes*). Extracellular polysaccharides probably have functions in the maintenance of the species other than acting as reserve compounds (summarized by Wilkinson, 1958).

The significance of these changes

The question may be raised whether the accumulation of these products in certain stationary phase cells represents (i) a pathological condition of unbalanced growth, (ii) a physiological adaptation useful for the economy and survival of the cell, or (iii) a valueless but harmless plasticity of growth processes in response to unusual environmental conditions.

In the first place, cells which are maximally or minimally loaded with polysaccharide, volutin, or lipid are generally not pathological to the point of being non-viable. Thus 50–100 % of *Klebsiella aerogenes* cells were found to be viable in carbon- and energy-limited, nitrogen-limited and acid-limited stationary phase cultures (Table 2), although the acid-limited cultures showed a progressive loss of viability after 16 hr. It is obvious that an indefinitely continued overproduction of an inclusion material would lead to disruption and death of the cell. Apparently the synthetic mechanisms are so regulated that this does not generally happen. In *K. aerogenes*, the levels of polysaccharide did not increase above certain maximal limits in the phosphorus-, nitrogen- or sulphur-limited cultures, although an excess of sugar was known to remain unconsumed in the medium (Duguid & Wilkinson, 1953). At the other extreme, in carbon- and energy-limited growth, the economizing of capsular polysaccharide is not carried to the extent of producing a final generation of entirely non-capsulate cells. The plasticity in the fluctuation of these materials is, therefore, generally regulated to within limits compatible with cell viability and is, in this sense, physiological.

The question remains as to whether or not this plasticity is valuable to the cell. Since polysaccharide, volutin, and lipid are only accumulated in excess when growth is limited with a supply of an energy source still available, it might appear that the accumulation of these substances has an energy-storage function (together with storage of carbon in lipid and polysaccharide, and phosphorus in volutin). On the other hand, it may be argued that these substances are 'overflow' or 'shunt' products analogous to the organic acids and other low molecular weight products which accumulate during growth in rich media (e.g. Foster, 1947). These products are the result of an incomplete oxidation of the carbohydrate which has been supplied in concentrations far exceeding those normally encountered in nature. They are characteristic of 'hot house' conditions and may be considered to result from the abnormally high level of carbohydrate metabolism intermediates. They may even effect a useful 'detoxification' under these conditions. It is possible that polymetaphosphate may be a normal intermediate occurring in large amounts as an overflow product, or as a means of removing any excess of energy produced during oxidative phosphorylation, particularly under conditions when carbohydrate metabolism is required to provide intermediates rather than energy. Polymetaphosphatase (Muhammed, Rodgers & Hughes, 1959) could also remove this excess energy as heat.

Intracellular polysaccharide and lipid, on the other hand, are produced to some extent even in the assimilation of only limiting amounts of the carbon and energy source. This may be seen in washed suspension experiments and in the still considerable level of these compounds occurring in stationary phase carbon- and energy-deficient cells. If they are solely 'overflow', 'shunt', or 'waste' products, it is surprising that bacteria have not evolved a means of suppressing the formation or activity of the synthesizing enzymes during starvation conditions.

It is difficult to prove a storage function and the only satisfactory way would be by experiments with mutants unable to synthesize these compounds in excess. Although polymetaphosphate may be a normal intermediate, and a mutant which had lost the ability to synthesize it might be lethal, this should not be the case with intracellular glycogen or poly-β-hydroxybutyrate. However, preliminary attempts in this laboratory and elsewhere (Holme, personal communication) to isolate a mutant of *Escherichia coli* which would no longer synthesize glycogen have failed.

If the substances considered do act as reserve storage materials, the excessive production in some stationary phase cells may or may not be a useful physiological adaptation. Perhaps the situation is analogous to the extra fat accumulated by some members of *Homo sapiens* when on a high carbohydrate diet. Although the extra fat may be of little value and even disadvantageous, this does not disprove a storage function of lipid in mammals. A reserve storage material, if present, would be useful to the cell in various ways (reviewed in detail by Wilkinson, 1959).

(1) *For cell growth in the absence of an exogenous carbon and energy source.* However, since intracellular polysaccharide and lipid will rarely amount, even under the most favourable circumstances, to more than 50% of the total bacterial dry weight, extensive growth cannot result from their utilization.

(2) *For the maintenance of cell integrity and viability.* The requirement of maintenance energy has sometimes been called in doubt. However, if the cell is to have the maximum degree of adaptability, it is advanta-

geous for its components, particularly its proteins, to be in a state of flux. Presumably this must entail a gradual loss of potential energy by the cell and an increased proneness to autolysis when placed in a deficient environment. The presence of intracellular polysaccharide and lipid which are fairly rapidly broken down to provide carbon intermediates and energy, will counteract this tendency. It might be expected therefore that stationary phase cells with high accumulations of possible reserve materials would have a greater viability than those with low accumulations; preliminary experiments in this department do not always bear this out.

(3) *For special phases of the division cycle.* There may be a special need for energy and intermediates during certain phases of the division cycle (e.g. nuclear or cell wall division). If this is so, reserve materials may be stored during periods when the requirement for energy and intermediates is minimal. Thus, volutin may undergo cyclical changes in level in synchronized cultures of *Corynebacterium diphtheriae* (Sall *et al.* 1958).

(4) *For adaptation to different media.* When bacteria are transferred to a new medium, they are often incapable of growth without some form of adaptation which may involve the formation of inducible enzymes or permease systems. The endogenous carbon and energy source may be important in this process, as has been shown in adaptation to galactose metabolism in yeasts (Spiegelman, Reiner & Cohnberg, 1947).

(5) *For special mechanisms of survival.* The two most obvious examples are: (*a*) in a motile organism, the presence of a reserve storage compound may allow sufficient motility to move from areas of insufficient nutrients to areas more suitable to growth; (*b*) in sporogenesis, energy and intermediates are required for the synthesis of the spore which often occurs under external conditions deficient in the carbon and energy source. Poly-β-hydroxybutyrate may function in this way in *Bacillus megaterium* (Tinelli, 1955), though it may also be required for direct incorporation as a major protective component of the spore coat which confers the property of acid-fastness (Yoneda & Kondo, 1959).

FIMBRIAL PHASE VARIATION

Bacteria, in changing between the fimbriate and non-fimbriate states, exhibit an environmentally determined variation between two phases having different physiological properties adapted to different conditions of growth. Fimbriae are fine filamentous appendages visible only by the electron microscope, 5–10 mμ in width, peritrichously arranged, and commonly numbering 100–300 per bacillus (Pl. 1, fig. 6). They occur in many strains of most genera of the Enterobacteriaceae and confer on the bacilli certain adhesive properties, including an easily demonstrable haemagglutinating activity (Duguid, Smith, Dempster & Edmunds, 1955; Duguid & Gillies, 1957, 1958; Duguid, 1959). A pure culture of a fimbriate strain usually contains a proportion of fully fimbriate cells, a proportion of non-fimbriate cells and only a small number of cells with scanty fimbriae. Brinton, Buzzell & Lauffer (1954) distinguished the fimbriate and non-fimbriate cells by their electrophoretic mobilities, respectively slow and fast. From the proportions of the two forms found in cultures derived from single cells of *Escherichia coli*, strain B, they estimated that the fimbriate to non-fimbriate variation occurred at the rate of about once per 1000 cells per generation, and the reverse variation at a similar rate.

Environmental control of the fimbrial phase variation was demonstrated in certain 'variable' strains of *Escherichia coli* by Duguid *et al.* (1955) and in all tested fimbriate strains of *Shigella flexneri* by Duguid & Gillies (1957). By serial cultivation under appropriate conditions it was possible to obtain cultures which were almost wholly in the fimbriate phase (e.g. over 95 % of cells fimbriate) or almost wholly in the non-fimbriate phase (e.g. under 0·2 % of cells fimbriate), and to change the strains repeatedly to and fro between the phases. *S. flexneri* was regularly converted to the fimbriate phase by serial cultivation in deep tubes or shallow layers of broth incubated aerobically without agitation or artificial aeration for 24–48 hr. periods at 37°. It was converted to the non-fimbriate phase by serial cultivation on aerobically incubated agar plates, or in broth aerated in continuously rotating bottles, or in static tubes of broth incubated anaerobically in hydrogen, or in tubes of 1 % glucose broth which became acid during growth. Since complete conversion to the new phase usually required two, three, four or more successive cultivations under the inducing conditions, Duguid & Gillies concluded that these conditions probably acted by selection of mutants and not by a direct influence on the mechanism of fimbrial synthesis.

Most of the enterobacteria resemble *Shigella flexneri* in showing a mixture of fimbriate and non-fimbriate cells in their cultures and in undergoing a similar environmentally controlled variation, but the organisms differ in the rapidity and extent of their response to particular conditions. Thus most strains of *Escherichia coli*, Klebsiella and Salmonella are reduced in degree of fimbriation by serial cultivation on agar, but not to the extent of yielding completely non-fimbriate cultures like *S. flexneri*. Some Klebsiella strains remain fully fimbriate on agar.

While there is evidence suggesting that the reversible fimbrial phase variation may depend on spontaneous changes of state in some heritable factor, a clearly different kind of genetic change can occur which causes an irreversible loss of the potentiality for fimbriation. Maccacaro, Colombo & Di Nardo (1959) have shown that the fimbriation gene (*Fim*) may be lost by mutation in *Escherichia coli*, strain K12, an organism which otherwise shows the normal, reversible variation (Maccacaro & Turri, 1959, fig. 2). Out of 61 strains in various mutational lines of K12, four strains in one line were found to be irreversibly non-fimbriate (*Fim*−). They could be rendered fimbriate again only after gene recombination by mating with the fimbriate phase or non-fimbriate phase cells of a *Fim*+ donor strain. These *Fim*− mutants are comparable to the minority of permanently non-fimbriate wild strains occurring in generally fimbriate species such as *E. coli*, *Shigella flexneri* and *Salmonella typhimurium*.

Change to fimbriate phase in aerobic static broth

The fimbrial phase change would be readily explicable as an environmentally impressed phenotypic modification if the conversion of cultures were completed more rapidly. Typical impressed changes, such as those involving the amount of capsular substance or volutin, are completed within a few generations and a few hours of growth under the inducing conditions, but completion of the fimbrial phase change in *Shigella flexneri* requires many generations of growth, e.g. 20–100, in the course of several 24–48 hr. subcultures. This is exemplified in the change to fimbriation effected during serial aerobic cultivation in tubes of static broth. From small inocula, each successive 48 hr. culture yielded nearly 10 generations of growth (1000-fold multiplication). The first broth culture was inoculated from a non-fimbriate phase colony on agar, and in most cases it remained wholly non-fimbriate. Some fimbriate cells were first seen usually in the second or third broth culture and the succeeding cultures were richly fimbriate. Results for a type 1a strain are shown in Table 3 and similar results were obtained for other *flexneri* strains.

Duguid & Gillies (1957) obtained evidence suggesting that the conditions of culture in aerobic static broth selectively favoured growth of the fimbriate phase cells and enabled them to outgrow the non-fimbriate. Non-fimbriate phase cultures mostly failed to form a pellicle on the broth, and in 48 hr. gave an amount of growth which was hardly greater than that in broth incubated anaerobically (e.g. 1·5-fold) and much less than that in artificially aerated broth (e.g. 0·1-fold). Growth thus seemed

Table 3. *Amount of growth, pellicle formation, haemagglutinating power and percentage of fimbriate cells in serial aerobic static broth cultures of* Shigella flexneri, *NCTC* 8192, *grown for* 48 *hr. periods at* 37°

First tube of broth (10 ml.) was inoculated from a single non-fimbriate phase colony on agar and later broths with cells from the top of the preceding culture, amounting to *c.* 0·1 % of the total crop. Different results are for two colonies picked at random.

	From colony 1			From colony 2		
Broth culture	Growth and pellicle*	HP	Percentage cells fimbriate	Growth and pellicle*	HP	Percentage cells fimbriate
1st	0·29	0	0	0·28	0	0
2nd	0·31	0	0	0·29	60	20
3rd	0·32	40	15	1·01 P	1500	95
4th	1·12 P	1500	95	1·16 P	1500	.
5th	1·29 P	2000	.	0·80 P	1500	.
6th	1·05 P	1500	.	0·87 P	2000	.
7th	0·94 P	2000	.	0·93 P	2000	.
8th	1·06 P	2000	.	0·91 P	2000	.

* Amount of growth given as opacity reading by Spekker photoelectric absorptiometer (1·0 = *c.* 10^9 bacilli/ml.); P = surface pellicle present.

HP = haemagglutinating power measured by method of Duguid & Gillies (1957). Percentage of cells bearing fimbriae was estimated approximately by inspection of electron microscope fields.

to be limited by deficiency of dissolved oxygen. Conversion to the fimbriate phase was marked by the appearance on the aerobic static broth of a thin surface pellicle consisting of closely packed bacilli and by a 3- to 4-fold increase in the amount of growth (Table 3). This suggested that the fimbriae played a part in concentrating the fimbriate bacilli as a pellicle at the broth surface and that the increased growth was due to the free supply of atmospheric oxygen available to these bacilli in the pellicle. In agreement with this interpretation it was found that the fimbriate cells gave no greater growth than the non-fimbriate when cultured in anaerobically incubated broth or artificially aerated broth, i.e. in conditions not permitting pellicle formation or any resulting advantage in access to oxygen.

These relationships are illustrated by the growth curves recorded in Fig. 3 for parallel fimbriate phase and non-fimbriate phase cultures of the type 1a strain. From equal inocula, the fimbriate and non-fimbriate cultures in aerated, aerobic static, and anaerobic broths all grew exponentially at very nearly the same rate for the first 4 hr. Between 4 and 6 hr. the growth was slowed almost to a standstill in the aerobic static and anaerobic broths, presumably due to exhaustion of fermentable substrates and dissolved oxygen; it continued exponentially for 2 hr. longer in the aerated broth. The only considerable difference

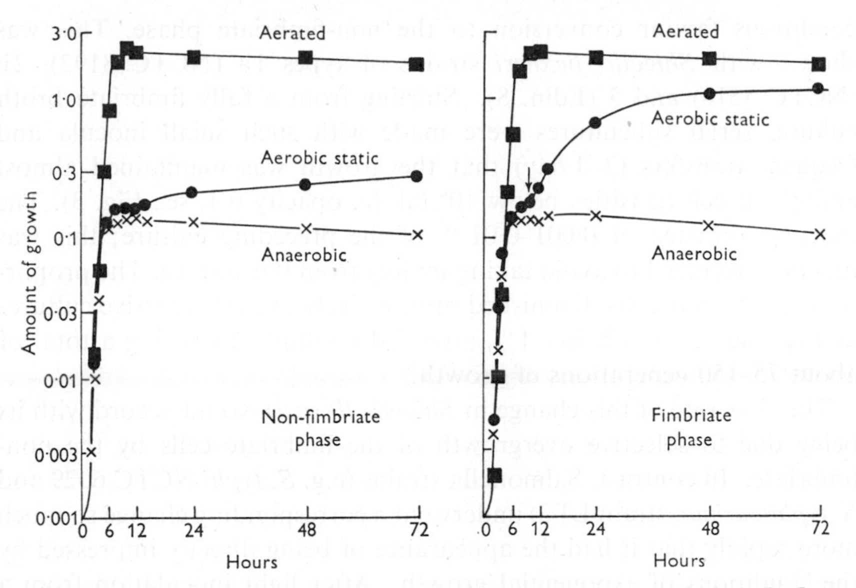

Fig. 3. Amount of growth of fimbriate phase and non-fimbriate phase *Shigella flexneri*, NCTC 8192, in broth incubated at 35°: (i) anaerobically in hydrogen, (ii) aerobically without agitation, and (iii) with aeration by continuous rotation (methods of Duguid & Gillies, 1957). Opacity readings of separate cultures grown for different times are plotted logarithmically, the values under 0·05 being opacity-equivalents calculated from cell counts. Inocula giving *c.* 10^6 bacilli/ml. (opacity equivalent = 0·001) were taken either from a 24 hr. fimbriate broth culture or from a non-fimbriate agar culture. A pellicle became visible on the fimbriate aerobic static cultures after 10 hr. (arrow).

between fimbriate and non-fimbriate growths occurred in the later stages in aerobic static broth; these fimbriate cultures underwent a large secondary increase of growth after forming a surface pellicle at about 10 hr., while the non-fimbriate showed little further increase. In some experiments the non-fimbriate cultures gave rise to fimbriate phase variants after 48, 72 or 96 hr. and then showed pellicle formation and late secondary growth.

It should not be concluded from this account that pellicle formation can occur only as a result of fimbriation. A minority of permanently non-fimbriate *flexneri* strains were found to form pellicles, thicker and more opaque than the fimbrial type, and these gave as much growth in aerobic static broth as fimbriate cultures.

Change to non-fimbriate phase during exponential growth in broth

Other evidence confirms that the conditions in aerobic static broth which favour conversion to fimbriation apply only in the later stages of culture, when exponential growth has been checked by exhaustion of dissolved oxygen and fermentable substrates. In the earlier stages the conditions favour conversion to the non-fimbriate phase. This was shown with *Shigella flexneri* strains of types 1a (NCTC 8192), 2b (NCTC 8518) and 3 (Edin. 8). Starting from a fully fimbriate broth culture, serial subcultures were made with such small inocula and frequent transfers (3–4/day) that the growth was maintained almost entirely at cell densities below 10^8/ml. (*c.* opacity 0·1, see Fig. 3). The inocula consisted of 0·001–0·01 % of the preceding culture; this was first homogenized to avoid taking mainly from the surface. The proportion of fimbriate cells diminished progressively in the successive cultures and reached a level below 1 % after 5–10 cultures involving a total of about 75–150 generations of growth.

The slow rate of this change in *Shigella flexneri* would accord with its being due to selective overgrowth of the fimbriate cells by the non-fimbriate. In contrast, Salmonella strains (e.g. *S. typhi* NCTC 6029 and *S. typhimurium*, strain LT2) underwent a corresponding change so much more rapidly that it had the appearance of being directly impressed by the conditions of exponential growth. After light inoculation from a fimbriate culture into a fresh tube of broth, the new culture became almost completely non-fimbriate within 10–20 generations growth in the first 4–8 hr. During the slower, later stages of growth the fimbriate cells increased again in number and became predominant after 24–48 hr.

The exponential cultures which are becoming non-fimbriate consist mainly of fully fimbriate cells and entirely non-fimbriate cells, together with a few cells which are scantily fimbriate (Pl. 2, fig. 7) and a very few which are fully fimbriate at one end and non-fimbriate at the other (Pl. 2, fig. 8). The existence of this last type might be considered evidence of unipolar growth as proposed by Bisset & Pease (1957), but the even distribution of fimbriae on both halves of most of the dividing cells argues for the occurrence of diffuse growth and binary fission (Pl. 1, fig. 6; Pl. 2, fig. 9).

Change to non-fimbriate phase on agar

The simplest means of converting cultures from the fimbriate to the non-fimbriate phase is by serial cultivation on nutrient agar plates incubated aerobically for 48 hr. at 37°. In most *flexneri* strains (e.g. NCTC 8192 and 8518) the change was fairly slow. When starting with a fully fimbriate broth culture and inoculating successive plates so as to obtain a confluent growth comprising 12–15 generations in 48 hr., usually 4–8 subcultures were needed to produce a completely non-fimbriate growth. The results of a typical experiment are shown in Table 4 (NCTC 8192). It is apparent that a proportion of the cells

Table 4. *Haemagglutinating power (HP) and approximate percentage of fimbriate cells in serial aerobic agar plate cultures of* Shigella flexneri, *strains NCTC* 8192 *and Edin.* 8, *grown for* 48 *hr. periods at* 37°

The first plate of each series was inoculated by covering it with 0·04 ml. of a fully fimbriate broth culture and later plates were inoculated similarly with *c.* 0·1 % of the cells of the preceding culture. The 48 hr. plate cultures of NCTC 8192 contained *c.* 5–25 % of viable cells and those of the strain Edin. 8 only 0·01–1·0 %; thus each subculture of the former involved *c.* 12–15 generations growth, and of the latter, *c.* 17–24 generations.

	NCTC 8192		Edin. 8	
Culture	HP	Percentage cells fimbriate	HP	Percentage cells fimbriate
Broth	2000	>95	2000	>95
1st plate	1000	90	500	50
2nd plate	150	30	1	0
3rd plate	30	7	0	0
4th plate	5	2	0	0
5th plate	1	0·2	0	.
6th plate	0	0·2	0	.
7th plate	0	0	0	.

HP = haemagglutinating power measured by method of Duguid & Gillies (1957). Percentage of cells bearing fimbriae was estimated approximately by inspection of electron microscope fields.

continue to inherit the fimbriate phase determinant through many generations growth under conditions favouring conversion to the non-fimbriate phase.

Certain type 3 and type 4a strains were liable to become non-fimbriate after only two or three cultures on agar (Table 4, strain Edin. 8). It was found that their 48 hr. plate cultures contained an exceptionally small proportion of viable cells and this suggests that the replacement of fimbriate by non-fimbriate cells may be due in part to a differential mortality in the later stages of growth.

Significance of the fimbrial phase variation

In assessing the biological significance of this variation, it may be assumed with fair certainty that the changes are physiological and not pathological. Both the fimbriate and the non-fimbriate cells are viable in favourable cultural conditions; both grow promptly from small inocula and their growth rates are similar. Brinton (1959) found only minor differences in growth rate, e.g. 10 %, between fimbriate and non-fimbriate strains of *Escherichia coli*, strain B, cultured under a variety of conditions.

Since the fimbriate and non-fimbriate cells differ in adhesive and other physiological properties, the variation mechanism may be of value to the organism in ensuring that part of its cell population assumes the properties best adapted for growth, survival or dissemination under the current conditions. Enterobacteria having the commonest kind of fimbriae (with the mannose-sensitive adhesive factor, 'MS adhesin', Duguid, 1959) are strongly adhesive for the surfaces of a wide variety of animal, plant and fungal cells to which they do not adhere when in the non-fimbriate phase. It may be advantageous for an intestinal or soil organism to vary between an adhesive form which will grow in aggregations on the surface of favourable substrates such as mucous membrane, food particles or plant root hairs, and a non-adhesive form which will readily disperse from the first colonized sites to reach fresh substrates. Saprophytic enterobacteria inhabiting oxygen-depleted water may benefit both from the ability of their fimbriate variants to grow as a surface pellicle with access to atmospheric oxygen and from the ability of their non-fimbriate variants to disperse from this pellicle. Maccacaro & Dettori's (1959) finding that non-fimbriate phase cells of *Escherichia coli*, strain K12, have greater fermentative activity than the fimbriate cells, seems to indicate that the non-fimbriate are better adapted for growth in anaerobic liquids. There may be further, yet unknown, physiological differences associated with the fimbrial phase change, and our assessment of its value is probably incomplete.

Since bacteria are liable to encounter widely varying conditions, they will benefit from a capacity for readily reversible variation in appropriate physiological characters. Such variation may occur as a direct phenotypic response to environmental conditions, e.g. the enzymic inductions and repressions which economically fit the cells to deal with changing supplies of nutrients (see Pardee in this Symposium). An alternative mechanism would be the control of a character by a nuclear gene or heritable cytoplasmic factor which was endowed with an exceptionally high degree of spontaneous reversible mutability between active and inactive states. This may be the mechanism of the fimbrial phase variation. There might indeed be an advantage in spontaneous as compared to induced variation. Thus, a bacterial clone growing in conditions favouring one phase would still continue to produce a few cells of the opposite phase, ready to take advantage of suddenly changing conditions.

CONCLUSIONS

The evidence reviewed suggests a different interpretation for each of these three groups of morphological changes.

(1) Gross variations in cell shape and size are involutionary and

result from unbalanced growth which is liable to proceed to cell disruption and death. In many cases, this involves a continuance of protoplasmic growth while either cell wall formation or cell division is selectively inhibited.

(2) Fluctuations in the amount of capsular and intracellular polysaccharides, and volutin and lipid inclusions, are commonly due to a nutrient imbalance causing disproportion between synthesis of the component and general protoplasmic growth. This unbalanced growth is generally regulated by the cell within limits compatible with viability and there is some evidence that the accumulation of intracellular polysaccharide, lipid and volutin may have a useful energy-storage function.

(3) The fimbrial phase variation provides for alternation between two viable forms having different physiological properties adapted to different conditions of growth. The mechanism of the change is still obscure, but the balance of evidence suggests that it involves spontaneous changes in a heritable determinant and environmental selection of mutants, rather than a direct phenotypic response to the influence of environmental conditions.

REFERENCES

ALMQUIST, E. (1922). Variation and life cycles of pathogenic bacteria. *J. infect. Dis.* **31**, 483.

AUSTRIAN, R. (1953). Morphologic variation in Pneumococcus. I. An analysis of the bases for morphologic variation in Pneumococcus and a description of a hitherto undefined morphologic variant. *J. exp. Med.* **98**, 21.

BERNHEIMER, A. W. (1953). Synthesis of type III pneumococcal polysaccharide by suspensions of resting cells. *J. exp. Med.* **97**, 591.

BISSET, K. A. (1949). The nuclear cycle in bacteria. *J. Hyg., Camb.* **47**, 182.

BISSET, K. A. & PEASE, P. (1957). The distribution of flagella in dividing bacteria. *J. gen. Microbiol.* **16**, 382.

BRINTON, C. C. (1959). Non-flagellar appendages of bacteria. *Nature, Lond.* **183**, 782.

BRINTON, C. C., BUZZELL, A. & LAUFFER, M. A. (1954). Electrophoresis and phage susceptibility studies on a filament-producing variant of the *E. coli* B bacterium. *Biochim. biophys. Acta*, **15**, 533.

CEDERGREN, B. & HOLME, T. (1959). On the glycogen in *Escherichia coli* B; electron microscopy of ultrathin sections of cells. *J. Ultrastruct. Res.* **3**, 70.

COHEN, S. S. & BARNER, H. D. (1954). Studies on unbalanced growth in *Escherichia coli*. *Proc. nat. Acad. Sci., Wash.* **40**, 885.

DAGLEY, S. & JOHNSON, A. R. (1953). The relation between lipid and polysaccharide contents of *Bact. coli*. *Biochim. biophys. Acta*, **11**, 158.

DAWES, E. A. & HOLMS, W. H. (1958). Metabolism of *Sarcina lutea*. III. Endogenous metabolism. *Biochim. biophys. Acta*, **30**, 278.

DAWSON, M. H. (1934). Variation in the Pneumococcus. *J. Path. Bact.* **39**, 323.

DIENES, L. & WEINBERGER, H. J. (1951). The L forms of bacteria. *Bact. Rev.* **15**, 245.

DIENES, L. & ZAMECNIK, P. C. (1952). Transformation of bacteria into L forms by amino acids. *J. Bact.* **64**, 770.

DUGUID, J. P. (1946). The sensitivity of bacteria to the action of penicillin. *Edinb. med. J.* **53**, 401.

DUGUID, J. P. (1948). The influence of cultural conditions on the morphology of *Bacterium aerogenes* with reference to nuclear bodies and capsule size. *J. Path. Bact.* **60**, 265.

DUGUID, J. P. (1959). Fimbriae and adhesive properties in Klebsiella strains. *J. gen. Microbiol.* **21**, 271.

DUGUID, J. P. & GILLIES, R. R. (1957). Fimbriae and adhesive properties in dysentery bacilli. *J. Path. Bact.* **74**, 397.

DUGUID, J. P. & GILLIES, R. R. (1958). Fimbriae and haemagglutinating activity in *Salmonella*, *Klebsiella*, *Proteus* and *Chromobacterium*. *J. Path. Bact.* **75**, 519.

DUGUID, J. P., SMITH, I. W., DEMPSTER, G. & EDMUNDS, P. N. (1955). Non-flagellar filamentous appendages ('fimbriae') and haemagglutinating activity in *Bacterium coli*. *J. Path. Bact.* **70**, 335.

DUGUID, J. P., SMITH, I. W. & WILKINSON, J. F. (1954). Volutin production in *Bacterium aerogenes* due to development of an acid reaction. *J. Path. Bact.* **67**, 289.

DUGUID, J. P. & WILKINSON, J. F. (1953). The influence of cultural conditions on polysaccharide production by *Aerobacter aerogenes*. *J. gen. Microbiol.* **9**, 174.

DUGUID, J. P. & WILKINSON, J. F. (1954). The influence of potassium deficiency upon production of polysaccharide by *Aerobacter aerogenes*. *J. gen. Microbiol.* **11**, 71.

EISLER, M. (1909). Ueber Wirkungen von Salzen auf Bakterien. *Zbl. Bakt.* (1. *Abt. Orig.*), **51**, 546.

FISCHER, A. (1900). Die Empfindlichkeit der Bakterienzelle und das baktericide Serum. *Z. Hyg. InfektKr.* **35**, 1.

FOSTER, J. W. (1947). Some introspections on mold metabolism. *Bact. Rev.* **11**, 167.

GARDNER, A. D. (1925). The growth of branching forms of bacilli ('three point multiplication'). *J. Path. Bact.* **28**, 189.

GARDNER, A. D. (1945). Microscopical effect of penicillin on spores and vegetative cells of bacilli. *Lancet*, **1**, 658.

GILLESPIE, H. B. & RETTGER, L. F. (1939). Bacterial variation: formation and fate of certain variant cells of *Bacillus megatherium*. *J. Bact.* **38**, 41.

GLICK, M. C., SALL, T., ZILLIKEN, F. & MUDD, S. (1960). Morphological changes of *Lactobacillus bifidus* var. *pennsylvanicus* produced by a cell-wall precursor. *Biochim. biophys. Acta*, **37**, 361.

GORDON, J. & GORDON, M. (1943). Involution forms of the genus *Vibrio* produced by glycine. *J. Path. Bact.* **55**, 63.

GRELET, N. (1952). Le détermination de la sporulation de *Bacillus megatherium*. II. L'effet de la pénurie des constituants minéraux du milieu synthétique. *Ann. Inst. Pasteur*, **82**, 66.

HOLME, T. (1957). Continuous culture studies on glycogen synthesis in *Escherichia coli* B. *Acta chem. scand.* **11**, 763.

HOLME, T. & PALMSTIERNA, H. (1956*a*). Changes in glycogen and nitrogen-containing compounds in *Escherichia coli* B during growth in deficient media. I. Nitrogen and carbon starvation. *Acta chem. scand.* **10**, 578.

HOLME, T. & PALMSTIERNA, H. (1956*b*). Changes in glycogen and nitrogen-containing compounds in *Escherichia coli* B during growth in deficient media. II. Phosphorus and sulphur starvation. *Acta chem. scand.* **10**, 1553.

HORT, E. C. (1920). The reproduction of aerobic bacteria. *J. Hyg., Camb.* **18**, 369.

HUGHES, W. H. (1953). The origin of the L-form variants in anaerobic cultures of *Bacterium coli*. *J. gen. Microbiol.* **8**, 307.

HUGHES, W. H. (1956). The structure and development of the induced long forms of bacteria. In *Bacterial Anatomy. Symp. Soc. gen. Microbiol.* **6**, 341.

JEENER, H. & JEENER, R. (1952). Cytological study of *Thermobacterium acidophilus* R36 cultured in absence of deoxyribonucleosides or uracil. *Exp. Cell Res.* **3**, 675.

KORNBERG, S. R. (1957). Adenosine triphosphate synthesis from polymetaphosphate by an enzyme from *Escherichia coli. Biochim. biophys. Acta*, **26**, 294.

LEDERBERG, J. (1956). Bacterial protoplasts induced by penicillin. *Proc. nat. Acad. Sci., Wash.* **42**, 574.

LEDERBERG, J. & ST CLAIR, J. (1958). Protoplasts and L-type growth of *Escherichia coli. J. Bact.* **75**, 143.

LIEBERMEISTER, K. (1960). Morphology of the PPLO and L forms of *Proteus. Ann. N.Y. Acad. Sci.* **79**, 326.

LIEBERMEISTER, K. & KELLENBERGER, E. (1956). Studien zur L-form der Bakterien. I. Die Umwandlung der bazillären in die globuläre Zellform bei *Proteus* unter Einfluss von Penicillin. *Z. Naturf.* **11**b, 200.

LÖHNIS, F. (1921). Studies upon the life cycles of bacteria. *Mem. nat. Acad. Sci.* **16**, 1.

LOMINSKI, I., CAMERON, J. & WYLLIE, G. (1958). Chaining and unchaining *Streptococcus faecalis*—a hypothesis of the mechanism of bacterial cell separation. *Nature, Lond.* **181**, 1477.

MACCACARO, G. A., COLOMBO, C. & DI NARDO, A. (1959). Studi sulle fimbrie batteriche. I. Lo studio genetico delle fimbrie. *Giorn. Microbiol.* **7**, 1.

MACCACARO, G. A. & DETTORI, R. (1959). Studi sulle fimbrie batteriche. IV. Metabolismo ossidativo e fermentativo in cellule fimbriate e sfimbriate. *Giorn. Microbiol.* **7**, 52.

MACCACARO, G. A. & TURRI, M. (1959). Studi sulle fimbrie batteriche. II. Osservazioni microelettroforetiche. *Giorn. Microbiol.* **7**, 21.

MACRAE, R. M. & WILKINSON, J. F. (1958*a*). Poly-β-hydroxybutyrate metabolism in washed suspensions of *Bacillus cereus* and *Bacillus megaterium. J. gen. Microbiol.* **19**, 210.

MACRAE, R. M. & WILKINSON, J. F. (1958*b*). The influence of cultural conditions on poly-β-hydroxybutyrate synthesis in *Bacillus megaterium. Proc. Roy. Phys. Soc. Edinb.* **27**, 73.

MATZUSCHITA, T. (1900). Die Einwirkung des Kochsalzgehaltes des Nährbodens auf die Wuchsform der Mikroorganismen. *Z. Hyg. InfektKr.* **35**, 495.

MCQUILLEN, K. (1958). Lysis resulting from metabolic disturbance. *J. gen. Microbiol.* **18**, 498.

MCQUILLEN, K. (1960). Bacterial protoplasts. In *The Bacteria*, vol. I, p. 249. Edited by I. C. Gunsalus and R. Y. Stanier. London: Academic Press.

MEADOW, P., HOARE, D. S. & WORK, E. (1957). Interrelationships between lysine and $\alpha\epsilon$-diaminopimelic acid and their derivatives and analogues in mutants of *Escherichia coli. Biochem. J.* **66**, 270.

MELLON, R. R. (1942). The polyphasic potencies of the bacterial cell; general biologic and chemotherapeutic significance. *J. Bact.* **44**, 1.

MUDD, S., YOSHIDA, A. & KOIKE, M. (1958). Polyphosphate as accumulator of phosphorus and energy. *J. Bact.* **75**, 224.

MUHAMMED, A., RODGERS, A. & HUGHES, D. E. (1959). Purification and properties of a polymetaphosphatase from *Corynebacterium xerosis. J. gen. Microbiol.* **20**, 482.

MURRAY, R. G. E., FRANCOMBE, W. H. & MAYALL, B. H. (1959). The effect of penicillin on the structure of staphylococcal cell walls. *Canad. J. Microbiol.* **5**, 641.

NORRIS, K. P. & GREENSTREET, J. E. S. (1958). On the infra-red absorption spectrum of *Bacillus megaterium*. *J. gen. Microbiol.* **19**, 566.

PARK, J. T. (1958). Selective inhibition of bacterial cell-wall synthesis: its possible applications in chemotherapy. In *The Strategy of Chemotherapy*. *Symp. Soc. gen. Microbiol.* **8**, 49.

PULVERTAFT, R. J. V. (1952). The effect of antibiotics on growing cultures of *Bacterium coli*. *J. Path. Bact.* **64**, 75.

QUADLING, C. & STOCKER, B. A. D. (1956). An environmentally induced transition from the flagellated to the non-flagellated state in *Salmonella*: the fate of parental flagella at cell division. *J. gen. Microbiol.* **15**, i.

REED, G. & ORR, J. H. (1923). The influence of H ion concentration upon structure. I. *H. influenzae*. *J. Bact.* **8**, 103.

RICHTER, K. (1934). Untersuchungen über den Einfluss von Lithiumchlorid auf *Bacterium coli*. *Zbl. Bakt.* (2. Abt.), **90**, 134.

ROBINOW, C. F. (1944). Cytological observations on *Bact. coli*, *Proteus vulgaris* and various aerobic spore-forming bacteria with special reference to the nuclear structures. *J. Hyg., Camb.* **43**, 413.

SALL, T., MUDD, S. & DAVIS, J. C. (1956). Factors conditioning the accumulation and disappearance of metaphosphate in cells of *Corynebacterium diphtheriae*. *Arch. Biochem. Biophys.* **60**, 130.

SALL, T., MUDD, S. & TAKAGI, A. (1958). Polyphosphate accumulation and utilization as related to synchronized cell division of *Corynebacterium diphtheriae*. *J. Bact.* **76**, 640.

SALTON, M. R. J. (1960). Surface layers of the bacterial cell. In *The Bacteria*, vol. I, p. 97. Edited by I. C. Gunsalus and R. Y. Stanier. London: Academic Press.

SHANKAR, K. & BARD, R. C. (1952). The effect of metallic ions on the growth and morphology of *Clostridium perfringens*. *J. Bact.* **63**, 279.

SHOCKMAN, G. D., KOLB, J. J. & TOENNIES, G. (1958). Relations between bacterial cell wall synthesis, growth phase, and autolysis. *J. biol. Chem.* **230**, 961.

SMITH, I. W., WILKINSON, J. F. & DUGUID, J. P. (1954). Volutin production in *Aerobacter aerogenes* due to nutrient imbalance. *J. Bact.* **68**, 450.

SPIEGELMAN, S., REINER, J. M. & COHNBERG, R. (1947). The relation of enzymatic adaptation to the metabolism of endogenous and exogenous substrates. *J. gen. Physiol.* **31**, 27.

TAKAHASHI, I. & GIBBONS, N. E. (1959). Effect of salt concentration on the morphology and chemical composition of *Micrococcus halodenitrificans*. *Canad. J. Microbiol.* **5**, 25.

TINELLI, R. (1955). Étude de la biochimie de la sporulation chez *Bacillus megaterium*. II. Modifications biochimiques et échanges gazeux accompagnant la sporulation provoquée par carence de glucose. *Ann. Inst. Pasteur*, **88**, 364.

TUTTLE, A. L. & GEST, H. (1960). Induction of morphological aberrations in *Rhodospirillum rubrum* by D-amino acids. *J. Bact.* **79**, 213.

WEBB, M. (1949). The influence of magnesium on cell division. 2. The effect of magnesium on the growth and cell division of various bacterial species in complex media. *J. gen. Microbiol.* **3**, 410.

WEBB, M. (1951). The influence of magnesium on cell division. 6. The action of certain hydrolytic enzymes on the filamentous and chain forms of Gram-positive rod-shaped organisms. *J. gen. Microbiol.* **5**, 496.

WEIBULL, C. (1956). Bacterial protoplasts; their formation and characteristics. In *Bacterial Anatomy*. *Symp. Soc. gen. Microbiol.* **6**, 111.

WELSCH, M. & OSTERREITH, P. (1958). A comparative study of the transformation of Gram-negative rods into 'Protoplasts' under the influence of penicillin and glycine. *Leeuwenhoek ned. Tijdschr.* **24**, 257.

PLATE 1

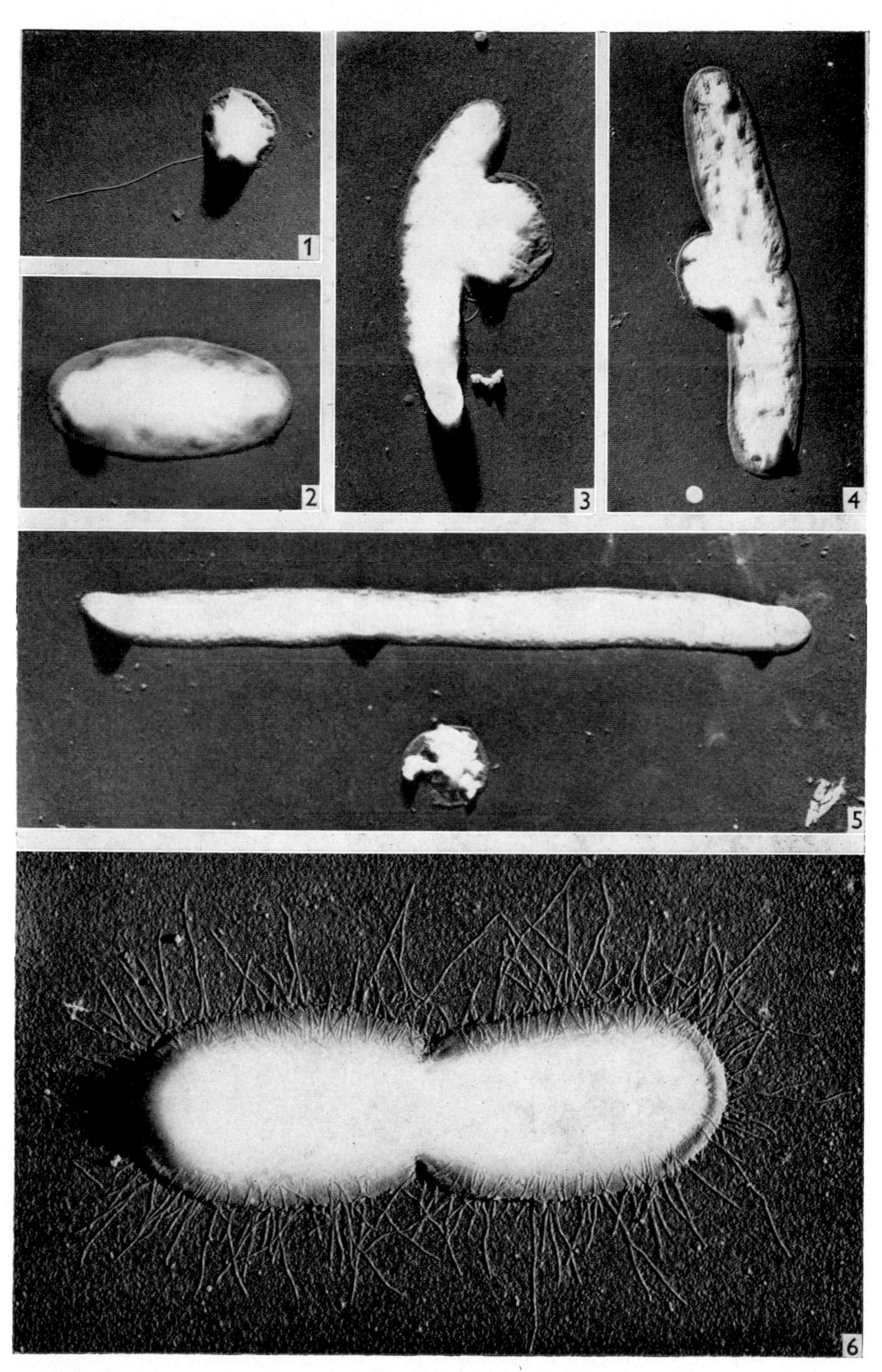

(*Facing page 98*)

PLATE 2

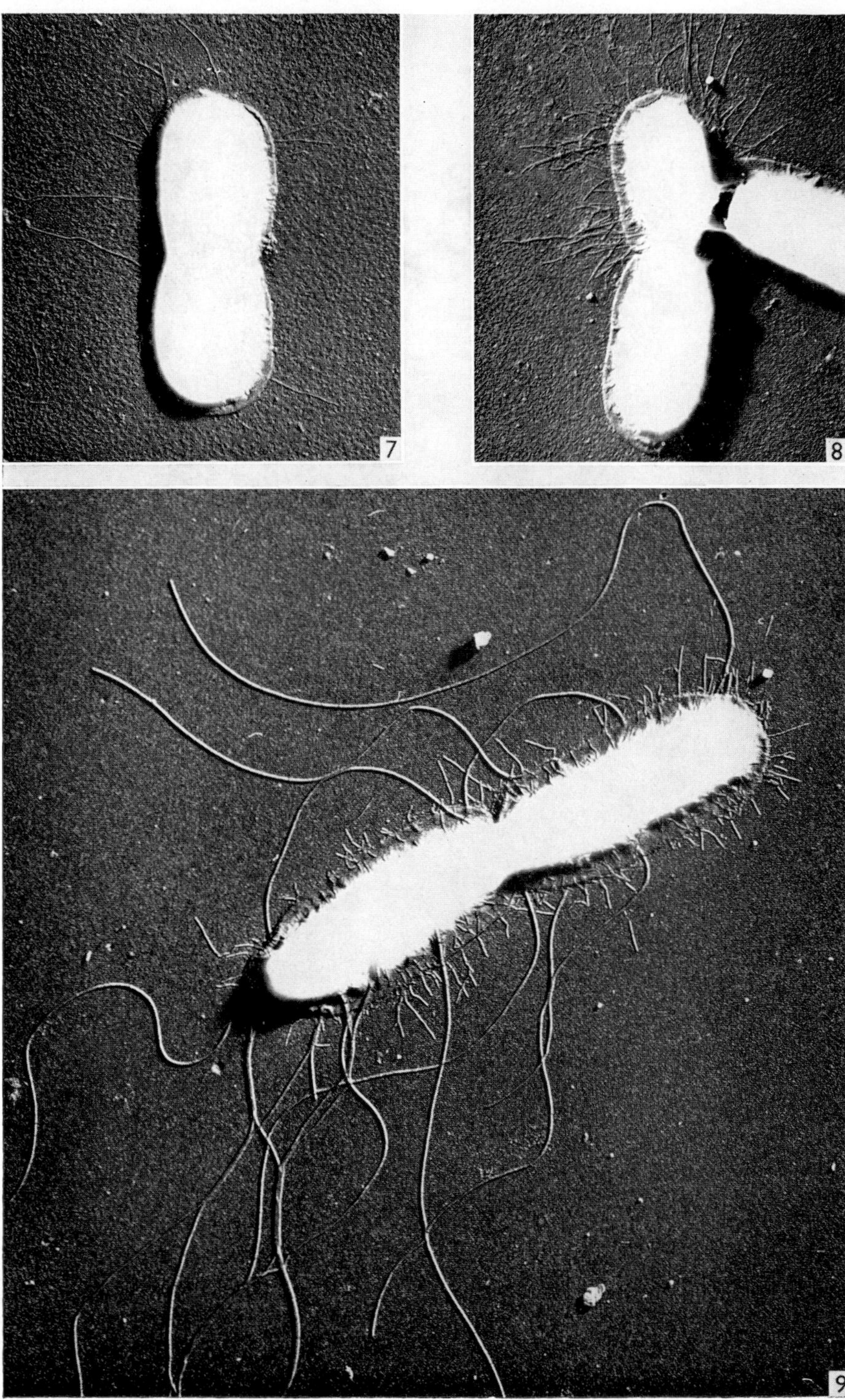

WIAME, J. M. (1946*a*). Remarque sur la métachromasie des cellules de levure. *C.R. Soc. Biol., Paris*, **140**, 897.

WIAME, J. M. (1946*b*). Basophilie et métabolisme du phosphore chez la levure. *Bull. Soc. chim. Biol., Paris*, **28**, 552.

WIAME, J. M. (1947*a*). Yeast metaphosphate. *Fed. Proc.* **6**, 302.

WIAME, J. M. (1947*b*). The metachromatic reaction of hexametaphosphate. *J. Amer. chem. Soc.* **69**, 3146.

WIDRA, A. (1959). Metachromatic granules of microorganisms. *J. Bact.* **78**, 664.

WILKINSON, J. F. (1958). The extracellular polysaccharides of bacteria. *Bact. Rev.* **22**, 46.

WILKINSON, J. F. (1959). The problem of energy-storage compounds in bacteria. *Exp. Cell Res.* (*Suppl.*), **7**, 111.

WILKINSON, J. F. & DUGUID, J. P. (1960). The influence of cultural conditions on bacterial cytology. *Int. Rev. Cytol.* **9**, 1.

WILKINSON, J. F. & STARK, G. H. (1956). The synthesis of polysaccharide by washed suspensions of *Klebsiella aerogenes*. *Proc. Roy. phys. Soc. Edinb.* **25**, 35.

WYCKOFF, R. W. G. (1933). The morphology of *Bacterium shigae* cultivated on various media favorable to the development of filterability and life cycle forms. *J. exp. Med.* **57**, 165.

YONEDA, M. & KONDO, M. (1959). Studies on poly-β-hydroxybutyrate in bacterial spores. I. Existence of poly-β-hydroxybutyrate in mature spores of a strain of *Bacillus cereus* and its relation to the acid-fast stainability. *Biken's J.* **2**, 247.

ZAMENHOF, S., DE GIOVANNI, R. & RICH, K. (1956). *Escherichia coli* containing unnatural pyrimidines in its deoxyribonucleic acid. *J. Bact.* **71**, 60.

EXPLANATION OF PLATES

PLATES 1 AND 2

Figs. 1–5. Abnormal forms of *Salmonella typhi* (NCTC 6029) found in an aerobic broth culture grown at 44° for 48 hr. Fig. 1, globular form; fig. 2, large form; figs. 3, 4, plasmoptysis forms; fig. 5, long form and lysed globular form; figs. 1–4, × 10,000; fig. 5, × 5000.

Figs. 6–8. *Shigella flexneri*, type 2b (NCTC 8518); dividing bacilli from an exponential growth in broth (8 hr. at 37°) undergoing conversion to non-fimbriate phase. Figure 6, typical dividing fimbriate bacillus shows even distribution of fimbriae on both daughter cells; × 24,000. Figure 7, uncommon form is scantily fimbriate; × 15,000. Figure 8, rare form bears fimbriae on one daughter cell only; × 15,000.

Fig. 9. *Salmonella typhi* (NCTC 6029); fimbriate and flagellate dividing bacillus from exponential growth in broth (6 hr. at 37°); × 15,000.

SPORE FORMATION AND GERMINATION AS A MICROBIAL REACTION TO THE ENVIRONMENT

W. G. MURRELL

C.S.I.R.O. Division of Food Preservation, Homebush, N.S.W., Australia

Bacterial spores are formed within vegetative cells, mainly of the genera *Bacillus* and *Clostridium*. A densely staining body forms, enlarges, and in a short time becomes refractive and impermeable to dilute cold stains. Eventually, the spore usually becomes free of the mother cell or sporangium. It may germinate immediately or remain in a resting state for long periods. The spore may be defined or recognized by its refractility, impermeability to stains and by its increased resistance to heat, chemicals, enzymes, radiation, and freezing and thawing. The resistance may vary widely, e.g. in a dilute aqueous system the rate of killing by heat of *B. stearothermophilus* spores is *c.* 10^5 times less than the rate for *C. botulinum*, type E spores (Table 1, Murrell, 1955). The spore is also well defined by its method of formation and its structure—the thick well-defined cortical region, and the several coat layers outside the cortex (Mayall & Robinow, 1957; Young & Fitz-James, 1959*b*; Hashimoto, Black & Gerhardt, 1960). Chemically, it also appears unique in containing up to 15% (dry wt.) dipicolinic acid (pyridine-2:6-dicarboxylic acid: Powell & Strange, 1953; Woese 1959*a*).

In this paper the processes of formation and germination of bacterial spores are considered as metabolic reactions or changes induced in the cell by some change in the environment. Alternatively, we may say that the absence of spore formation and germination is the result of environments inadequate for these processes.

SPORE FORMATION: INTRODUCTION

Cytologically, sporogenesis begins when the nuclear material of the cell, in the form of two condensed chromatin bodies, fuses and forms into an axial thread. A portion of the chromatinic material then becomes positioned at one end of the sporangium and a thin septum, growing centripetally from opposite sides of the cell and very close to one end, confines the chromatin and a small amount of cytoplasm. The process of sporogenesis is believed to become irreversible or committed at this stage which may be reached in *c.* 35 min. after the last cell division. The septum finally encloses the spore chromatin completely by growing, from its slightly thickened base at the inside of the cell wall, round inside the end of the cell. The spore body then becomes free in the sporangium. This description is based on the important studies of Young & Fitz-James (1959*b*–*d*) with *Bacillus cereus* and *B. cereus* var. *alesti.* During maturation of the spore, first a cortical region (*c.* 0·1 μ thick) and then at least two coat layers, and in some organisms an exosporium, are formed peripheral to the spore body enclosed by the initial septum. The initial spore septum or membrane eventually becomes the cell wall of the germ cell during germination. The development of refractility of the spore appears to coincide with the formation of the cortex and the first spore coat. The stage, from the completion of the septum to the development of refractility, which appears as a densely staining body in the light microscope, is known as 'forespore' formation.

Vegetative growth does not always lead to spore formation since sporulation is greatly affected by the environment. Many media apparently give excellent vegetative growth, but are often unsuitable for spore formation. Two types of observation should be considered before studying the effect of environment on sporulation. First, only cells grown under suitable conditions become sporangia, and secondly, these sporangial cells are capable of completing sporogenesis independently of the chemical environment of the culture in which they are grown and without an external source of nutrients. These observations are derived largely from studies on sporulation of washed cells suspended in distilled water. This valuable technique, due to Hardwick & Foster (1952), has enabled considerable advances to be made in our knowledge of spore formation, and will continue to do so, even though, in some organisms, lytic products contribute significantly to the environment (Powell & Hunter, 1953; Perry & Foster, 1954; Young & Fitz-James, 1959*c*; Black, Hashimoto & Gerhardt, 1960).

The evidence suggesting that vegetative cells grown under many conditions are not sporangia is as follows: (1) Cells grown on media not favouring sporogenesis due to either omission of glucose, deficiency of nitrogen, or the presence of inhibitors of sporogenesis do not form spores in the washed cell method (Hardwick & Foster, 1952). Cells growing on media deficient in manganese, which is essential for sporulation, do not spore when Mn is supplied (Charney, Fisher & Hegarty, 1951; Amaha, Ordal & Touba, 1956). (2) The cultural age of the cells is important. Miwatani (1957) found that washed lag phase cells of *Bacillus cereus* containing smaller amounts of free amino acids than logarithmic phase cells and no cytochromes did not sporulate, but cells in the logarithmic growth phase did. Gollakota & Halvorson (1960) observed that spore formation of washed cells of *B. cereus* depended on the age of the cells since cells harvested at the end of the logarithmic phase did sporulate. Black *et al.* (1960) found that only the cells appearing granulated (the cytological stage preceding forespore formation) and cells with forespores could form spores in aerated suspensions of washed cells of *B. cereus.* Ungranulated cells preceding these stages would not form spores. Further, cells of cultures of *B. anthracis* and *B. globigii* in the late vegetative growth phase, when transferred to a fresh medium, sporulated and began a new growth cycle, while those in the early stages of growth on transfer continued to grow vegetatively (Roth, Lively & Hodge, 1955).

The intracellular environment or the organization of the cell, not the extracellular medium during sporogenesis, is the main variable in these observations. These observations indicate that the cell irreversibly committed to sporogenesis, the sporangium, must actually be produced and go through its pre-spore development in a suitable environment. That is, not only is the spore a product of a suitable environment but the sporangium is also. The possibility of lag or early log phase cells or cells grown on asporogenic media becoming sporangia in a suitable environment in the absence of cell division is not excluded by the above experiments.

Washed cells of *Clostridium* spp. also form spores. Collier (1957) obtained complete sporulation in 6–7 hr. of cells of *C. roseum* washed under anaerobic conditions and incubated anaerobically in distilled water. Lund (1957*a*), however, failed to obtain spores with washed cells of *C. sporogenes*, incubated in either buffered thioglycollate or glucose with thioglycollate, which sporulated in a peptone medium.

The ability of sporangia to form spores after removal from the previous environment and washing will depend on the adequacy of the intracellular reserves, the growth medium and the organism. Perry & Foster (1954) showed that it was possible to produce heat-stable spores of *Bacillus mycoides* in the absence of lysis of cells that did not form spores. Black *et al.* (1960) found that spores of *B. cereus* produced from washed cells may be deficient in some properties compared to those produced in the growth medium unless the growth medium was enriched with calcium. The calcium could be added at the time of reincubation of the washed cells.

In sporulation studies with growing cultures or washed cells, viable or microscopic counts of spores in conjunction with total counts of all cell forms should be made in order to assess the true spore yield and the importance of lysis (Powell & Hunter, 1953; Perry & Foster, 1954). Synchronously sporing cultures are likely to provide more valuable information than asynchronous ones (Collier, 1957; Young & Fitz-James, 1959*b*). Washed cell cultures should not be dividing.

The intracellular environment during sporogenesis depends not only on the medium but on the equilibrium between endogenous reserves and the free solute pool in the cell and the products of lysis. In washed cell studies, lysis assumes much greater importance. The effect of added substances and of variations in the external environment can be studied much more conclusively with washed cells than with growing cultures.

EFFECT OF ENVIRONMENTAL CONDITIONS ON THE GROWTH OF SPORANGIA AND ON SPORE FORMATION

Sporangia are recognizable by special microscopic techniques, especially electron micrographs of ultra-thin sections, before or very early in sporogenesis (Young & Fitz-James, 1959*b*; Hashimoto *et al.* 1960). However, as the main criterion of a sporangium until recently was the production of a spore, the information in this section refers largely to studies on spore formation in growing cultures. During the process of sporulation, which may take several hours, the growth medium and the intracellular environment are changing. Further, many of the environmental factors are interdependent. Another complication is that in some instances the response to an environmental change during spore formation may be due to selection of cells which are more capable of growing and sporulating under the new conditions.

Physical environment

The limits for pH, temperature, water availability (water activity), oxygen pressure and surface tension are usually narrower for spore formation than for growth, although the optima are generally similar (Knaysi, 1948; Wynne, 1948, 1952).

Temperature. Temperature affects both the rate and amount of spore formation, and the properties of the spores. Quantitative observations on the effect of temperature on the rate and amount of spore formation for either growing or washed cells are not available. Qualitative observations show that sporogenesis occurs most rapidly at or near the optimum growth temperature and the amount is reduced by unfavourable temperatures (Migula, 1904).

Schreiber (1896) observed that transfer of 14 hr. cultures of *Bacillus anthracis* from 37° to 18° produced irregular cell forms and abnormal sporulation, but when the temperature change was less (30° to 20°) no effect other than a reduced growth rate was observed. Christian (1931) also observed, that at the maximum temperature for spore formation of a species of *Bacillus*, the cells concentrated protoplasm in a terminal position but no spores formed. These 'abortive' spores were freed spontaneously and persisted for long periods, but were not shown to be viable. These examples suggest that temperature affects some of the biochemical processes in sporogenesis differently. Interruption of sporulation by cooling to 4° did not prevent prompt sporulation of washed cells of *B. mycoides* when returned to 30° after a month at 4° (Hardwick & Foster, 1952).

Temperature may also affect the properties of the spores which are produced. Spores formed at high temperatures are usually more heat-resistant than those formed at lower temperatures (Williams, 1929; Theophilis & Hammer, 1938; Williams & Robertson, 1954; El-Bisi & Ordal, 1956). The thermal death rate varied 2 to 5 times for growth temperature differences of 13–15°. Sugiyama (1951) obtained spores of *Clostridium botulinum* of greater resistance at 37° than either at 41° or at 24° and 29°. Selection of cells capable of forming more resistant spores is not excluded in these experiments.

Lechowich (1959) found that in spores of *Bacillus subtilis* the heat resistance, dipicolinic acid and cation (Ca^{++}, Mg^{++}, Mn^{++}) content increased with the sporulation temperature; but in spores of *B. coagulans* (*thermoacidurans*), although the heat resistance increased, the dipicolinic acid decreased and the cation content changed little with sporulation temperature. However, in both organisms the ratio of the total cation concentration to dipicolinic acid was greater in spores grown at high temperatures.

pH. The difficulties of stabilizing pH during growth have prevented accurate studies of its effect. It is evident, however, that pH values not affecting vegetative growth completely prevent spore formation (Leifson, 1931). Some species appear to have a low optimum, e.g. *Clostridium pasteurianum*, pH 5·4–5·7 (Bowen & Smith, 1955) and *Bacillus coagulans* var. *thermoacidurans*, pH *c.* 5·5 (Amaha *et al.* 1956).

With some organisms, e.g. *Bacillus cereus* and *B. subtilis*, the pH of the medium falls rapidly from 7 to 5, and then rises to over pH 8 during the final stages of spore formation and lysis (Williams, 1929; Knaysi, 1945*a*; Roth, Lively & Metcalfe, 1958). Halvorson (1957*b*) observed that the initial fall in pH usually closely followed the first peak in the oxygen demand curve of the culture. Gollakota & Halvorson (1960) found that certain cations (Zn^{++}, Co^{++}, Ni^{++}), phosphate or yeast extract (but not the ash of yeast extract) were essential for this rise in pH and for the occurrence of sporulation.

In washed cell studies with *Bacillus mycoides* in phosphate buffer, Hardwick & Foster (1952) found that sporulation decreased sharply below pH 6, and above pH 6 was not affected up to 8·5, the highest value tested. In similar studies, Ordal (1957) observed a sharp optimum at pH 7, but the range and optimum was about 0·5 unit higher than with growing agar cultures.

Water activity (a_w). The effect of a_w (the ratio of the vapour pressure of the solution to that of the solvent) on spore formation is not clear. Although observations by several early workers suggested that some degree of drying or slow drying increased spore formation, Knaysi (1945*a*) found that storage of agar slants over sodium hydroxide or sulphuric acid solutions of known a_w reduced spore formation by *Bacillus mycoides*. Sporulation of some species occurs readily in dilute liquid media of a_w *c.* 0·999.

Radiations. Sunlight gave variable effects which were sometimes detrimental with several aerobic species (Schreiber, 1896; Holzmüller, 1909). Wynne (1948), however, obtained no effect with artificial light on *Clostridium botulinum*. Certain stages of spore formation are, however, affected by other radiations. Hardwick & Foster (1952) observed that sporogenesis in washed cells was more sensitive to ultraviolet light than were endogenous respiration and viability. Romig & Wyss (1957) found in synchronously sporing washed cell cultures of *Bacillus cereus* that the enhanced ultraviolet resistance of spores developed at the time of forespore appearance, and when the cells also ceased to be photo-reactivated after ultraviolet irradiation. This stage occurred *c.* 2 hr. before the appearance of refractive, heat-stable spores. McDonald & Wyss (1959) showed that cells of *B. cereus* irradiated with ultraviolet light and X-rays in the period immediately preceding forespore appearance had double the frequency of mutation to streptomycin resistance as washed vegetative cells irradiated 2–3 hr. before the appearance of forespores. The cells with the high mutation rate were also more sensitive to ultraviolet light than the cells committed to sporogenesis. The increased sensitivity in this period just before the cells became refractory to doses mutagenic for vegetative cells, indicated genetic instability.

Oxygen. First, *Bacillus* spp. Oxygen is essential for the production of sporangia, but its necessity in the final stages of the metabolic processes of sporogenesis is not yet clear. Knaysi (1945*a*) observed that good aeration increased the percentage of cells sporing and the rate of onset of sporulation in *B. mycoides*. Wund (1906) attempted to define the limiting oxygen requirements for a large group of organisms. The range of oxygen concentrations for spore formation was generally less than for germination and growth. Holzmüller (1909) questioned the necessity of oxygen for sporulation of facultative anaerobes. However, Leifson (1931) found that of four species able to grow at 20 mm. O_2, including one which grew at < 10 mm., not one formed spores at 20 mm. Several organisms, including some that grew quite well under strict anaerobic conditions, failed to form spores in growing cultures in periods up to 3 months under anaerobic conditions (Knaysi, 1945*a*; Roth & Lively, 1956; Puziss & Rittenberg, 1957). Grelet (1952*b*, 1957) found that a mutant of *B. megaterium* selected by insufficient aeration, did not sporulate unless oxygen became limiting for growth. The growth of the mutant and its parent strain were limited to the same degree by reduced aeration, but only the mutant sporulated.

Roth *et al.* (1955) observed with *Bacillus anthracis* and *B. globigii* that the type of inoculum affected the oxygen uptake during spore formation. The minimum requirements for maximum spore formation of *B. anthracis* were 0·7, 1·0, 0·28 and 0·13 mM O_2/l./min. for a heat-activated (65°, 30 min.) spore inoculum, and for vegetative inocula obtained from cultures inoculated with the spore inoculum and incubated for 6, 8 and 12 hr. respectively. A similar variation occurred with *B. globigii*.

The oxygen demand during sporogenesis has been studied to a limited extent. Hardwick & Foster (1952) failed to obtain spores with washed cells of *Bacillus mycoides* and *B. lacticola* in 24 hr. in an atmosphere of nitrogen. These cells formed spores in 8–10 hr. under aeration. However, exposure of the washed cells to anaerobic conditions for 4 hr. did not prevent normal sporulation on return to aerobic conditions. The oxygen uptake of these two organisms during spore formation showed no marked decline, except in *B. lacticola* during lysis of the sporangia. Roth *et al.* (1955), however, noted in growing cultures of *B. anthracis* and *B. globigii* a marked change in the oxygen requirements at 6–8 hr. (see above), *c.* 4 hr. before forespores were observed. These authors suggested that spores continue to develop under extremely low levels of oxygen once the precursors for sporulation have been formed. Halvorson (1957*b*), also, observed in broth cultures of *B. cereus* var. *terminalis* that the oxygen demand, which reached a peak of 600 μl./l./min. at 4 hr. coinciding with the maximum cell population and exhaustion of glucose, fell to 100 μl./l./min. by 9 hr. when the first heat-stable spores had formed.

Secondly, *Clostridium* spp. There is little precise information available on the effect of oxygen on sporogenesis in these organisms. Migula (1904) suggested that some oxygen hastened spore formation by making the medium unsuitable for further growth. Leifson (1931), however, found that spore formation in five species was inhibited by 20 mm. oxygen pressure and in two species by 10 mm.; and in all cases sporulation was less at 10 mm. oxygen pressure than at *c.* 0 mm.

Chemical environment

Inorganic constituents.

The increasing knowledge of the role of cations and anions in sporulation indicates that some are essential and have a specific role in biochemical reactions involved in spore formation (Curran, 1957). Others probably simply contribute to the ionic environment of the cells.

(*a*) *Essential minerals.* Certain ions, notably potassium and manganese, are required either for the maximum production of sporangia or for the process of sporogenesis. They are required in much greater amounts than for vegetative growth, and many common complex media adequate for vegetative growth contain too low a concentration of these ions for spore formation to occur (Table 1).

A deficiency of some ions may limit growth and induce spore formation. Grelet (1946, 1950, 1952*a*, *c*, 1957) observed that when iron, zinc,

Table 1. *Specific ion requirements for production of sporangia and spores*

Organism	Effective concentration (μg./ml.)								
	K	Mn	Ca	Mg	Fe	Zn	Co	PO_4	SO_4
Bacillus cereus	360[6]	6[12]	.	.	.	.	.	.	.
B. cereus var. *lacticola*	.	.	.	.	.	(+)[10]	.	.	.
B. cereus var. *anthracis*	(+)[3]	(+)[3]	20[3], (−)[16]	.	(+)[3]	(−)[3]	(−)[3]	.	.
B. cereus var. *mycoides*	.	(−)[5]	.	(+)[8]	.	.	.	.	.
B. megaterium	(+)[7]	0·6[7], 6[12], (−)[5]	2[7,14]	(+)[7]	0·6[7]	3[7,14]	.	(+)[7]	20[7]
B. subtilis	(+)[11]	0·05–2,[4,5,18], 6[12]	(+)[11,19]	(+)[11,19]	(+)[19]	.	.	(+)[19]	.
B. brevis	.	2[5]	.	.	.	.	.	.	.
B. sphaericus	.	(−)[12]	.	.	.	.	.	.	.
B. laterosporus	.	2[5]	.	.	.	.	.	.	.
B. coagulans var. *thermoacidurans*	.	(−)[5], 1[1,2]	50[2,17]	50[1,2]	.	(+)[17]	1[1]	.	.
B. stearothermophilus and other thermophiles	.	(−)[5]	(−)[13]	.	.	.	.	.	.
Clostridium botulinum	.	.	40[15]	.	50[15]	.	.	(+)[9]	(+)[9]

(+) = stimulation; (−) = no response.

[1] Amaha *et al.* 1956; [2] Amaha & Ordal, 1957; [3] Brewer *et al.* 1946; [4] Charney *et al.* 1951; [5] Curran & Evans, 1954; [6] Foster & Heiligman, 1949*a*; [7] Grelet, 1946, 1950, 1952*a*, *c*; [8] Knaysi, 1945*a*; [9] Leifson, 1931; [10] Lundgren & Beskid, 1958; [11] Perdue, 1933; [12] Powell, 1957*c*; [13] Schmidt, 1950; [14] Slepecky & Foster, 1959; [15] Sugiyama, 1951; [16] Vanini, 1957; [17] Ward, 1947; [18] Weinberg, 1955; [19] Williams, 1929.

sulphate and phosphate became scarce sporulation occurred even though carbon and nitrogen nutrients were still adequate. Where the ion was involved in sporogenesis itself, e.g. phosphate and zinc, the amount of spore formation was reduced by the deficiency.

The concentration of certain ions in the growth medium affects the properties of the spore. Sugiyama (1951) found that the lower the iron and calcium concentration, the lower the heat stability of spores of *Clostridium botulinum.* Grelet (1952*c*) observed that the omission of calcium reduced sporulation in *Bacillus megaterium* and the spores produced were less refractive and less heat resistant. Amaha & Ordal (1957) showed that manganese and calcium increased the heat resistance of *B. coagulans* var. *thermoacidurans* and maximum heat resistance was only obtained with adequate amounts of both cations in the medium.

Slepecky & Foster (1959) found that from 3 to >100 times more metal ions (calcium, cobalt, manganese, zinc, copper and nickel in ascending order) were incorporated into spores, the higher their concentration. The concentration in the spores was higher than in the medium suggesting a concentration mechanism. Competitive effects occurred. Spores with high calcium content had a high resistance, but with a high content of zinc, manganese and nickel (low calcium), they had a much lower resistance. The metal content of the spores did not appreciably affect their resistance to ultraviolet light, desiccation, phenol, or their refractility or impermeability to stains.

(*b*) *Electrolyte environment.* Organisms appear to vary greatly in their response to the ionic environment. This is possibly related to their ability to exercise some control of the internal ionic environment. Various salts (1–5 %) such as sodium and potassium chloride, sodium sulphate, sodium carbonate, magnesium sulphate and certain bi-, tri- and tetravalent metal chlorides had little effect on spore formation of *Clostridium botulinum*, *C. tetani*, *Bacillus anthracis*, *B. subtilis*, *B. tumescens*, *B. cereus* and *B. mesentericus* (Schreiber, 1896; Fitzgerald, 1911; Williams, 1929; Cook, 1931; Leifson, 1931; Fabian & Bryan, 1933; Schmidt, 1950).

Schreiber (1896), however, reported that when vegetative cells of *Bacillus anthracis*, *B. subtilis* and *B. tumescens* were suspended in 2 % solutions of sodium chloride, sodium carbonate and magnesium sulphate, but not potassium nitrate or glycerol, sporulation was slightly accelerated. Fabian & Bryan (1933) increased spore formation of *B. subtilis*, *B. cereus*, *B. mesentericus* and *B. megaterium* with the cations sodium, potassium, ammonium and lithium added as chlorides to the growth medium. Wynne (1948), however, found that sodium chloride, potassium chloride and sodium sulphate (>M/30) reduced sporulation of *Clostridium botulinum*. Sodium chloride (2 %) reduced spore formation by 50 % without affecting vegetative growth.

Organic constituents.

The nature of the carbon and nitrogen nutrients is very important in the production of sporangia and the carbon-nitrogen balance can determine when spore formation will occur. High concentrations of organic nutrients usually retard the onset of spore formation and reduce the percentage of sporulating cells (Williams, 1929; Tarr, 1932; Brunstetter & Magoon, 1932; Wynne, 1948). These effects are related to increased vegetative growth and associated changes in the environment during growth, such as reduction in the oxygen supply, pH, and mineral availability, and increases in the concentrations of inhibitors (Knaysi, 1945*a*; Foster, Hardwick & Guirard, 1950).

The following separate considerations of the carbon and nitrogen nutrition may over-simplify the position as some compounds are involved in both.

(*a*) *Carbon source.* Numerous carbon compounds are reported to affect sporulation rather than vegetative growth. A separation of the effects of these has been attempted.

(i) Group 1 comprises compounds such as fermentable sugars which reduce sporulation. These compounds increase the carbon:nitrogen ratio in the growth medium, and the excess of acid metabolic products produced lowers the pH to unfavourable levels (Fitzgerald, 1911; Knaysi, 1948). The degree of diversion of the cell metabolism depends on the amount of available nitrogen (Kaplan & Williams, 1941; Hardwick & Foster, 1952) and the ability of the organism adaptively to metabolize the acid metabolic products (Halvorson, 1957*b*; Gollakota & Halvorson, 1960).

(ii) A number of compounds which might be expected to fall into group 1 actually increase sporulation markedly (Table 2). Foster & Heiligman (1949*b*) suggested that a low glucose concentration (0·1 %) in the growth medium may provide extra energy for spore formation without affecting growth. The effect was observed in high and low glutamate concentrations, but not when DL-alanine was present.

(iii) Depletion of the carbon source results in initiation of spore formation. Grelet (1946, 1950, 1951) observed with *Bacillus megaterium* growing in a simple chemically defined medium that sporulation occurred when the single carbon source was almost exhausted. Sporulation resulted from exhaustion of the following carbon sources: glucose, glutamate, D- and L-alanine, malate, fumarate, succinate and acetate. If more glucose was added to the culture which had sporulated, germination and regrowth occurred. This has also been observed with *Clostridium sporogenes* (Ordal, 1957).

(iv) If growth is limited by another factor, such as the supply of either

Table 2. *Some carbon compounds that increase spore formation*

Compound*	Concentration (%, w/v)	Organism	Spore formation† (% increase)	Reference
Glucose	0·1	*Bacillus cereus*	63	Foster & Heiligman, 1949*b*
Oxalate	0·06	*B. megaterium*	++‡	Powell, 1951
Maleic acid	0·05	*B. coagulans* var. *thermoacidurans*	65	Amaha *et al.* 1956
α-Ketoglutaric acid	0·05		65	
Xylose	0·5	*B. cereus*	21	Majumder & Padma, 1957
		B. subtilis	25	
		B. megaterium	35	
Mannitol[11]	1·0	*Clostridium perfringens*	+++§	Fitzgerald, 1911
Amygdalin	1·0		+++	
Raffinose	1·0		+++	
Inulin	1·0		++	
Dulcitol	1·0		++	
Iso-dulcitol	1·0		++	
Lactose‖	1·0	*C. sporogenes*	26	Kaplan & Williams, 1941

* Added to the complex growth medium before growth.

† % increase = % spore formation in presence of compound minus % spore formation in control medium.

$$\frac{\%\text{ spore formation}}{\text{(estimated microscopically)}} = \frac{\text{No. spores}}{\text{Total no. cells (free spores, cells with spores, vegetative cells)}} \times 100.$$

‡ Spore formation increased from a few to many stained, unrefractive spores up to *c.* 100% spores of which *c.* 20% were still densely stained.

§ Spore formation increased from odd spores in the cultures to very many spores (= ++), to abundant (= +++).

‖ Not fermented.

nitrogen or zinc, sporulation occurs without depletion of the carbon source (Grelet, 1950, 1952*a*, 1957). Powell & Strange (1956) also observed that in a complex broth medium when spore formation of *Bacillus cereus* was complete, up to 20% of the original carbohydrate was still present.

Limitation of growth by some factor may help explain some of the effects of the compounds in Table 2. If growth was limited before the cells had built up enough endogenous reserves, an exogenous supply of readily available energy may enable spore formation to proceed.

The carbon sources of the growth medium can also affect the properties of the spore. Williams (1929) showed that the addition of utilizable carbohydrates such as glucose, lactose, dextrose and starch, to a peptone medium increased the heat-resistance of the spores of *Bacillus subtilis*. Of various organic acids tested, only acetate and tartrate increased the heat resistance.

Spores of *Bacillus megaterium* produced in media with different carbon sources had the same amounts of lipid, polysaccharide, nitrogen and ash. However, if spore formation was induced by nitrate deficiency rather than by a glucose deficiency, the spores contained more ash, lipid, and twice the amount of polysaccharide (Tinelli, 1955*a*, 1957).

(*b*) *Nitrogen source.* In a group of organisms for which nitrogen requirements vary from simple inorganic sources to complex amino acid mixtures and possibly peptides, the nitrogen nutrition will be difficult to assess. Further, it is likely that cells which can grow on a simple medium will synthesize a pool of intracellular substrates and reserve compounds as complicated as cells growing in a complex medium.

The significant observations on the nitrogen status of the environment during formation of the sporangia include the following:

(i) Nitrogen depletion as a cause of initiation of sporulation is evident with organisms with either simple or complex nutritional requirements (Grelet, 1946, 1955, 1957).

(ii) In complex amino acid media, sporulation is induced by a deficiency of one or more amino acids. These may either be at a low level initially or become depleted during growth. This is evident from the studies of Grelet (1955, 1957) with a defined medium containing L-glutamate, L-alanine, L(+)-valine, L(−)-leucine and L(+)-isoleucine (50, 50, 15, 8·6 and 8·2 mM respectively), salts and glucose in which *Bacillus cereus* var. *mycoides* formed few spores. The glutamate was important in growth for the reduction of the lag period, and annulling the inhibitory effects of the other amino acids, but was not appreciably used by the cells until the other amino acids became depleted. When the alanine concentration was reduced to < 5 mM, sporulation occurred just before growth ceased; the glutamate was still at its initial concentration and the other amino acids were not exhausted. Variations in the concentrations of valine, leucine and isoleucine inhibited growth due to amino acid disequilibria, but this did not result in sporulation. Omission of all three amino acids, however, resulted in *c.* 50% spore formation; after a lag, growth restarted during which the alanine was used, its depletion limited growth again and further spore formation occurred. Later still, a similar series of events occurred wih depletion of glutamate.

(iii) Some amino acids may have specific roles in spore formation. Krask (1953) observed that spore formation of *Bacillus subtilis* in a simple glucose-salts-glutamate medium required more glutamate than was required for maximum vegetative growth. Methionine sulphoxide, an inhibitor of glutamic acid-glutamine conversion, inhibited spore formation but not vegetative growth. Krask suggested that a specific step, possibly glutamine synthesis, in a chemical sequence in sporulation was interrupted.

It is likely that these observations will apply equally to the clostridia, although the carbon-nitrogen balance must vary considerably for the saccharolytic and proteolytic species. Lund (1957*a*) has shown that *Clostridium sporogenes* depleted the media of certain amino acids, while the concentration of glutamate, aspartate and tryptophan remained the same.

The nature and concentration of the nitrogen compounds can affect the properties of the spore. A medium with more nitrogen and less minerals yielded spores of *Bacillus cereus* with a higher nitrogen content, a slightly greater average weight, smaller volume and less ribonucleic acid. The spores also tended to germinate more rapidly during storage in wet suspensions (Fitz-James, 1957; Fitz-James & Young, 1959*a*). The deoxyribonucleic acid content, however, was approximately constant for each of 24 species regardless of the growth medium (Fitz-James & Young, 1959*a*). Williams & Harper (1951) found that the heat resistance of spores of *B. cereus* grown in a chemically defined medium was not affected by any one amino acid, and was independent of the amount of growth and sporulation.

(*c*) *Miscellaneous nutrients*. Very little is known of the effect of growth factors and coenzymes on spore formation. Williams & Harper (1951) observed that *p*-aminobenzoic acid increased sporulation of *Bacillus cereus* by 30–40 %, but the heat resistance of the spores was not affected by any one growth factor or by yeast nucleic acid. Ordal (1957) found that omission of folic acid reduced spore formation of *B. coagulans*; *p*-aminobenzoic acid, but not vitamin B_{12}, replaced this requirement. Adenosine or adenine slightly increased sporulation, but not inosine, guanine, uracil or thymine. Lund (1957*a*) reported that thiamin (0·1 μg./ml.) increased spore formation in a *Clostridium* sp. At 10 μg./ml., 100 % incipient spore formation occurred, but the cells later lysed. Other growth factors had no effect (Lund, Janssen & Anderson, 1957).

Various associations with other organisms, or the addition of culture filtrates of other organisms, have increased spore formation of some organisms (Mellon, 1926; Powell & Hunter, 1955*a*; Lund *et al.* 1957). In *Bacillus sphaericus* this was due to the higher CO_2 concentration in mixed culture (Powell & Hunter, 1955*a*).

Low concentrations of certain fatty acids affect spore formation and the properties of the spores. Meisel & Rymkiewicz (1954) found that oleic and sometimes formic acid increased spore formation of *Clostridium tetani*. Sugiyama (1951) showed that the presence of certain saturated and unsaturated fatty acids in the growth medium increased the heat resistance of spores of *C. botulinum*: the longer the carbon chain, the greater the increase. Linoleic acid, however, decreased the heat resistance.

(*d*) *Inhibitors*. Substances occur in media which specifically inhibit sporulation and not growth. Frequently these inhibitors may be removed by treating media with adsorbents such as soluble starch or activated carbon (Roberts & Baldwin, 1942; Foster *et al.* 1950). Low concentrations (15–300 μg./ml.) of the $C_{(1)}$–$C_{(14)}$ saturated fatty acids, oleic, linoleic and ricinoleic acids, inhibited sporulation of several organisms (Hardwick, Guirard & Foster, 1951; Meisel & Rymkiewicz, 1954). Hardwick *et al.* (1951) also observed that *Bacillus cereus* and *B. mycoides* produced considerable ether-extractable inhibitors during growth in a peptone medium. Hardwick & Foster (1952) found that fatty acid inhibition could be partly reversed by washing the cells.

This suggests that fatty acids inhibit sporogenesis but not production of sporangia. Butyric and heptylic acids inhibited sporogenesis of washed cells committed to sporogenesis (Hardwick & Foster, 1952).

The effect of changing environmental conditions during growth and sporulation is well illustrated by the effects of fatty acids. The media may initially contain acids inhibitory to spore formation, some of which may be essential to growth, while during growth, others may be produced in concentrations which will favour vegetative growth at the expense of sporogenesis. Some affect the properties of the spore, others inhibit germination. The effect of fatty acids on sporulation and possibly germination will be determined by the actual concentrations at the time of formation of each spore. This concentration will depend on the medium, and on the consumption and production of fatty acids by the organism.

Some amino acids inhibit spore formation. Tarr (1932) observed that asparagine (2%, w/v) in a casein digest medium almost completely inhibited sporulation of several aerobic organisms. Foster & Heiligman (1949*b*) reported inhibition by either α- or β-alanine, L or DL-alanine or DL-aspartic acid.

THE INTRACELLULAR ENVIRONMENT AND INTRACELLULAR EVENTS OCCURRING DURING SPORE FORMATION

From the recent cytological studies of Young & Fitz-James (1959*b–d*) and Hashimoto *et al.* (1960) it is evident that after the initial enclosure of the spore chromatin and a relatively small part of the sporangial cytoplasm by a thin septum, the spore-sporangium complex consists of a cell growing within the mother cell. Further spore walls are deposited as the spore develops. The main source of substrates for the growth of this cell is, therefore, likely to be the intracellular pool of low molecular weight solutes and reserve materials of the sporangium.

Some information is available on the nature of the intracellular substrates and the stages at which these are incorporated into the spore. Our knowledge of the metabolic reactions involved in sporogenesis, however, is far from complete.

Amino acid, purine and pyrimidine metabolism

The concentration of these substances in the intracellular pool falls during sporulation, and some have been shown to be incorporated into the spore. Hardwick & Foster (1952) showed that amino acid, purine and pyrimidine antimetabolites completely inhibited sporulation, and the inhibition was reversed by the corresponding metabolite. Foster & Perry (1954) found that cells inhibited by metabolite analogues showed no changes in their content of free amino acids, purines and pyrimidines. Also, inhibition by any one analogue prevented the utilization of all the amino acids, purines and pyrimidines in the pool. This 'all-or-nothing' inhibition indicated *de novo* synthesis of protein and nucleic acids from these low molecular weight substances during sporogenesis. ^{35}S-Methionine experiments showed that methionine was fixed into an insoluble form during sporogenesis. These experiments do not exclude synthesis in other parts of the sporangium.

Glucose (but not glucose + ammonia) prevented sporulation if added before the cells were committed to spore formation (Hardwick & Foster, 1952). The interpretation of these authors was that glucose diverted the metabolism in the direction of growth, which competed for the low molecular weight nitrogenous solutes required for sporogenesis. Ammonia prevented depletion of these metabolites.

Vinter (1959*a–c*) in studies with ^{35}S-labelled amino acids found that practically all the cysteine + cystine of the sporangium-spore system was incorporated into the protein of spores of *Bacillus megaterium*. Incorporation of cysteine was highest during and for a short time after sporogenesis, the cysteine + cystine content of the spores reaching five times that of the vegetative cells. Cysteine was fixed preferentially to methionine, and no further methionine was taken up during sporulation. During sporulation, methionine was probably converted to cysteine and cystine to meet the increased requirement for these amino acids.

Millet & Aubert (1960) found that glutamic acid formed the greater part of the amino acids and about one-fifth of the free solute pool of *Bacillus megaterium*. During the period 3–2 hr. before forespores were visible, the amount of glutamate increased threefold. Studies with ^{14}C-glucose in the medium showed that the increase in glutamate was not due to synthesis from glucose or its degradation products. Glutamic acid did not provide tricarboxylic acid intermediates and was not itself derived from protein breakdown as no other free amino acids appeared. Apparently the glutamate was derived from glucidic and lipid reserves known to disappear during sporogenesis. During sporulation, glutamate decreased and experiments with ^{14}C-glutamate added at various times indicated that it was metabolized in relatively large amounts from the end of the exponential growth phase up to the appearance of visible spores. About two-thirds of the glutamate was lost as CO_2, and the remainder of the radioactivity appeared in the spore proteins mainly in glutamate, aspartate, alanine and proline in descending order, and also in dipicolinic acid. Thus, the results showed that up to the visible spore stage, the sporangium + spore still actively metabolized glutamate, incorporating it, as well as other amino acids derived from it, into the spore protein.

Young & Fitz-James (1959*c*) found that the free amino acids of *Bacillus cereus* and *B. cereus* var. *alesti* were mainly glutamate and alanine. The alanine concentrations did not change during sporulation. In *B. cereus* var. *alesti* the glutamate concentration fell during sporogenesis and formation of the protein crystal (associated with the spore of this organism). In *B. cereus* a sharp increase in glutamate concentration occurred at the beginning of sporogenesis followed by a fall during spore formation. This incorporation may be either into the thick cortex which comprises *c.* 30–40 % of the spore volume or into the outer spore walls, rather than into the spore core proper.

Metabolism of reserve substances

Hardwick & Foster (1952) considered that pre-existing sporangial proteins, such as enzymes that were no longer needed, supplemented the amino acid pool as a source of substrates. They showed that certain enzymic activities were lost, and that the sporangia became unable to adapt to certain substrates. Foster & Perry (1954) with ^{35}S-labelled cells obtained evidence of protein degradation during sporogenesis.

Tinelli (1955*b*) found that during sporogenesis of cells of *Bacillus megaterium*, the polysaccharides and β-hydroxybutyric lipids progressively disappeared from the sporangia without appearing in the medium. If only partial sporulation occurred, β-hydroxybutyrate and acetate were liberated into the medium. The dry weight of the cells with spores was *c.* 30 % less than that of the vegetative cells; the loss was accounted for by CO_2 and small amounts of volatile acids.

Dipicolinic acid formation and calcium incorporation

Dipicolinic acid (DPA) first appears at the onset of sporulation, presumably always inside the outer spore walls as it is not detectable in vegetative cells. Its rate of formation is directly related to that of spore formation (Perry & Foster, 1955; Powell & Strange, 1956). In *Clostridium roseum*, Halvorson (1957*b*) observed that heat-resistant cells were formed 1–2 hr. after the maximum amount of DPA appeared, but some cells with the staining properties and refractility of spores were produced before heat-resistant cells were present. Hashimoto *et al.* (1960), working with synchronized sporing cultures of *Bacillus cereus*, found that the onset of DPA synthesis occurred at the stage when the second coat began to form outside the first coat at the time the cortex first appeared. These events occurred before an increase in heat-resistance was detected.

DPA and calcium are both concerned in heat-resistance, and calcium in the synthesis of DPA. Powell & Strange (1956) found that the uptake of calcium and also iron followed the synthesis of DPA, the metal content of the spores reaching up to 9 and 4 times respectively that of the vegetative cells. Vinter (1956*a*, *b*) made similar observations on the calcium content. Vinter (1957) also showed that cysteine (10^{-4}M) added after the start of sporogenesis disorganized spore formation and reduced calcium uptake; the effect was reversible by heavy metals. Black *et al.* (1960) observed that spores formed by cells of *B. cereus* suspended in water were smaller than spores formed in broth cultures and although both types of spores were equally resistant to staining, radiation and phenol, the former spores were not resistant to 80° for 30 min. These spores contained only a quarter of the DPA found in spores formed in broth. Higher concentrations of calcium (1 mg./ml.) in the growth medium, or 0·1 mg./ml. added to the cell suspension during sporulation in water, gave heat-resistant spores with a normal amount of DPA. The calcium was most effective when added to the sporangia before commitment to sporogenesis, although some effect was obtained by adding it during sporogenesis, but not after the spores became partly refractile. The amount of DPA formed depended on the concentration of calcium.

The location of DPA in the spores is unknown. Its ready and practically complete release in germination exudates (Powell & Strange, 1953); the cytological evidence of the disappearance of the cortex in the first few minutes of germination (Mayall & Robinow, 1957); the late stage of its formation and its association with the development of the cortex (Hashimoto *et al.* 1960) suggest the cortex as its location. Ultraviolet light microscopy failed to locate the position of the DPA (Hashimoto & Gerhardt, 1960).

The biosynthesis of DPA is unsolved. Perry & Foster (1955) showed with *Bacillus mycoides* that ^{14}C from ^{14}C-labelled $\alpha\epsilon$-diaminopimelic acid (DAP) was incorporated in DPA, to a much greater extent than from 14 other amino acids. Powell & Strange (1956) obtained no evidence of synthesis of DPA from DAP using either extracts of sporulating cells or whole cells. They observed, however, a transfer of DAP from the insoluble fraction of the cell to the soluble fraction of the spore + sporangium and the spore. In *B. sphaericus*, Powell & Strange (1957) found no free DAP before sporulation, but during sporogenesis the free DAP concentration increased until all the DAP was in the soluble fraction. Martin & Foster (1958) showed with *B. megaterium* and ^{14}C-labelled metabolites that glutamate, aspartate, alanine, proline and serine were the most efficient precursors of DPA. Assimilated CO_2 formed 6·5 % of the carbon of DPA. Experiments in which the labelling in the pyridine and carboxyl groups of DPA was determined suggested that the carbon skeleton of DPA was formed from a C_4 and C_3 condensation of either (i) aspartate and pyruvate or their derivatives, or (ii) alanine and oxalacetate or their derivatives.

It is not known whether calcium is required either in reactions in the synthesis of DPA or to combine with the end product. Perry & Foster (1955) and Slepecky & Foster (1959) found that the calcium present was less than equivalent with DPA. However, Powell & Strange (1956) recorded *c.* 5 % calcium in spores of one organism which would be approximately equivalent. Lund (1959) observed a 1:1 molar ratio of calcium and DPA in spores and in exudates from germinated spores.

Nuclear material

Recent studies of the nucleic acid composition and of the movement of ^{32}P-labelled phosphate compounds, in conjunction with parallel cytological observations by phase contrast, bright field and electron microscopy, of synchronously sporing cultures of two strains of *Bacillus cereus*, has given a much better understanding of the nature of sporogenesis. Spores were found to contain a characteristic and constant amount of deoxyribonucleic acid (DNA) for each organism. This amount was approximately half that of rapidly growing vegetative cells of the same species. The vegetative cell content, however, varied from *c.* 1·5 to 3 times the average amount per spore, depending on the state of division in the synchronous cultures (Fitz-James, 1955*a*, 1957; Fitz-James & Young, 1959*a*, *b*; Young & Fitz-James, 1959*a*, *b*).

When the DNA content of the sporangium reached twice the average amount per spore, and the two condensed chromatin bodies had formed into an axial thread, sporogenesis commenced (Young & Fitz-James, 1959*b*–*d*). Growth and the net synthesis of both nucleic acids ceased about this time. Part of the chromatin became enclosed by the thin septum of the forespore. Tracer studies confirmed that the chromatin of the spore was derived from the vegetative cell (Young & Fitz-James, 1959*c*). Further synthesis of DNA occurred until *c*. 3 times the average amount per spore was contained by the spore-sporangium system when spore formation was complete. This DNA was apparently synthesized in the part of the sporangium outside the spore, as ^{32}P added to the system after the beginning of sporulation was not incorporated into the spore proper. Although ribonucleic acid (RNA) synthesis ceased during sporulation and even a slight decrease in the net amount in *Bacillus cereus* occurred during the latter stages, the studies indicated that there was a turnover of RNA occurring in the sporangium (Young & Fitz-James, 1959*b*–*d*).

Labelled phosphorus was also incorporated from the medium into a bound organic form in the spore coat fraction (Young & Fitz-James, 1959*b*). Young & Fitz-James (1959*c*) also observed in electron micrographs that the second spore coat formed outside the original thin coat.

THE INITIATION AND CAUSE OF SPORE FORMATION

Spore formation has been considered as: (1) an intermediate stage in normal development which may be inhibited by any partial physiological damage short of prevention of growth (Behring, 1889); (2) the result of a factor necessary for vegetative growth becoming limiting (Grelet, 1946, 1950, 1951, 1952*a*, 1957); (3) the response of a 'healthy cell facing starvation' (Knaysi, 1948); (4) 'the end product of a series of enzymatic reactions in the proper balance and integrated toward the generation of the spore' and 'subject to environmental and particularly nutritional, influences' (Foster & Heiligman, 1949*b*).

If we accept sporogenesis as a normal metabolic process, then the second and third views suggest causes for the initiation or induction of the process. We have seen that unfavourable growth conditions prevent rather than cause spore formation, and that when nutrient conditions are satisfactory for the growth of sporangia, the only factor which is definitely connected with the initiation of sporulation is a deficiency of some growth factor leading to cessation of vegetative cell growth (Grelet). However, the effect of radiations on mutations in the pre-spore stage (Romig & Wyss, 1957; McDonald & Wyss, 1959) and recent cytological evidence (Young & Fitz-James, 1959*b–d*; Hashimoto *et al.* 1960) very strongly suggest that the initiation of sporogenesis depends on a genetic event. Jacob, Schaeffer & Wollman (1960) suggest that sporulation may be under the control of genetic elements (episomes), non-functional during vegetative growth. Genetic control of the cellular system results in the initiation of a peculiar type of cell division which leads to 'the biological goal of the sporulating cell—to parcel in an insulated chamber the functioning unit of chromatin' (Fitz-James, 1957, p. 87).

There are no observations proving that this nuclear event is initiated by a threshold concentration of a specific endogenous substance, but there are some suggestions that this is likely. For instance, the sporangium might be the type of cell produced in the environment when this specific substance reaches the threshold level. The unknown factor in the spent cultures that Lund (1957*b*) has studied may be relevant, as also may some of the effects observed in mixed cultures. Further, some of the carbohydrates listed in Table 2 may be involved in the synthesis of (or directly related to, or contaminated by) such a factor. It is possible that limitation of growth by factors such as those observed by Grelet results in the production of such a substance due to the disturbance of a steady-state reaction in some metabolic sequence.

The origin of the process of sporulation can be visualized as a mutation—a peculiar cell division taking place at the extreme end of the cell, perhaps by single layer septum development instead of the usual cell division. After the initial surrounding of the nuclear element, the new cell grows and excretes cell wall material inside or through the first thin wall, and this may cause the mother cell to deposit wall material to confine the new cell. This could result in the complex multi-layered spore coat seen in electron micrographs of thin sections (cf. Mayall & Robinow, 1957). The laying down of the walls may be an encystment of the foreign cell by the mother cell—the result of antagonistic reactions. This would not be unexpected if the sporangium has a negative and positive polarity* and the polarity of the spore cell is opposite to that of the mother cell as observed by Streshinskiĭ (1955). Such a mutant with an increased survival value could account for the origin of the spore.

* During division the ends of the cells near the division are the positive poles; the opposite ends are the negative ones. The positive poles are poles of growth. Spores are formed at or close to the negative pole, i.e. at the point where growth proceeds least intensely (Streshinskiĭ, 1955).

SPORE GERMINATION

Spores stored in conditions unsuitable for germination may remain viable and heat-stable for many years (Evans & Curran, 1960). However, within a few minutes of placing in a favourable environment the spores may swell, become permeable to dilute stains, and lose their resistance to heat (Powell, 1957*a*). These primary changes which occur more or less simultaneously are often described as 'germination', but may better be described as 'initiation of germination' or more briefly as 'initiation'. In favourable conditions initiation is followed by swelling of the spore, and either rupture or absorption of the spore coats. The germ cell elongates and divides to become two vegetative cells. The germinal development after initiation until the beginning of the first cell division has been defined as 'outgrowth' by O'Brien & Campbell (1956). In this review, initiation and outgrowth up to the beginning of the first cell division will be considered as germination.

Dormancy

Viable spores failing to germinate in apparently favourable conditions are said to be 'dormant'. Many instances of prolonged dormancy have been recorded. The fraction of spores which exhibits dormancy may be small or large. Dormancy is now believed to be due to previously unsuspected inhibitors of germination or to unfavourable conditions of the medium. During dormancy, changes may occur both in the spore and in the environment. The prospects for germination may depend on both.

Morrison & Rettger (1930*a*, *b*) showed that a suitable medium completely eliminated dormancy of spores of *Bacillus megaterium*, *B. cereus* and *B. vulgatus* even in spores surviving a drastic heat treatment. Knaysi (1948) was able to vary the percentage germination in spores of *B. mycoides* by varying either the amount of yeast extract or the oxidation-reduction potential of the medium. Curran & Evans (1937) and Nelson (1943) showed that heat-treated spores had more exacting nutritional requirements than unheated ones and that the percentage spores germinating depended on the recovery media. Metal ions and certain fatty acids in low concentrations which do not affect vegetative growth can completely prevent germination (see later).

Recent observations suggest that metabolic changes occur in resting spores which affect both the cultural requirements for germination and the organism's chances of survival. Knaysi (1948) observed that normal spores of *Bacillus mycoides* germinated in glucose and acetate, and that germination was increased by heated glucose or traces of potassium phosphate. Spores aged in broth or water, however, were dormant in glucose and acetate, and as the storage period increased, the response to heated sugar or phosphate was gradually lost. Knaysi suggested that this was due to leakage or destruction of a substance(s) needed for growth. Powell (1950) found that freshly harvested spores initiated germination slowly, but if stored in water at 20° the initiation rate gradually increased to a maximum after 20 days storage. This increase occurred more slowly at lower temperatures, and was not reduced by further storage. Spores aged in the growth medium behaved as young cells. Murty & Halvorson (1957*b*) observed that *B. terminalis* spores stored in frozen distilled water lost the need for L-alanine for the activation of oxidative enzymes, an indication of initiation. Church & Halvorson (1957) showed that in stored spores (4 months, −20°) the endogenous and glucose oxidative activity was five times higher than in freshly harvested spores. No further change in activity occurred in frozen storage for four years. Extracts of freshly harvested spores were as active as those of aged spores. The extracts were supplemented with diphosphopyridine nucleotide in both cases.

During storage, spores also die (Williams, 1929; Evans & Curran, 1960), a fact sometimes disregarded in relation to the effect of storage on the properties of the spore. Nevertheless, the nature of the environment and the changes occurring in the dormant spore affect its germination.

REQUIREMENTS FOR THE INITIATION OF GERMINATION

Initiation is quantitatively studied by measuring the loss in heat resistance; permeability to dilute stains; loss in turbidity or dry matter; and 'darkening' under phase contrast. These methods give good agreement (Powell, 1951; Levinson & Sevag, 1953; Brown, 1957). Enhanced metabolic activity, such as O_2 uptake, probably occurs only as a result of initiation, and is therefore less satisfactory for such measurements (Murrell, 1955, pp. 29, 51). The majority of spores initiate germination within a short time of being placed in a favourable environment, the number of uninitiated spores decreasing exponentially with time (Wynne & Foster, 1948*b*; Mehl & Wynne, 1951). During initiation, therefore, the spores initiating germination in the later stages will be affected by substances released from spores that have already initiated germination and commenced outgrowth. These substances may be either stimulatory or inhibitory. Recent studies have shown that initiation is usually brought about by chemical substances, and that even though it may occur most rapidly in optimal growth conditions, it can occur even in environmental conditions unfavourable to germinal development.

Chemical initiation

Germination is initiated by simple specific substances such as L-alanine, glucose and adenosine. Initiation in solutions of these substances can be as rapid as in complex nutrient media (Powell, 1957*a*). The array of compounds reported to initiate germination is, however, somewhat bewildering. Since Hills (1949*a*, *b*, 1950) showed that L-alanine was specific for *Bacillus subtilis* spores and adenosine (2 μM)+L-alanine (500 μM)+DL-tyrosine (500 μM) for *B. anthracis*, *c.* 30 papers have reported at least fifty substances with some activity for one or more species. It is not possible to discern any clear principles from these studies because they have been limited to very few strains in each of a small number of species. Further, in many cases, only a few arbitrarily chosen substances were studied. Many metabolic breakdown products of initiators were only recently found to be active and then only under special conditions (Church & Halvorson, 1957). These have been tested with very few organisms.

Table 3 lists some of the substances known to be most active for 15 species, and illustrates that the compounds commonly involved are simple sugars, amino acids and ribosides, either singly or in combination. More comprehensive reviews of initiators exist (Stedman, 1956*a*, *b*; Heiligman, Desrosier & Broumand, 1956; Schmidt, 1957). The following observations are apparent from a review of chemical initiation. None of the substances tested has been shown to be active for all types of spores with which it was tested. L-Alanine has most commonly shown activity, but it has also been tested more frequently. Substances, e.g. adenosine, which are effective in very low concentrations (2 μM; Hills, 1949*a*, *b*) in some organisms may be inactive at much higher concentrations in others (750 μM; Levinson & Sevag, 1953). Spores of a few organisms initiate germination in distilled water (Henrici, 1928; Curran & Evans, 1946). This may occur only after heat activation (*Bacillus megaterium*; Powell & Hunter, 1955*b*). These spores apparently either contain adequate endogenous substrates or adsorbed initiating substances. The nature of the exogenous initiators may also differ after heat activation (see later).

Many substances chemically related to initiators inhibit initiation. Hills (1949*b*, 1950) observed that D-alanine at 0·03 times the concentration of L-alanine completely inhibited its action. Glycine, methionine, valine and cysteine also inhibited, but at much higher concentrations. Guanosine inhibited the action of inosine (Powell & Hunter, 1958).

Table 3. *Examples of substances that initiate germination in various organisms*

Organism	Substances	References
Bacillus anthracis	Inosine, adenosine + tyrosine + L-alanine	Hills, 1949*a*, *b*; Powell & Hunter, 1955*b*
B. cereus	Inosine, adenosine + tyrosine + L-alanine	Powell & Hunter, 1955*b*
B. cereus var. *mycoides*	Glucose	Knaysi, 1945*b*
B. cereus var. *terminalis*	L-Alanine + adenosine	Stewart & Halvorson, 1953
B. cereus var. *thuringiensis*	L-Alanine	Wolf & Mahmoud, 1957*b*
B. megaterium	Glucose, Mn^{++} + glucose, mannose	Powell, 1951; Levinson & Sevag, 1953; Powell & Hunter, 1958
B. subtilis	L-Alanine, L-alanine + glucose	Hills, 1950; Powell, 1950
B. circulans	L-Alanine	Wolf & Mahmoud, 1957*b*
B. licheniformis	DL-Alanine	Wolf & Mahmoud, 1957*b*
B. globigii	L-Alanine + glucose	Church *et al.* 1954
B. polymyxa	L-Alanine + adenosine	Church *et al.* 1954
B. sphaericus	L-Alanine	Wolf & Mahmoud, 1957*b*
B. stearothermophilus	Glucose	O'Brien & Campbell, 1957
Clostridium botulinum	Glucose, L-alanine + L-arginine + L-phenylalanine, yeast extract	Wynne, Mehl & Schmeiding, 1954; Halvorson, 1957*a*; Treadwell *et al.* 1958
C. perfringens	Glucose	Wynne *et al.* 1954
C. chauvei	Glucose	Wynne *et al.* 1954
C. sporogenes	Glucose	Wynne *et al.* 1954
C. roseum	L-Arginine + L-alanine + L-phenylalanine	Hitzman, Halvorson & Ukita, 1957

Caracò, Falcone & Salvatore (1958) showed that glycine, DL-serine, DL-threonine and β-alanine inhibited the action of L-alanine. Glycine initiated germination at 10–50 mM but at 100 mM inhibited the action of L-alanine.

Woese, Morowitz & Hutchison (1958) studied with *Bacillus subtilis* the activity of numerous compounds related to L-alanine. The results showed the molecular configuration necessary for initiation. Esterification of the carboxyl group reduced its effectiveness but by no means eliminated it; its removal, reduction to alcohol or replacement by a sulphonyl group destroyed the activity. The amino group was essential, its replacement by hydrogen, hydroxyl or sulphydryl leading to a loss in activity. The hydrogen was not essential, but all compounds which inhibited initiation had at least one hydrogen attached to the central carbon atom. Each extra carbon addition to the methyl group reduced its effectiveness. A positive charge on this part also made the compound inactive. D-Alanine was bound more strongly than L-alanine. D-Isomers of the compounds tested were not initiators and D-α-amino acids were mainly inhibitors.

Wolf & Thorley (1957) observed no correlation in initiation requirements and the nitrogen requirements for vegetative growth.

Levinson & Sevag (1953) found that manganese ions in the presence of glucose initiated germination in *Bacillus megaterium* but not in *B. subtilis* or *B. cereus*. Manganese (*c.* 10^{-4}M) + glucose gave over 75% initiation, but glucose, none. Heat activation was not necessary with manganese present. Cobalt and zinc (0·8 mM) slightly increased initiation.

Mechanical initiation

Rode & Foster (1960) found that crushing or abrasion with glass particles (*c.* 40μ dia.) under certain conditions caused 80–90% of *Bacillus megaterium* spores, while still remaining viable, to become heat-sensitive, stainable, and sensitive to various lethal agents. Spores so treated oxidized glucose, had lost dipicolinate and spore peptides, and appeared similar to physiologically germinated spores.

Heat activation

This is an important phenomenon shown by many but not all types of spores (Evans & Curran, 1943; Curran & Evans, 1944, 1945; Reynolds & Lichenstein, 1949; Treadwell, Jann & Salle, 1958). It is sometimes needed to induce germination in part of the spore population even in complex media. It can vary the initiation requirements, metabolic activities and chances of survival in many organisms.

The magnitude and type of response varies a great deal and is affected by many factors. Curran & Evans (1945) showed that the lower the temperature of spore formation, the greater the response. The response varies according to the sporulation medium (Murrell, unpublished). The amount of heating required for maximum activation varies greatly for different organisms—from a few minutes at 60° to a short time at 110° (Curran & Evans, 1945). Powell & Hunter (1955*b*) found with *Bacillus megaterium* that the time for maximum activation varied non-linearly from 110 min. at 65° to *c.* 10 min. at 80°. At 45°, 18 hr. gave *c.* 100% initiation, but no initiation occurred after 18 hr. at 44°. Figure 1 illustrates an extreme case. *B. coagulans* showed maximum activation in *c.* 6 min. at 110° or after *c.* 80 min. at 95° (Q_{10} *c.* 5·5).

The magnitude of the response depends on the heating medium. Curran & Evans (1945) found the following decreasing order of effectiveness for different media: glucose (0·1%) and lactose (0·5%), peptone (0·5%), skim milk, glucose nutrient agar, beef extract, glucose nutrient broth, distilled water, NaCl (0·5 %). Greater responses to heat activation were also observed, the lower the temperature of subcultivation. They attributed this to some activation during incubation at the higher temperatures. The response also depends on transfer to a suitable environment within a certain time after heating otherwise deactivation or germination changes occur. The type and rapidity of the change depend on the medium. Curran & Evans (1947) found that activated spores

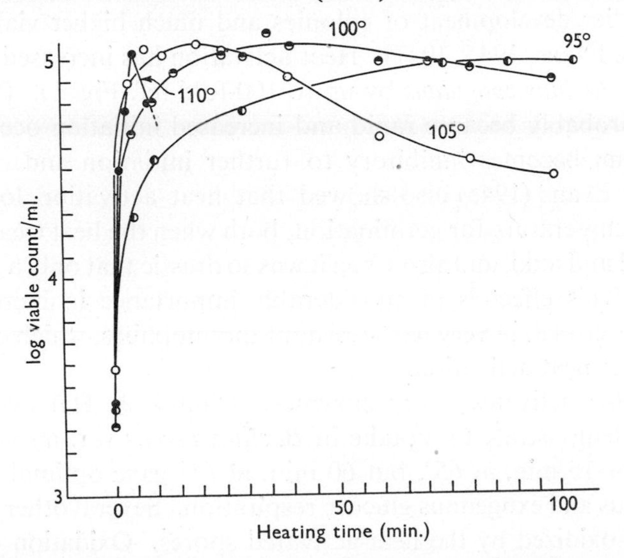

Fig. 1. The effect of heat activation on viable counts of spores of *Bacillus coagulans* (strain 320). Spores, derived from cells grown for 8 days on nutrient agar slopes at 50°, were washed 4 times with distilled water, heated in phosphate buffer (0·05 M, pH 7), at the temperature shown in each curve, and survivors counted on nutrient agar after 48 hr. incubation at 50°. Count at 0 min. equals number of spores germinating without heat treatment. (Murrell, unpublished.)

stored in nutritionally incomplete media (distilled water ± glucose) either germinated and died rapidly or became deactivated with retention of viability. When activation was conducted in a nutritionally complete medium (milk) with storage at low temperatures, the spores died more slowly and little deactivation occurred. Poststorage heating caused some reactivation in both types of medium even where considerable loss in viability had occurred.

Heat activation can vary the initiation requirements and increase the rate and amount of initiation. Powell & Hunter (1955*b*) obtained spontaneous initiation in the absence of exogenous substances in *Bacillus megaterium* spores after heat activation. Freshly harvested spores of *B. cereus* required either inosine or L-alanine + tyrosine + adenosine for optimal initiation, but the heat-activated spores initiated germination rapidly and completely in adenosine (cf. Powell, 1957*a*). Powell & Hunter (1956) were unable to demonstrate the activation of adenosine deaminase. Riemann (1957) observed that heat activation increased the rate of initiation rather than outgrowth in *Clostridium sporogenes* and *C. botulinum*. Heat activation may increase the amount of initiation by up to 40-fold (Curran & Evans, 1945).

By increasing the rate and amount of initiation, heat activation results in the earlier development of colonies and much higher viable counts (Curran & Evans, 1945, 1947). Heat activation has increased the viable counts of *Bacillus coagulans* by up to 100-fold (cf. Fig. 1). Part of this effect is probably because rapid and increased initiation occurs before the medium becomes inhibitory to further initiation and outgrowth. Curran & Evans (1945) also showed that heat activation lowered the minimal temperature for germination, both when the heat treatment was non-lethal and mild, and also when it was so drastic that only a few spores survived. This effect is of considerable importance in increasing the chances of growth in very heat-resistant thermophiles, which often show the greatest heat activation.

Heat also activates many enzymes. Church & Halvorson (1957) failed to demonstrate O_2 uptake in *Bacillus cereus* var. *terminalis* after heating for 15 min. at 65°, but 60 min. at 65° gave optimal activity of endogenous and exogenous glucose respiration. Several other substrates were also oxidized by the heat-activated spores. Oxidation of glucose by extracts was increased 3 times by heat activation of the spores before disruption. Falcone & Caracò (1958*a*, *b*) obtained transaminase activity of intact spores only after heat activation. The transaminase activities were similar to those in spores initiated in L-alanine. Levinson, Sloan & Hyatt (1958) found 60% greater pyrophosphatase activity in extracts of heat-activated spores than in extracts from unheated spores. Heat activation releases dipicolinate and possibly L-alanine from spores (Harrell & Mantini, 1957; Falcone & Caracò, 1958*a*; Levinson & Hyatt, 1955).

What is the mechanism of heat activation? The temperature coefficient (Q_{10}) appears to be *c.* 5. No initiation occurs during heating, and spores of the strain of *Bacillus megaterium* showing spontaneous initiation after activation could be heated and ice-cooled alternately many times without spontaneous initiation occurring, but when placed at growth temperatures initiated germination completely (Powell & Hunter, 1955*b*). The activation is lost on storage, but the spores can be reactivated. As solutes are lost, there is presumably a limit to the number of cycles during which spores will continue to respond. Church & Halvorson

(1957) observed that the glucose oxidizing capacity of heat-activated spores, which was largely lost in 24 hr. and was fully restored by a second heat treatment, was irreversibly lost after 1 week of storage at 5°. Levinson & Hyatt (1955) found that D-alanine inhibited only part of the increased activity, possibly that due to L-alanine release. They, therefore, suggested that heat activation may act as a more general metabolic stimulant. O'Connor & Halvorson (1959) found that although most of the substrates metabolized in the first few minutes of initiation are endogenous, exogenous alanine is preferentially utilized in this period and possibly has a catalytic role. Heat activation by releasing an endogenous initiator in the free state may therefore catalyse and accelerate catabolic activity. The activation of enzymes and parallel release of dipicolinate may be evidence of such catabolism.

ENVIRONMENTAL FACTORS AFFECTING INITIATION

Although germination may sometimes be initiated in environmental conditions unfavourable to outgrowth and vegetative growth, initiation does not always occur when conditions are favourable for vegetative growth. The environmental limits for initiation may generally be narrower than for growth, but there are insufficient data yet to decide this.

Temperature

Even though initiation sometimes occurs at temperatures unfavourable for outgrowth, optimal initiation occurs at temperatures near the growth optimum and at the lower and upper temperature limits, initiation may be restricted before vegetative growth. Ohye & Scott (1953, 1957) observed that at the lowest and highest temperature at which multiplication of *Clostridium botulinum* types A, B and E occurred, spore inocula did not grow and at least some of the spores retained heat resistance suggesting that initiation had not occurred. The spores of several strains of type E which grew vegetatively at 5° and 42·5° did not initiate germination at 2·5° in 20 weeks and at 45° in 1 week.

Mundt, Mayhew & Stewart (1954) obtained 90% initiation of *Clostridium sporogenes* spores in 2 weeks at 4·4°. No growth occurred in 6 weeks at 10°. Mol (1957) observed 85, 30 and 0% initiation in spores of *Bacillus cereus* in 6 hr. at 8°, 6° and 4° respectively, although the minimum temperature for growth was above 8°. Similar decreases in heat-stable spore counts have been found at temperatures as low as 0° in some strains of *B. subtilis* but not in spores of *B. licheniformis* (Wolf & Mahmoud, 1957*a*; Williams, Clegg & Wolf, 1957). Wynne, Galyen &

Mehl (1955) and Wynne (1957) observed that spores of several *Clostridium* spp. which did not grow above 50° showed loss in heat resistance in caramelized glucose at 75°. Their interpretation from this and other unpublished data was that initiation had occurred although no loss in turbidity of the spore suspensions had taken place. Vas & Prozt (1957*a*) found that the effect of temperature on *B. cereus* spores varied with the pH. At pH 7 the rate, but not the amount of initiation, changed much more with temperature (Q_{10} = 3·1) than at pH 5·1 (Q_{10} = *c.* 2). The maximum amount of initiation occurred at pH 5·1, but at pH 7 the amount showed little variation between 30° and 40°, while the highest rate of initiation occurred at 35° at pH 5·1 and 40° at pH 7. Keynan, Halman & Avi-Dor (1958) could not obtain initiation of *B. subtilis* spores in either L-alanine or a mixture of L-leucine, L-norleucine and DL-phenylalanine below 18°, but if the process began above 18°, it continued even at 0°.

Water activity

The evidence on whether spores initiate germination at a_w values too low for growth is conflicting: this may vary with the species. Bullock & Tallentire (1952) found that *c.* 70% of the dried spores of *Bacillus subtilis* present in dry material at an a_w of 0·78 lost their heat resistance but remained viable. At 0·8–0·9 a_w, both heat resistance and viability were lost. At higher levels in the presence of nutrients germination and growth occurred. In canned cheese spread containing 36% moisture and 0·2% salt, Wagenaar & Dack (1955) observed a 65% fall in the spore count of *Clostridium botulinum* spores in 6 months at *c.* 32°, but no growth occurred. Under these conditions the effect of inhibitors was not excluded.

Beers (1956, 1957) measured the lower limit for initiation of germination in several species. The limit in *Bacillus cereus* var. *terminalis* was slightly lower for freshly harvested unrefrigerated spores than for old spores. A slight reduction in the a_w limit was also observed after frozen storage of the spores for 10 months (Beers, 1957, p. 54). The lower limits for initiation in *B. cereus* var. *terminalis*, *B. subtilis*, *B. megaterium* and *Clostridium botulinum* type B were 0·968, 0·970, 0·963 and 0·975 a_w respectively, and for vegetative growth 0·947, 0·944, 0·937 and 0·976 respectively (Beers, 1957). Ohye & Scott (unpublished) found the lower limit for growth of *C. botulinum* to be *c.* 0·94 a_w. Mundt *et al.* (1954) observed that vegetative cells of *C. sporogenes* grew sparingly in broth with 6% (w/v) NaCl, but not with 8% (w/v) NaCl or 60% (w/v) glucose in 6 weeks; 95% of the spores however initiated germination in 6 hr. in broth with 60% glucose or 8% NaCl.

pH

Maximum initiation usually occurs between pH 6 and 8 but the optimum pH may vary with the initiating substance. Powell (1951) observed little difference in initiation at pH 6 and 8 for *Bacillus megaterium* spores. Church *et al.* (1954) working with *B. terminalis* found little effect on the rate and amount of initiation between pH 7 and 10. Below pH 7 the rate fell rapidly but the amount only below pH 6. Above pH 10 both the rate and amount fell sharply. Lawrence (1955*b*) found an optimum near pH 8 for *B. cereus* var. *terminalis*. Wolf & Mahmoud (1957*b*) obtained more initiation of *B. subtilis* spores at pH 8·5 than pH 7 or 6. Vas & Proszt (1957*a*) found that the rate and extent of initiation in *B. cereus* spores exposed to any one of five acids depended on the pH. Below pH 6 the amount of initiation fell only slightly until it decreased suddenly at pH 5. The rate of initiation decreased linearly below pH 6 to zero at *c*. 4·0. Acetic acid completely prevented initiation at pH 4·5.

Wolf & Thorley (1957) observed with *Bacillus subtilis* spores that more germination was initiated by glucose at pH 5·5 than at pH 7·5 or 8·5, whereas in L-alanine more initiation occurred at pH 8·5. In *B. megaterium*, Levinson & Hyatt (1956) however observed greater initiation at pH 6 than 7 both in heated (15 min., 60°) and unheated spores in glucose with and without L-alanine or manganese.

Oxygen

Initiation has been observed in spores of *Bacillus anthracis*, *B. subtilis* var. *niger*, *B. megaterium* and *B. cereus* var. *terminalis* in N_2 atmospheres free of O_2, even in the presence of 0·5 % thioglycollate (Roth & Lively, 1956; Murty & Halvorson, 1957*a*; Hyatt & Levinson, 1959). Roth & Lively (1956) showed that the oxidation-reduction potential (E_h) value in their experiments was *c*. −0·3 V. and that resazurin remained colourless throughout the tests. Harrell & Halvorson (1955) observed that spores of *B. terminalis* exposed to L-alanine for 45 sec. under anaerobic conditions, after washing, did not initiate germination in buffered adenosine at 30°. Control spores exposed under aerobic conditions initiated 40 % germination. Little work has been reported on the effect of oxygen or E_h on the initiation of germination by anaerobes. Spores of *Clostridium roseum* and *C. botulinum* did not initiate germination in L-alanine, arginine and L-phenylalanine unless the last traces of oxygen were removed (Halvorson, 1957*a*).

Carbon dioxide

Wynne & Foster (1948*a*) found that initiation in spores of *Clostridium botulinum* was markedly increased by 1 % CO_2. This requirement disappeared in the presence of oxalacetate or a mixture of L-malic, fumaric and succinic acids in a complex medium, but not in a synthetic medium. Four other anaerobes showed no effect. Treadwell *et al.* (1958) confirmed that CO_2 greatly increased initiation in *C. botulinum* in casein hydrolysate or yeast extract solution.

Inorganic environment

(*a*) *Mineral requirements.* Initiation requires trace metals but higher concentrations of the essential metals and other metals are inhibitory (see later). In experiments with *Bacillus subtilis*, Keilin & Hartree (1947) found that germination was inhibited completely by low concentrations of 8-hydroxyquinoline (2×10^{-5} M) and related compounds. The spores germinated normally when washed 3 times with water and suspended in phosphate. Powell (1950, 1951) obtained similar results with *B. megaterium*, the inhibition being partly reversible by Zn, Mg, Cu and Fe. Viability was again unaffected. Murty & Halvorson (1957*a*) completely inhibited initiation in *B. cereus* var. *terminalis* by oxine and high phosphate concentrations (M/3). Although added phosphate is not necessary for initiation (Powell, 1951; Murty & Halvorson, 1957*a*; Hyatt & Levinson, 1959), it may actually reduce initiation by combining with essential trace metals. This is probably the mechanism by which phosphate reduced viable counts of unheated and heated spores (Johnson, 1952; Williams & Hennessee, 1956). Conversely, in the presence of toxic concentrations of metals, phosphate increased initiation and germination (Murty & Halvorson, 1957*a*).

(*b*) *Ionic environment.* The results discussed below suggest that for one species maximum initiation requires an ionic balance in the medium. This was not confirmed in one other organism tested. Whether this is a difference due to the species or other factors is not known. Levinson & Sevag (1953) observed that omission of chloride ions from a salt solution containing Mn^{++} and glucose reduced initiation of *Bacillus megaterium* spores from 92 to 23 %. Omission of sulphate or non-inhibitory cations of the salt solution had no effect. Four univalent anions (0·05 M) but not fluoride and formate replaced the chloride requirement. With a less concentrated phosphate buffer the 'anion effect' was not so marked. Adenosine but not related compounds relieved the need for anions.

Murty & Halvorson (1957*a*) found that chloride (0·01 M) and sulphate ions had no effect on initiation of *B. cereus* var. *terminalis* and high concentrations of chloride (M/3) inhibited.

Density of suspension

In dense suspensions of spores of *Bacillus subtilis* and *B. terminalis* initiation of germination is slower and less complete than in dilute suspensions (Powell, 1950; Halvorson, 1959). This is not always so. Powell & Hunter (1955*b*) observed that spores closely packed in the pellet during centrifugation of *B. megaterium* germinated more rapidly than suspensions. This appeared to be due to stimulatory substances released from a small proportion of spores which initiated germination spontaneously. On the other hand, with spores requiring an exogenous substrate for initiation, the rate in very dense suspensions was slower due to a limiting supply of the active substance (Murrell, 1955, p. 31). Vas & Proszt (1957*b*) have confirmed this in a detailed study with spores of *B. cereus*. In thick suspensions, the medium became unsuitable for initiation but not for germinal development. The content of solids of the medium showed no change, but the analyses were not adequate to detect qualitative changes in the cell exudate and the medium.

The reduced initiation in thick suspensions is a reaction to the changing environment. Not only is the concentration of initiating substances changing, but inhibitors may be produced by the activity of the germinating spores on the initiating substance; for example, the production of D-alanine by alanine racemase (Church *et al.* 1954; Halvorson, 1957*a*). Stedman, Kravitz, Anmuth & Harding (1956) and Stedman, Kravitz, Harding & King (1957) described a similar phenomenon in *Bacillus globigii* as 'auto-inhibition': alanine racemase was probably not involved. Half the activity of the inhibitory substances was lost on heating. DL-Serine, L-tyrosine and pyruvate, substances likely to be in the exudate, competitively inhibited spore initiation by L-alanine. Dipicolinic acid did not inhibit in the presence of calcium.

Powell (1957*a*) found that increasing the L-alanine concentration (from 2–5 mM to 20–40 mM) or adding a copper salt (1 mM) abolished this retardation of initiation in suspensions with 10^9/ml. of *Bacillus subtilis* spores. She suggested that the higher L-alanine concentration delayed the time to reach the inhibitory D:L ratio, and that the copper inhibited alanine racemase.

Inhibitors

Some substances inhibit initiation without having much effect on vegetative growth. Foster & Wynne (1948) observed that in *Clostridium botulinum* spores, loss of heat stability was strongly inhibited by low concentrations of oleic, linoleic and linolenic acids. Spores of four *Bacillus* spp. were not inhibited and other clostridia were only slightly affected. Oleic acid at 100 μg./ml. prevented initiation of large inocula, but vegetative growth was unaffected. The saturated acids were inactive. Soluble starch (0·1 %) in the medium neutralized the effects of the fatty acids. Roth & Halvorson (1952) found that unsaturated fatty acids only became inhibitory after oxidation. Severely heated spores were much more sensitive than unheated spores to unsaturated fatty acids and also to lauric and myristic acids (Murrell, 1955, p. 60). Adsorbents increased the viable counts on many complex media and reversed the action of added fatty acids (Olsen & Scott, 1946, 1950; Murrell, Olsen & Scott, 1950; Murrell, 1955). In these experiments it was not shown that initiation was the stage inhibited.

Inhibition by D-alanine is not necessarily permanent. In several aerobes, Wolf & Mahmoud (1957*b*) showed that organisms possessing alanine racemase were able to initiate germination in D-alanine although germination was sometimes considerably delayed. Several other substances related to initiators inhibit or antagonize their activity (see p. 122).

Initiation of germination in spores of several organisms was not prevented by penicillin, subtilin or streptomycin, but the ensuing outgrowth was inhibited (Wynne & Harrell, 1951; Wynne, Collier & Mehl, 1952; Treadwell *et al.* 1958). Kosaki (1959) observed the same effect with tetracycline, chloramphenicol and carbomycin.

Metal ions frequently inhibit initiation and their action is reversible by chelating compounds. Levinson & Sevag (1953) and Hyatt & Levinson (1957) observed that Fe^{++} and Cu^{++} (0·8 mM) inhibited initiation in *Bacillus megaterium*, but not Na^+, K^+, NH_4^+ (50 mM), Mg^{++} (5 mM), Ni^{++}, Cu^{++}, Fe^{++} (0·2 mM). Murty & Halvorson (1957*a*) showed that a crop of spores of *B. cereus* var. *terminalis* grown in a pilot plant made of metal did not initiate germination until washed with versene, phosphate, thioglycollate and other metal-binding compounds. Fe^{++}, Cu^{++}, Cr^{++}, Hg^{++} at 0·1 mM were inhibitory, the inhibition being reversible by washing with phosphate or versene.

MECHANISM OF INITIATION

The conditions under which initiation can occur have been described. Usually some specific substances are required. Strict temperature, pH and E_h limits operate. Heat, storage, and freezing have some effect, possibly indicating permeability changes. Mechanical rupture of one or more of the outer integuments can also initiate germination. Under optimal conditions the change occurs within 5 min. Certain metal ions are essential, others inhibit. What type of reaction causes the initiation process? Studies of this problem have proceeded vigorously along different lines at several centres, and although the puzzle has not yet been solved some interesting mechanisms have been postulated.

(i) *The lytic system*

Dr Joan Powell put forward the hypothesis that the key reactions in initiation were the activation of a lytic enzyme and its attack on the spore coat, causing the simultaneous release of certain spore constituents and hydration of the spore. A considerable amount of solutes is released during initiation (Powell & Strange, 1953; Falcone, 1954). Powell & Strange (1953) showed that this exudate contained calcium dipicolinate (*c.* 50–60 %), a characteristic peptide (20 %), free amino acids and proteins. The exudates from several organisms were similar. The peptide (mol. wt. *c.* 10,000) was composed chiefly of $\alpha\epsilon$-diaminopimelic acid (DAP), alanine, glutamic acid and hexosamines in the ratio 1:3:1:8 and smaller amounts of other amino acids (Strange & Powell, 1954). The peptide was obtained from crushed spores; a similar peptide occurred in vegetative cell walls. Strange & Dark (1957*a*, *b*) showed that spores and sporulating cells of several *Bacillus* spp. contained a lytic enzyme which released hexosamine from vegetative cell walls, spore coat preparations and autoclaved spores. Its activity was increased by various metal ions (Co^{++}, Mg^{++}, Cu^{++}, Ni^{++} and Mn^{++}). Spore coat preparations contained some bound hexosamine although most was released on disintegration. This bound hexosamine was slowly released in water or buffer in the form of the DAP-hexosamine peptide.

Powell (1957*a*) obtained some evidence of a connexion between the lytic enzyme and initiation. In crushed spores of *Bacillus cereus* and in one strain of *B. subtilis* the enzyme was very active, but in another strain of *B. subtilis* which was dormant in tryptic meat broth, the enzyme was less active. Further, it was found that although the DAP and hexosamine contents of *B. cereus* were fairly constant during growth and spore formation, there was a change in its distribution. In non-sporulating cells most was found in the insoluble fraction, but in spores and sporulating cells only in the soluble fraction. Some other evidence does not support this hypothesis. Berger & Marr (1960) reported that when the exosporia (outermost loose layer of the spore coat in some spores) was removed from *B. cereus* spores, the spores contained no hexosamine and were still heat-resistant. Secondly, Powell & Hunter (1958) reported only partial success in relating the activity of initiation inhibitors with an effect on the spore lytic system.

(ii) Mn^{++} *activated system*

Levinson and his co-workers found two enzymes in spores of *Bacillus megaterium* which were activated by Mn^{++}. They therefore suggested two systems for explaining their relationship to Mn^{++} initiation of this organism. The first enzyme was a protease which Levinson & Sevag (1954*a*) showed to attack spore protein and to release various substances including L-alanine. These substances increased initiation and respiration of spores of this organism. Mn^{++} activation of this enzyme thus would cause initiation by the released L-alanine. Exogenous L-alanine gave more rapid initiation and onset of O_2 uptake than Mn^{++}. Levinson & Hyatt (1955) also found that although D-alanine inhibited initiation by L-alanine, it only partly reduced (by 40 %) the initiation by Mn^{++}. They deduced that D-alanine reversed that part of the activity due to L-alanine release, and that Mn^{++} both activated the proteolytic enzyme and raised the overall metabolic level (Mn^{++} caused greater O_2 uptake than L-alanine). The absence of Mn^{++} initiation in two other organisms, and the lack of information on Mn^{++} activated proteolytic enzymes in spores of other organisms at present preclude a general application of this hypothesis.

The second enzyme activated by Mn^{++} was a pyrophosphatase. Murrell (1952, 1955) found a pyrophosphatase in spores of *Bacillus subtilis* and *B. coagulans*, and Levinson *et al.* (1958) demonstrated the enzyme in spores of *B. megaterium*. The enzyme in the latter organism was Mn^{++} activated, and its activity was greatly increased by heat activation of the spores before disintegration. Further, Levinson *et al.* (1958) showed that the enzyme activity was reduced after initiation and that the pyrophosphatase of the vegetative cells was activated by Co^{++} but not by Mn^{++}. Both acid- and alkali-insoluble phosphorus was found by Fitz-James (1955*a*, *b*) in the spore coat residues of *B. megaterium* and *B. cereus* and in the latter during initiation the concentration of acid-insoluble phosphorus decreased.

Levinson (1957) and Levinson *et al.* (1958) suggested that the insoluble residue of the spore coats may form a lattice making the spore coats impermeable to nutrients and that the function of the Mn^{++}activated pyrophosphatase in initiation may be the breakdown or partial breakdown of this lattice permitting the entry of nutrients. They were unable, however, to detect release of phosphate from the spore coats exposed to spore extract. Neither the release of the substrate of the enzyme during disintegration of the spores, nor the immediate use of the products in synthetic reactions was excluded in these experiments. Fitz-James (1955*b*) detected an increase in soluble phosphorus and RNA synthesis within 10 min. of initiation.

(iii) *Alanine racemase system*

The discovery of this enzyme in spores by Stewart & Halvorson (1953, 1954) suggested a relationship with initiation. However, Church *et al.* (1954) were unable to detect the enzyme in some organisms in which germination was initiated by L-alanine; other organisms possessing the enzyme initiated germination without the addition of L-alanine.

(iv) *Transamination system*

Levinson & Sevag (1954*b*) reported a transaminase in spores of *Bacillus megaterium*. Falcone & Caracò (1958*a*, *b*) found transamination involving L-alanine to occur in *B. subtilis* spores after initiation or heat activation, but not in unheated intact spores. The activity in heat-activated spores increased in parallel with the release of substances absorbing at 2700 Å. (probably dipicolinic acid). It was concluded that heat activation or the initiating substances activated the transaminase system, which was possibly of great importance in the transformation of spore proteins into enzyme proteins during germination. However, Falcone, Salvatore & Covelli (1959) found that isonicotinic hydrazide, an inhibitor of transaminase reactions, did not inhibit initiation by L-alanine.

(v) *Activation of an energy-yielding or respiratory system*

Foster (1957) suggested that initiators make energy available and that adaptive syntheses or synthetic activities may occur within seconds of the initiating effect and before heat resistance is lost. Hardwick (1957) reported that ^{32}P or ^{35}S-methionine was fixed by *Bacillus* spores within 1 min. of being placed in a germination medium.

Early studies on the metabolism of initiators and the initiating activity of obvious breakdown products of glucose and L-alanine were disappointing, pyruvate being ineffective as an initiator (Hills, 1949*a*; Powell,

1951) and alanine apparently not metabolized (Harrell & Halvorson, 1955). Spores contain a ribosidase which releases ribose from many ribosides active in initiation and also from adenosine triphosphate, adenosine diphosphate, and ribose-5-phosphate (Lawrence, 1955*a*, *b*; Nakata, 1957; Krask & Fulk, 1959*a*, *b*). Also, L-alanine is metabolized by intact heat-activated spores to ammonia, pyruvate and hydrogen peroxide (Falcone, 1955; Halvorson & Church, 1957; O'Connor & Halvorson, 1959). Pyruvate + adenosine triphosphate and hydrogen peroxide initiated germination in spores of *Bacillus subtilis* (Falcone, 1955; Falcone *et al.* 1958). These authors also showed that alanine initiation was inhibited by atebrin and *p*-chloromercuribenzoate, but not by cyanide; pyruvate + ATP initiation was inhibited by phosphoenolpyruvate, and hydrogen peroxide initiation by cyanide but not by atebrin or *p*-chloromercuribenzoate. They, therefore, concluded that the Krebs cycle and cytochromes were involved. Halvorson & Church (1957) showed that alanine initiation was inhibited by BEP* which inhibits utilization of pyruvate. They also showed that both pyruvate and acetate initiated germination in aged spores or in freshly harvested spores which had been pre-incubated with these substances at pH 5.

A glucose oxidation system exists in heat-activated spores of *Bacillus cereus* var. *terminalis* (Church & Halvorson, 1957; Halvorson, 1957*c*; Doi, Halvorson & Church, 1959). This involved a complete hexose monophosphate shunt system and a di- and tri-carboxylic acid system. Extracts of the heat-activated spores were shown to oxidize many of the substrates involved in these systems, including ribose-5-PO_4 and further that these substrates could initiate germination in aged or heat-activated spores. Halvorson & Church (1957) suggested that initiation involved the immediate participation of energy-yielding reactions. Halvorson, Doi & Church (1958) found that oxidation of glucose was stimulated by DPA and inhibited by metal ions, this inhibition being reversed by DPA. They suggested that initiation might lead to an increase in chelating potential by releasing bound DPA. Such removal of an inhibitory metal may promote a flow of electrons from substrate to molecular oxygen. Heat activation may affect such a reversible system, deactivation occurring in the absence of suitable conditions.

(vi) *Permeability*

The effect of various physical and mechanical treatments on initiation and initiation requirements suggests that a permeability change is involved. Whether spores of one organism are permeable only to par-

* bis-1:3-β-ethylhexyl-5-methyl-5-aminohexahydropyrimidine.

ticular initiators at germination temperatures has not been determined. It is difficult to locate permeability barriers in spores and to decide whether an observed change in permeability is due to initiation or *vice versa*. Perhaps the initiator need only reach the cortex to bring about the necessary changes.

EFFECT OF ENVIRONMENT ON GERMINAL DEVELOPMENT

Most of the information available before 1948 was derived from studies of the over-all germination process, and indicated that the environment most suited to germination was very similar to the optimum for vegetative growth. This information has been reviewed elsewhere (Knaysi, 1948; Wynne, 1952; Murrell, 1955). Recent studies, however, have revealed some of the particular requirements for development of the germ cell, after initiation of germination has occurred. These requirements and their detection will depend on the nutritional status of the mature spore, and this in turn depends on the sporulation medium. The detection of these requirements for outgrowth does not preclude similar requirements during initiation or during vegetative growth.

Requirements for outgrowth

(*a*) *Phosphate.* Hyatt & Levinson (1959) showed that, although added phosphate was not required for initiation, no outgrowth occurred unless phosphate was supplied. A minimum concentration of 0·5–1 mM was necessary to support the outgrowth of *c.* 10^8 spores/ml. of *Bacillus megaterium*. Several inorganic and organic phosphate compounds permitted outgrowth and although initiated spores hydrolysed pyrophosphate this was possible only under conditions inhibiting outgrowth and phosphate uptake.

(*b*) *Sulphate.* Hyatt & Levinson (1957) found that *Bacillus megaterium* spores required a source of sulphur for germinal development, but not for initiation. It was active at 0·01 mM, but the optimal concentration was *c.* 0·1 mM. Normal emergence of the germ cell occurred only in the presence of sulphate. A wide variety of sulphur compounds supported outgrowth, but selenium did not replace sulphur. Woese (1959*b*) did not observe a sulphate requirement for germinal development in *B. subtilis*.

(*c*) *Amino acids and growth factors.* O'Brien & Campbell (1957) showed that *Bacillus stearothermophilus* spores would initiate germination in a glucose-mineral salts medium, but required amino acids and growth factors for outgrowth. Leucine and nicotinic acid were essential for outgrowth but not for vegetative growth. With *B. coagulans*, methionine, thiamin, biotin and folic acid were required for growth at 35° and 55°. Spores needed, in addition, glutamic acid, histidine, isoleucine, leucine and valine for outgrowth at 55°. Leucine was not essential at 35°. *B. cereus* var. *terminalis* spores required isoleucine, leucine, valine and methionine; vegetative growth did not need leucine.

In *Bacillus subtilis*, Woese (1959*b*) found that glutamate and asparagine were needed for the rapid incorporation of phosphate into acid-insoluble phosphate in the germinating cell. About 110 min. after initiation the phosphate in the germinating spore had recovered from the loss in the germination exudate.

(*d*) *pH*. In *Bacillus megaterium* the optimum pH for germinal development was higher (pH 7·5–8) than for initiation (pH 6–7) (Hyatt & Levinson, 1959).

Inhibitors

A few substances are known to inhibit outgrowth. Hachisuka, Sugai & Asano (1958) found that while DL- or L-serine increased initiation in spores of *Bacillus subtilis*, they inhibited their outgrowth in a synthetic medium (containing L-glutamic acid, L-alanine and L-asparagine) and in meat infusion broth. Hyatt & Levinson (1957) observed that although cobalt and nickel ions (0·2 mM) had no effect on initiation, they inhibited outgrowth of *B. megaterium* spores. Cobalt ions inhibited before rupture, and nickel ions inhibited during swelling, emergence and elongation of the germ cell.

THE ROLE OF THE SPORE IN SURVIVAL AND EVOLUTION

No generally acceptable theory of the role of the spore is evident in the literature. Various theories have been discussed by Lamanna (1952) and Foster (1956). The usefulness of the spore's resistance in survival is accepted, but other organisms survive without spores. A great variation in properties of the spores may occur in natural environments due to changes in the environment during their formation, life and germination. The spores of a single species may vary widely in cultural age, composition, size, resistance and their requirements for germination. The first-formed spores in a culture may have high metal contents, more adsorbed inhibitors and greater amounts of endogenous substrates than those formed at other stages of the culture. As the temperature of sporulation varies, so do the properties. Some spores will be better adapted to immediate germination, others to periods of dormancy. Products of germination and growth operate to prevent germination of all the spores, tending to reserve spore inocula for seeding the environment at later stages.

The wide variation in resistance and germination capacity suggests that one effective role of the spore is the seeding of the environment with spores adapted to survive and germinate under a wide range of conditions at appropriate periods when the environment may by chance be suitable for growth. Although the production of one spore per cell has no reproductive value, a cell population arising from a small inoculum may produce a vast number of spores adapted to provide the continual seeding of the environment.

CONCLUSION

The effect of the many factors on spore germination and formation suggests that although germination is a direct metabolic response to the environment, sporulation may not be. To affirm that sporogenesis is a direct response to the environment, it will be necessary to show that the initiation of the genetic changes involved in sporogenesis is a response to a nutrient deficiency or to a threshold concentration of a hormone-like substance.

Although considerable advances have been made in the recent decade in our understanding of the spore there are still great deficiencies in our knowledge. Nutrient conditions—particularly mineral requirements for the growth of sporangia and conditions for the maintenance of these during sporulation, the production of heat-resistant spores, the basis of dormancy, and the requirements for initiation of germination—are aspects in which a considerable advance in knowledge occurred. There is a great deficiency of precise quantitative data on the effect of factors such as temperature, pH, a_w, and O_2 pressure, on the rate and amount of spore formation and initiation of germination and on the properties of spores, particularly in the clostridia. Much of these data need to be obtained with washed suspensions of sporangia under conditions uncomplicated by changing environmental conditions. The mechanisms of heat activation and initiation are unsolved but considerable advances in the knowledge of these processes can be expected. The intracellular events during spore formation are just beginning to take shape and will need to be extended and confirmed in more species. The biochemical changes and metabolic processes involved in the growth of the spore within the cell and in the development of the intraspore state of the mature spore are fascinating problems for the future. The determination of the biosynthesis of dipicolinate, its location and function in the spore needs to be ascertained. Experiments revealing the site of labelled dipicolinate, the composition and method of formation of each coat layer are needed. Finally, the resistant properties of the spore need to be related to the spore structure by determining whether they are due to the coat layers, the state of the spore core, or both. When the state of the protoplasm in the spore core, whether wet or dry, and the basis of heat resistance are known, then the extremely large differences between spores of different species will need to be explained.

REFERENCES

AMAHA, M. & ORDAL, Z. J. (1957). Effect of divalent cations in the sporulation medium on the thermal death rate of *Bacillus coagulans* var. *thermoacidurans*. *J. Bact.* **74**, 596.

AMAHA, M., ORDAL, Z. J. & TOUBA, A. (1956). Sporulation requirements of *Bacillus coagulans* var. *thermoacidurans* in complex media. *J. Bact.* **72**, 34.

BEERS, R. J. (1956). Effect of moisture level on germination of bacterial endospores. Thesis Publ. no. 18,112, University of Illinois.

BEERS, R. J. (1957). Effect of moisture activity on germination. *Publ. Amer. Inst. biol. Sci.* no. 5, 45.

BEHRING, E. (1889). Beiträge zur Aetiologie des Milzbrandes. *Z. Hyg. Infektkr.* **6**, 117.

BERGER, J. A. & MARR, A. G. (1960). Sonic disruption of spores of *Bacillus cereus*. *J. gen. Microbiol.* **22**, 147.

BLACK, S. H., HASHIMOTO, T. & GERHARDT, P. (1960). Calcium reversal of the heat susceptibility and dipicolinate deficiency of spores formed 'endotrophically' in water. *Canad. J. Microbiol.* **6**, 213.

BOWEN, J. F. & SMITH, E. S. (1955). Sporulation in *Clostridium pasteurianum*. *Food Res.* **20**, 655.

BREWER, C. R., MCCULLOUGH, W. G., MILLS, R. C., ROESSLER, W. G., HERBST, E. J. & HOWE, A. F. (1946). Studies on the nutritional requirements of *Bacillus anthracis*. *Arch. Biochem. Biophys.* **10**, 65.

BROWN, W. L. (1957). Informal discussion of activators and inhibitors of germination. *Publ. Amer. Inst. biol. Sci.* no. 5, p. 70.

BRUNSTETTER, B. C. & MAGOON, C. A. (1932). Studies on bacterial spores. III. A contribution to the physiology of spore production in *Bacillus mycoides*. *J. Bact.* **24**, 85.

BULLOCK, K. & TALLENTIRE, A. (1952). Bacterial survival in systems of low moisture content. Part IV. The effects of increasing moisture content on heat resistance, viability and growth of spores of *Bacillus subtilis*. *J. Pharm., Lond.* **4**, 917.

CARACÒ, A., FALCONE, G. & SALVATORE, G. (1958). Sull'azione di alcuni aminoacidi sulla germinazione da l-alanina di spore di *Bacillus subtilis*. *Giorn. Microbiol.* **5**, 127.

CHARNEY, J., FISHER, W. P. & HEGARTY, C. P. (1951). Manganese as an essential element for sporulation in the genus *Bacillus*. *J. Bact.* **62**, 145.

CHRISTIAN, M. I. (1931). A contribution to the bacteriology of commercial sterilized milk. Part II. The coconut or carbolic taint. A study of the causal organism and the factors governing its spore-formation. *J. Dairy Res.* **3**, 113.

CHURCH, B. D. & HALVORSON, H. (1957). Intermediate metabolism of aerobic spores. I. Activation of glucose oxidation in spores of *Bacillus cereus* var. *terminalis*. *J. Bact.* **73**, 470.

CHURCH, B. D., HALVORSON, H. & HALVORSON, H. O. (1954). Studies on spore germination: its independence from alanine racemase activity. *J. Bact.* **68**, 393.

COLLIER, R. E. (1957). An approach to synchronous growth for spore production in *Clostridium roseum. Publ. Amer. Inst. biol. Sci.* no. 5, p. 10.

COOK, R. P. (1931). Some factors influencing spore formation in *B. subtilis* and the metabolism of its spores. *Zbl. Bakt.* (1. *Abt. Orig.*), **122**, 329.

COOK, R. P. (1932). Bacterial spores. *Biol. Rev.* **7**, 1.

CURRAN, H. R. (1957). Mineral requirements for sporulation. *Publ. Amer. Inst. Biol. Sci.* no. 5, p. 1.

CURRAN, H. R. & EVANS, F. R. (1937). The importance of enrichments in the cultivation of bacterial spores previously exposed to lethal agencies. *J. Bact.* **34**, 179.

CURRAN, H. R. & EVANS, F. R. (1944). Heat activation inducing germination in the spores of thermophilic aerobic bacteria. *J. Bact.* **47**, 437 (Abstr.).

CURRAN, H. R. & EVANS, F. R. (1945). Heat activation inducing germination in the spores of thermo-tolerant and thermophilic aerobic bacteria. *J. Bact.* **49**, 335.

CURRAN, H. R. & EVANS, F. R. (1946). The viability of heat-activatable spores in distilled water or glucose solution as influenced by prestorage or poststorage heating. *J. Bact.* **51**, 567.

CURRAN, H. R. & EVANS, F. R. (1947). The viability of heat-activatable spores in nutrient and non-nutrient substrates as influenced by prestorage or poststorage heating and other factors. *J. Bact.* **53**, 103.

CURRAN, H. R. & EVANS, F. R. (1954). The influence of iron or manganese upon the formation of spores of mesophilic aerobes in fluid organic media. *J. Bact.* **67**, 489.

DOI, R., HALVORSON, H. & CHURCH, B. (1959). Intermediate metabolism of aerobic spores. III. The mechanism of glucose and hexose phosphate oxidation in extracts of *Bacillus cereus* spores. *J. Bact.* **77**, 43.

EL-BISI, H. M. & ORDAL, Z. J. (1956). The effect of sporulation temperature on the thermal resistance of *Bacillus coagulans* var. *thermoacidurans. J. Bact.* **71**, 10.

EVANS, F. R. & CURRAN, H. R. (1943). The accelerating effect of sub-lethal heat on spore germination in mesophilic aerobic bacteria. *J. Bact.* **46**, 513.

EVANS, F. R. & CURRAN, H. R. (1960). Influence of preheating, pH, and holding temperature upon viability of bacterial spores stored for long periods in buffer substrates. *J. Bact.* **79**, 361.

FABIAN, F. W. & BRYAN, C. S. (1933). The influence of cations on aerobic sporogenesis in a liquid medium. *J. Bact.* **26**, 543.

FALCONE, G. (1954). Spectrophotometric tests on substances liberated from bacterial spores during germination. *Boll. Ist. sieroter. Milano*, **33**, 460.

FALCONE, G. (1955). Metabolismo della 1-alanina e germinazione della spore. *Giorn. Microbiol.* **1**, 185.

FALCONE, G. & CARACÒ, A. (1958*a*). Attività transaminasiche in spore quiescenti attivate al calore e germinanti di *B. subtilis. Giorn. Microbiol.* **5**, 80. (*Biol. Abstr.* **35**, 10,769, 1960.)

FALCONE, G. & CARACÒ, A. (1958*b*). Transaminase activities in germinating and heat-activated spores of *Bacillus subtilis. Congr. int. Microbiol.* (Abstr.), 7th, 3d, p. 34.

FALCONE, G., SALVATORE, G. & COVELLI, I. (1958). On the mechanism of spore germination by L-alanine in *Bacillus subtilis. Congr. int. Microbiol.* (Abstr.), 7th 3c, p. 33.

FALCONE, G., SALVATORE, G. & COVELLI, I. (1959). Mechanism of induction of spore germination in *Bacillus subtilis* by L-alanine and hydrogen peroxide. *Biochim. biophys. Acta*, **36**, 390.

Fitzgerald, M. P. (1911). The induction of sporulation in the bacilli belonging to the *Aerogenes capsulatus* group. *J. Path. Bact.* **15**, 147.

Fitz-James, P. C. (1955*a*). The phosphorus fractions of *Bacillus cereus* and *Bacillus megaterium*. I. A comparison of spores and vegetative cells. *Canad. J. Microbiol.* **1**, 502.

Fitz-James, P. C. (1955*b*). The phosphorus fractions of *Bacillus cereus* and *Bacillus megaterium*. II. A correlation of the chemical with the cytological changes occurring during spore germination. *Canad. J. Microbiol.* **1**, 525.

Fitz-James, P. C. (1957). Discussion on cytological changes during germination. *Publ. Amer. Inst. biol. Sci.* no. 5, p. 85.

Fitz-James, P. C. & Young, I. E. (1959*a*). Comparison of species and varieties of the genus *Bacillus*. Structure and nucleic acid content of spores. *J. Bact.* **78**, 743.

Fitz-James, P. C. & Young, I. E. (1959*b*). Cytological comparisons of spores of different strains of *Bacillus megaterium*. *J. Bact.* **78**, 755.

Foster, J. W. (1956). Morphogenesis in bacteria: some aspects of spore formation. *Quart. Rev. Biol.* **31**, 102.

Foster, J. W. (1957). Discussion of chemical changes occurring during spore germination. *Publ. Amer. Inst. biol. Sci.* no. 5, p. 76.

Foster, J. W. & Heiligman, F. (1949*a*). Mineral deficiencies in complex organic media as limiting factors in the sporulation of aerobic bacilli. *J. Bact.* **57**, 613.

Foster, J. W. & Heiligman, F. (1949*b*). Biochemical factors influencing sporulation in a strain of *Bacillus cereus*. *J. Bact.* **57**, 639.

Foster, J. W., Hardwick, W. A. & Guirard, B. (1950). Antisporulation factors in complex organic media. 1. Growth and sporulation studies on *Bacillus larvae*. *J. Bact.* **59**, 463.

Foster, J. W. & Perry, J. J. (1954). Intracellular events occurring during endotrophic sporulation in *Bacillus mycoides*. *J. Bact.* **67**, 295.

Foster, J. W. & Wynne, E. S. (1948). Physiological studies on spore germination with special reference to *Clostridium botulinum*. IV. Inhibition of germination by unsaturated C_{18} fatty acids. *J. Bact.* **55**, 495.

Gollakota, K. G. & Halvorson, H. O. (1960). Biochemical changes occurring during sporulation of *Bacillus cereus*. Inhibition of sporulation by α-picolinic acid. *J. Bact.* **79**, 1.

Grelet, N. (1946). Sporulation d'une souche de *Bacillus megaterium* par épuisement soit du fer, soit du carbone, soit de l'azote en milieu synthétique. *C.R. Acad. Sci., Paris*, **222**, 418.

Grelet, N. (1950). Culture d'une souche de *Bacillus megatherium* en milieu synthétique glucosé sporulation par pénurie de zinc en présence de calcium. *Ann. Inst. Pasteur*, **78**, 423.

Grelet, N. (1951). Le déterminisme de la sporulation de *Bacillus megatherium*. I. L'effet de l'épuisement de l'aliment carboné en milieu synthétique. *Ann. Inst. Pasteur*, **81**, 430.

Grelet, N. (1952*a*). Le déterminisme de la sporulation de *Bacillus megatherium*. II. L'effet de la pénurie des constituants minéraux du milieu synthétique. *Ann. Inst. Pasteur*, **82**, 66.

Grelet, N. (1952*b*). Le déterminisme de la sporulation de *Bacillus megatherium*. III. L'effet d'une aération limitante sur les cultures agitées de deux variants d'une même souche. *Ann. Inst. Pasteur*, **82**, 310.

Grelet, N. (1952*c*). Le déterminisme de la sporulation de *Bacillus megatherium*. IV. Constituants minéraux du milieu synthétique nécessaires à la sporulation. *Ann. Inst. Pasteur*, **83**, 71.

Grelet, N. (1955). Nutrition azotée et sporulation de *Bacillus cereus* var. *mycoides*. *Ann. Inst. Pasteur*, **88**, 60.

GRELET, N. (1957). Growth limitation and sporulation. *J. appl. Bact.* **20,** 315.

HACHISUKA, Y., SUGAI, K. & ASANO, N. (1958). Inhibitory effect of serine on the growth of the germinated spores of *Bacillus subtilis*. *Jap. J. Microbiol.* **2,** 317. (*Chem. Abstr.* **53,** 10,382e.)

HALVORSON, H. O. (1957*a*). The germination of bacterial spores. *Proc. Res. Conf. Amer. Meat Inst., Univ. Chicago,* **9,** 1.

HALVORSON, H. O. (1957*b*). Rapid and simultaneous sporulation. *J. appl. Bact.* **20,** 305.

HALVORSON, H. (1957*c*). Oxidative enzymes of bacterial spore extracts. *Publ. Amer. Inst. Biol. Sci.* no. 5, 144.

HALVORSON, H. O. (1959). Symposium on initiation of bacterial growth. *Bact. Rev.* **23,** 267.

HALVORSON, H. & CHURCH, B. D. (1957). Intermediate metabolism of aerobic spores. II. The relationship between oxidative metabolism and germination. *J. appl. Bact.* **20,** 359.

HALVORSON, H. O., DOI, R. & CHURCH, B. (1958). Dormancy of bacterial endospores: regulation of electron transport by dipicolinic aid. *Proc. nat. Acad. Sci., Wash.* **44,** 1171.

HARDWICK, W. A. (1957). Quoted by Foster (1957).

HARDWICK, W. A. & FOSTER, J. W. (1952). On the nature of sporogenesis in some aerobic bacteria. *J. gen. Physiol.* **35,** 907.

HARDWICK, W. A., GUIRARD, B. & FOSTER, J. W. (1951). Antisporulation factors in complex organic media. II. Saturated fatty acids as antisporulation factors. *J. Bact.* **61,** 145.

HARRELL, W. K. & HALVORSON, H. (1955). Studies on the role of L-alanine in the germination of spores of *Bacillus terminalis*. *J. Bact.* **69,** 275.

HARRELL, W. K. & MANTINI, E. (1957). Studies on dipicolinic acid in the spores of *Bacillus cereus* var. *terminalis*. *Canad. J. Microbiol.* **3,** 735.

HASHIMOTO, T., BLACK, S. H. & GERHARDT, P. (1960). Development of fine structure, thermostability, and dipicolinate during sporogenesis in a *Bacillus*. *Canad. J. Microbiol.* **6,** 203.

HASHIMOTO, T. & GERHARDT, P. (1960). Monochromatic ultraviolet microscopy of microorganisms: preliminary observations on bacterial spores. *J. biophys. biochem. Cytol.* **7,** 195.

HEILIGMAN, F., DESROSIER, N. W. & BROUMAND, H. (1956). Spore germination. I. Activators. *Food Res.* **21,** 63.

HENRICI, A. T. (1928). *Morphologic Variation and the Rate of Growth of Bacteria*. London: Baillière, Tindall and Cox.

HILLS, G. M. (1949*a*). Chemical factors in the germination of spore-bearing aerobes. The effect of yeast extract on the germination of *Bacillus anthracis* and its replacement by adenosine. *Biochem. J.* **45,** 353.

HILLS, G. M. (1949*b*). Chemical factors in the germination of spore-bearing aerobes. The effect of amino acids on the germination of *Bacillus anthracis,* with some observations on the relation of optical form to biological activity. *Biochem. J.* **45,** 363.

HILLS, G. M. (1950). Chemical factors in the germination of spore-bearing aerobes: observations on the influence of species, strain and conditions of growth. *J. gen. Microbiol.* **4,** 38.

HITZMAN, D. O., HALVORSON, H. O. & UKITA, T. (1957). Requirements for production and germination of spores of anaerobic bacteria. *J. Bact.* **74,** 1.

HOLZMÜLLER, K. (1909). Die Gruppe des *Bacillus mycoides* Flügge. Ein Beitrag zur Morphologie und Physiologie der Spaltpilze. *Zbl. Bakt.* (2 *Abt. Orig.*), **23,** 304. (Quoted by Knaysi, 1948.)

HYATT, M. T. & LEVINSON, H. S. (1957). Sulfur requirements for postgerminative development of *Bacillus megaterium* spores. *J. Bact.* **74**, 87.
HYATT, M. T. & LEVINSON, H. S. (1959). Utilization of phosphates in the postgerminative development of spores of *Bacillus megaterium*. *J. Bact.* **77**, 487.
JACOB, F., SCHAEFFER, P. & WOLLMAN, E. L. (1960). Episomic elements in bacteria. In *Microbial Genetics. Symp. Soc. gen. Microbiol.* **10**, 67.
JOHNSON, D. E. (1952). The effect of phosphate on the germination of heated and unheated bacterial spores. Thesis for M.A. Degree, University of Texas.
KAPLAN, I. & WILLIAMS, J. W. (1941). Spore formation among the anaerobic bacteria. I. The formation of spores by *Clostridium sporogenes* in nutrient agar media. *J. Bact.* **42**, 265.
KEILIN, D. & HARTREE, E. F. (1947). Comparative study of spores and vegetative forms of *Bacillus subtilis*. *Leeuwenhoek ned. Tijdschr.* **12**, 115.
KEYNAN, A., HALMAN, M. & AVI-DOR, Y. (1958). The influence of temperature on the different stages of germination of *Bacillus subtilis* spores. *Congr. int. Microbiol.* (Abstr.), 7th, 3h, p. 37.
KNAYSI, G. (1945*a*). A study of environmental factors which control endospore formation by a strain of *Bacillus mycoides*. *J. Bact.* **49**, 473.
KNAYSI, G. (1945*b*). Investigation of the existence and nature of reserve material in the endospore of a strain of *Bacillus mycoides* by an indirect method. *J. Bact.* **49**, 617.
KNAYSI, G. (1948). The endospore of bacteria. *Bact. Rev.* **12**, 19.
KOSAKI, N. (1959). Studies on bacterial spores. IV. Effect of various antibiotics on the germination and growth of some endospores of bacteria. *Jap. J. Bact.* **14**, 161.
KRASK, B. J. (1953). Methionine sulfoxide and its specific inhibition of sporulation in *Bacillus subtilis*. *J. Bact.* **66**, 374.
KRASK, B. J. & FULK, G. E. (1959*a*). The utilization of inosine, adenosine and ribose by spores of *Bacillus cereus* var. *terminalis*. *Arch. Biochem. Biophys.* **79**, 86.
KRASK, B. J. & FULK, G. E. (1959*b*). The non-enzymic formation of ribose oxine upon cleavage of adenine nucleotides and purine nucleosides by spores of *Bacillus cereus* var. *terminalis* in the presence of NH_2OH. *Arch. Biochem. Biophys.* **85**. 131,
LAMANNA, C. (1952). Biological role of spores. In *Symposium on the biology of bacterial spores. Bact. Rev.* **16**, 90.
LAWRENCE, N. L. (1955*a*). The cleavage of adenosine by spores of *Bacillus cereus*. *J. Bact.* **70**, 577.
LAWRENCE, N. L. (1955*b*). The relationship between the cleavage of purine ribosides by bacterial spores and the germination of the spores. *J. Bact.* **70**, 583.
LECHOWICH, R. V. (1959). Studies on thermally induced changes in the bacterial endospore and on the relationship of its chemical composition to thermal resistance. Thesis for Ph.D. in Food Tech. University of Illinois.
LEIFSON, E. (1931). Bacterial spores. *J. Bact.* **21**, 331.
LEVINSON, H. S. (1957). Non-oxidative enzymes of spore extracts. *Publ. Amer. Inst. biol. Sci.* no. 5, p. 120.
LEVINSON, H. S. & HYATT, M. T. (1955). The stimulation of germination and respiration of *Bacillus megaterium* spores by manganese, L-alanine and heat. *J. Bact.* **70**, 368.
LEVINSON, H. S. & HYATT, M. T. (1956). Correlation of respiratory activity with phases of spore germination and growth in *Bacillus megaterium* as influenced by manganese and L-alanine. *J. Bact.* **72**, 176.
LEVINSON, H. S. & SEVAG, M. G. (1953). Stimulation of germination and respiration of the spores of *Bacillus megatherium* by manganese and mono-valent anions. *J. gen. Physiol.* **36**, 617.

LEVINSON, H. S. & SEVAG, M. G. (1954*a*). Manganese and the proteolytic activity of spore extracts of *Bacillus megaterium* in relation to germination. *J. Bact.* **67**, 615.

LEVINSON, H. S. & SEVAG, M. G. (1954*b*). The glutamic-aspartic transaminase of extracts of vegetative cells and of spores of *Bacillus megatherium. Arch. Biochem. Biophys.* **50**, 507.

LEVINSON, H. S., SLOAN, J. D. & HYATT, M. T. (1958). Pyrophosphatase activity of *Bacillus megaterium* spore and vegetative cell extracts. *J. Bact.* **75**, 291.

LUND, A. J. (1957*a*). Discussion on the effect of nutritional and environmental conditions of sporulation. *Publ. Amer. Inst. biol. Sci.* no. 5, p. 26.

LUND, A. J. (1957*b*). Factors influencing food sterilization and preservation: sporulation in the genus *Clostridium. Rep. Hormel Inst. Univ. Minn.* 1956, 72.

LUND, A. J. (1959). Physiology of bacterial spores. *Rep. Hormel Inst. Univ. Minn.* 1959, 14.

LUND, A. J., JANSSEN, F. W. & ANDERSON, L. E. (1957). Effect of culture filtrates on sporogenesis in a species of *Clostridium. J. Bact.* **74**, 577.

LUNDGREN, D. G. & BESKID, G. (1958). Induced asporogenic mutants of *Bacillus cereus* var. *lacticola. Bact. Proc.* p. 46.

MCDONALD, W. G. & WYSS, O. (1959). Mutagenic response during sporulation of *Bacillus cereus. Radiation Res.* **11**, 409.

MAJUMDER, S. K. & PADMA, M. C. (1957). Screening of carbohydrates for sporulation of bacilli in fluid medium. *Canad. J. Microbiol.* **3**, 639.

MARTIN, H. H. & FOSTER, J. W. (1958). Biosynthesis of dipicolinic acid in *Bacillus megaterium. J. Bact.* **76**, 167.

MAYALL, B. H. & ROBINOW, C. F. (1957). Observations with the electron microscope on the organization of the cortex of resting and germinating spores of *B. megaterium. J. appl. Bact.* **20**, 333.

MEHL, D. A. & WYNNE, E. S. (1951). A determination of the temperature characteristic of spore germination in a putrefactive anaerobe. *J. Bact.* **61**, 121.

MEISEL, H. & RYMKIEWICZ, D. (1954). Spore formation by *Clostridium tetani. Med. dosw. Mikrobiol.* **6**, 191. (*Chem. Abstr.* **48**, 10,838 b).

MELLON, R. R. (1926). Studies in microbic heredity. X. The agglutinin-absorption reaction as related to the newer biology of bacteria, with special reference to the nature of spore formation. *J. Immunol.* **12**, 355.

MIGULA, W. (1904). Allgemeine Morphologie, Entwicklungsgeschichte, Anatomie und Systematik der Schizomyceten. In Lafar: Handb. techn. Mykol. **1**, 29. (Quoted by Knaysi, 1948.)

MILLET, J. & AUBERT, J. P. (1960). The métabolisme de l'acide glutamique au cours de la sporulation chez *Bacillus megaterium. Ann. Inst. Pasteur*, **98**, 282.

MIWATANI, T. (1957). Studies on the development of *Bacillus cereus. Jap. J. Bact.* **12**, 283. (*Biol. Abstr.* (1958), **32**, 9325.)

MOL, J. H. H. (1957). The temperature characteristics of spore germination and growth of *Bacillus cereus. J. appl. Bact.* **20**, 454.

MORRISON, E. W. & RETTGER, L. F. (1930*a*). Bacterial spores. I. A study in heat resistance and dormancy. *J. Bact.* **20**, 299.

MORRISON, E. W. & RETTGER, L. F. (1930*b*). Bacterial spores. II. A study of bacterial spore germination in relation to environment. *J. Bact.* **20**, 313.

MUNDT, J. O., MAYHEW, C. J. & STEWART, G. (1954). Germination of spores in meats during cure. *Food Tech., Champaign*, **8**, 435.

MURRELL, W. G. (1952). Metabolism of bacterial spores and thermophilic bacteria. Thesis for D.Phil. University of Oxford.

MURRELL, W. G. (1955). The bacterial endospore. Mimeo, University of Sydney.

MURRELL, W. G., OLSEN, A. M. & SCOTT, W. J. (1950). The enumeration of heated bacterial spores. II. Experiments with *Bacillus* species. *Aust. J. Sci. Res.* **B, 3,** 234.

MURTY, G. G. K. & HALVORSON, H. O. (1957*a*). Effect of enzyme inhibitors on the germination and respiration of and growth from *Bacillus cereus* var. *terminalis* spores. *J. Bact.* **73**, 230.

MURTY, G. G. K. & HALVORSON, H. O. (1957*b*). Effect of duration of heating, L-alanine and spore concentration on the oxidation of glucose by spores of *Bacillus cereus* var. *terminalis. J. Bact.* **73**, 235.

NAKATA, H. M. (1957). Discussion on enzymes active in the intact spore. *Publ. Amer. Inst. biol. Sci.* no. 5, p. 97.

NELSON, F. E. (1943). Factors which influence the growth of heat treated bacteria. I. A comparison of four agar media. *J. Bact.* **45**, 395.

O'BRIEN, R. T. & CAMPBELL, L. L. (1956). The nutritional requirements for germination and outgrowth of spores of thermophilic bacteria. *Bact. Proc.* 1956, 46.

O'BRIEN, R. T. & CAMPBELL, L. L. (1957). The nutritional requirements for germination and outgrowth of spores and vegetative cell growth of some aerobic spore forming bacteria. *J. Bact.* **73**, 522.

O'CONNOR, R. & HALVORSON, H. (1959). Intermediate metabolism of aerobic spores. IV. Alanine deamination during the germination of spores of *Bacillus cereus. J. Bact.* **78**, 844.

OHYE, D. F. & SCOTT, W. J. (1953). The temperature relations of *Clostridium botulinum* types *A* and *B. Aust. J. biol. Sci.* **6**, 178.

OHYE, D. F. & SCOTT, W. J. (1957). Studies in the physiology of *Clostridium botulinum* type E. *Aust. J. biol. Sci.* **10**, 85.

OLSEN, A. M. & SCOTT, W. J. (1946). Influence of starch in media used for the detection of heated bacterial spores. *Nature, Lond.* **157**, 337.

OLSEN, A. M. & SCOTT, W. J. (1950). The enumeration of heated bacterial spores. 1. Experiments with *Clostridium botulinum* and other species of *Clostridium. Aust. J. sci. Res.* B, **3**, 219.

ORDAL, Z. J. (1957). Effect of nutritional and environmental conditions of sporulation. *Publ. Amer. Inst. biol. Sci.* no. 5, p. 18.

PERDUE, L. (1933). The effect of inorganic salts in spore formation. Master's Thesis, University of Texas. Quoted Foster & Heiligman (1949*a*).

PERRY, J. J. & FOSTER, J. W. (1954). Non-involvement of lysis during sporulation of *Bacillus mycoides* in distilled water. *J. gen. Physiol.* **37**, 401.

PERRY, J. J. & FOSTER, J. W. (1955). Studies on the biosynthesis of dipicolinic acid in spores of *Bacillus cereus* var. *mycoides. J. Bact.* **69**, 337.

POWELL, J. F. (1950). Factors affecting the germination of thick suspensions of *Bacillus subtilis* spores in L-alanine solution. *J. gen. Microbiol.* **4**, 330.

POWELL, J. F. (1951). The sporulation and germination of a strain of *Bacillus megatherium. J. gen. Microbiol.* **5**, 993.

POWELL, J. F. (1957*a*). Biochemical changes occurring during spore germination in *Bacillus* species. *J. appl. Bact.* **20**, 349.

POWELL, J. F. (1957*b*). Chemical changes occurring during spore germination. *Publ. Amer. Inst. biol. Sci.* no. 5, p. 72.

POWELL, J. F. (1957*c*). Effect of manganese on sporulation of laboratory strains of *Bacillus cereus, B. subtilis* and *B. megatherium. Publ. Amer. Inst. biol. Sci.* no. 5, p. 163.

POWELL, J. F. & HUNTER, J. R. (1953). Sporulation in distilled water. *J. gen. Physiol.* **36**, 601.

POWELL, J. F. & HUNTER, J. R. (1955*a*). The sporulation of *Bacillus sphaericus* stimulated by association with other bacteria: an effect of carbon dioxide. *J. gen. Microbiol.* **13**, 54.

POWELL, J. F. & HUNTER, J. R. (1955*b*). Spore germination in the genus *Bacillus*: the modification of germination requirements as a result of preheating. *J. gen. Microbiol.* **13**, 59.

POWELL, J. F. & HUNTER, J. R. (1956). Adenosine deaminase and ribosidase in spores of *Bacillus cereus*. *Biochem. J.* **62**, 381.

POWELL, J. F. & HUNTER, J. R. (1958). The stimulation and inhibition of spore germination in *Bacillus* species. *Congr. int. Microbiol.* (Abstr.), **7**, 30, p. 43.

POWELL, J. F. & STRANGE, R. E. (1953). Biochemical changes occurring during the germination of bacterial spores. *Biochem. J.* **54**, 205.

POWELL, J. F. & STRANGE, R. E. (1956). Biochemical changes occurring during sporulation in *Bacillus* species. *Biochem. J.* **63**, 661.

POWELL, J. F. & STRANGE, R. E. (1957). $\alpha\epsilon$-Diaminopimelic acid metabolism and sporulation in *Bacillus sphaericus*. *Biochem. J.* **65**, 700.

PUZISS, M. & RITTENBERG, S. C. (1957). Studies on the anaerobic metabolism of *Bacillus anthracis* and *Bacillus cereus*. *J. Bact.* **73**, 48.

REYNOLDS, H. & LICHENSTEIN, H. (1949). Germination of anaerobic spores induced by sublethal heating. *Bact. Proc.* p. 9.

RIEMANN, H. (1957). Some observations on the germination of *Clostridium* spores and the subsequent delay before the commencement of vegetative growth. *J. appl. Bact.* **20**, 404.

ROBERTS, J. L. & BALDWIN, I. L. (1942). Spore formation by *Bacillus subtilis* in peptone solutions altered by treatment with activated charcoal. *J. Bact.* **44**, 653.

RODE, L. J. & FOSTER, J. W. (1960). Mechanical germination of bacterial spores. *Proc. nat. Acad. Sci., Wash.* **46**, 118.

ROMIG, W. R. & WYSS, O. (1957). Some effects of ultra-violet radiation on sporulating cultures of *Bacillus cereus*. *J. Bact.* **74**, 386.

ROTH, N. G. & HALVORSON, H. O. (1952). The effect of oxidative rancidity in unsaturated fatty acids on the germination of bacterial spores. *J. Bact.* **63**, 429.

ROTH, N. G. & LIVELY, D. H. (1956). Germination of spores of certain aerobic bacilli under anaerobic conditions. *J. Bact.* **71**, 162.

ROTH, N. G., LIVELY, D. H. & HODGE, H. M. (1955). Influence of oxygen uptake and age of culture on sporulation of *Bacillus anthracis* and *Bacillus globigii*. *J. Bact.* **69**, 455.

ROTH, N. G., LIVELY, D. H. & METCALFE, S. N. (1958). Correlation of environmental and biological changes occurring in a complex medium during growth and sporulation of *Bacillus* species. *J. Bact.* **75**, 436.

SCHMIDT, C. F. (1950). Spore formation by thermophilic flat sour organisms. 1. The effect of nutrient concentration and the presence of salts. *J. Bact.* **60**, 205.

SCHMIDT, C. F. (1957). Activators and inhibitors of germination. *Publ. Amer. Inst. biol. Sci.* no. 5, p. 56.

SCHREIBER, O. (1896). Ueber die physiologischen Bedingungen der endogenen. Sporenbildung bei *Bacillus anthracis*, *subtilis* und *tumescens*. *Zbl. Bakt.* (1.*Abt Orig.*), **20**, 353. (Quoted by Knaysi, 1948.)

SLEPECKY, R. & FOSTER, J. W. (1959). Alterations in metal content of spores of *Bacillus megaterium* and the effect on some spore properties. *J. Bact.* **78**, 117.

STEDMAN, R. L. (1956*a*). Biochemical aspects of bacterial endospore formation and germination. *Amer. J. Pharm.* **128**, 84.

STEDMAN, R. L. (1956*b*). Biochemical aspects of bacterial endospore formation and germination. II. Chemical changes during sporulation and germination. *Amer. J. Pharm.* **128**, 114.

STEDMAN, R. L., KRAVITZ, E., ANMUTH, M. & HARDING, J. (1956). Autoinhibition of bacterial endospore germination. *Science*, **124**, 403.

STEDMAN, R. L., KRAVITZ, E., HARDING, J. & KING, J. D. (1957). Further studies on bacterial endospore germination relative to autoinhibition. *Food Res.* **22**, 396.

STEWART, B. T. & HALVORSON, H. O. (1953). Studies on the spores of aerobic bacteria. I. The occurrence of alanine racemase. *J. Bact.* **65**, 160.

STEWART, B. T. & HALVORSON, H. O. (1954). Studies on the spores of aerobic bacteria. II. The properties of an extracted heat-stable enzyme. *Arch. Biochem. Biophys.* **49**, 168.

STRANGE, R. E. & DARK, F. A. (1957*a*). A cell-wall lytic enzyme associated with spores of *Bacillus* species. *J. gen. Microbiol.* **16**, 236.

STRANGE, R. E. & DARK, F. A. (1957*b*). Cell-wall lytic enzymes at sporulation and spore germination in *Bacillus* species. *J. gen. Microbiol.* **17**, 525.

STRANGE, R. E. & POWELL, J. F. (1954). Hexosamine-containing peptides in spores of *Bacillus subtilis, B. megatherium, B. cereus*. *Biochem. J.* **58**, 80.

STRESHINSKIĬ, M. O. (1955). Spore formation and polarity of the bacterial cell. *J. gen. Biol., Moscow*, **16**, 480.

SUGIYAMA, H. (1951). Studies on factors affecting the heat resistance of spores of *Clostridium botulinum*. *J. Bact.* **62**, 81.

TARR, H. L. A. (1932). The relation of the composition of the culture medium to the formation of endospores by aerobic bacilli. *J. Hyg., Camb.* **32**, 535.

THEOPHILIS, D. R. & HAMMER, B. W. (1938). Influence of growth temperature on the thermal resistance of some bacteria from evaporated milk. *Res. Bull. Ia. agric. Exp. Sta.* no. 224.

TINELLI, R. (1955*a*). Étude de la biochimie de la sporulation chez *Bacillus megaterium*. I. Composition des spores obtenues par carence de différents substrats carbonés. *Ann. Inst. Pasteur*, **88**, 212.

TINELLI, R. (1955*b*). Étude de la biochimie de la sporulation chez *Bacillus megaterium*. II. Modifications biochimiques et échanges gazeux accompagnant la sporulation provoquée par carence de glucose. *Ann. Inst. Pasteur*, **88**, 364.

TINELLI, R. (1957). Quoted by GRELET (1957).

TREADWELL, P. E., JANN, G. J. & SALLE, A. J. (1958). Studies on factors affecting the rapid germination of spores of *Clostridium botulinum*. *J. Bact.* **76**, 549.

VANINI, G. C. (1957). Influenza dei sali di calcio sullo sviluppo e sulla sporogenesi del bacillo del carbonchio. *Ateneo parmense*, **28**, 700.

VAS, K. & PROSZT, G. (1957*a*). Effect of temperature and hydrogen-ion concentration on the germination of spores of *Bacillus cereus*. *Nature, Lond.* **179**, 1301.

VAS, K. & PROSZT, G. (1957*b*). Correlation between cell count and nutrient concentration in the germination of *Bacillus cereus* spores. *J. appl. Bact.* **20**, 413.

VINTER, V. (1956*a*). Sporulation of Bacilli. Consumption of calcium by the cells and decrease in the proteolytic activity of the medium during sporulation of *Bacillus megatherium*. *Folia biol. (Prague)*, **2**, 216.

VINTER, V. (1956*b*). Sporulation of Bacilli. III. Transference of calcium to cells and decrease in proteolytic activity in the medium in the process of sporulation of *Bacillus megatherium*. *Čsl. Mikrobiol.* **1**, 145.

VINTER, V. (1957). Sporulace bacilů IV. Cvilvěni tvorby spor *Bacillus megatherium* cysteinem a cystinem. *Čsl. Mikrobiol.* **2**, 80. (*Biol. Abstr.* (1958), **32**, 16,672.)

VINTER, V. (1959*a*). Differences in cyst(e)ine content between vegetative cells and spores of *Bacillus cereus* and *Bacillus megaterium*. *Nature, Lond.* **183**, 998.

VINTER, V. (1959*b*). Sporulation of bacilli. VI. The incorporation of ^{35}S-labelled amino acids into sporulating cells of *Bacillus megatherium* inhibited by cystine. *Folia microbiol.* **4**, 1.

VINTER, V. (1959*c*). Sporulation of Bacilli. VII. The participation of cysteine and cystine in spore formation by *Bacillus megatherium*. *Folia microbiol.* **4**, 216.

WAGENAAR, R. O. & DACK, G. M. (1955). Studies on canned cheese spread experimentally inoculated with spores of *Clostridium botulinum*. *Food Res.* **20**, 144.

WARD, B. Q. (1947). Studies on the sporulation of *Bacillus thermoacidurans*. Thesis, University of Texas. (Quoted by Curran, 1957.)

WEINBERG, E. D. (1955). The effect of Mn^{++} and antimicrobial drugs on sporulation of *Bacillus subtilis* in nutrient broth. *J. Bact.* **70**, 289.

WILLIAMS, D. J., CLEGG, L. F. L. & WOLF, J. (1957). The germination of spores of *Bacillus subtilis* and *Bacillus licheniformis* at temperatures below the minimum for vegetative growth. *J. appl. Bact.* **20**, 167.

WILLIAMS, O. B. (1929). Heat resistance of bacterial spores. *J. infect. Dis.* **44**, 421.

WILLIAMS, O. B. & HARPER, O. F. Jr. (1951). Studies on heat resistance. IV. Sporulation of *Bacillus cereus* in some synthetic media and the heat resistance of the spores produced. *J. Bact.* **61**, 551.

WILLIAMS, O. B. & HENNESSEE, A. D. (1956). Studies on heat resistance. VII. The effect of phosphate on the apparent heat resistance of spores of *Bacillus stearothermophilus*. *Food Res.* **21**, 112.

WILLIAMS, O. B. & ROBERTSON, W. J. (1954). Studies on heat resistance. VI. Effect of temperature of incubation at which formed on heat resistance of aerobic thermophilic spores. *J. Bact.* **67**, 377.

WOESE, C. (1959*a*). Further studies on the ionizing radiation inactivation of bacterial spores. *J. Bact.* **77**, 38.

WOESE, C. R. (1959*b*). Effect of withholding glutamic acid and asparagine on the germination of spores of *Bacillus subtilis*. *J. Bact.* **77**, 690.

WOESE, C. R., MOROWITZ, H. J. & HUTCHISON, C. A. (1958). Analysis of action of L-alanine analogues in spore germination. *J. Bact.* **76**, 578.

WOLF, J. & MAHMOUD, S. A. Z. (1957*a*). The germination and enzymic activities of *Bacillus* spores at low temperatures. *J. appl. Bact.* **20**, 124.

WOLF, J. & MAHMOUD, S. A. Z. (1957*b*). The effects of L- and D-alanine on the germination of some *Bacillus* spores. *J. appl. Bact.* **20**, 373.

WOLF, J. & THORLEY, C. M. (1957). The effects of various germination agents on the spores of some strains of *B. subtilis*. *J. appl. Bact.* **20**, 384.

WUND, M. (1906). Festellung der Kardinalpunkte der Säuerstoffkonzentration für Sporenkeimung und Sporenbildung einer Reihe in Luft ihren ganzen Entwicklungsgang durchführender, sporenbildender Bakterienspecies. *Zbl. Bakt.* (1. *Abt. Orig.*), **42**, 97, 193, 289, 385, 481, 577, 673. (Quoted by Knaysi, 1948.)

WYNNE, E. S. (1948). Physiological studies on spore formation in *Clostridium botulinum*. *J. infect. Dis.* **83**, 243.

WYNNE, E. S. (1952). Some physiological aspects of bacterial spore formation and spore germination. In *Symposium on the biology of bacterial spores*. *Bact. Rev.* **16**, 101.

WYNNE, E. S. (1957). Discussion of bacterial spore germination. *Publ. Amer. Inst. biol. Sci.* no. 5, p. 38.

WYNNE, E. S., COLLIER, R. E. & MEHL, D. A. (1952). Locus of action of streptomycin in the development of *Clostridia* from spore inocula. *J. Bact.* **64**, 883.

WYNNE, E. S. & FOSTER, J. W. (1948*a*). Physiological studies on spore germination with special reference to *Clostridium botulinum*. III. Carbon dioxide and germination, with a note on carbon dioxide and aerobic spores. *J. Bact.* **55**, 331.

WYNNE, E. S. & FOSTER, J. W. (1948*b*). Physiological studies on spore germination with special reference to *Clostridium botulinum*. II. Quantitative aspects of the germination process. *J. Bact.* **55**, 69.

WYNNE, E. S., GALYEN, L. I. & MEHL, D. A. (1955). Thermophilic germination of spores of mesophilic *Clostridium*. *Bact. Proc.* 1955, 39.

WYNNE, E. S. & HARRELL, K. (1951). Germination of spores of certain *Clostridium* species in the presence of penicillin. *Antibiot. and Chemother.* **1**, 198.

WYNNE, E. S., MEHL, D. A. & SCHMEIDING, W. R. (1954). Germination of *Clostridium* spores in buffered glucose. *J. Bact.* **67**, 435.

YOUNG, I. E. & FITZ-JAMES, P. C. (1959*a*). Pattern of synthesis of deoxyribonucleic acid in *Bacillus cereus* growing synchronously out of spores. *Nature, Lond.* **183**, 372.

YOUNG, I. E. & FITZ-JAMES, P. C. (1959*b*). Chemical and morphological studies of bacterial spore formation. I. The formation of spores in *Bacillus cereus*. *J. biophys. biochem. Cytol.* **6**, 467.

YOUNG, I. E. & FITZ-JAMES, P. C. (1959*c*). Chemical and morphological studies of bacterial spore formation. II. Spore and parasporal protein formation in *Bacillus cereus* var. *Alesti*. *J. biophys. biochem. Cytol.* **6**, 483.

YOUNG, I. E. & FITZ-JAMES, P. C. (1959*d*). Chemical and morphological studies of bacterial spore formation. III. The effect of 8-Azaguanine on spore and parasporal protein formation in *Bacillus cereus* var. *Alesti*. *J. biophys. biochem. Cytol.* **6**, 499.

EXTERNAL FACTORS INFLUENCING STRUCTURE AND ACTIVITIES OF *STREPTOCOCCUS PYOGENES*

H. GOODER AND W. R. MAXTED

Streptococcus Reference Laboratory, Central Public Health Laboratory, Colindale, London, N.W. 9

Haemolytic streptococci of Lancefield's group A form an appropriate subject for a discussion of the effect of environment, partly because we have a considerable amount of information about the structure and composition of their cell walls; partly because the changes induced by variations in culture media have been widely studied in connexion with the identification of their serotypes and the recognition of virulent strains; and partly because of our ability by simple external treatment to produce aberrant forms.

The reactions of the streptococcus to its environment are of several varieties, ranging from the simplest situation in which an enzyme present in the culture medium removes some part of the coccus without killing it, through the selective action of enzymes and bacteriophages, to the apparently permanent changes that can be induced by exposure to enzymes or antibiotics that destroy or inhibit the production of the bacterial cell wall.

The environmental influences that we shall discuss are practically all those of laboratory cultures, and because of its current interest we shall give especial prominence to recent work in our own laboratory on the induction of L-type growth.

THE STRUCTURE OF GROUP A STREPTOCOCCI

Present concepts regarding the structure, composition and serological activity of the surface of streptococci have their origins in the extensive studies of Lancefield and McCarty at the Rockefeller Institute (see Lancefield, 1954). The principal serologically active components of the cocci are: (*a*) the 'group' polysaccharide hapten, sometimes called the 'C' substance, and (*b*) three series of more or less type-specific protein antigens known as 'M', 'T' and 'R'. Particular cultures of streptococci may or may not contain all these antigens; for convenience strains containing them will be referred to as M+, T+, etc., and strains lacking them as M−, etc.

Mechanical disruption of the streptococcus liberates the cell wall essentially free of cytoplasmic constituents (Salton, 1952). This cell wall material contains the group hapten and the M antigen. After digestion by trypsin (during which M protein is destroyed), the residue contains rhamnose and a restricted range of hexosamines and amino acids. In 30 strains of group A streptococci studied by Cummins & Harris (1956) and by Michel & Gooder (unpublished), the principal amino acids found in acid hydrolysates of the trypsin-treated cell wall material proved to be alanine, lysine and glutamic acid; the sugars were glucosamine, muramic acid and rhamnose. Southard, Hayashi & Barkulis (1959) have characterized the rhamnose as L(−)-rhamnose.

The structure of the cell wall has also been studied with the use of the enzymes prepared from a culture filtrate of a streptomyces strain that is able to render soluble the group hapten present in whole streptococcal cells (Maxted, 1948). McCarty (1952) studied seven strains in this way and found the ratio of rhamnose to total hexosamine to be approximately 1·6:1. Formamide-extracted group hapten had a slightly higher ratio, 2·5:1, for the same constituents (Schmidt, 1952).

Thus, the basal cell wall matrix of group A streptococci appears to be a mucopeptide complex composed of a small number of relatively simple chemical substances. The group hapten seems to be an integral part of this matrix. Variation in the ratio of the components present in the mucopeptide has been observed in a naturally occurring variant of a group A streptococcus originally isolated by mouse passage (McCarty & Lancefield, 1955). This variant appears to be a mutant which differs from the normal organisms by the loss of β-glucosamine-linked material (McCarty, 1956).

Located at the cell wall of group A streptococci is the type-specific M antigen (Lancefield, 1928). This is an alcohol-soluble protein which resists heating to 100° at pH 2·0 and is destroyed by proteolytic enzymes. A purified preparation of the M antigen from a type 1 organism was found by Lancefield & Perlmann (1952) to be almost homogeneous electrophoretically and to have an isoelectric point at pH 5·3. Similarly, located on the surface of the organism are two other serologically and chemically distinct proteins designated T and R respectively, but comparatively little work falling within the scope of this Symposium has been performed on these antigens.

Most strains of group A streptococci also have, at least in some stage of their growth, a capsular polysaccharide, hyaluronic acid, which is composed of *n*-acetyl glucosamine and glucuronic acid. Underlying the cell wall proper is a cytoplasmic membrane, which is presumably concerned with osmotic functions since the protoplasts remaining after removal of the cell wall rapidly lyse in physiological salt solutions (Gooder & Maxted, 1958).

This, then, is the current picture of the structure of a group A streptococcus which can serve as a standard for estimating environmental changes. Such a picture is based upon cells derived from growth in normal laboratory media, such as a simple nutrient glucose broth or on a blood agar plate, at a pH around neutrality. Growth in such media may be summarized as follows. In general, a lag phase of about 3 hr. is followed by a logarithmic growth phase of about 6 hr. In the latter phase the mean generation time is about 20–30 min. The normal chain length varies between 4 and 30 cocci per chain. Enzymes of varying specificity are liberated into the surrounding medium and in certain cases, which will be described, have a pronounced effect on the surface composition of the organism. The general aspects of enzyme formation and activity as affected by environment are discussed elsewhere in this Symposium (Pardee, p. 19).

THE INFLUENCE OF THE CULTURE MEDIUM

Changes in the external surface during the growth cycle

Many investigators have studied the variation during growth of a single identifiable portion of the cell surface, such as M protein, hyaluronic acid, etc. The techniques employed to study these well-known substances necessarily fail to indicate changes among hitherto undetected components on the surface. By following the change in electrophoretic mobility of an organism during growth in a laboratory medium it is possible to discern the changes taking place in surface properties. Such a study has recently been made in our laboratory with a type 6 M+ group A streptococcus. This strain was grown in Todd Hewitt medium at 37°, and samples of the culture removed at intervals for the estimation of growth and electrophoretic mobility (Fig. 1). The mobility measurements were made on thoroughly washed organisms in M/150 phosphate buffer (pH 7·0, ionic strength = 0·012), and growth followed by the change in optical density at a wavelength of 600 mμ (Plummer, D., unpublished results). The electrophoretic mobility is directly related to the zeta potential and hence to the electrical charge density of the bacterial surface (Abramson, Moyer & Gorin, 1942). Thus, the variation in observed mobility is the result of actual alterations in the chemical nature of the bacterial surface during growth and represents the response of the organism to its changing environment. Figure 1 shows that at neutral pH the organisms present in the inoculum carried a net negative charge and hence had a negative mobility value. During the logarithmic phase of growth, this mobility value increased to a maximum and then, as the organisms entered the stationary phase, rapidly decreased. The mobility value after 18 hr. growth was similar to that of the original inoculum.

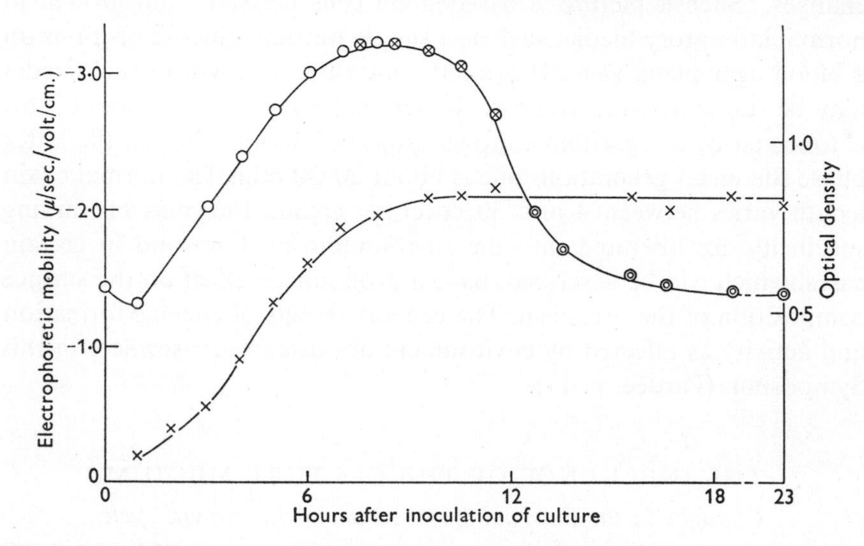

Fig. 1. Type 6 M+ group A streptococcus grown in Todd Hewitt broth pH 7·8 at 37°. At the intervals shown samples were removed for growth and mobility measurements. The optical density of the suspension was determined at 600 mμ. The sample for mobility determinations was washed 3 times in M/150 phosphate buffer pH 7·0 (ionic strength = 0·012) and finally suspended in this buffer. The mobility values were determined in the microelectrophoresis apparatus as described by Loveday & James (1957). ○, ⊗, ◎, Mobility (different experiments); ×, optical density (composite growth curve).

It is tempting to try to interpret these findings in the light of knowledge available from the use of other techniques. Many investigators (Morison, 1940; Pike, 1948) had noted from studies on capsular morphology and staining, the tendency for group A streptococci to produce hyaluronic acid during the most active phase of growth. Hyaluronic acid is a large negatively charged polymer and the laying down of such a polymer on the bacterial surface could account for the observed increase in mobility. Similarly, the decrease in electrophoretic mobility could be explained by the action of hyaluronidase produced by the streptococcus during the later stages of growth. McClean (1941) and Rogers (1945) reported the appearance of this enzyme in culture fluids, possibly as an adaptive response to hyaluronic acid or its breakdown products. Crowley (1944),

Pike (1948), Faber & Rosendal (1954) and MacLennan (1956) have all shown that some strains of streptococci produce both hyaluronic acid and hyaluronidase during growth. In the experiments quoted above, hyaluronidase was found to be present in the culture fluid when growth of the type 6 M+ strain ceased. Growth of this strain in the presence of impure hyaluronidase was not accompanied by the same large increase in mobility during the logarithmic phase of growth, thus supporting the idea that the changes in mobility reflect the formation and subsequent destruction of the hyaluronic acid capsule. However, until more purified preparations of the enzyme have been used it would be premature to ascribe the effects noted solely to the hyaluronic acid-hyaluronidase system.

Growth conditions and their effects

The hyaluronic acid-hyaluronidase balance can be altered in a number of ways. Both anaerobiosis (Sallman, 1951) and the presence of cysteine in the medium (Crowley, 1951) appear to favour increased hyaluronidase production. Rosendal & Faber (1955) reported a decrease in hyaluronidase production when the organisms had been trained to acquire an increased tolerance to penicillin. This decrease in enzyme production may have been due to a reduced ability to release the enzyme in the logarithmic growth phase. Serum was reported by many earlier workers to inhibit hyaluronidase action and, although the mechanism of this inhibition is not yet clear, streptococci grown on a medium containing serum and glucose certainly appear to produce greater quantities of hyaluronic acid.

It has always been recognized that the colonial appearance of bacteria may vary considerably on solid media, depending on the constituents of that medium. The streptococcus is no exception, and Wilson (1959) has drawn attention to some of these differences. It is easy to understand that strains synthesizing hyaluronic acid will, given the necessary substrates, form large mucoid colonies on moist plates. When these colonies dry out, they collapse to form flat disks with a matt surface, now designated 'post-mucoid forms'. These are not to be confused with the matt colonies first described by Todd & Lancefield (1928), which are smaller domed colonies with a dull matt surface and are not necessarily composed of capsulated organisms. As Todd & Lancefield (1928) showed, these matt forms often contain the M antigen, while 'glossy' colonies from the same stock are composed of streptococci lacking the M antigen. If, however, these matt, M+, strains are grown on a serum glucose agar medium, and produce capsules, they will form large mucoid colonies with a glossy surface. The same type of colonies would be produced on this medium by M –, capsulated, strains. When hyaluronidase is added to the medium the obscuring capsule is removed and the colonies recognized again as true matt and glossy forms. Often these differences may be detected more easily by using transmitted light, when the M + strains then appear much more opaque than the translucent M – variants (Pl. 1, fig. 1).

Striking differences in colony form may occur with some strains when grown on blood agar made up from different broth bases, e.g. enzyme digests, infusion broth, Difco or Oxoid Blood Agar Base. On some of these media strong hyaluronidase-producing M + organisms form extremely rough dry colonies, while on others they produce the more typical smooth matt colonies.

One further effect of the environment on the ability of group A streptococci to synthesize high molecular weight polysaccharide should be mentioned. In the presence of maltose and an unidentified component of animal plasma, some streptococci are capable of synthesizing a starch-like polysaccharide. This polysaccharide can be synthesized in the presence of plasma in media containing starch, glycogen or maltose but not when these substances are replaced by glucose, glucose-1-PO_4, cellobiose, melibiose, saccharose or lactose (Crowley & Jevons, 1955). The starch-like material has been characterized as probably either a mixture of amylose and amylopectin, or a single substance with a composition between that of the two. Synthesis of starch by any one strain appears to be markedly affected by cultural conditions, the relative proportions of maltose and serum being of prime importance (Crowley, 1959). Subsequent loss of starch from the bacterial cells can be attributed to amylase production. The comparatively minor alterations in environment which influence this phenomenon emphasize the importance of *in vivo* studies and the possible relation of this activity to pathogenicity.

The M antigen endows the streptococcus with serological type specificity and is also one of the main virulence factors (Rothbard, 1948). The destruction of this antigen would thus not only prevent precise identification of the organism in the laboratory but would also affect its infectivity and chance of survival *in vivo*. Since the antigen is susceptible to proteolysis, any environment which stimulated the appearance of proteolytic enzymes might well have far reaching results.

In fact, a simple manipulation of laboratory media promotes the synthesis and activation of just such an agent by the streptococcus itself; i.e. streptococcal proteinase, an extracellular enzyme activated by reducing agents, destroys the M protein (Elliott, 1945). An enzyme precursor (Elliott & Dole, 1947) is liberated into the culture fluid and is converted into active enzyme by an autocatalytic reaction brought about by reducing conditions or traces of proteolytic enzymes. Elliott (1950) could not detect the precursor in cultures maintained either above 40° or at 4°, nor when the pH of the culture fluid was greater than 7·0. Since the precursor is produced during the most active phases of growth, there may be in many culture media a competition between the synthesis of the M antigen and its destruction by proteinase. The balance of these factors can be manipulated by effecting a change in the surrounding environment. Elliott (1945) had noted that with some strains the M antigen could be identified in cultures grown at 22° but not in cultures grown at 37°. This is apparently due to destruction of M antigen by proteinase action. The use of certain peptones (e.g. neopeptone) in the preparation of the media inhibits the transformation of precursor protein into active enzyme, and advantage can be taken of this fact when growing organisms for serological typing.

In any natural infection it may be assumed that some of the streptococci will come into contact with proteolytic enzymes in the tissue fluids of the host. This might result in loss of M antigen and hence bring about a decrease in virulence of the invading organisms. While the normal content of proteinases in blood and skin is probably low, the increase in amount of certain peptidases of plasma following fever, shock, etc. may be important in this respect. Rothbard & Watson (1948) found that streptococci isolated from convalescent patients often produced less type-specific M substance than streptococci isolated during the active phase of the infection; they were also often susceptible to the bacteriostatic action of normal children's blood, whereas the organisms obtained in the acute phase were invariably resistant. It could not be shown that these changes resulted from the action of streptococcal proteinase but it may be argued that they arose under the influence of convalescent antibodies.

Wormald (1956) also investigated the influence of the environment on the loss of M substance and found that such a loss in a strain of type 3 in a tonsillar carrier was associated with the gradual replacement of M+ forms by M− variants. The same change took place *in vitro* and could be attributed to different growth rates of the two forms. However, in our experience, repeated subculture and storage in the laboratory does not necessarily eliminate M+ variants; nor do strains from either convalescent patients or healthy carriers lack M antigen more often than strains obtained from acute infections (Report, 1957). In one instance prolonged storage in blood broth resulted in the identification of a variant carrying a second M antigen (Wiley, 1960).

Influence of antibody in the environment

Issaeff (1893) found that in the presence of homologous antiserum *Streptococcus pneumoniae* grew in long chains and not as diplococci. The work of Stollerman and his associates (Stollerman & Ekstedt, 1957; Ekstedt & Stollerman, 1960) affords a striking example of the effect of adding type-specific M antibody to highly virulent streptococcal populations in liquid culture. With heterologous antiserum there is no change in morphology, but homologous antiserum causes the organism to grow in the form of very long chains. The chain length may increase from 4–8 cocci per chain to 100 or more. The importance of the type-specific globulin was neatly confirmed by the fact that addition of excess type-specific antigen led to a break-up of the long chains into smaller units. The phenomenon may prove similar to the long chain formation in *S. faecalis* observed by Lominski, Cameron & Wyllie (1958) and might provide an additional useful technique in the search for type-specific antibody in human sera.

An important effect in blood containing homologous antibody is the opsonization of the streptococcus. In such blood the streptococci undergo phagocytosis and destruction by polymorphonuclear leucocytes, whereas in normal heparinized human blood streptococci containing M antigen do not undergo phagocytosis but continue to grow. However, in many animal bloods the organism continues to grow whether antibody is present or not (see also Meynell, this Symposium). Organisms lacking the M antigen are destroyed in normal human blood, but again, in most animal bloods with or without antibody, they are able to grow. Thus the particular blood environment is of importance to the survival of the organism. The presence of type-specific antibody would exert a selective effect on the homologous strain should the blood be infected with a mixed population of streptococcal types. Similarly, in a culture of one type containing M+ and M− variants, an environment of normal human blood would exert a selective effect allowing multiplication of only the M+ strain. Whole human blood is a common enough environment even outside the laboratory and may be of considerable importance in the natural selection of streptococcal variants.

Environments containing bacteriophage

The presence in the culture medium of certain virulent bacteriophages derived from sewage has been shown to have a visible selective effect on a streptococcal population, and this may also be related to antigenic differences (Maxted, 1955). Briefly, the effect is to select capsulated variants from a mixed population of group A streptococci; capsulated cocci do not absorb the virulent phages and they survive, while the non-capsulated organisms, which absorb phage, are subsequently lysed (Pl. 1, fig. 2). The incubation of a predominantly non-mucoid streptococcal population with the appropriate phage therefore resulted in a culture yielding almost entirely mucoid colonies. In some strains, capsulated organisms were also M+ and virulent, whereas the non-capsulated organisms were M− and avirulent. Thus the selection of the capsulated forms resulted in the selection of virulent variants (Maxted, 1955).

Bacteriophages obtained from lysogenic streptococci do not show the same selective action because they have a mechanism for penetrating the capsular material (Kjems, 1958) and are absorbed equally well by capsulated and non-capsulated strains.

There is, however, another quite different way in which bacteriophages can influence the streptococci, namely by inducing the formation of bacterial lysins. Evans (1934) observed that a phage, ordinarily lytic for group C but not group A streptococci, lysed the group A cocci when sensitive group C cocci were also present in the culture. No phage active on group A cocci could be recovered from the lysate. This phenomenon was termed 'nascent lysis'.

The effect was found to be due to the production, during phage lysis of the group C streptococci, of a factor that has a powerful lytic effect on group A streptococci (Maxted, 1957). Lysis of the cells liberates type-specific M antigen and the group polysaccharide which, as already described, is a major constituent of the cell wall. The activity of the lytic agent was unaffected by the removal of the phage by high-speed centrifugation or neutralization by antiphage serum. The properties of the lytic agent were consistent with its being a protein which under reducing conditions possessed enzyme activity directed against a portion of the streptococcal cell wall. More recent work has supported this view and we shall subsequently refer to the lytic factor as the 'cell wall-attacking enzyme'. Partial purification of this enzyme yielded preparations which were only active in the presence of reducing substances; maximum lytic activity on streptococcal cells was found with 10^{-3} M-sodium thioglycollate or potassium cyanide.

Experiments with isolated cell walls prepared by Mickle disintegration confirmed these findings. Among the constituents released by the enzyme was a non-dialysable muco-complex with the same rhamnose-hexosamine ratio as the original cell wall substrate (Maxted & Gooder,

1958). Essentially similar findings have been reported from the Rockefeller Institute (Krause, 1958; Krause & McCarty, 1960). The exact nature of the point of attack of the enzyme is uncertain.

INFLUENCE OF ENZYME ATTACKING THE CELL WALL

Weibull (1953) had observed that the removal of the cell wall of *Bacillus megaterium* by lysozyme in media containing added sucrose (6·7 %, w/v) resulted in the formation of protoplasts of the organism. Similarly, in an environment containing the streptococcal cell wall-attacking enzyme and either sodium chloride (6–12 %, w/v) or sucrose (30–50 %, w/v), the group A streptoccocal cells were freed of all cell wall mucopeptide and were converted to protoplasts (Gooder & Maxted, 1958) conforming to the criteria of Brenner *et al.* (1958).

Initial attempts to grow normal streptococci from these protoplasts by subculture into nutrient liquid media, containing up to 10 % (w/v) added sodium chloride, failed. However, pour plate and surface subculture of enzyme-treated suspensions using suitable solid media allowed up to 10 % or more of the original cocci to grow as stable L-form colonies (see later and Table 1). Similar findings have been reported by Freimer, Krause & McCarty (1959). The L-form growth might have arisen from the streptococcal population either: (*a*) by the selection of pre-existing organisms devoid of cell wall material which are unable to grow on *normal* laboratory media, or (*b*) during enzyme treatment by the production of cells which cannot regenerate cell wall material after its removal by enzyme. These possibilities are considered again in the discussion.

ENZYME-INDUCED L-FORM GROWTH OF *STREPTOCOCCUS PYOGENES*

(*a*) *Growth on solid media*

The method first found successful consisted in treating the streptococci, suspended in 11·7 % (w/v) sodium chloride, with the cell wall-attacking enzyme for 5 min. at 37°. The suspension was then diluted in molten Hartley agar containing 4·3 % (w/v) sodium chloride, 0·5 % (w/v) glucose and 10 % (w/v) horse serum, which was then poured into Petri dishes already containing a 4 mm. thick layer of solidified agar of the same composition. The L-type colony forms appeared after 2–5 days incubation at 37°. The high salt concentration in the agar tended to suppress the growth of any remaining normal streptococci, especially in the early stages of incubation. Penicillin was not necessary for the production of the L-type colony forms but in order to facilitate the differentiation of these colonies we usually incorporated it in low concentration, 5 units/ml.

The L-type colonies were very tenacious and since they grew partly in the agar were difficult to move or break up. They were, however, transferable to the surface of a similar medium by pushing an agar block containing exposed colonies over the surface of a new plate. Phase-contrast microscopy of the L-type colonies revealed a rather amorphous mass which showed large round or oval bodies at the periphery. No normal bacterial forms were seen.

Although different batches of media varied considerably in their yields of growth, no L-type growth occurred, either in the presence or absence of penicillin, without preliminary treatment of the organisms with the cell wall-attacking enzyme. Digestion times between 5 and 60 min., and cultures between 4 and 18 hr. old were used without notable difference in yield of L-type colonies.

With this system it was found essential to have a bottom layer of uninoculated agar with a thin layer of inoculated medium upon it. There is no explanation for this finding at present, but its effectiveness could easily be shown by using gradient plates with a bottom layer of uninoculated agar and a top layer of inoculated agar medium. The number of colonies developing was greatest in the thinnest area of the top layer and became smaller as the depth of the top layer increased and that of the bottom layer decreased; in the portion of the plate with no bottom layer there were usually no colonies.

Another rather surprising effect was also noted. A good batch of medium would grow many hundreds of colonies per plate while another would yield perhaps only 10 or 20. However, it was possible to obtain a good growth of L-type colonies even on a poor medium by the simple expedient of laying a piece of sterile filter-paper on the surface of the agar before incubation. The mass of growth arising deep in the agar layer was sharply limited at the edge of the paper, so giving areas of growth the exact shape of the paper used (Pl. 2, fig. 1). Filter-paper or blotting paper produced the best effect, glossy or waxed paper having little effect, while dialysing membrane or high density polythene gave intermediate results. The paper had to be placed on the surface of the medium shortly after inoculation and left on for the greater part of the incubation time. When papers were removed after 3–18 hr. incubation few colonies were seen. It appeared that the paper improved the local conditions enough to permit growth of the latent cells beneath but no adequate explanation for this effect has been found. The deep situation of the growth shows that the effect could not be due to the spreading beneath the paper of a few surface colonies. Such a sharply defined surface action was unlikely to be due to the adsorption of inhibitory substances from the medium. There was some evidence that the effect might be partly due to the retention of water on the agar surface because growth, even deep in the agar, could also be improved by flooding the plate with 0·9 % (w/v) sodium chloride (Pl. 2, fig. 2).

Table 1. *Yield of L-forms obtained after enzyme induction*

Streptococci suspended for 5 min. at 37° in	Inoculated into	Yield* (colony counts/ml.)
Nutrient broth	Nutrient agar	$10^8/2\times10^8$ streptococcal colonies
Broth containing 11·7 % (w/v) NaCl	Serum agar medium containing 4·3 % (w/v) NaCl	$4\times10^6/4\times10^6$ streptococcal colonies
Broth containing 11·7 % (w/v) NaCl	Serum agar medium containing 4·3 % (w/v) NaCl and penicillin 5 units/ml.	No growth
Broth containing 11·7 % (w/v) NaCl and enzyme	Serum agar medium containing 4·3 % (w/v) NaCl and penicillin 5 units/ml.	$7\times10^7/3\times10^6$ L-form colonies

* The figures before and after the oblique stroke were obtained in separate experiments.

It was later found that more consistent growth could be obtained by using tryptone soya agar instead of Hartley agar as the base; with this medium containing salt, serum and glucose, it was possible to determine the approximate proportion of the original streptococcal cells that give rise to L-type colonies (Table 1). It seemed that 10 % or more of the inoculated organisms that had been exposed to the short enzyme treatment were capable of giving rise to L-type colonial forms. This is in sharp contrast to the induction of L-forms by penicillin. To some extent the very high apparent yield is fallacious because the control count of streptococci is necessarily based on colony-forming units composed of several streptococcal cells and the first action of the enzyme is to break up these chains; this alone would increase the viable count. However, even when this is considered the yield would appear very high. On the improved agar medium it was found possible to obtain good surface growth of L-type colonies directly from the enzyme-treated streptococcal suspensions. Microscopically, the surface colonies were similar to the L-forms described by other workers with a great network of processes growing out from each colony, often merging with a similar outgrowth from neighbouring colonies (Pl. 3, fig. 1). The colonies had the usual characteristics of tenacity and the ability to burrow into the agar. The addition of penicillin to the medium produced no morphological differences in the L-form colony surface growth but chemical analysis may be more revealing.

This surface growth could be propagated on agar plates, with or without penicillin, by inverting a block of growth-covered agar on to the new surface and moving it across the plate at 24 hr. intervals. When the inverted block was moved on, the original well-grown colonies left an easily discernible depression on the new surface. Around the periphery of these depressions appeared tiny white specks of new growth which increased in size to form a complete ring (Pl. 3, fig. 2). This suggests that new growth arises from the network of outgrowth rather than from the dense centre of the old colony.

(*b*) *Grown in liquid media*

In most of our work penicillin (5 units/ml.) was incorporated in the plates in order to reduce the risk of contamination. Since penicillin has been used to induce L-form growth it seemed essential to know whether a stable line of L-forms could be obtained and established in liquid media without exposure to penicillin.

Three systems were used. For the first two, the standard enzyme-treated streptococcal suspension, for a type 6 M+ strain was inoculated on to the standard medium (*a*) without and (*b*) with penicillin (5 units/ml.). For the third system (*c*) the parent streptococcus selected was a type 12 M+ strain which was resistant to 100 μg./ml. (w/v) tetracycline. L-forms of this strain induced in the usual way were grown on the standard medium except that tetracycline 100 μg./ml. (w/v) was included. Tetracycline resistance provided another marker for subsequent identification. Tetracycline alone does not induce L-form growth and all the L-forms that we have made from tetracycline-sensitive strains have been inhibited by this antibiotic.

The initial L-form growth obtained from systems (*a*), (*b*) and (*c*) was subcultured continuously on a solid medium similar to that used to establish the line. Subcultures were made at intervals of 5–6 days by the block technique. The plates were incubated at 37° in polythene bags to retard evaporation and the entire process was pursued over 5–6 months to a total of 20–25 subcultures. The growth was then well removed from the original enzyme treatment used for induction; two of the lines had never been in contact with penicillin. No gross morphological changes were detected during this prolonged subculture.

After the repeated subcultures, attempts were made to obtain growth in fluid media by the transfer of blocks of agar covered with growth to tryptone soya broth (containing 4·3 %, w/v, NaCl; 0·5 %, w/v, glucose;

and 10%, w/v, horse serum) but without success. When, however, subcultures were first made to plates containing the appropriate standard media, except that the concentration of sodium chloride was lowered to 2·5% (w/v), growth occurred on transfer to broth with this lower salt concentration. Subsequently, the broth cultures of lines (*a*) and (*b*) could be transferred in the conventional way. Normal streptococcal growth did not arise and at the time of writing both lines (*a*) and (*b*) are liberating homologous type 6 M antigen so relating the L-form growth to its original streptococcal origin. Agar block to broth transfer of line (*c*) has on two occasions out of five given rise to streptococcal growth identical with the parent strain in type and tetracycline resistance. This reversion only occurred in 2·5% (w/v) salt broth and could not be demonstrated by inoculating similar blocks from the same plate on to blood agar or into normal Todd Hewitt medium; this suggests that the medium must be capable of supporting L-form growth for reversion to occur.

(*c*) *Metabolic activities*

Free type-specific antigen, but no cell wall polysaccharide antigen, is found in the supernatant obtained after growth in liquid medium or in the liquid on the surface of plates growing L-form colonies. These colonies liberate haemolysin and a serologically active streptococcal deoxyribonuclease similar to that produced by the parent streptococcus (Freimer *et al.* 1959). Starch synthesis occurs when the glucose in the medium is replaced by maltose (Gooder, H. & Maxted, W. R., unpublished). The starch is presumably stored internally by the L-form since it is protected from active amylase present in the medium.

EFFECTS DUE TO THE PRESENCE OF CERTAIN ANTIBIOTICS IN GROWTH MEDIA

Growing cultures of haemolytic streptococci are rapidly killed in the presence of penicillin. By passing strains of virulent streptococci through media containing penicillin over a prolonged period of time, Rosendal (1958) isolated penicillin-resistant variants which were relatively avirulent for mice and had lost their M antigen. Mouse-passaged resistant strains regained the M antigen. Such relatively penicillin-resistant strains have never been reported as occurring after a single subculture of a penicillin-sensitive *Streptococcus pyogenes* on to penicillin-containing medium. However, under suitable environmental conditions, streptococcal L-form growth can be induced by penicillin in this manner (Sharp, 1954). Petri dishes containing nutrient agar with added horse serum 10% (w/v) and sodium chloride (1·5–6·0%, w/v) were seeded with a heavy inoculum of group A streptococci. 1000 units of penicillin were deposited in a small trough cut in the agar and the plates incubated anaerobically at 36°. The bacterial growth of streptococci was inhibited for several centimetres around the deposited penicillin; L-form colonies appeared near the margin of the zone of bacterial inhibition. Similar results using this method have been reported by others (Crawford, Frank & Sullivan, 1958; Freimer *et al.* 1959). The ease with which any given strain can be induced to give L-form growth is dependent both on the medium and the added inorganic salts (Dienes & Sharp, 1956). Induction of L-form growth of group A streptococci by D-cycloserine (Michel & Hijmans, 1960) and bacitracin (Gooder, H. & Maxted, W. R., unpublished) is also possible. Morphologically the antibiotic-induced L-forms are similar to those arising from enzyme induction. These L-form cultures have been propagated by the block technique and ultimately in broth culture. Chemical analysis of the L-forms grown in liquid culture revealed the absence of rhamnose and a reduced content of hexosamines, substances normally present in the streptococcal cell wall. Serologically-active group carbohydrate is absent but the L-forms liberate haemolysin and type-specific M antigen (Sharp, Hijmans & Dienes, 1957; Panos, Barkulis & Hayashi, 1959; Freimer *et al.* 1959).

DISCUSSION

The recognized changes in the structure and activities of the surface layers of group A streptococci fall into three broad divisions. First, there are the changes, produced by alterations in the growth conditions in laboratory media, that result in the destruction or synthesis of surface structures but which, in this environment, do not result in the death of the organism. Secondly, there are changes due to either bacteriophage or type-specific antibody which, *in vitro* or *in vivo*, select particular forms of the organism; survivors from the encounter with these agents owe their resistance to the possession of particular structures. Finally, there are changes due to destruction of the cell wall or interference with its formation; in normal laboratory media, the structural damage is severe enough to kill the streptococcus but under special environmental conditions viability is maintained by the development of a completely new and characteristic form.

In the first category, the effects described include those resulting from the interaction of enzymes in the culture medium with their substrates on the cell surface giving rise to many characteristic cultural forms. The best example is offered by the changes occurring in colonial form that result from the synthesis, first, of the hyaluronic acid capsule and, second, of the hyaluronidase that destroys the capsule, resulting in the mucoid appearance of young colonies and a rough appearance in the old. Another easily demonstrable effect of changes in culture media is the promotion or inhibition of proteinase production. The obvious opportunity that streptococcal proteinase has for destroying the M protein suggests a number of ways in which this enzyme might affect both pathogenicity and the laboratory identification of serological types. However, numerous studies have failed to discover any role played by the proteinase in the production of streptococcal disease, nor does it seem to raise any practical difficulties in the routine type identification of group A streptococci. On the other hand, the ability of some strains, in appropriate culture media, to synthesize starch is a characteristic that has been found to have some relation to the production of rheumatic fever and nephritis (Crowley, 1959).

The changes in surface structure, just discussed, do not affect the survival of streptococci in laboratory media; in whole blood, however, they can have a pronounced effect. The presence of a capsule may delay phagocytosis and the type-specific M antigen ordinarily makes the streptococci quite resistant to phagocytosis. Factors in the environment that influence capsule or M antigen production can thus lead to a selective action when the bacteria are exposed to leucocytes. Destruction of the capsule or M antigen naturally removes the resistance; the resistance due to the M antigen can also be overcome by addition of type-specific antibody. This promotion of type-specific bactericidal action of blood by antibody has a practical laboratory application: it affords a method for the identification of type-specific antibody in serum.

The selective action of certain bacteriophages in promoting the survival of the more virulent strains can prove a help in laboratory identification. Whether the effect is of importance *in vivo* is difficult to assess, especially since it is only the virulent phages that have the selective action and hitherto virulent phages have only been isolated from sewage. The occurrence of lysogenic streptococci in the throat, especially of children, is not uncommon (Kjems, 1960). An investigation of the effects of the temperate phages from such strains on the properties of streptococci, and consequently on the host, may prove rewarding.

It is, however, the survival of aberrant forms of the streptococcus following severe localized damage to the cell wall that offers great scope for controversy and even more for the imagination. In the so-called

L-form the organism can survive and grow in circumstances that would otherwise certainly destroy it. It is of considerable interest, therefore, to compare the two known systems for the production of stable L-form cultures of group A streptococci. Induction by exposure to penicillin, or other antibiotics that interfere with cell wall metabolism, usually results in the appearance of only a few L-form colonies from a relatively large number of streptococci. Even under the best known conditions, where the action of the antibiotic is increased by the addition of glycine or other amino acids, it appears unlikely that the conversion of the coccus to its L-form is ever more frequent than about 1 in 100,000. In contrast to this, the cell wall-attacking enzyme can convert from 1 % to approximately 100 % of the original colony-forming units of streptococci present.

However, the L-forms induced by the two systems appear similar, both morphologically and in the absence of cell wall material. Chemical analysis of the enzyme-induced L-forms, similar to that already described for those induced by penicillin, should now be possible if the L-form line recently obtained in liquid culture remains stable. The enzyme-induced L-forms resemble the parent streptococcus in haemolysin formation, the production of serologically active deoxyribonucleases and the ability to synthesize starch. Only haemolysin has as yet been shown to be produced by the antibiotic-induced growth.

Perhaps the most striking phenomenon is the continued production, by both the enzyme- and the antibiotic-induced L-forms, of the type-specific M antigen which on the original coccus appears to be located on the cell wall but is now liberated into the culture medium. If the L-forms are thought of as structures differing from their parents chiefly in the inability to synthesize a complete cell wall (see Klieneberger-Nobel, 1960), the liberation of the M antigen suggests that this antigen is held on to the cells of normal streptococci by the presence of the cell wall. The absorption by streptococci of other macromolecules from culture media has been reported: e.g. polysaccharide haptens (Glynn & Holborow, 1952), and a myocardial antigen, probably lipoidal in nature (Kaplan, 1958). It is not yet known whether streptococcal L-forms can synthesize cell wall precursor material similar to the complex *n*-acetyl-amino sugar compounds found to accumulate in cultures of *Staphylococcus aureus* incubated with penicillin or bacitracin (Park, 1958), or with D-cycloserine (Ciak & Hahn, 1959). In this connexion it is of great interest that the enzyme-induced L-form cultures, although never having been in contact with these antibiotics, are immediately on subculture resistant to high concentrations of them. They remain sensitive to chloromycetin, terramycin, aureomycin and streptomycin. These latter antibiotics are not known to affect cell wall metabolism of normal sensitive organisms.

The continued growth of the penicillin-induced L-forms in media containing penicillin is reasonably interpreted as a direct evasive response of the bacteria to the adverse environment. It is possible to think of the cell wall-attacking enzyme as assisting the streptococcus to convert into the penicillin-resistant form and so to enable it to survive in the presence of penicillin. But the establishment of stable L-form cultures capable of subculture in the absence of the antibiotic presumes a genetic or continuing environmental effect.

The low yield obtained by penicillin induction might be the result of the selection of mutants although there appears to be no evidence that penicillin itself has any direct mutagenic effect. Stabilization of the once selected L-forms might be due to environmental factors. The high frequency of conversion of the cocci to L-forms by the enzyme system seems impossible to explain on the basis of the selection of mutants. Moreover, the L-forms resulting from the conversion remain stable in the absence of any stronger selective agent than high salt concentration.

It is pertinent at this point to consider the nature of the inducing enzyme system. Pure cell wall-attacking enzyme is not yet available and all the experiments reported have been performed with relatively crude material. In addition to the enzyme, such preparations contain intracellular and extracellular products of group C streptococci and also some group C bacteriophage. The enzyme preparations still induce L-form growth even when considerably diluted but it is conceivable that, after the initial attack by the enzyme on the coccal cell wall, some other factor in this mixture is able to enter the damaged coccus and bring about a permanent change in the internal genetic material. It would be surprising, however, for such a change to be so strongly directional, possibly affecting only cell wall synthesis, and at the same time so efficient. It will be interesting to see whether other similar cell wall-attacking enzymes that are available in a purified form, e.g. lysozyme, can be used in a similar manner to effect L-form induction.

Another possible explanation of the permanence of the transformation would be that the damage to the cell wall allows leakage from the cell of the genetic material concerned with the control of cell wall synthesis. Such genetic material might be located centrally in the nuclear region or might be close to the cytoplasmic membrane and behave like an episome in its free phase (cf. Jacob, Schaeffer & Wollman, 1960). It may even be that the cell wall synthesizing systems lie between the wall proper and the cytoplasmic membrane and that this system (or its regulatory mechanism) is lost as a result of the enzyme damage. There is as yet no conclusive evidence regarding the probable location of cell wall synthesizing enzymes in bacteria. It may be that a 'primer' of cell wall material is needed on the cell surface before more can be added. Such a mechanism would be analogous to the synthesis of polyribonucleotides by the polyribonucleotide phosphorylase system which requires oligonucleotides as a primer (Mii & Ochoa, 1957).

The construction of any morphologically identifiable macromolecule requires not only the presence of the various building blocks but also their assembly in order at the right moment, probably on specific sites. Should any one of these functions be impaired then the final morphological unit may not be completed. There are many ways in which these highly specialized functions could be upset, e.g. disturbance of the required spatial relationships on the cytoplasmic membrane, or the separation of the process of cell wall synthesis which may be geared in a synchronous manner to other cellular functions. Should any of these changes occur after cell wall removal and the organism then stabilize itself in its new environment, the process of reversing the change might require techniques and laboratory manipulations which are yet to be discovered.

Superficially, the mass conversion of the bacterial form to L-type growth is similar to the phenomenon described by Lacey (this Symposium, p. 343) under the term 'modulation'. There is, however, the very important difference that the conversion to L-forms appears to be irreversible. This irreversibility is unlikely to be due to the continued presence of the enzyme added in the initial treatment; the dilution factor alone would rule out this explanation unless new enzyme is being synthesized.

The nature of the changes involved in the process of L-form formation may be revealed if a reproducible technique for obtaining mass reversion to the original bacterial form is found. On those occasions when, after manipulation of the environment, it has been possible to obtain a streptococcus from a previously stable L-form line, the occurrence could always be explained by mutation. Crawford *et al.* (1958) suggested that the incorporation of yeast nucleic acid in the medium had a beneficial effect on the reversion to streptococci of their L-form cultures which had been subcultured for 6 months without previously giving rise to streptococci.

The surface changes in group A streptococci stimulated by variation in the medium, e.g. hyaluronic acid and M protein synthesis, would certainly have a useful function in the survival of the organism *in vivo*. *In vitro*, however, such changes confer no obvious advantage. The L-form, on the other hand, is a mechanism of survival under otherwise adverse conditions *in vitro*. It is the *in vivo* function of the L-form which invites speculation. The persistence of antibiotic-sensitive bacteria *in vivo* despite massive antibiotic treatment is not uncommon and it has been suggested that L-forms may be implicated in such microbial persistence (McDermott, 1958). The change to the bacterial form could then occur in the absence of the antibiotic due to an environmental stimulus.

ACKNOWLEDGMENT

We are indebted to Dr Armine T. Wilson for permission to use the photograph represented in Pl. 1, fig. 1.

REFERENCES

Abramson, H. A., Moyer, L. S. & Gorin, M. H. (1942). *Electrophoresis of Proteins.* New York: Rheinold Publishing Co.

Brenner, S., Dark, F. A., Gerhardt, P., Jeynes, M. H., Kandler, O., Kellenberger, E., Klieneberger-Nobel, E., McQuillen, K., Rubio-Huertos, M., Salton, M. R. J., Strange, R. E., Tomcsik, J. & Weibull, C. (1958). Bacterial protoplasts. *Nature, Lond.* **181**, 1713.

Ciak, J. & Hahn, F. E. (1959). Mechanisms of action of antibiotics. II. Studies on the mode of action of cycloserine and its L-isomer. *Antibiot. and Chemother.* **9**, 47.

Crawford, Y. E., Frank, P. F. & Sullivan, B. (1958). Isolation and reversion of L-forms of β-haemolytic streptococci. *J. infect. Dis.* **102**, 44.

Crowley, N. (1944). Hyaluronidase production by haemolytic streptococci of human origin. *J. Path. Bact.* **56**, 27.

Crowley, N. (1951). The effect of amino acids and other substances on the activity of streptoccocal hyaluronidase. *J. gen. Microbiol.* **5**, 906.

Crowley, N. (1959). The association of starch-accumulating strains of group A streptococci with acute nephritis and acute rheumatic fever. *J. Hyg., Camb.* **57**, 235.

Crowley, N. & Jevons, M. P. (1955). The formation of a starch-like polysaccharide from maltose by strains of *Streptococcus pyogenes*. *J. gen. Microbiol.* **13**, 226.

Cummins, C. S. & Harris, H. (1956). The chemical composition of the cell wall in some gram positive bacteria and its possible value as a taxonomic character. *J. gen. Microbiol.* **14**, 583.

Dienes, L. & Sharp, J. T. (1956). The role of high electrolyte concentration in the production and growth of L-forms of bacteria. *J. Bact.* **71**, 208.

Ekstedt, R. D. & Stollerman, G. H. (1960). Inhibition of the chain splitting mechanism of group A streptococci by antibody to M protein. *Fed. Proc.* **19**, 243.

Elliott, S. D. (1945). A proteolytic enzyme produced by group A streptococci with special reference to its effect on type-specific M antigen. *J. exp. Med.* **81**, 573.

Elliott, S. D. (1950). The crystallization and serological differentiation of a streptococcal proteinase and its precursor. *J. exp. Med.* **92**, 201.

Elliott, S. D. & Dole, V. P. (1947). An inactive precursor of streptococcal proteinase. *J. exp. Med.* **85**, 305.

Evans, A. C. (1934). Streptococcus bacteriophage: A study of four serological types. *Publ. Hlth Rep., Wash.* **49**, 1386.

Faber, V. & Rosendal, K. (1954). Streptococcal hyaluronidase. II. Studies on the production of hyaluronidase and hyaluronic acid by representatives of all types of hemolytic streptococci belonging to group A. *Acta path. microbiol. scand.* **35**, 159.

Freimer, E. H., Krause, R. M. & McCarty, M. (1959). Studies of L-forms and protoplasts of group A streptococci. I. Isolation, growth and bacteriologic characteristics. *J. exp. Med.* **110**, 853.

Glynn, L. E. & Holborow, E. J. (1952). The production of complete antigens from polysaccharide haptens by streptococci and other organisms. *J. Path. Bact.* **64**, 775.

Gooder, H. & Maxted, W. R. (1958). Protoplasts of group A β-haemolytic streptococci. *Nature, Lond.* **182**, 808.

Issaeff, B. (1893). Contribution à l'étude de l'immunité. Acquise contre le pneumocoque. *Ann. Inst. Pasteur*, **7**, 260.

Jacob, F., Schaeffer, P. & Wollman, E. L. (1960). Episomic elements in bacteria. In *Microbiol Genetics. Symp. Soc. gen. Microbiol.* **10**, 63.

Kaplan, M. H. (1958). Immunologic studies of heart tissue. I. Production in rabbits of antibodies reactive with an autologous myocardial antigen following immunization with heterologous heart tissue. *J. Immunol.* **80**, 254.

Kjems, E. (1958). Studies on streptococcal bacteriophage. 2. Adsorption, lysogenization and one step growth experiments. *Acta path. microbiol. scand.* **42**, 56.

Kjems, E. (1960). Studies on streptococcal bacteriophages. 4. The occurrence of lysogenic strains among group A haemolytic streptococci. *Acta path. microbiol. scand.* **49**, 199.

Klieneberger-Nobel, E. (1960). L-forms of bacteria. In *The Bacteria*, Vol. I, p. 361. Edited by I. C. Gunsalus and R. Y. Stanier. New York and London: Academic Press.

Krause, R. M. (1958). Studies on the bacteriophages of haemolytic streptococci. II. Antigen released from the streptococcal cell-wall by a phage associated lysin. *J. exp. Med.* **108**, 803.

Krause, R. M. & McCarty, M. (1960). Lysis of streptococcal cell-wall by a phage associated lysin. *Fed. Proc.* **19**, 244.

Lancefield, R. C. (1928). The antigenic complex of streptococcus haemolyticus. III. Chemical and immunological properties of the species specific substance. *J. exp. Med.* **47**, 481.

Lancefield, R. C. (1954). Cellular constituents of Group A streptococci concerned in antigenicity and virulence. In *Streptococcal Infections*, p. 3. Edited by M. McCarty. New York: Columbia University Press.

Lancefield, R. C. & Perlmann, G. E. (1952). Preparation and properties of type-specific M antigen isolated from a group A type 1 haemolytic streptococcus. *J. exp. Med.* **96**, 71.

Lominski, I., Cameron, J. & Wyllie, G. (1958). Chaining and unchaining *Streptococcus faecalis*—a hypothesis of the mechanism of bacterial cell separation. *Nature, Lond.* **181**, 1477.

Loveday, D. E. E. & James, A. M. (1957). An improved micro electrophoresis electrode assembly for use in microbiological investigations. *J. Sci. Instrum.* **34**, 97.

McCarty, M. (1952). The lysis of group A haemolytic streptococci by extracellular enzymes of *Streptomyces albus*. II. Nature of the cellular substrate attacked by the lytic enzymes. *J. exp. Med.* **96**, 569.

McCarty, M. (1956). Variation in the group-specific carbohydrate of group A streptococci. II. Studies on the chemical basis for serological specificity of the carbohydrates. *J. exp. Med.* **104**, 629.

McCarty, M. & Lancefield, R. C. (1955). Variation in the group-specific carbohydrates of group A streptococci. I. Immunochemical studies on the carbohydrate of variant strains. *J. exp. Med.* **102**, 11.

McClean, D. (1941). Studies on diffusing factors. The hyaluronidase activity of testicular extracts, bacterial culture filtrates and other agents that increase tissue permeability. *Biochem. J.* **35**, 159.

McDermott, W. (1958). Microbial persistence. *Yale J. Biol. Med.* **30**, 257.

MacLennan, A. P. (1956). The production of capsules, hyaluronic acid, and hyaluronidase by group A and group C streptococci. *J. gen. Microbiol.* **14**, 134.

Maxted, W. R. (1948). Preparation of streptococcal extracts for Lancefield grouping. *Lancet*, **2**, 255.

Maxted, W. R. (1955). The influence of bacteriophage on *Streptococcus pyogenes*. *J. gen. Microbiol.* **12**, 484.

Maxted, W. R. (1957). The active agent in nascent phage lysis. *J. gen. Microbiol.* **16**, 584.

Maxted, W. R. & Gooder, H. (1958). Enzymic lysis of group A streptococci. *J. gen. Microbiol.* **18**, xiii.

Michel, M. F. & Hijmans, W. (1960). The additive effect of glycine and other amino acids on the induction of the L-phase of group A β-haemolytic streptococci by penicillin and D-cycloserine. *J. gen. Microbiol.* **23**, 35.

Mii, S. & Ochoa, S. (1957). Polyribonucleotide synthesis with highly purified polynucleotide phosphorylase. *Biochem. biophys. Acta*, **26**, 445.

Morison, J. E. (1940). Capsulation of haemolytic streptococci in relation to colony formation. *J. Path. Bact.* **51**, 401.

Panos, C., Barkulis, S. S. & Hayashi, J. A. (1959). Streptococcal L-forms. II. Chemical composition. *J. Bact.* **78**, 863.

Park, J. T. (1958). Inhibition of cell-wall synthesis in *Staphylococcus aureus* by chemicals which cause accumulation of wall precursors. *Biochem. J.* **70**, 2P.

Pike, R. M. (1948). Streptococcal hyaluronic acid and hyaluronidase. I. Hyaluronidase activity of non-capsulated group A streptococci. *J. infect. Dis.* **83**, 1.

Report (1957). Serotypes of *Str. pyogenes*. Their relative prevalence in England and Wales 1952–1956. *Mon. Bull. Minist. Hlth Lab. Serv.* **16**, 163.

Rogers, H. J. (1945). The conditions controlling the production of hyaluronidase by micro-organisms grown in simplified media. *Biochem. J.* **39**, 435.

Rosendal, K. (1958). Investigations of penicillin-resistant streptococci belonging to group A. *Acta path. microbiol. scand.* **42**, 181.

Rosendal, K. & Faber, V. (1955). Streptococcal hyaluronidase. III. The effect of penicillin on the production of hyaluronic acid and hyaluronidase by haemolytic streptococci (type 24 group A). *Acta path. microbiol. scand.* **36**, 263.

Rothbard, S. (1948). Protective effect of hyaluronidase and type-specific anti-M serum on experimental group A infections in mice. *J. exp. Med.* **88**, 325.

Rothbard, S. & Watson, R. F. (1948). Variation occurring in group A streptococci during human infection. Progressive loss of M substance correlated with increasing susceptibility to bacteriostasis. *J. exp. Med.* **87**, 521.

Sallman, B. (1951). The process of hyaluronidase formation by haemolytic streptococci. *J. Bact.* **62**, 741.

PLATE 1

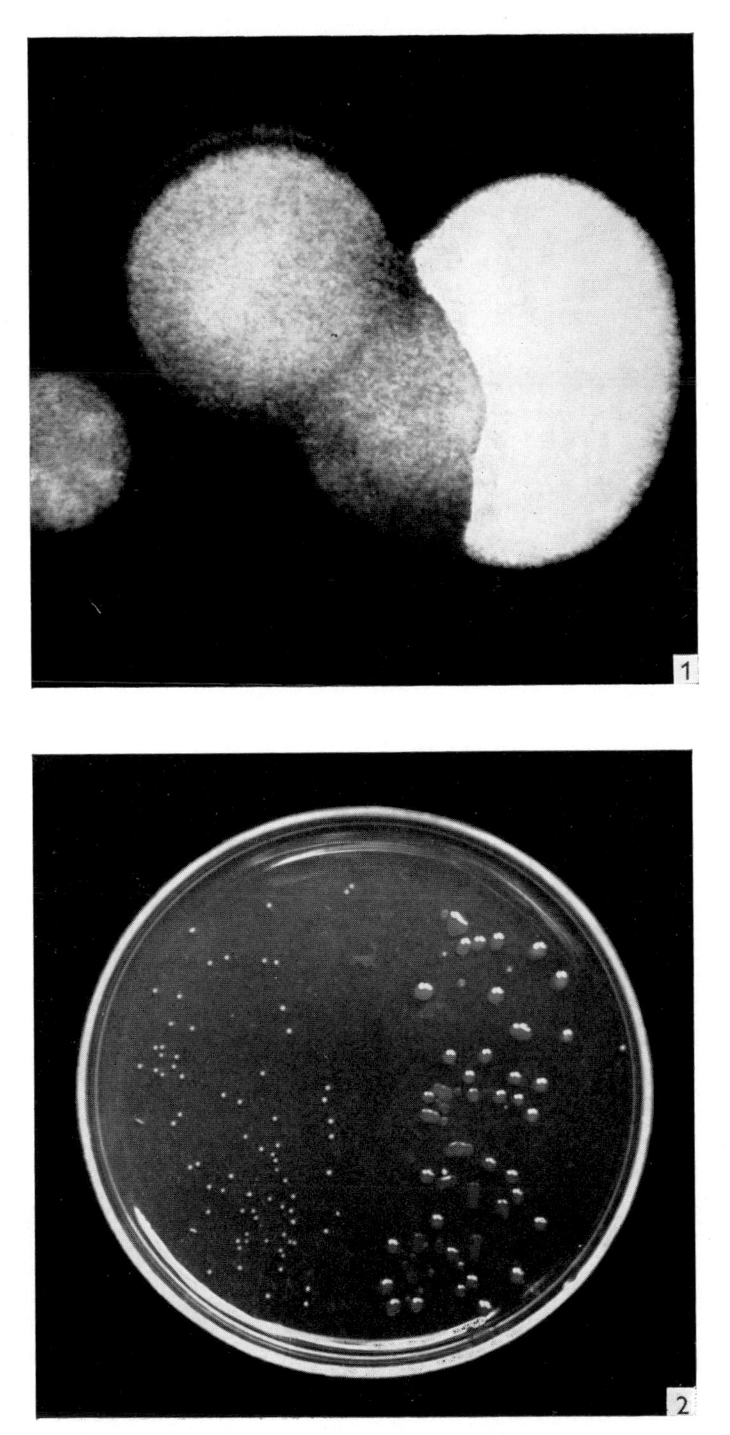

(*Facing page 172*)

PLATE 2

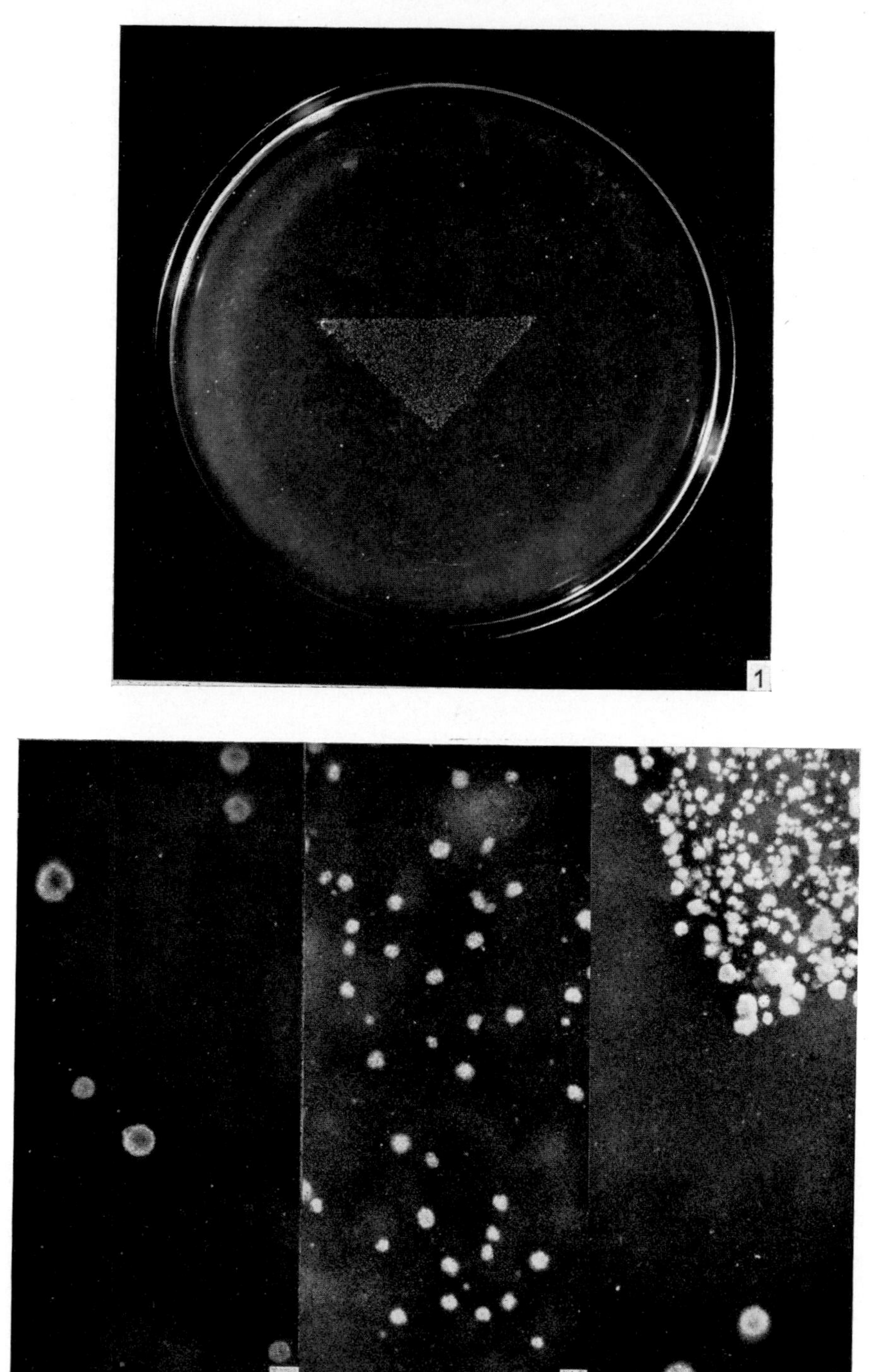

PLATE 3

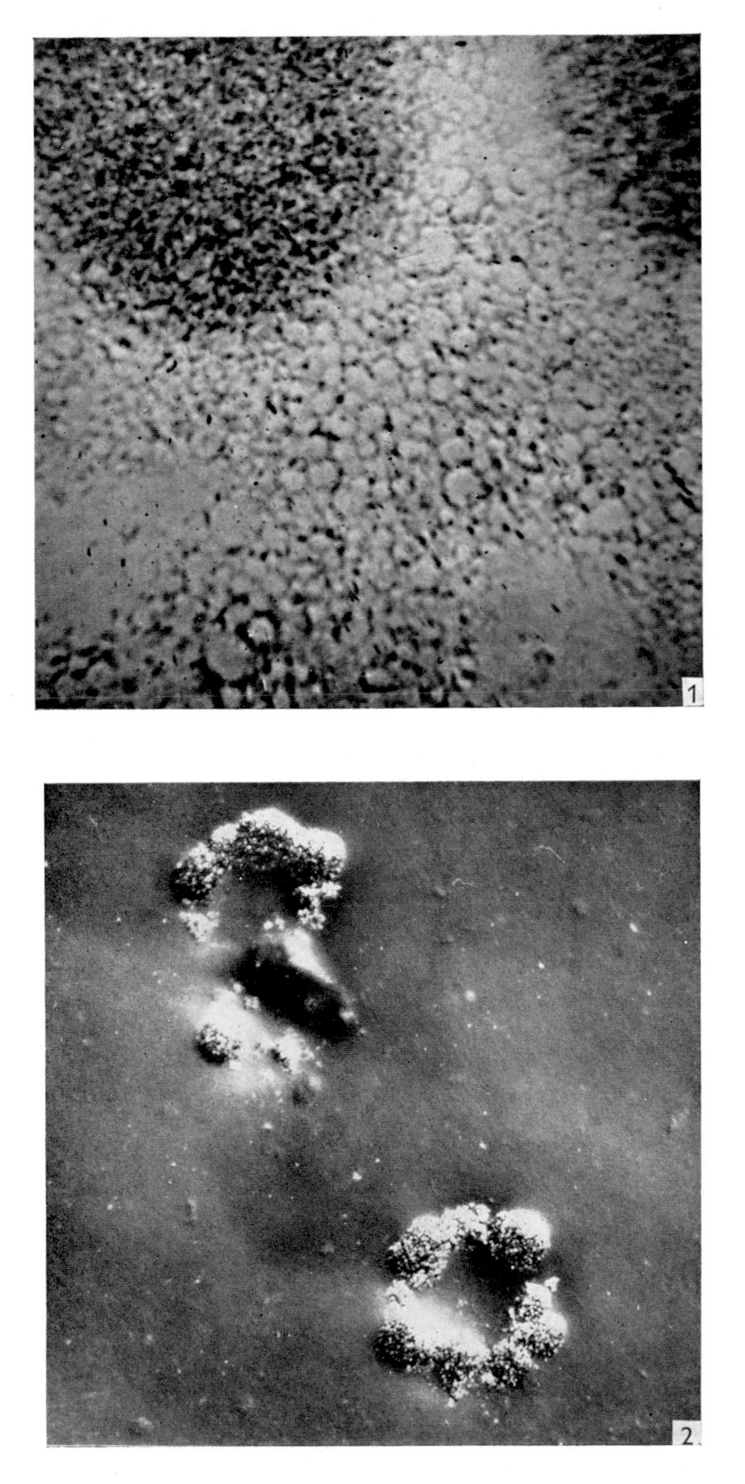

SALTON, M. J. R. (1952). The nature of the cell walls of some Gram-positive and Gram-negative bacteria. *Biochim. biophys. Acta*, **9**, 334.

SCHMIDT, W. C. (1952). Group A streptococcus polysaccharide: Studies on its preparation, chemical composition and cellular localization after intravenous injection into mice. *J. exp. Med.* **95**, 105.

SHARP, J. T. (1954). L-colonies from β-haemolytic streptococci. New technique in the study of L-forms of bacteria. *Proc. Soc. exp. Biol., N.Y.* **87**, 94.

SHARP, J. T., HIJMANS, W. & DIENES, L. (1957). Examination of the L-form of group A streptococci for the group-specific polysaccharide and M protein. *J. exp. Med.* **105**, 153.

SOUTHARD, W. H., HAYASHI, J. A. & BARKULIS, S. S. (1959). Studies on streptococcal cell walls. IV. The conversion of D-glucose to cell wall L-rhamnose. *J. Bact.* **78**, 79.

STOLLERMAN, G. H. & EKSTEDT, R. (1957). Long chain formation by strains of group A streptococci in the presence of homologous antiserum: A type-specific reaction. *J. exp. Med.* **106**, 345.

TODD, E. W. & LANCEFIELD, R. C. (1928). Variants of haemolytic streptococci; their relation to type-specific substance, virulence, and toxin. *J. exp. Med.* **48**, 751.

WEIBULL, C. (1953). The isolation of protoplasts from *Bacillus megaterium* by controlled treatment with lysozyme. *J. Bact.* **66**, 688.

WILEY, G. G. (1960). The M antigen composition of group A type 14, streptococci. *Fed. Proc.* **19**, 204.

WILSON, A. T. (1959). The relative importance of the capsule and the M antigen in determining colony form of group A streptococci. *J. exp. Med.* **109**, 257.

WORMALD, P. J. (1956). Some observations on the influence of the micro-environment on the loss of M substance in strains of *Streptococcus pyogenes*. *J. Hyg., Camb.* **54**, 89.

EXPLANATION OF PLATES

PLATE 1

Fig. 1. Colonies of *Streptococcus pyogenes*, growing on agar medium containing hyaluronidase, viewed with transmitted light. Opaque colony M+. Translucent colony M−. *c.* ×60.

Fig. 2. Mucoid colonies derived from a non-mucoid strain by passage through media containing phage. ×$\frac{3}{4}$. Mucoid colonies shown on right of Plate.

PLATE 2

Fig. 1. Pour plate of L-form growth in Hartley salt serum agar. A mass of colonies arise beneath a triangular piece of blotting paper placed on the agar surface immediately after inoculation. ×$\frac{3}{4}$.

Figs. 2, 3 and 4. Growth of L-form colonies in pour plates of Hartley salt serum agar showing (2) normal conditions, (3) continued excess of liquid on the surface during incubation and (4) the effect of blotting paper as shown in Pl. 2, fig. 1. (2) *c.* ×7. (3) and (4) ×5.

PLATE 3

Fig. 1. L-form colony of *Streptococcus pyogenes* growing on the surface of tryptone salt serum agar. ×60.

Fig. 2. New L-form growth arising from the transfer to fresh medium of an inverted block carrying surface L-form colonies similar to that shown in Pl. 3, fig. 1. ×10.

PHENOTYPIC VARIATION AND BACTERIAL INFECTION

G. G. MEYNELL

Lister Institute of Preventive Medicine, Chelsea Bridge Road, London, S.W. 1

It seems likely that none of the serious consequences of bacterial infection can occur unless the bacteria multiply in their host. Many generations presumably elapse between the first moment of infection and the end of the disease, and during this time it is extremely unlikely that the environment of the infecting organisms is constant. Clearly the organisms may pass from one part of the host to another; but even if they are confined to one site, such as an abscess, its state will continually be altering. Environmental changes can affect the virulence of a bacterial population in several ways. For example, by *selection*, resulting in the predominance of certain mutants in the population; by supplying substances such as antibody which act directly on the organisms and, in doing so, alters their behaviour (e.g. by making them more susceptible to phagocytosis); and lastly, through *phenotypic variation*, by causing the organisms themselves to change their overt properties (their phenotype) without alteration in their genetic structure (genotype). Only phenotypic variation is discussed here since direct effects and selection deserve separate treatment (Braun, 1956).

The most serious consequence of infection is death: yet only four reports describe systems in which virulence (measured by the LD 50, i.e. the number of viable bacteria killing 50 % of inoculated hosts) is affected by the media on which the bacteria are grown before inoculation. Most of the literature deals with bacterial behaviour under artificial conditions such as phagocytic experiments using mixtures of washed bacteria and phagocytic host cells. Moreover, the underlying environmental changes are usually not known so that the phenotype is described, not as that produced by a given factor but, for example, as that exhibited by the cells in the stationary phase of growth, since the cause of cessation of growth is often unknown. Exotoxin production is an exception in having a large literature devoted to phenotypic control of toxin production *in vitro*. Most of this is not referred to here, as it is often reviewed and because preference has been given to observations that appear to bear more directly on the course of infection in the intact host. One section deals with phenotypic resistance to therapeutic agents for, besides being of practical interest, the work concerned provides good examples of how a genotypically uniform bacterial population may become phenotypically heterogeneous even when grown in a uniform environment. The various ways in which this can occur, and its possible significance in infection, form the main topic of the Discussion.

EFFECT OF CULTURAL CONDITIONS ON THE VIRULENCE OF INOCULATED BACTERIA

Cultural conditions are well known to have a great effect on the formation of substances related to pathogenicity: for example, diphtheria toxin (Pappenheimer, 1955), the capsule of *Bacillus anthracis* (Chu, 1952) and the VW antigens of *Pasteurella pestis* (Burrows & Bacon, 1956*b*). Nevertheless, there are only a few examples where the LD 50 is altered by the conditions under which the organisms are grown before inoculation.

Effect of the medium

Bordetella pertussis can grow in either the X or the C mode, depending on the composition of the medium (Lacey, 1960, and this Symposium). The LD 50's of the two modes differ by a factor of *c*. 10^3 when determined by titrations in mice by either the intranasal or the intraperitoneal routes.

Mouse-virulent strains of *Pasteurella pestis* owe their virulence in part to the formation of the antigens, V and W, which are formed during growth at 37° but not at 28° (Burrows & Bacon, 1956*b*). The virulence of organisms grown at 37° is greater than that of those grown at 28°, for the fraction of mice killed after intraperitoneal injection of similar doses was 36/80 and 8/80 respectively.

The LD 50 of spores of *Bacillus anthracis* for guinea-pigs infected by inhalation is affected by the medium in which sporulation occurs (Lincoln *et al.* 1946). When corn steep liquor was added to the growth medium, the LD 50 of the spores by the respiratory route was increased *c*. 3-fold but that by subcutaneous injection was decreased. The increased LD 50 by the respiratory route was therefore thought to be due, not to lesser ability to multiply *in vivo*, but to reduced ability to invade the lung. Colony counts on the lungs and peribronchial lymph nodes of the challenged animals showed this to be so, for the viable count of organisms grown in corn steep liquor fell more rapidly in the lung and far fewer viable organisms were found in the lymph nodes. This effect was not due to the mere presence of corn steep liquor in the inhaled material. Nor was it due to selection of more virulent mutants during multiplication *in vitro*, as shown by experiments in which the organisms were alternately grown in medium either containing or lacking corn steep liquor, 11–18 generations elapsing between inoculation of the medium and collection of the spores. Moreover, no selection occurred during experiments using known mixtures of virulent and avirulent mutants. The effect of corn steep liquor was attributed to its high lactate content, for adding lactate to the growth medium in place of corn steep liquor had the same effects on the LD 50. The differing virulence of *Bacillus anthracis* grown *in vitro* and *in vivo* (Webb, Williams & Barber, 1909) is discussed in the next section.

Effect of growth phase of inoculated bacteria

Various instances are given in the next section of organisms which are more virulent if they are actively growing at the time of inoculation (Felty & Bloomfield, 1924; Felix & Pitt, 1951). The virulence of cultures of *Bordetella pertussis* falls progressively after 24 hr. incubation when titrated in mice by intranasal, but not by intracerebral, injection (Gray, 1946; Standfast, 1958).

The scarcity of examples of environmental influence on the LD 50 is not perhaps surprising, since bacteria may change their phenotype so quickly after inoculation that they rapidly become much the same *in vivo* whatever their state beforehand. This is true of haemolytic streptococci. For example, a strain lacking M antigen has an LD 50 for the mouse of 10^5–10^7 by intraperitoneal injection, compared to an LD 50 of 1 for a strain forming large amounts of this antigen (Foley, Smith & Wood, 1959). Yet removal of the M protein from fully virulent strains by growth in trypsin has no effect on the LD 50 (Lancefield, 1943). There is evidence that this antigen can be synthesized even in the absence of multiplication (Fox & Krampitz, 1954).

RESISTANCE TO PHAGOCYTOSIS

Two factors, growth phase and the nature of the growth medium, affect bacterial resistance to ingestion by phagocytic host cells. The bearing of most of the observations on the course of natural infections was thrown open to discussion by the discovery of 'surface phagocytosis' by Wood, Smith & Watson (1946). The essence of this phenomenon is that although an organism may resist phagocytosis in the environment provided by the usual phagocytic test, where both phagocyte and bacterium are in suspension, ingestion may rapidly occur when the experiment is carried out with phagocyte and bacterium placed on an irregular surface such as filter-paper or the cut surface of lung. Some of the following examples of phenotypic resistance to ingestion may therefore not show that the organisms would have been resistant to ingestion *in vivo* where suitable surfaces are known to be present (Wood *et al.* 1946).

Effect of growth phase

Haemolytic streptococci will be discussed in detail since they show most of the general features; the observations on other species will be summarized.

Two factors determine the resistance of haemolytic streptococci to phagocytosis, namely, the type-specific M protein on the surface of the cell wall and a capsule of hyaluronic acid (Lancefield, 1954; Wiley & Wilson, 1956). Foley *et al.* (1959) showed that a group A strain forming large amounts of both factors was highly virulent for mice (LD 50 = 1 for a growing culture given by intraperitoneal injection) and in the absence of antibody was not phagocytosed either in suspension or on rough surfaces. However, phagocytosis of such strains can occur either if the M antigen is even partially removed with trypsin (Morris & Seastone, 1955) or if the capsule is either removed with hyaluronidase (Rothbard, 1948) or exposed to an unknown factor in human serum (Hirsch & Church, 1960).

When the resistance of haemolytic streptococci to phagocytosis is measured during the growth cycle, virulent strains are found to be most resistant at the end of the exponential phase, being more susceptible both before and after this phase, whereas avirulent strains are susceptible throughout (Hare, 1931). Microscopy and chemical analysis of the cells at the corresponding times show that virulent strains, especially of group C, form a capsule which enlarges greatly during the exponential phase and tends to disintegrate during the stationary phase (Seastone, 1934, 1939, chart 1), its stability depending on the strain (Hirst, 1941) and on the medium (Bazeley, 1940). These changes account, at least in part, for the changes in susceptibility to phagocytosis of virulent streptococci during the growth cycle.

The size of the LD 50 is said to be poorly correlated with resistance to phagocytosis in the conventional phagocytic test using organisms in suspension (Foley *et al.* 1959) although phagocytosis is generally thought to be the principal defence against streptococcal infection. However, when phagocytic tests are performed on rough surfaces, such as filter-paper, resistance to phagocytosis was found to be clearly correlated with the size of the LD 50, so far as could be judged from the behaviour of four strains differing in virulence for mice and rats (Foley *et al.* 1959).

In view of these findings, it is not surprising that actively growing cultures of haemolytic streptococci have been reported to be more virulent than stationary phase cultures. Felty & Bloomfield (1924) reported that the LD 50 of 5 hr. cultures given to mice by intraperitoneal injection was *c.* 100 times less than that of 24 hr. cultures (*c.* $10^{5.5}$ colony-forming units compared to *c.* $10^{7.5}$).* Chains of organisms were commoner in the 24 hr. than in the 5 hr. cultures so that the difference between the LD 50's in terms of the number of individual bacteria may be even greater than these estimates suggest. Similar experiments with group C streptococci gave the same general result (Bazeley, 1940) although Seastone (1934) mentions that he failed to find any difference between growing and stationary phase cultures with other strains of group C streptococci.

Many other species increase substantially in resistance to phagocytosis during active growth. For example, the only difference between two strains of *Streptococcus pneumoniae*, differing in virulence for the rabbit, was that, during growth in broth, the less virulent lost its capsule and its resistance to ingestion earlier in the stationary phase than the more virulent (Shaffer, Enders & Wu, 1936). Type III pneumococci form large capsules during active growth which inhibit even surface phagocytosis in the absence of antibody (Wood & Smith, 1949). Virulent strains of *Pasteurella pestis* become resistant to phagocytosis during active growth either *in vivo* or *in vitro* when grown on a suitable medium at 37° owing to the formation of a new antigen which is not visible as a capsule. This antigen is not formed at 28° so that cells growing at this temperature remain susceptible to phagocytosis. Avirulent strains are susceptible both in the stationary and the exponential phases (Burrows & Bacon, 1956*a*; Burrows, 1955). Virulent strains of *Salmonella typhi* produce Vi antigen, visible as a capsule (Ando & Nakumara, 1951) which inhibits phagocytosis (Felix & Pitt, 1951). Since most genotypically Vi-positive strains only produce Vi antigen during multiplication (Craigie & Brandon, 1936), changes in resistance might be expected during the growth cycle. Phagocytic experiments with certain virulent strains, such as Ty 2 (Bensted, 1940), would presumably not show these changes

* The titrations used four-fold dilutions of the cultures, only one mouse being inoculated with each dilution, the final result being expressed as the smallest dose found to kill. The LD 50 cannot be estimated by the usual methods, such as probit analysis, in this case where only one host/dose was used, but an estimate can be obtained by a method described by Dr P. Armitage who kindly analysed the above data (Armitage, 1959). The estimates are derived on the assumption that the dose–response curves were Poissonian (Meynell, 1957).

because they are exceptional in forming Vi antigen which is stable during the stationary phase. The LD 50 of strain Ty 2 falls slightly during active growth (Felix & Pitt, 1951). No change occurs in the LD 50 of *Salmonella typhimurium* during the growth cycle (Wilson, 1926). *Staphylococcus aureus* forms a capsule while actively dividing and, at the same time, becomes resistant to phagocytosis (Lyons, 1937). Virulent strains of *Bacillus anthracis* may be non-capsulated and susceptible to phagocytosis when grown *in vitro* on certain media, but when grown either on other media or *in vivo*, develop a capsule and become resistant (Smith, Keppie & Stanley, 1953). The reverse change was probably observed by Webb *et al.* (1909) who isolated *B. anthracis* from the blood of infected mice and measured its virulence immediately after isolation and after various periods of growth on agar, the organisms being isolated by micromanipulation and administered by subcutaneous injection. Immediately after isolation from an infected host, one chain of bacilli usually caused a fatal infection in mice but even 100–200 chains failed to do so after 1–3 days growth *in vitro*.

These changes in resistance to phagocytosis account, at least in part, for the finding that after intravenous injection of virulent strains of *Streptococcus pneumoniae* into rabbits, when the organisms are exposed to phagocytosis by the reticulo-endothelial system, the viable count in the blood begins to rise immediately if the organisms are actively dividing; but falls initially and only later rises if the cells are in the stationary phase (Wright, 1927; Enders, Shaffer & Wu, 1936). The strain used by Enders and his colleagues was capsulated in the exponential phase but not after overnight growth. The secondary rise in the viable count in the blood after inoculation of overnight cultures was presumably due to the joint effects of the onset of bacterial division (which would tend to counteract the fall in count) and concurrent capsule formation (which would inhibit phagocytosis). Another strain of pneumococcus described by these authors differed in remaining capsulated in the stationary phase and after inoculation of its cultures, the viable count of the blood did not fall sharply as with the other strain, but stayed virtually constant until it began to rise at the usual time after inoculation.

Effect of the medium

Virulent group A streptococci growing in a medium of low oxidation-reduction potential (E_h) form an active protease which destroys their M antigen (Elliott, 1945; Stamp, 1953) and presumably makes them susceptible to phagocytosis.

Cultural conditions affect the stability of the capsule of group C streptococci, the capsule disappearing after the pH falls below 6·8. This may not be due to a direct effect of acid on the hyaluronic acid capsule but to potentiation of hyaluronidase formed by most strains of group C streptococci (MacLennan, 1956) which, if it resembles that tested by Rogers (1948), is most active at *c.* pH 6, judging from its viscosity-reducing power. Adding glucose to the medium hastens the fall in pH so that the organisms become non-capsulated earlier during growth of the culture (Bazeley, 1940). These changes may be of importance *in vivo* since inflammatory areas may be acid (Menkin, 1956) and so cause capsulated streptococci to become susceptible to phagocytosis.

RESISTANCE TO ANTIBACTERIAL AGENTS

Killing by serum

The bactericidal action of serum is greatly affected by the growth phase of the organisms and by the growth medium. For example, *Pasteurella multocida*, the cause of haemorrhagic septicaemia in cattle, resists the bactericidal action of serum taken from immune cattle when the organisms are grown *in vivo*, but killing occurs when they are grown *in vitro* (Bain, 1960).

Killing of Gram-negative organisms now appears to be due to a weakening of the cell wall following combination with antibody and complement (Amano *et al.* 1957). When the reaction occurs in an isotonic medium such as 0·85 % (w/v) saline, plasmoptysis and lysis may follow; but when the medium is made hypertonic (e.g. by adding sucrose to 20 %, w/v) spheroplasts are formed (Muschel, Carey & Baron, 1959; Michael & Braun, 1959*a*). The effect is partly controlled by the growth phase of the organisms and Michael & Braun (1959*b*) have suggested that death is dependent on the occurrence of growth at the time the cell is damaged. Most of the observations on phenotypic changes in resistance to killing by serum can be interpreted in this way.

For example, Rowley & Wardlaw (1958) showed that *Escherichia coli* is only killed by normal human serum once the culture begins to multiply. Morris (1943, table 8) described a similar effect with the Vi-negative strain of *Salmonella typhi*, 0901, exposed to O antibody and rabbit complement. This change would not be expected with most genotypically Vi-positive strains of *S. typhi* exposed to O antibody since these become phenotypically Vi-positive during active growth (Craigie & Brandon, 1936) and are then resistant to this antibody (Felix & Pitt, 1951). Killing of *Shigella dysenteriae* by normal human serum is prevented by the presence of amino acid analogues which inhibit multiplication (Michael & Braun, 1959*b*). A discrepant observation is that this organism is said to be most susceptible in the stationary phase (Michael & Braun, 1959*a*).

Maaløe (1948*a*, *b*) analysed the effect of the medium on the resistance of *Salmonella typhimurium* to fresh human serum. Growth in diluted broth at pH $< 6{\cdot}2$ lowered resistance without lowering the growth rate; the drop was prevented at any pH by adding glucose to the growth medium. The organisms were then grown in dilute broth at pH 5·5–6·0 containing various supplements in order to determine their effect on resistance. Hexoses, pentoses, lactate and pyruvate (in descending order of efficiency) were all found to increase resistance. Phlorhizin (M/200) reduced resistance and the cell yield without affecting the growth rate, and its effect was counteracted by glucose (cf. Rowley & Wardlaw, 1958). The composition of a simple chemically defined medium affected resistance which became lower, the higher the concentration of citrate and ammonium sulphate. Michael & Braun (1959*b*) describe experiments giving similar results with *Shigella dysenteriae*.

Bacteria appear to die initially at a constant rate in serum but the rate soon falls to zero (Morris, 1943; Adler, 1953). The organisms surviving at this time are not genotypically resistant, for their descendants have the same resistance as the original culture (Adler, 1953); they must therefore be phenotypically resistant (cf. survival after exposure to antibiotics).

Other agents present in host tissues

Many agents besides serum have been described, but their importance in infection is still uncertain (Skarnes & Watson, 1957). Most of the examples concern lysozyme which is probably the factor of greatest potential importance.

Until recently, lysozyme was not thought to be of great value to an infected host because its distribution was thought to be restricted to surface secretions and as it was thought to attack only a few non-pathogenic bacterial species. However, its significance has been increased by detection of relatively very high concentrations in one of the commonest types of phagocyte, the polymorphonuclear leucocyte (Myrvik & Weiser, 1955), and by finding that many bacterial species, including pathogens, which were formerly regarded as resistant to lysozyme, can be made susceptible by simple means. For example, *Bacillus anthracis* grown on ordinary media *in vitro* is resistant but 5/13 strains became susceptible when grown in 10–20 % CO_2 with 0·1 M-$NaHCO_3$ (Gladstone & Johnston, 1955). Increasing the concentration of NaCl in the growth medium reduces the susceptibility of the classic test organism, *Micrococcus lysodeikticus* (Litwack & Pramer, 1956). *Escherichia coli* and *Salmonella typhi* are susceptible after treatment with chelating agents or with antibody+complement respectively (Repaske, 1958; Muschel *et al.* 1959). Both these reports mention that the growth phase affects susceptibility.

Two other reports of naturally occurring agents mention effects due to variation in phenotype. The growth phase does not affect the susceptibility of *Escherichia coli* and *Shigella sonnei* to 'phagocytin', a bactericidal agent extracted from rabbit polymorphs (Hirsch, 1956). The susceptibility of *E. coli* to killing by histone is partly determined by the medium in which the organisms are grown before test but not by their growth phase (Hirsch, 1958).

Therapeutic agents

Many antibacterial drugs kill organisms of sensitive strains only when they are actively growing. This is true of sulphonamides (Wolff & Julius, 1939), penicillin, streptomycin (Bigger, 1944; Hobby, Meyer & Chaffee, 1942; Garrod, 1948), isoniazid (Schaefer, 1954) and erythromycin (Haight & Finland, 1952). It has often been pointed out that during treatment of an infected patient, some organisms might chance to be phenotypically resistant because they were not multiplying; and consequently, being alive when treatment stopped, would be able to cause fresh disease. This was soon shown to be a serious possibility when genotypically sensitive organisms were found in patients relapsing after treatment. The significance of such findings has been discussed by many authors (e.g. McDermott, 1958).

One of the earliest observations was made by Bigger (1944) on the effect of penicillin on growing cultures of *Staphylococcus aureus*. After several hours exposure, a small proportion (*c.* 10^{-6}) of the population was still viable although, on subculture, its resistance was no greater than that of the original culture. Furthermore, the culture became resistant to the bactericidal action of penicillin when multiplication was stopped by dilution in cold broth or by the addition of boric acid. Phenotypically resistant organisms do not always form only a small part of a growing culture, as Bigger suggested, and the size of this fraction can change considerably in a few minutes during growth *in vitro* (Mitchison & Selkon, 1956; Meynell, 1958). Moreover, sensitive organisms are now thought to become resistant to some agents of this kind before they stop growing, as shown by comparing the survival of samples from the same culture treated either by cold shock or by streptomycin (Meynell, 1958).

The difficulty in applying such observations to the interpretation of either natural or experimental infections lies, not in showing that genotypically sensitive organisms can survive *in vivo* during antibacterial therapy, but in proving that effective concentrations of the drug are in contact with the organisms for a period as long as that necessary for killing of the organism *in vitro* during exponential growth. Nevertheless, convincing results were obtained by Eagle (1952) in his study of the effect of penicillin on the course of group A streptococcal infections in the mouse.

Mice inoculated in the thigh muscle with a known number of cocci were also injected at varying times in the shoulder or opposite leg with either a single dose of procaine penicillin or with several doses of aqueous penicillin. The number of viable organisms in the thigh muscle was measured at intervals by colony counts. After inoculation of 10^3 organisms, penicillin cured the infection, provided it was not given more than 6 hr. after challenge, and rapidly reduced the viable count. But when given more than 12 hr. after inoculation of the cocci, penicillin failed to kill a substantial fraction of the organisms and to prevent death. The failure of penicillin treatment was due neither to a general failure of host defence mechanisms, nor to the large number of organisms present; nor to poor diffusion of penicillin into the infected muscle, which contained up to 10,000 times the smallest concentration effective *in vitro*. The onset of penicillin resistance coincided with the time at which the rate of increase of the viable count began to fall, just in the same way that a phenotypically resistant fraction appears in broth cultures as the stationary phase approaches (Mitchison & Selkon, 1956; Meynell, 1958). Eagle's conclusion that bacterial survival in this case is due to phenotypic resistance is probably correct. Several other authors have shown that genotypically sensitive organisms can survive *in vivo* during antibacterial therapy of experimental infections (Mitchison & Selkon, 1956; McCune & Tompsett, 1956; Fukai & Ishizuka, 1958). During infection of mice by *Salmonella enteriditis*, the onset of phenotypic resistance coincided with a diminution in size of the organisms as determined by differential centrifugation (Mitsuhashi *et al.* 1959).

In all these examples, phenotypic resistance is believed to arise because the bacteria cease to grow. A different form of phenotypic resistance was described by Henderson, Peacock & Belton (1956) in their study of penicillin prophylaxis in monkeys infected by spores of *Bacillus anthracis* by inhalation. Penicillin kills the vegetative organisms, not the spores, of this species. However, penicillin prophylaxis is generally unsuccessful even when continued for 20 days after challenge, for many animals later die from the growth of penicillin-sensitive organisms. The reason is that some retained spores may not germinate until 100 days after infection so that some organisms, being sporulated, will survive a course of penicillin given for any shorter time and will later germinate to cause a fatal infection.

EXOTOXIN-FORMING ORGANISMS

Only three organisms whose exotoxins have been shown to determine pathogenicity are considered: *Corynebacterium diphtheriae* (van Heyningen, 1955) and the two anaerobes, *Clostridium tetani* and *C. welchii* (van Heyningen, 1955; Macfarlane, 1943, 1955).

Corynebacterium diphtheriae

There are no data concerning variation in phenotype and the initiation of infection. However, toxin production is known to be extremely dependent on the growth medium, particularly its iron content. The cells neither liberate nor contain toxin during exponential growth while iron is available (Barksdale, 1959), but when growth stops with exhaustion of the available iron, bacterial lysis occurs with liberation of toxin (Pappenheimer, 1955). All potentially toxigenic strains of *C. diphtheriae* are thought to be lysogenized by temperate phage whose vegetative growth can be induced, resulting in bacterial lysis and release of toxin (Barksdale, 1959). Barksdale suggests that the low concentration of iron optimal for toxin production may act by allowing accumulation of substances inducing phage development and incidentally, toxin production.

The optimal iron concentration *in vitro* is *c.* 0·14 μg./ml., whereas the iron concentration encountered by the organism in a diphtheritic membrane may be much higher. Mueller (1941) found 43 μg. iron/g. moist weight in one membrane and suggested that the differing severity of the disease produced by the three types of diphtheria bacillus, *gravis*, *intermedius* and *mitis*, might be due to their differing abilities to produce toxin in media whose iron content is above optimal. However, although Mueller obtained confirmatory results in tests on four strains grown for 8 days in media containing up to 4 μg. iron/ml., Zinnemann (1943), using sixteen strains of each type grown for 24 hr. in broth with 2–7 μg. iron/ml., found no correlation between yield of toxin and bacterial type.

Pappenheimer (1955) tentatively suggested that the three types differ, not in toxin yield per cell, but in their growth rates; since fast-growing strains would exhaust the available iron and become toxigenic sooner than if growth occurred slowly. Another factor may be the number of cells that can be produced from the amount of iron available *in vivo*; the classical toxigenic strain, P.W. 8, owes its unusual toxigenic power to its ability to continue multiplying at iron concentrations too low to support growth of the majority of strains (Pappenheimer, 1955).

Clostridium tetani

Virtually all cases of clinical tetanus are thought to arise from wound infection, not by vegetative organisms, but by spores which are not toxigenic. Spore germination *in vitro* is dependent on the degree of anaerobiosis of the medium, being delayed indefinitely when the E_h is more than +0·11 V. at pH 7·15–7·65. The E_h also affects the rapidity of germination which takes, for example, 8 hr. when the E_h is +0·10 V., but only 4 hr. at an E_h of −0·05 V. (Knight & Fildes, 1930). These findings account for the observation that washed tetanus spores injected subcutaneously into a normal guinea-pig remain viable for months but do not cause tetanus (Vaillard & Rouget, 1892), the reason being that germination cannot occur in normal subcutaneous tissue because its E_h (*c.* +0·12 V.) is too high (Fildes, 1929). However, tetanus frequently occurs when the tissues are either damaged or treated with various adjuvants (reviewed by Wilson & Miles, 1955). One of these, $CaCl_2$, is known to reduce the tissue E_h to that of methylene blue (−0·05 V.) which is low enough for germination to occur (Fildes, 1929). Tetanus will follow even when spores are injected 7 days before the adjuvant or if both are given simultaneously at different sites (Bullock & Cramer, 1919).

Multiplication of vegetative organisms also requires anaerobiosis. Miller, Eaton & Gray (1959) have shown that subcutaneous injection of washed bacilli into normal mice produces tetanic symptoms solely by virtue of preformed intracellular toxin and that no more toxin is formed *in vivo*. However, even multiplying organisms may not be toxigenic because exotoxin is an intracellular product formed, but not secreted, by living cells which is only liberated by cell lysis (Mueller & Miller, 1948; Raynaud, 1951; Miller *et al.* 1959). Furthermore, multiplication of potentially toxigenic cells can occur without formation of intracellular toxin if the medium lacks certain compounds, notably histidine peptides (Mueller & Miller, 1956). Toxigenesis *in vivo* therefore presumably only begins when the environment causes the cells to lyse following multiplication on suitable substrates.

Clostridium welchii

The exotoxin concerned is the α-toxin or lecithinase (Macfarlane, 1943, 1955). The factors determining the occurrence of gas-gangrene following wound infection are similar to those involved in tetanus. For example, infection is again initiated by spores whose germination is dependent on an E_h less than $+0{\cdot}074$ V. As in tetanus, the LD 50 of spores injected subcutaneously in mice is greatly reduced by the injection of adjuvants, such as $CaCl_2$ (Bullock & Cramer, 1919) or adrenaline (Evans, Miles & Niven, 1948; see Discussion). Multiplication of vegetative cells is controlled by E_h and the highest E_h value compatible with growth is markedly affected by pH, the highest value being $+0{\cdot}16$ V. at pH 6·5 (Hanke & Bailey, 1945). This figure of $+0{\cdot}16$ V. is higher than the E_h of normal guinea-pig tissue but spores presumably cannot germinate because the pH is too high. Oakley (1954, fig. 2) points out that tissue damage may potentiate the virulence of this organism, not only by lowering E_h, but also by reducing tissue pH from the normal value of 7·2, where the highest E_h compatible with multiplication is $+0{\cdot}04$ V., to pH 6·6 where the maximum is $+0{\cdot}12$ V.

The yield of free toxin/unit weight of organism bears a different relation to growth phase than with *Clostridium tetani*, being greatest at the end of the exponential phase and staying constant thereafter (Gale & van Heyningen, 1942). As usual, the toxin yield is affected by the medium (see, for example, Adams & Hendee, 1945) and it may be of significance in the aetiology of gas gangrene that substances promoting toxin production can be extracted from muscle (Macfarlane & Knight, 1941).

DISCUSSION

Phenotypic variation intervenes in most of the processes believed to control the growth of bacteria *in vivo* and may have obvious importance in the preparation of vaccines (Bazeley, 1940; Stamp, 1953; Magheru *et al.* 1957) and other bacterial products (reviewed by Smith, 1958). It is also clear that phenotypic changes may explain many observations made on experimental infections. For example, the LD 50 of *Clostridium welchii* and *C. septicum* given to guinea-pigs by intramuscular injection is lowered at least 100,000-fold if 2 μg. adrenaline is simultaneously injected at the same site (Evans *et al.* 1948, table 4). This may act by delaying an inflammatory response long enough for the organisms to alter their phenotype in such a way that they can resist host antibacterial mechanisms. Another instance where delay may account for differences in LD 50 occurs when the LD 50 is estimated for the same culture given by either intravenous or intraperitoneal injection (Dutton, 1955, using stationary phase cultures). The LD 50 for many species given to mice by intravenous injection is far greater than after intraperitoneal injection. The smaller LD 50 by the intraperitoneal route can be ascribed to the cells having sufficient time during their stay in the peritoneal cavity before entering the blood to alter in phenotype so as to become resistant to destruction by the reticulo-endothelial system, to which cells inoculated intravenously are exposed immediately after injection.

Phenotypic heterogeneity in bacterial populations growing in a uniform environment

The numerous kinds of phenotypic variation open to micro-organisms suggest that the bacterial population in an infected host is likely to appear extremely heterogeneous, if only because the population is scattered amongst different sites, each providing a different environment. However, a bacterial population is often heterogeneous even when growing in an environment which appears uniform, such as a stirred, aerated broth culture.

For example, heterogeneity becomes apparent when a culture is exposed to either cold shock by sudden dilution in saline at 4° or streptomycin, both of which kill only growing cells. When the cells are dividing exponentially at their maximum rate and the viable count is relatively low, all the cells are killed, with the exception of negligible numbers of resistant mutants. On the other hand, no cells die in the stationary phase. Between the exponential and the stationary phases, the population is increasing in resistance but does so, not uniformly, but heterogeneously. This is evident when samples from a growing culture are exposed to the bactericidal agent at different times (Fig. 1). If the culture was always uniform, all the cells would die at the same rate and the survival curve of each sample would be exponential. The change from maximum susceptibility would then have to consist of a progressive fall in death rate as the culture neared the stationary phase so that each survival curve would have a smaller slope than its predecessor (Fig. 2). In practice, however, the survival curves are not exponential but more or less discontinuous (Fig. 3), indicating that the population is heterogeneous and consists of a sensitive and a resistant fraction, the size of the latter steadily increasing as the stationary phase approaches.

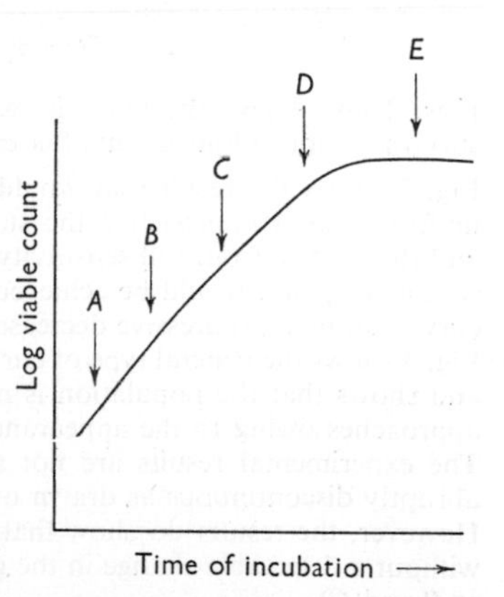

Fig. 1. Samples of the culture whose growth curve is shown in this figure are exposed at times *A*, *B*, *C*, *D* and *E* to a bactericidal agent, such as streptomycin, which kills only growing organisms.

Another example of heterogeneity in a population growing in a supposedly uniform environment is provided by Fraser's (1957) observation that as a culture nears the stationary phase, a growing fraction of cells lyse late after infection by phage T3 and yield an unusually high proportion of mutant phage particles.

Two probable examples of phenotypic heterogeneity affecting pathogenicity were met with. These are the detection of phenotypic resistance

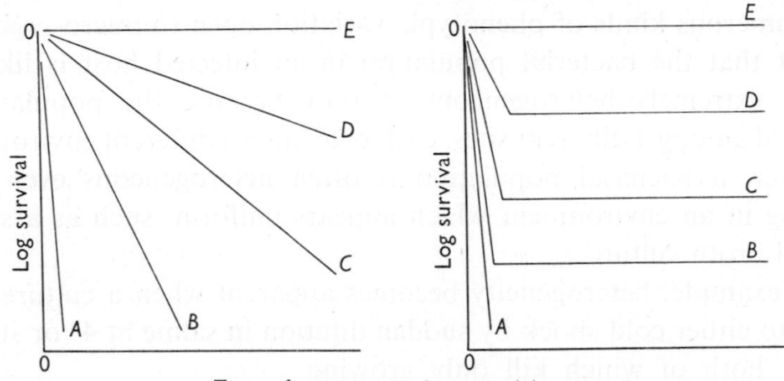

Figs. 2 and 3 give the possible extreme results of the experiment shown in Fig. 1. Log survival = log colony count after exposure for a given time minus log initial viable count. Fig. 2 shows the result that would be obtained if the population increased in resistance uniformly as it approached the stationary phase. Each curve is necessarily exponential and the change from full sensitivity in the logarithmic phase to complete resistance in the stationary phase would be achieved by a progressive decrease in the slope of each survival curve (i.e. by a progressive decrease in the killing rate constant).
Fig. 3 shows the general type of curve obtained with penicillin, streptomycin or cold shock, and shows that the population is not uniform but heterogeneous as the stationary phase approaches owing to the appearance of a resistant fraction whose size steadily increases. The experimental results are not sufficiently precise to show either that the curves are abruptly discontinuous as drawn or the exact slope of the descending part of each curve. However, the results do show that the size of the resistant fraction can change markedly without a detectable change in the growth rate of the culture as a whole (cf. samples taken at *B* and *C*).

to the bactericidal action of serum (Adler, 1953) and the behaviour of virulent strains of *Pasteurella pestis* exposed to phagocytosis by polymorphonuclear leucocytes. The organisms are uniformly susceptible to ingestion in the stationary phase and become most resistant during exponential growth although, even then, *c.* 5% of the cells remain susceptible (Burrows & Bacon, 1956*a*). The susceptible fraction may consist of dead cells, as the authors suggest, but it could also consist of cells of stationary phase phenotype persisting in the growing culture.

In all the above examples, the proportion of resistant organisms can change very rapidly (e.g. 100-fold in one generation time). But phenotypic heterogeneity can also arise when stable cell components are either formed or lost during growth in a uniform environment. Examples occur during the formation of β-galactosidase induced under certain conditions (Benzer, 1953, fig. 6C; Novick & Weiner, 1957), and during the segregation of enzymically active particles (Campbell & Spiegelman, 1956) and motility conferring particles (Quadling, 1958) that occurs during multiplication of organisms in environments in which such particles are not formed. In all cases, the cell phenotype may take several generations to become approximately stable (Rogers, 1957): the kinetics of these changes are discussed in detail by Pollock (1958, 1960). There is no reason to think that cell characters concerned in pathogenicity cannot behave in similar ways but unfortunately, so far, none seem to have been studied from this point of view.

The significance of phenotypic heterogeneity in infection

Phenotypic heterogeneity may partly account for one of the general problems of immunity: namely, the cause of fatal infections in partially resistant hosts. The LD 50 for such hosts may have any value between 0·7 organisms, the LD 50 for fully sensitive hosts (Meynell, 1957), and some much greater figure, such as 10^{10} for *Salmonella*, the number of organisms which can cause death simply by their toxin content without having to multiply *in vivo*. The problem is why mortality in partially resistant hosts rises as the number of viable organisms inoculated is increased between these limits.

One explanation is that advanced by Meynell & Stocker (1957), Meynell (1957) and by Meynell & Meynell (1958). This postulates in the case of a partially resistant host, that each inoculated organism has a probability between 0 and 1 ($1 > p > 0$) of multiplying to a lethal extent after inoculation. The value of this probability is equal, very approximately, to the reciprocal of the LD 50 (Meynell, 1957). For example, the probability for *Salmonella paratyphi* B given to mice by intraperitoneal injection is *c*. 10^{-7} since the LD 50 is 10^7. Conversely, there must be a probability of $(1-10^{-7})$ per inoculated organism of *not* multiplying to a lethal extent. Therefore, by postulating a probability between 0 and 1 for each inoculated organism of multiplying to a lethal extent, the hypothesis implies that the fates of individual organisms usually differ after inoculation. The hypothesis does not specify the cause of these differences but several possibilities can be distinguished.

In some systems, the bacterial population is genotypically heterogeneous in respect of virulence and contains a minority of mutants of greater virulence than the majority of organisms. Increase in the number of organisms inoculated will then increase mortality merely by increasing the chance that 1 or more such mutants will be inoculated into a given host. This situation can be detected by isolating the organisms from fatally infected hosts and by re-measuring the LD 50 after inoculation into fresh hosts, when it will be found to be smaller than for either the original culture or the organisms isolated from surviving hosts.

However, in many systems the bacterial population is genotypically homogeneous for, on re-measuring the LD 50, it is the same as that of the original culture. Examples are *Salmonella paratyphi* B and Vi-positive strains of *S. typhi* given to mice by intraperitoneal injection where the lowest observed LD 50's are 10^7 and 4×10^7 organisms respectively and are not reduced however many times the organisms are grown *in vivo*.

Thus, in these systems, the fates of individual bacteria *in vivo* must differ either because the organisms are phenotypically heterogeneous, despite their uniform genotype, or because the organisms are scattered amongst host sites differing in their ability to kill bacteria and to support bacterial multiplication. (An interesting possibility is that most sites are identical in these respects and that a 'favourable' site is one causing the phenotype to change so it can resist host antibacterial agents: the host would then be playing an active part in the aetiology of its own fatal infection.) In any event, the organisms isolated from fatally infected hosts would have the same LD 50 as the original culture because even if heterogeneity in phenotype, and not host sites, led to death, the phenotype of the organisms would revert to its initial state during growth *in vitro* before re-inoculation. In a system in which phenotypic heterogeneity was important, the LD 50 might be much smaller if the organisms were taken directly from one animal to the next, as in the micromanipulator experiments with *Bacillus anthracis* described by Webb *et al.* (1909).

Phenotypic heterogeneity existing before inoculation seems unlikely to be of much importance in determining the fates of inoculated organisms, considering the relatively small effect on the LD 50 produced by either the medium or the growth phase of inoculated organisms. Differences arising *in vivo* after inoculation are likely to be far more significant. In a partially resistant host, the organisms are more likely to have an avirulent than a virulent phenotype (or to be more likely to occupy an unfavourable than a favourable site). All the inoculated organisms are likely to succumb *in vivo* if only a small number are inoculated, for the odds of survival are against them; but if more and more organisms are inoculated, it becomes more and more likely that at least one inoculated organism and sufficient of its progeny will chance to form a virulent phenotype (or to occupy a favourable site) and so become able to multiply to a lethal extent. We do not know whether differences between sites or differences in phenotype are of most importance; but the examples given here show that the range of phenotypic variation open to bacterial pathogens is already known to be so great that it must be considered one of the major factors determining the course of infection.

REFERENCES

ADAMS, M. H. & HENDEE, E. D. (1945). Methods for the production of the alpha and theta toxins of *Clostridium welchii*. *J. Immunol.* **51**, 249.

ADLER, F. L. (1953). Studies on the bactericidal reaction. I. Bactericidal action of normal sera against a strain of *Salmonella typhosa*. *J. Immunol.* **70**, 69.

AMANO, T., INOUE, K., TANIGAWA, Y., MORIOKA, T. & UTSUMI, S. (1957). Studies on the immune bacteriolysis. XII. Properties of coccoid-form bacteria produced by immune bacteriolysis. *Med. J. Osaka Univ.* **7**, 819.

ANDO, K. & NAKAMURA, Y. (1951). Studies on the Vi antigen of typhoid bacilli. II. On a method of staining Vi antigen as a capsule. *Jap. J. exp. Med.* **21**, 41.

ARMITAGE, P. (1959). An examination of some experimental cancer data in the light of the one-hit theory of infectivity titrations. *J. nat. Cancer Inst.* **23**, 1313.

BAIN, R. V. S. (1960). Mechanism of immunity in haemorrhagic septicaemia. *Nature, Lond.* **186**, 734.

BARKSDALE, L. (1959). Symposium on the biology of cells modified by viruses or antigens. I. Lysogenic conversions in bacteria. *Bact. Rev.* **23**, 202.

BAZELEY, P. L. (1940). Studies with equine streptococci. 2. Experimental immunity to *Str. equi*. *Aust. vet. J.* **16**, 243.

BENSTED, H. J. (1940). Bacterium typhosum. The development of Vi antigen and Vi antibody. *J. Roy. Army med. Cps*, **74**, 19.

BENZER, S. (1953). Induced synthesis of enzymes in bacteria analysed at the cellular level. *Biochim. biophys. Acta*, **11**, 383.

BIGGER, J. W. (1944). Treatment of staphylococcal infections with penicillin by intermittent sterilization. *Lancet*, **2**, 497.

BRAUN, W. (1956). Cellular products affecting the establishment of bacteria of different virulence. *Ann. N.Y. Acad. Sci.* **66**, 348.

BULLOCK, W. E. & CRAMER, W. (1919). On a new factor in the mechanism of bacterial infection. *Proc. roy. Soc.* B, **90**, 513.

BURROWS, T. W. (1955). The basis of virulence for mice of *Pasteurella pestis*. In *Mechanisms of Microbial Pathogenicity*. *Symp. Soc. gen. Microbiol.* **5**, 152.

BURROWS, T. W. & BACON, G. A. (1956*a*). The basis of virulence in *Pasteurella pestis*: the development of resistance to phagocytosis *in vitro*. *Brit. J. exp. Path.* **37**, 286.

BURROWS, T. W. & BACON, G. A. (1956*b*). The basis of virulence in *Pasteurella pestis*: an antigen determining virulence. *Brit. J. exp. Path.* **37**, 481.

CAMPBELL, A. M. & SPIEGELMAN, S. (1956). The growth kinetics of elements necessary for galactozymase formation in 'long term adapting' yeasts. *C.R. Lab. Carlsburg* (Ser. Physiol.), **26**, 13.

CHU, H. P. (1952). Variation of *Bacillus anthracis* with special reference to the non-capsulated avirulent variant. *J. Hyg., Camb.* **50**, 433.

CRAIGIE, J. & BRANDON, K. F. (1936). The identification of the V and W forms of *B. typhosus* and the occurrence of the V form in cases of typhoid fever and in carriers. *J. Path. Bact.* **43**, 249.

DUTTON, A. A. C. (1955). The influence of the route of injection on lethal infections in mice. *Brit. J. exp. Path.* **36**, 128.

EAGLE, H. (1952). Experimental approach to the problem of treatment failure with penicillin. *Amer. J. Med.* **13**, 389.

ELLIOTT, S. D. (1945). A proteolytic enzyme produced by Group A streptococci with special reference to its effect on the type-specific M antigen. *J. exp. Med.* **81**, 573.

ENDERS, J. F., SHAFFER, M. F. & WU, C-J. (1936). Studies on natural immunity to pneumococcus Type III. III. Correlation of the behaviour *in vivo* of pneumococci Type III varying in their virulence for rabbits with certain differences observed *in vitro*. *J. exp. Med.* **64**, 307.

EVANS, D. G., MILES, A. A. & NIVEN, J. S. F. (1948). The enhancement of bacterial infections by adrenaline. *Brit. J. exp. Path.* **29**, 20.

FELIX, A. & PITT, R. M. (1951). The pathogenic and immunogenic activities of *Salmonella typhi* in relation to its antigenic constituents. *J. Hyg., Camb.* **49**, 92.

FELTY, A. R. & BLOOMFIELD, A. L. (1924). The relation of vegetative activity of bacteria to pathogenicity. *J. exp. Med.* **40**, 703.

FILDES, P. (1929). Tetanus. IX. The oxidation-reduction potential of the subcutaneous tissue fluid of the guinea-pig; its effect on infection. *Brit. J. exp. Path.* **10**, 197.

FOLEY, M. J., SMITH, M. R. & WOOD, W. B. (1959). Studies on the pathogenicity of group A streptococci. I. Its relation to surface phagocytosis. *J. exp. Med.* **110**, 603.

FOX, E. N. & KRAMPITZ, L. O. (1954). Synthesis of M antigen protein in non-proliferating group A haemolytic streptococci. *Bact. Proc.* p. 66.

FRASER, D. K. (1957). Host range mutants and semitemperate mutants of bacteriophage T3. *Virology*, **3**, 527.

FUKAI, K. & ISHIZUKA, H. (1958). Studies on the experimental typhoid. IV. The status of the presence of bacteria in the organs of infected mice with *S. enteritidis*. *Jap. J. Bact.* **13**, 704.

GALE, E. F. & VAN HEYNINGEN, W. E. (1942). The effect of the pH and the presence of glucose during growth on the production of α and θ toxins and hyaluronidase by *Clostridium welchii*. *Biochem. J.* **36**, 624.

GARROD, L. P. (1948). The bactericidal action of streptomycin. *Brit. med. J.* **1**, 382.

GLADSTONE, G. P. & JOHNSTON, H. H. (1955). The effect of cultural conditions on the susceptibility of *Bacillus anthracis* to lysozyme. *Brit. J. exp. Path.* **36**, 363.

GRAY, D. F. (1946). Some factors influencing the virulence of *Haemophilus pertussis* phase 1. *Aust. J. exp. Biol. med. Sci.* **24**, 301.

HAIGHT, T. H. & FINLAND, M. (1952). Observations on mode of action of erythromycin. *Proc. Soc. exp. Biol., N.Y.* **81**, 188.

HANKE, M. F. & BAILEY, J. H. (1945). Oxidation-reduction potential requirements of *Cl. welchii* and other Clostridia. *Proc. Soc. exp. Biol., N.Y.* **59**, 163.

HARE, R. (1931). Studies on immunity to haemolytic streptococci. III. Observations on the variations in resistance to phagocytosis displayed by broth cultures of strains of high and low virulence. *Brit. J. exp. Path.* **12**, 261.

HENDERSON, D. W., PEACOCK, S. & BELTON, F. C. (1956). Observations on the prophylaxis of experimental pulmonary anthrax in the monkey. *J. Hyg., Camb.* **54**, 28.

HIRSCH, J. G. (1956). Phagocytin: a bactericidal substance from polymorphonuclear leucocytes. *J. exp. Med.* **103**, 589.

HIRSCH, J. G. (1958). Bactericidal action of histone. *J. exp. Med.* **108**, 925.

HIRSCH, J. G. & CHURCH, A. B. (1960). Studies of phagocytosis of group A streptococci by polymorphonuclear leucocytes *in vitro*. *J. exp. Med.* **111**, 309.

HIRST, G. K. (1941). The effect of a polysaccharide-splitting enzyme on streptococcal infection. *J. exp. Med.* **73**, 493.

HOBBY, G. L., MEYER, K. & CHAFFEE, E. (1942). Observations on the mechanism of action of penicillin. *Proc. Soc. exp. Biol., N.Y.* **50**, 281.

KNIGHT, B. C. J. G. & FILDES, P. (1930). Oxidation-reduction studies in relation to bacterial growth. III. The positive limit of oxidation-reduction potential required for the germination of *B. tetani* spores *in vitro*. *Biochem. J.* **24**, 1496.

LACEY, B. W. (1960). Antigenic modulation of *Bordetella pertussis*. *J. Hyg., Camb.* **58**, 57.

LANCEFIELD, R. C. (1943). Studies on the antigenic composition of Group A hemolytic streptococci. I. Effects of proteolytic enzymes on streptococcal cells. *J. exp. Med.* **78**, 465.

LANCEFIELD, R. C. (1954). Cellular constituents of group A streptococci concerned in antigenicity and virulence. In *Streptococcal Infections*. Edited by M. McCarty. New York: Columbia University Press.

LINCOLN, R. E., ZELLE, M. R., RANDLES, C. I., ROBERTS, J. L. & YOUNG, G. A. (1946). Respiratory pathogenicity of *Bacillus anthracis* spores. III. Changes in pathogenicity due to nutritional modifications. *J. infect. Dis.* **79**, 254.

LITWACK, G. & PRAMER, D. (1956). Growth of *Micrococcus lysodeikticus* as substrate for lysozyme. *Proc. Soc. exp. Biol., N.Y.* **91**, 290.

LYONS, C. (1937). Antibacterial immunity to *Staphylococcus pyogenes*. *Brit. J. exp. Path.* **18**, 411.

MAALØE, O. (1948*a*). Pathogenic-apathogenic transformation of *Salmonella typhimurium*. I. Induced change of resistance to complement. *Acta path. microbiol. scand.* **25**, 414.

MAALØE, O. (1948*b*). Pathogenic-apathogenic transformation of *Salmonella typhimurium*. II. Induced change of resistance to complement (continued). *Acta path. microbiol. scand.* **25**, 755.

MACFARLANE, M. G. (1943). The therapeutic value of gas-gangrene antitoxin. *Brit. med. J.* **2**, 636.

MACFARLANE, M. G. (1955). On the biochemical mechanism of action of gas-gangrene toxins. In *Mechanisms of Microbial Pathogenicity*. *Symp. Soc. gen. Microbiol.* **5**, 57.

MACFARLANE, M. G. & KNIGHT, B. C. J. G. (1941). The biochemistry of bacterial toxins. 1. The lecithinase activity of *Cl. welchii* toxins. *Biochem. J.* **35**, 884.

MACLENNAN, A. P. (1956). The production of capsules, hyaluronic acid and hyaluronidase by 25 strains of group C streptococci. *J. gen. Microbiol.* **15**, 485.

MCDERMOTT, W. (1958). Microbial persistence. *Yale J. Biol. Med.* **30**, 257.

MCCUNE, R. M. & TOMPSETT, R. (1956). Fate of *Mycobacterium tuberculosis* in mouse tissues as determined by the microbial enumeration technique. I. The persistence of drug-susceptible tubercle bacilli in the tissues despite prolonged antimicrobial therapy. *J. exp. Med.* **104**, 737.

MAGHERU, G., MAGHERU, A., FAUR, Y., OLINICI, N., RIMNICEANU, I. & TIGOIU, V. (1957). Recherches concernant la préparation d'un vaccin contre la fièvre typhoide. *Arch. roum. Path. exp. Microbiol.* **16**, 517.

MENKIN, V. (1956). *Biochemical Mechanisms in Inflammation*, 2nd ed. Springfield, Ill.: Charles Thomas.

MEYNELL, G. G. (1957). Inherently low precision of infectivity titrations using a quantal response. *Biometrics*, **13**, 149.

MEYNELL, G. G. (1958). The effect of sudden chilling on *Escherichia coli*. *J. gen. Microbiol.* **19**, 380.

Meynell, G. G. & Meynell, E. W. (1958). The growth of micro-organisms *in vivo* with particular reference to the relation between dose and latent period. *J. Hyg., Camb.* **56**, 323.
Meynell, G. G. & Stocker, B. A. D. (1957). Some hypotheses on the aetiology of fatal infections in partially resistant hosts and their application to mice challenged with *Salmonella paratyphi*-B or *Salmonella typhimurium* by intraperitoneal injection. *J. gen. Microbiol.* **16**, 38.
Michael, J. G. & Braun, W. (1959*a*). Serum protoplasts of *Shigella dysenteriae*. *Proc. Soc. exp. Biol., N.Y.* **100**, 422.
Michael, J. G. & Braun, W. (1959*b*). Modification of bactericidal effects of human sera. *Proc. Soc. exp. Biol., N.Y.* **102**, 486.
Miller, P. A., Eaton, M. D. & Gray, C. T. (1959). Formation of tetanus toxin within the bacterial cell. *J. Bact.* **77**, 733.
Mitchison, D. A. & Selkon, J. B. (1956). The bactericidal activities of antituberculous drugs. *Amer. Rev. Tuberc.* **74**, 109.
Mitsuhashi, S., Kawakami, M., Goto, S., Yoshimura, T. & Hashimoto, H. (1959). Studies on the experimental typhoid. IV. The microorganisms in the organs of infected mice. *Jap. J. exp. Med.* **29**, 311.
Morris, M. C. (1943). The validity of the 'percentage law' in bactericidal reactions. *J. Immunol.* **47**, 359.
Morris, M. & Seastone, C. V. (1955). The relationship of M protein and resistance to phagocytosis in the beta hemolytic streptococci. *J. Bact.* **69**, 195.
Mueller, J. H. (1941). Toxin-production as related to the clinical severity of diphtheria. *J. Immunol.* **42**, 353.
Mueller, J. H. & Miller, P. A. (1948). Unidentified nutrients in tetanus toxin production. *J. Bact.* **56**, 219.
Mueller, J. H. & Miller, P. A. (1956). Essential role of histidine peptides in tetanus toxin production. *J. biol. Chem.* **223**, 185.
Muschel, L. H., Carey, W. F. & Baron, L. S. (1959). Formation of bacterial protoplasts by serum components. *J. Immunol.* **82**, 38.
Myrvik, Q. N. & Weiser, R. S. (1955). Studies on antibacterial factors in mammalian tissues and fluids. I. A serum bactericidin for *Bacillus subtilis*. *J. Immunol.* **74**, 9.
Novick, A. & Weiner, M. (1957). Enzyme induction as an all-or-none phenomenon. *Proc. nat. Acad. Sci., Wash.* **43**, 553.
Oakley, C. L. (1954). Gas gangrene. *Brit. med. Bull.* **10**, 52.
Pappenheimer, A. M. (1955). The pathogenisis of diphtheria. In *Mechanisms of Microbial Pathogenicity*. *Symp. Soc. gen. Microbiol.* **5**, 40.
Pollock, M. R. (1958). Enzymic 'de-adaptation'; the stability of an acquired character on withdrawal of the external inducing stimulus. *Proc. roy. Soc.* B, **148**, 340.
Pollock, M. R. (1960). Drug resistance and mechanisms for its development. *Brit. med. Bull.* **16**, 16.
Quadling, C. (1958). The unilinear transmission of motility and its material basis in Salmonella. *J. gen. Microbiol.* **18**, 227.
Raynaud, M. (1951). Extraction de la toxine tétanique à partir des corps microbiens. *Ann. Inst. Pasteur*, **80**, 356.
Repaske, R. (1958). Lysis of gram-negative organisms and the role of versene. *Biochim. biophys. Acta*, **30**, 225.
Rogers, H. J. (1948). The complexity of the hyaluronidases produced by microorganisms. *Biochem. J.* **42**, 633.
Rogers, H. J. (1957). The preferential suppression of hyaluronidase formation in cultures of *Staphylococcus aureus*. *J. gen. Microbiol.* **16**, 22.

ROTHBARD, S. (1948). Protective effect of hyaluronidase and type-specific anti-M serum on experimental group A streptococcus infections in mice. *J. exp. Med.* **88**, 325.

ROWLEY, D. & WARDLAW, A. C. (1958). Lysis of Gram-negative bacteria by serum. *J. gen. Microbiol.* **18**, 529.

SCHAEFER, W. B. (1954). The effect of isoniazid on growing and resting tubercle bacilli. *Amer. Rev. Tuberc.* **69**, 125.

SEASTONE, C. V. (1934). Capsules in young cultures of *Streptococcus hemolyticus*. *J. Bact.* **28**, 481.

SEASTONE, C. V. (1939). The virulence of Group C hemolytic streptococci of animal origin. *J. exp. Med.* **70**, 361.

SHAFFER, M. F., ENDERS, J. F. & WU, C-J. (1936). Studies on natural immunity to pneumococcus type III. II. Certain distinguishing properties of two strains of pneumococcus type III varying in their virulence for rabbits, and the reappearance of these properties following R → S reconversion of their respective rough derivatives. *J. exp. Med.* **64**, 281.

SKARNES, R. C. & WATSON, D. W. (1957). Antimicrobial factors of normal tissues and fluids. *Bact. Rev.* **21**, 273.

SMITH, H. (1958). The use of bacteria grown *in vivo* for studies on the basis of their pathogenicity. *Annu. Rev. Microbiol.* **12**, 77.

SMITH, H., KEPPIE, J. & STANLEY, J. L. (1953). The chemical basis of the virulence of *Bacillus anthracis*. I. Properties of bacteria grown *in vivo* and preparation of extracts. *Brit. J. exp. Path.* **34**, 477.

STAMP, LORD (1953). Studies on O/R potential, pH and proteinase production in cultures of *Streptococcus pyogenes*, in relation to immunizing activity. *Brit. J. exp. Path.* **34**, 347.

STANDFAST, A. F. B. (1958). Some factors influencing the virulence for mice of *Bordetella pertussis* by the intracerebral route. *Immunology*, **1**, 123.

VAILLARD, L. & ROUGET, J. (1892). Contribution à l'étude du tétanos. *Ann. Inst. Pasteur*, **6**, 385.

VAN HEYNINGEN, W. E. (1955). The role of toxins in pathology. In *Mechanisms of Microbial Pathogenicity*. *Symp. Soc. gen. Microbiol.* **5**, 17.

WEBB, G. B., WILLIAMS, W. W. & BARBER, M. A. (1909). Immunity production by inoculation of increasing numbers of bacteria beginning with one living organism. *J. med. Res.* **20**, 1.

WILEY, G. G. & WILSON, A. T. (1956). The ability of Group A streptococci killed by heat or mercury arc irradiation to resist ingestion by phagocytes. *J. exp. Med.* **103**, 15.

WILSON, G. S. (1926). The relation between the age and the virulence of cultures of *B. aertrycke* (Mutton). *J. Hyg., Camb.* **25**, 142.

WILSON, G. S. & MILES, A. A. (1955). *Topley and Wilson's Principles of Bacteriology and Immunity*, 4th ed. London: Edward Arnold.

WOLFF, L. K. & JULIUS, H. W. (1939). Action du sulfanilamide *in vitro* et *in vivo*. *Ann. Inst. Pasteur*, **62**, 616.

WOOD, W. B. & SMITH, M. R. (1949). The inhibition of surface phagocytosis by the capsular 'slime layer' of pneumococcus type III. *J. exp. Med.* **90**, 85.

WOOD, W. B., SMITH, M. R. & WATSON, B. (1946). Studies on the mechanism of recovery in pneumococcal pneumonia. IV. The mechanism of phagocytosis in the absence of antibody. *J. exp. Med.* **84**, 387.

WRIGHT, H. D. (1927). Experimental pneumococcal septicaemia and anti-pneumococcal immunity. *J. Path. Bact.* **30**, 185.

ZINNEMANN, K. (1943). Toxin production by the three types of *C. diphtheriae*. *J. Path. Bact.* **55**, 275.

SOME ASPECTS OF THE INFLUENCE OF ENVIRONMENT ON THE RADIO-SENSITIVITY OF MICRO-ORGANISMS

F. J. DE SERRES

*Biology Division, Oak Ridge National Laboratory,**
Oak Ridge, Tennessee

One of the problems of great concern to the early radiologists was the fundamental problem of radiosensitivity. It was known that there was a wide range in the sensitivity of various cells of about $10^4 \times$ r., but little was known about the factors responsible for this wide range of variation. Since the beneficial results obtained with radiotherapy depended upon the differential susceptibility of tumor cells and normal cells, the most challenging problems to these investigators were to try to learn more about the factors responsible for this difference and to try to determine whether the radiosensitivity of particular cells might be subject to experimental control.

In some of the early experiments with X-irradiation, definite correlations with the degree of a particular effect were found with increased exposure time; however, in some cases, the same effects were reproducible only if the experimental conditions were duplicated precisely. In other words, many types of environmental variations appeared to have an important influence on the relative radiosensitivity of particular types of cells. Deliberate attempts to explore the influence of various biological and environmental factors showed that the relative radiosensitivity of certain cells could be influenced markedly. The experiments of Henshaw & Henshaw (1933) showed that changes in the shape of survival curves and the sensitivity of *Drosophila* eggs within the first 3 hr. after fertilization could be correlated with changes in the stages of development. Crabtree & Cramer (1933), working with mouse tumor cells, and Mottram (1935), studying the broad bean, found that sensitivity to ionizing radiation could be either increased by irradiating at ice-water temperatures instead of at room temperature, or decreased by irradiating under conditions of anaerobiosis at high temperatures. Zirkle (1936) showed that the effect of anaerobiosis could be mimicked in spores of the fern *Pteris longifolia* by irradiating in atmospheres of 20–100 % H_2S or CO_2. Other investigations (Stadler, 1928; Henshaw &

* Operated by Union Carbide Corporation for the U.S. Atomic Energy Commission.

Francis, 1935) showed that the radiosensitivity of seeds varied with water content, germinated seeds being much more sensitive to ionizing radiation than dry seeds. From these early experiments some of the unknown factors influencing relative radiosensitivity became more clearly defined, and it became apparent that more careful study should be made of the influence of such environmental factors as water content, oxygen tension, cultural conditions and temperature in addition to such biological factors as age, division stage and strain differences.

The advantages of micro-organisms for studies of this type were recognized by a number of investigators in the late thirties and, in the twenty years since the yeast *Torula cremoris* was used by Anderson & Turkowitz (1941) to investigate radiosensitivity in the presence and absence of atmospheric oxygen, a substantial body of information has been accumulated describing the conditions under which various factors exert their influence. The literature on the experimental modification of the effects of radiation is voluminous and a comprehensive review is beyond the scope and purpose of this paper. Excellent reviews of the work with micro-organisms, to mention only a few appearing since 1951, can be found in the chapters by Luria (1955), by Zelle & Hollaender (1955), and Pomper & Atwood (1955) in volume II of Hollaender's *Radiation Biology*, and in the chapters by Stapleton and by Conger in *Radiation Protection and Recovery* (1960), and in the reviews by Mortimer & Beam (1955), Kimball (1957), and Wood (1958). In what follows I have attempted a general and by no means complete review of some of the environmental and biological factors that alter the sensitivity of various groups of micro-organisms to ionizing radiations.

OXYGEN TENSION

The effect of oxygen removal by mechanical displacement

In accord with the early work on tumors and root tips, a marked reduction in radiosensitivity under conditions of anaerobiosis has been shown with a variety of micro-organisms. Anderson & Turkowitz (1941) showed that the radiosensitivity of the yeast *Torula cremoris* could be altered by bubbling suspensions of cells with hydrogen before irradiation. Despite the fact that the hydrogen was not oxygen-free, a dose-reduction factor (DRF) of from 2 to 3 was obtained with the use of relatively more anoxic environmental conditions. More detailed experiments by Hollaender, Stapleton & Martin (1951) on *Escherichia coli*, strain B/r, showed a pronounced effect on suspensions of cells when irradiated under anaerobic conditions. The oxygen tension was lowered by bubbling the suspension either with pure nitrogen, hydrogen, helium or carbon dioxide, and, although none of these treatments affected the survival of the unirradiated bacterial cells, a striking decrease in radio-sensitivity was obtained as compared with fully oxygenated cell suspensions. In these experiments at least three times the X-ray dose had to be used to bring the anaerobic suspensions to the same levels of survival that were obtained with aerobic suspensions. Protection under conditions of anoxia was also obtained with haploid strains of *Saccharomyces cerevisiae* by Birge & Tobias (1954), a DRF of *c.* 2 being obtained if cells were irradiated anaerobically rather than in air. The same twofold dose reduction was also found by Beam (1955) when irradiating dividing and interdivision cells of haploid, diploid and tetraploid strains of *S. cerevisiae*, showing that this effect is not influenced by division stage or ploidy. The experiments of Wood & Taylor (1957*a*) show that a greater resistance to the lethal effect is obtained with anaerobiosis even when the cells are frozen. The dose reduction factors for two different strains of *S. cerevisiae* were determined at $-9{\cdot}4°$, the cells being either supercooled or frozen. The DRF values vary somewhat, but of particular interest is the strain difference in the response to anoxic conditions. Whereas essentially the same dose in air is required to kill 90 % of the cells, the Berkeley strain is 2·26 times more resistant in the liquid phase under conditions of anoxia whereas the Chicago strain is only 1·41 times more resistant. The observations on the protective effect of anoxia were extended to fungi in the experiments of Stapleton & Hollaender (1952) on *Aspergillus terreus*. They observed that the effect of irradiating spores under anoxic conditions were much more pronounced if the spores were irradiated in aqueous suspension rather than dry. Kølmark (1961*b*), in experiments on *Neurospora crassa*, found that if ice-cold suspensions of conidia were bubbled with nitrogen before irradiation with a dose of 48 kr., a survival of 73 % was obtained in contrast to 33 % survival if the ice-cold conidial suspensions were in equilibrium with air.

The effect of anaerobic growth conditions

The experiments of Hollaender *et al.* (1951) with *Escherichia coli*, strain B/r, showed that cells cultured under anaerobic conditions were more radioresistant than cells grown aerobically. Quite a pronounced effect was observed even though the cells were irradiated in oxygen-saturated buffer, a treatment that completely reverses the effect of anoxia produced by endogenous respiration (discussed in more detail below). However, no modification of either the radiation sensitivity or the shape of the survival curve was found with anaerobic growth conditions in the experiments of Birge & Tobias (1954) on *Saccharomyces cerevisiae*, even though anaerobiosis was maintained for periods up to several days prior to X-irradiation.

The effect of endogenous anoxia as related to cell concentration and temperature

The effect of variation in the concentration of cells/unit volume of suspending medium was reported by Hollaender *et al.* (1951) for *Escherichia coli*, strain B/r, and by Biagini (1955) for *E. coli*, strain B. It was found that cells became more sensitive with decreasing cell concentration. Biagini's interpretation of these results was that with low concentrations of cells (below 10^{-4} g./ml.), the increased effectiveness of the radiation was due to an increased importance of indirect effects produced by the free radicals resulting from the ionization of water. However, it soon became apparent to Hollaender's group (Hollaender & Stapleton, 1953; Zelle & Hollaender, 1955) that the protective effect of high cell concentrations was not a concentration effect *per se*, but rather a modification of radiosensitivity due to the development of a state of endogenous anoxia that developed in direct proportion to cell concentration/unit volume of medium during the period the suspensions were kept at room temperature before X-irradiation. In the experiments of Gunter & Kohn (1956*b*) the effect of variation in cell concentration on radiosensitivity was studied in more detail with several different micro-organisms. The resistance (DRF values varied from 2 to 4) conferred upon the bacterial cells with concentrations above 2×10^9/ml. was completely eliminated, however, by aerating the suspensions either by shaking or by bubbling with oxygen before irradiation. A marked alteration of levels of survival obtained with a constant dose of X-rays on conidial suspensions of *Neurospora crassa* was observed by Kølmark (1961*a*) with change in preincubation temperature. A DRF of 2·6 was observed with a dose of 48 kr., if suspensions of spores were incubated at 30° for 165 min. before irradiation as compared with suspensions kept at 0°. As in the experiments of Gunther & Kohn, the resistance conferred in this manner could be eliminated by bubbling with oxygen before irradiation showing that the resistance was due to the development of anoxia as a result of endogenous metabolism during the pretreatment period. Kølmark also showed that the development of anoxia by endogenous metabolism at 25° is dependent on cell concentration and time. Somewhat surprisingly such changes in resistance were also found in cells kept at ice-water temperature for 75 min. prior to irradiation. One might have expected endogenous metabolism to be virtually non-existent at this temperature, so that variation in cell concentration would be without effect. However, it was found that the survival levels obtained with preincubation at 25° can also be obtained by pre-incubation at ice-water temperatures if the concentrations of conidia per unit volume used are above 10^8/ml.

Not all changes in radiosensitivity with temperature variation can be attributed to the development of endogenous anoxia and temperature-dependent changes will be discussed in a later section.

The concentration of oxygen required for maximum radiosensitivity

Very careful studies have been made to determine the concentrations of oxygen required to bring about these marked changes in various organisms. In experiments of Burnett reported by Hollaender & Stapleton (1953), it was found that the survival of *Escherichia coli*, B/r, after 80 kr. changed from 10 to 0·1 % as the oxygen dissolved in the suspending buffer changed from 1 to 10 mg./l. No striking additional effect was observed in the range of oxygen concentrations from 10 to 100 mg./l. With the use of very dilute suspensions (6 to $7{\cdot}5 \times 10^3$ cells/ml.) to avoid any possible influence of endogenous respiration, and very sensitive methods of measuring oxygen concentration, Howard-Flanders & Alper (1957) found marked changes in radiosensitivity with partial pressures of oxygen of less than 1 %. Dose-survival curves obtained with *Shigella flexneri* showed that half-maximum radiosensitivity was obtained with very low oxygen concentrations of about 4 μM./l. With this organism, the variation in radiosensitivity with change in oxygen concentration was found to fit closely to the curve represented by the equation

$$\frac{S}{S_N} = \frac{m(O_2)+K}{(O_2)+K},$$

where O_2 = the concentration of dissolved oxygen, S_N = the radiosensitivity in the absence of oxygen, S = the radiosensitivity at the oxygen concentration O_2, K = a constant 4·0 μM./l., and m = a constant 2·92. The constant, m, is the ratio of the effectiveness of a given dose at high oxygen concentrations to that under conditions of anoxia, and the constant K, the concentration of oxygen required for half-maximum radiosensitivity. It has been a point of considerable interest to try to obtain m and K values for different types of cells to determine whether this marked increase in radiosensitivity with low concentrations is of widespread occurrence. Values for m of *c.* 3 and for K of from 4 to 6 have been obtained with X-rays for a number of different bacterial strains, plant tissues and various mammalian cells and tissues (Gray,

1959). These calculations have shown that the concentration of oxygen required to produce a marked change in radiosensitivity is very small and nearly the same in a variety of different organisms.

Oxygen removal by means of various chemical agents

The rationale behind many of the attempts to modify the effects of ionizing radiations has been to use chemicals that would bring about a reduction both in the level of intracellular dissolved oxygen and that dissolved in the suspending medium. This can be accomplished in a variety of different ways, e.g. with the use of chemicals with a high affinity for oxygen, or with the use of chemicals that will promote oxidative metabolism, etc. The former approach was used in the experiments of Burnett *et al.* (1952) on *Escherichia coli*, strain B/r. It was found that sodium hydrosulphite and various other related sodium salts of inorganic sulphur compounds which have the ability to remove dissolved oxygen from the suspending medium, bring about a marked increase in the survival after irradiation. The DRF of 3·7 with sodium hydrosulphite was somewhat higher than that obtained by bubbling with nitrogen, and it seems likely that the difference can be attributed either to the greater efficiency of oxygen removal with the chemical or to the reaction of sodium hydrosulphite with other cellular components that affect radiosensitivity. Protection against the lethal effects of X-irradiation was obtained when cell suspensions of *E. coli*, B/r, were pre-incubated with pyruvate, formate, succinate, serine or ethanol (Stapleton, Billen & Hollaender, 1952). Protection was not obtained with incubation at 2°, suggesting that metabolism is required for a protective effect of these compounds, the most likely mechanism of action being the removal of oxygen from the cell and suspending medium by oxidative metabolism. Protection may be afforded by other mechanisms, however, since ethanol at high concentrations was found to give protection without pre-incubation. Moreover, pyruvate has been shown to protect against radiation effects in *E. coli* without preincubation, even if the suspension was aerated during irradiation (Thompson, Mefferd & Wyss, 1951). The variety of mechanisms that have been suggested to account for the protection against radiation conferred by such well-known agents such as cysteamine have been discussed in detail by Kølmark (1961*a*).

Mechanism of the oxygen effect

The work of Alper & Howard-Flanders (1956), Howard-Flanders & Alper (1957), and Howard-Flanders & Moore (1958) has led to the modification of the direct action model for primary radiation damage developed by Lea (1947) (see Wood, 1958, pp. 344–350 for an excellent review). These papers stress the point that the oxygen effect is unlikely to result from primary damage to biological targets caused by oxidation products of water irradiated in the presence of oxygen. The alternative hypothesis developed by these authors to account for the oxygen effect postulates that the lethal effect of radiation results from chemical change in some vital target molecule in the cell in accord with the 'target theory' of Lea (1947). They assume that the cell will retain the capacity to multiply only if a certain number of vital molecules, R, remain intact within the cell. They visualize two different types of damaged molecules, R' and R''. The first is short-lived and highly reactive, and the fate of the cell is dependent on whether R' is restored to its original functional state or whether it reacts with oxygen to be converted to a form that is non-functional and lethal to the cell. The second kind of damaged molecule, R'', is assumed to be lethal, whether oxygen is present or not, in order to account for death after irradiation in the absence of oxygen. Howard-Flanders & Moore point out that essential features of the effect of oxygen on radiosensitivity can be explained by the assumption that R' results from the production of a carbon radical within the large molecule R. Calculations of the mean lifetime of R' were made, based on the diffusion rate of oxygen in bacterial cytoplasm, it being assumed that the diffusion rate of oxygen in bacterial cytoplasm is not different from that in water. Estimates of 10^{-4} sec. were derived as a lifetime for R' in cells free from oxygen and 10^{-6} sec. in cells equilibrated with air. These estimates are consistent with Howard-Flanders & Moore's (1958) finding that oxygen provided 5–10 msec. after irradiation does not increase the lethal effect above the level obtained under complete anoxia. It is especially noteworthy at this point that support for the modifiable direct action model has come only from indirect lines of evidence and from experiments with negative results. The indirect action model that attributes a certain portion of the radiation damage to free radicals produced by ionization within water molecules has certainly not been eliminated from further consideration either by the experiments under consideration or by any others.

CULTURAL CONDITIONS

The influence of growth medium variation on radiosensitivity

The effect of cultural conditions on X-ray sensitivity was also studied in the experiments of Hollaender *et al.* (1951) on *Escherichia coli*, strain B/r. A comparison was made between cells grown with and without glucose both aerobically and anaerobically and irradiated in oxygen-saturated buffer. The greatest sensitivity was shown by cells grown under aerobic conditions without glucose, whereas cells grown anaerobically without glucose or aerobically with glucose were more resistant in that order. The greatest resistance was shown by cells grown anaerobically with glucose. With an X-ray dose of 60 kr. the difference in survival of the most resistant and the most sensitive cells was of the order of 10^5. The nature of the resistance conferred by growth in the presence of glucose has been studied more extensively by Stapleton & Engel (1960). It was found that the resistance was unrelated to anaerobiosis, multicellularity, clumping or a multinuclear condition, and that the level of the resistance of such cells was directly related to the final pH of the culture in the stationary phase of growth. The development of radioresistance is probably not an effect of pH *per se*, since the pH of the suspending medium during irradiation does not alter the sensitivity of either sensitive or resistant cell populations. Moreover, cells cultured in a glucose medium kept at a constant pH (5·6) during the growth cycle, do not show the radioresistance characteristic of that pH. Cells only become radioresistant if the pH is allowed to decrease as a result of acid production in the course of metabolism of the substrate by the organism. The nature of the resistance is still unknown, but one possibility considered was that protection is afforded by some product of metabolism present in the cell in low concentration that acts as an internal protective agent. It is also of interest that the small differences found in the chemical composition of the resistant and sensitive cells with regard to DNA, RNA, protein or dry weight cannot account for the marked differences in radiosensitivity observed. The observation that resistant cells have about twice the cell volume of the sensitive cells is being investigated further (Engel, personal communication).

The influence of post-irradiation growth conditions

Another interesting phenomenon uncovered in the work on *Escherichia coli*, strain B/r, is the influence of post-irradiation plating medium on the recovery of irradiated cells reported by Stapleton, Sbarra & Hollaender (1955). It was found that broth-grown cells were more sensitive to the effects of irradiation when plated on synthetic medium than cells grown on synthetic medium. The explanation offered to account for this observation was that broth-grown cells lack certain enzymes and complex metabolic intermediates necessary for growth on synthetic medium. Because of this, radiation-induced inhibition of protein synthesis leads to unbalanced growth and results in cell inactivation. However, if broth-grown cells are plated instead on a broth medium, the supplements present in this medium help carry the cell through some temporary, but crucial, post-irradiation period and a larger proportion survive.

The effect of post-irradiation conditions on survival and/or mutation has been the subject of extensive study in the experiments of Witkin (1959), Alper & Gillies (1958, 1960) and Doudney & Haas (1958, 1959) on *Escherichia coli* and by Kimball, Gaither & Perdue (1960) on *Paramecium*. They have found that incubation of irradiated cells in the presence of metabolic inhibitors such as caffeine, chloramphenicol, iodoacetate and streptomycin considerably lessens the radiation effects. Such investigations have shown that only a portion of the initial damage is subject to post-irradiation modification and that there is a time limit (the 'terminal event' of Kimball) beyond which further modification is not possible. Whether factors that alter intracellular metabolism act by changing the time of the terminal event or by changing the mutational processes *per se* has not as yet been resolved.

Hollaender, Billen & Doudney (1955) and Hollaender & Stapleton (1956) have shown that post-irradiation treatment can have quite different effects on survival and mutation. The addition of certain factors (glutamic acid, guanine and uracil) to the synthetic medium promotes increased survival but a decrease in the rates of mutation. After a dose of 60 kr., plating the cells on the supplemented medium gave a 100-fold increase in the number of survivors but only a tenfold increase in the number of mutations.

TEMPERATURE

The effect of changes in temperature before and during irradiation

The lethal effects of X-irradiation were reported to be independent of temperature during irradiation in the range 0–37° (Lea, 1947), but since that time, a number of experiments have shown a marked influence of temperature in this range and in higher and lower ranges. Stapleton & Edington (1956) found no change in the sensitivity of aerobic cells (oxygen-saturated) or anaerobic cells (sodium hydrosulphite-treated) of *Escherichia coli*, strain B/r, in the range from 0° to 40°. However, in the range from 0° to about −196°, a striking reduction of about threefold was found on going from the liquid (0°) to the solid state (−22°) with a further twofold reduction in sensitivity on going from −22° to −196°. These changes in radiosensitivity were found to be oxygen-dependent. If oxygen was removed from the suspensions either by nitrogen bubbling or sodium hydrosulphite treatment, the change in sensitivity at the freezing-point was considerably less and no significant change in sensitivity occurred with decreasing temperature below the freezing-point. Wood (1954) also found no temperature-dependent change in radiosensitivity with *Saccharomyces cerevisiae* in the range 0° to 40°. However, a sixfold change in sensitivity was found in the range 45–55°. More detailed experiments on yeast by Wood & Taylor (1957*b*) showed that the relative radiosensitivity in the low temperature range is very much dependent on the rate of freezing. For fast frozen yeast, radiosensitivity decreased by a factor of 2·5 with a decrease in temperature from 0° to −33°, whereas from −33° to −72° the radiosensitivity was found to increase to a value typical of 0°. However, if the yeast is cooled slowly to −72°, the relative radiosensitivity found was that characteristic of −33°. This peculiar effect was found to be correlated with the level of intracellular dehydration. With slow freezing the cell is dehydrated by the freezing process, whereas with rapid freezing the radiobiologically significant water is frozen within the cell. To account for these differences in sensitivity found with dehydration, it was assumed that the free radicals produced by irradiation of water are trapped within the ice crystals. Upon thawing, if the frozen cellular water is largely exterior to the cell, the free radicals will be diluted by the suspending medium and the probability of their effective interaction with sensitive sites in the cell will be much lower than if the cellular water is frozen within the cell.

The modification of the radiosensitivity of dry spores of *Bacillus megaterium* irradiated under anaerobic conditions was studied over a wide range of temperatures by Powers, Webb & Ehret (1959). No temperature-dependent change in radiosensitivity was found with temperatures between −268° and −145°. At temperatures higher than this, the relative sensitivity increases, with constant slope. No marked change in sensitivity was found at 0°, as in the work on yeast and *Escherichia coli* cited above, and a steady increase in sensitivity was found with a peak at 36°. In the range 40–80° there was a marked drop in sensitivity to 55 % of the peak value, to a point lower than that found with the lowest temperatures tested. Beyond 80° the sensitivity began to increase again with rising temperatures up to 100°. It was also found that spores heated for 20 min. at 80° and then returned to 20°, have the sensitivity characteristic of 20°. This type of reversibility indicates that the resistance of these cells when irradiated at high temperatures is not due to some type of permanent change, or to a loss of residual oxygen or moisture.

The effect of changes in post-irradiation temperature

Experiments on the effect of post-irradiation temperature on the survival of X-irradiated cells have shown that the effects vary greatly depending on the strains and organisms used. In the experiments of Stapleton, Billen & Hollaender (1953) three different strains of *Escherichia coli* were irradiated and held at various temperatures for 24 hr. on nutrient agar and at 37° for 24 hr. before counts were made. Although the variation in post-incubation temperature had no effect on viability of unirradiated cells, a marked dose-dependent effect was found on survival of irradiated cells. For *E. coli*, strain B/r, maximum survival with a dose of 80 kr. was obtained if cells were incubated at 18°, and much lower levels of survival were obtained with temperature variation up to 42° and down to 6°. The optimum recovery temperature and the shape of the survival–temperature curve was found to be strain-dependent. Of particular interest in these experiments was the finding that the relative radiosensitivities of these strains of *E. coli* were essentially the same with some constant X-ray dose when each strain was incubated at its own optimum recovery temperature.

Gunter & Kohn (1958) have performed similar experiments on the effect of post-incubation temperature on the survival of various bacterial and yeast strains (incubated at various temperatures from 6° to 37° for 24 hr. and then at 30° for 24–48 hr. before counts were made). Using the comparison, LD37* at the test temperature/LD37 at 30°, the following ranges of ratios were found: *Pseudomonas fluorescens*, 1·00–1·22; *S. cerevisiae*, 0·98–1·08; *Escherichia coli*, strain W-1485, 0·80–1·00; *E. coli*, strain B/r, 0·91–1·03. It seems likely that a more striking demonstration of the effect of post-irradiation temperature might have been possible with these organisms if they were made on the survivors of much larger doses of irradiation. In the earlier experiments by Stapleton and his associates, the most striking effect was observed with the high doses (80 kr.) that give 0·1–0·001 % survival depending on the post-irradiation temperature chosen.

WATER CONTENT

The role of water content on the modification of the effects of ionizing radiation has been studied by Stapleton & Hollaender (1952) in *Aspergillus terreus*. A comparison was made between the relative radiosensitivities of (1) desiccated spores (about 25 % water by weight),

* LD37, LD50, etc., are the doses killing 37 %, 50 % of the cells.

(2) dry spores (about 42 % water by weight), and (3) spores in water suspension (about 80 % water by weight). A direct relationship was found between the X-ray sensitivity and water content. The doses killing 99 % of cells were 126, 74 and 52 kr., respectively. In these same experiments the water content of spores was shown to influence the magnitude of the oxygen effect. With a dose of 80 kr. the ratio of survivors in nitrogen survivors in oxygen for dry spores was 7·4 in contrast to 25·3 for spores in suspension.

Changes in radiosensitivity with alteration of water content were also studied by Schwinghamer (1958) with urediospores of the flax rust, *Melampsora lini*. The sensitivity was found to be constant up to 52 % relative humidity with a twofold increase at 98 % humidity where the urediospores contained about 70 % water by weight.

In the experiments of Rosenberg (reported in Wood, 1959) very high concentrations of various chemicals (methanol, ethanol, glycerol, glucose, etc.) were used to dehydrate the cells of *Saccharomyces cerevisiae* and bring about changes in the relative water content. A marked progressive decrease in radiosensitivity was found with increase in glycerol concentration; a DRF of 4 was obtained with 6·9 M glycerol, a concentration found non-injurious to the cell. As in the experiments with *Aspergillus terreus*, additional protection was afforded by anoxic conditions; a combination of 6·9 M glycerol and anoxia gave a maximum DRF of about 5.

BIOLOGICAL FACTORS

In the previous sections, it has been shown that alterations in radiosensitivity can be brought about by various environmental factors that modify the intracellular environment indirectly. However, there are a number of biological factors that modify the intracellular environment more directly, and some of these also have been found to bring about marked changes in the radiosensitivity of certain groups of cells.

Alteration in radiosensitivity correlated with changes in division stage

In the experiments of Stapleton (1955) the variation in X-ray sensitivity of *Escherichia coli*, strain B/r, was correlated with various phases of the growth cycle. Stationary phase cells were inoculated into broth and the changes in the fraction surviving at 30 kr. were determined with increasing age of the culture for a period of 12 hr. Cells in the lag phase were characterized by an increase in resistance to X-rays (from about 0·8 to 10 % survival); cells in the phase of logarithmic growth showed a slow but steady decay in resistance with increasing time (from about 10 %

to 0·04 % survival) reaching a state of maximum sensitivity after 8 hr. at the end of this phase; the stationary phase was distinguished by a gradual return to the initial sensitivity (from about 0·04 to 0·8 % survival) displayed by cells after 24 hr. incubation at 37°. A DRF of about 3 was obtained regardless of the age of the culture when a comparison was made of oxygen- and nitrogen-saturated suspensions showing that the alteration in sensitivity was not due to changes in the concentration of oxygen during the course of growth.

A similar effect of division stage on radiosensitivity was reported for *Saccharomyces cerevisiae* by Beam *et al.* (1954). The survival curve for a haploid strain was found to have two distinct components; the first with an LD50 of 3100 r. and consisting of 90 % of the cells, and the second with an LD50 of 60,000 r. consisting of the remaining 10 % of the cells. The radioresistant component was found to be correlated with the survival of budding, not resting, yeast cells by direct microscopic examination. Starvation experiments confirmed these observations. When cells were incubated in a buffered dextrose medium in the absence of a source of nitrogen, the frequency of budding cells is very much reduced and this resulted in an essentially homogeneous population (99·999 %) with reference to radioresistance. However, when fresh nutrients were added, the onset of fresh budding gave approximate synchrony in division (about 60%) and the more radioresistant component of the survival curve reappeared. These observations were extended in the experiments of Beam (1955) and the radioresistant component was demonstrated in the survival curves of various haploid, diploid and tetraploid strains. Furthermore, it was shown that the radioresistance of the budding cells in all of these strains was not due to the development of a condition of endogenous anoxia, since the same DRF values of about 2 were obtained for both budding and resting cells irradiated in the presence and absence of oxygen. The nature of the greater resistance of the budding yeast cell is still unknown (Beam, 1959), but the hypothesis put forth to account for this behaviour proposes that the pronounced resistance is due to a nuclear rather than a cytoplasmic change. The various alternatives considered for this change in the sensitivity of the nucleus to radiation damage are (1) that an alteration in radioresistance might be brought about with change in the division stage during mitosis, (2) that the processes involved in genetic replication may also serve for repair of X-ray damage, or (3) because the genetic specificity usually residing in DNA undergoes a temporary transposition during duplication to a system of different susceptibilities.

The effect of changes in ploidy on radiosensitivity

Studies on the relative radiosensitivities of haploid and diploid strains of *Saccharomyces cerevisiae* (Latarjet & Ephrussi, 1949) showed marked differences in the shape of the survival curves and in resistance. An exponential type curve was found for the haploid strain and a sigmoidal curve for the more radioresistant diploid strain. Cell death and the difference in sensitivity in the haploid and diploid strains were attributed primarily to recessive lethal damage. Damage of this type can kill the haploid cell since only one chromosome is present, and could kill the diploid cell only if lethals were paired, i.e. were induced in both alleles of a homologous pair of chromosomes.

More extensive experimentation to determine the basis of the differential sensitivity of haploid and diploid strains of yeast (Mortimer, 1955) showed that death of haploid cells cannot be explained on the basis of recessive lethal damage alone. Mortimer found that many haploid cells receive damage which is lethal in the haploid cells but which is non-lethal to the zygotes formed when such cells are mated to a non-irradiated haploid cell of the opposite mating type. The type of damage 'covered up' by the presence of normal alleles in the zygote corresponds to recessive lethal mutation, whereas the damage expressed in the zygote corresponds to dominant lethal mutation. At 50 % survival, the relative rates of recessive lethal mutation to dominant lethal mutation was about 15:1.

Further evidence to determine the nature of the radiation damage leading to zygote inviability in yeast was obtained by irradiating haploid or diploid cells (homozygous for mating type) with haploid cells of opposite mating type to determine the relative frequencies of dominant lethals in these two types of cells (Owen & Mortimer, 1956). The yield of dominant lethals was considerably higher in the diploid cells and was found to depend on the product of ploidy and dose. A DRF of about 2 was found between haploid and diploid dominant lethal curves.

On the model attributing cellular inactivation primarily to a pairing of recessive lethal damage, one would predict that increasing ploidy would lead to a progressive increase in radioresistance due to shielding of recessive lethals by normal alleles. However, if cellular inactivation were primarily due to dominant lethal damage, one would predict a decrease in radioresistance with increase in ploidy, with inactivation due to pairing of recessive lethals playing a minor role.

The effect of changes in ploidy on radiosensitivity was studied by Mortimer (1958) in *Saccharomyces cerevisiae*. By crosses between various haploid, diploid and tetraploid strains, a complete polyploid series was developed from haploid to hexaploid. An increase in resistance to X-ray inactivation was found going from haploid to diploid. In accord with the model attributing cellular inactivation to the production of dominant lethals, a progressive decrease in resistance was found with further increase in ploidy from diploid to hexaploid. The doses giving 90 % survival for diploid, tetraploid and hexaploid cultures under comparable conditions were 47·5, 30 and 16·8 k.rad. From these studies it appears that recessive lethal mutations are more efficient in killing haploid cells whereas dominant lethals are more efficient in cells of higher ploidy.

Strain differences in radiosensitivity

One of the classic strain differences in radiosensitivity is that of the radiation-resistant mutant B/r of *Escherichia coli*, induced and analysed in the experiments of Witkin (1946). This strain, one of four radiation-resistant mutants recovered after ultraviolet irradiation of *E. coli*, strain B, with a dose of 1000 ergs/mm.^{2}, shows a marked heritable decreased sensitivity to both ultraviolet and X-rays. The LD99 was *c.* 16 kr. for strain B compared to 60 kr. for strain B/r; whereas with ultraviolet irradiation, a dose of 50 ergs/mm.^{2} gives 100 % survival of strain B/r but only 10 % survival for cells of strain B.

Studies on a radiation-resistant strain of *Escherichia coli*, strain B, were reported by Hill (1958). The new mutant strain was one of 22 survivors that appeared on a plate spread with $2{\cdot}2 \times 10^6$ cells irradiated with ultraviolet light. Subcultures were made of 12 of these and the survivals of each were compared after exposure to ultraviolet light. A radiosensitivity comparable to that of strain B was found with 10 isolates that gave about 1·2 % survival. Another isolate appeared to be like B/r since 18 % of the cells survived, and one that gave less than 0·01 % survival was designated B_s. A study of this mutant showed that the change was heritable and stable after repeated subculture in the same manner as B/r. Comparative survival curves obtained after 10 sec. ultraviolet light gave about 90 % survival for B/r, about 10 % survival for strain B and less than 0·001 % survival for B_s. Comparative survival curves obtained after X-irradiation showed that B_s is more sensitive than strain B or B/r, but that the differences are not as marked as with ultraviolet light.

The radiosensitivity of various strains of cultures labelled *Escherichia coli*, strain B/r, and cultures labelled *E. coli*, strain *B*, from various laboratories in the United States and France were compared by Adler & Haskins (1960) with rather surprising results. Their data show that cultures labelled *E. coli*, strain B/r, were a phenotypically and genotypically heterogeneous group, whereas all *E. coli*, strain B, cultures were relatively homogeneous. *E. coli*, strain B/r, cultures can be characterized by a greater resistance to ultraviolet light, X-rays and penicillin than strain B; by a difference in the rate of growth; and by the fact that they only occasionally produce the snake-like forms frequently found after ultraviolet irradiation of strain B. Three of the four B/r cultures tested had the X-ray sensitivity characteristic of strain B, but none of them produced the snake-like forms characteristic of strain B. Furthermore, two of the three that have the same X-ray sensitivity can be distinguished from one another on the basis of final optical density when grown in liquid medium and sensitivity to penicillin. The precise nature of the factors responsible for this variation in sensitivity of B/r cultures are not known, but it indicates a very important source of variation to be considered in the evaluation of work done in different laboratories on supposedly the same mutant form of strain B.

A systematic approach to the problem of variation in radiosensitivity of various diploid strains of *Saccharomyces cerevisiae* was taken in the experiments of Mortimer (1958). On the hypothesis that heterogeneity in the constitution of homologous chromosomes was an important source of variation, a comparison was made of the sensitivity of two different homozygous diploid strains and their diploid hybrid. It was found that both homozygous diploids were more radiosensitive (LD 10 *c.* 30 kr.) than the diploid hybrid (LD 10 *c.* 50 kr.) between them, in addition to a number of other non-related diploid hybrids. Mortimer accounts for these differences on the basis of a differential sensitivity to recessive lethal damage in various strains. He makes the assumption that particular chromosome regions are more sensitive to recessive lethal damage and that they are not in the same location on the chromosomes of different haploid strains. Thus the diploid hybrid of two unrelated haploid strains would be less sensitive to this type of damage than the homozygous diploids derived from either strain.

An apparent single gene difference that affects the shape of survival curves and the levels of survival obtained by X-irradiating at various times during meiosis was studied by Magni (1959). The shape of the survival curve was found to be correlated with the genotype of the diploid strain such that strains of genotype *rs/rs* gave a sigmoidal inactivation curve, strains of genotype *Rs/Rs* gave an exponential survival curve, whereas strains of genotype *Rs/rs* gave either one or other of the two types of curves.

There is a wide range in the sensitivity of various micro-organisms to X-irradiation and also considerable variation in the shape of the survival curves. In an attempt to evaluate these differences, Gunter & Kohn (1956*a*) obtained dose–survival curves on resting cells after X-irradiation under standardized and comparable conditions. The LD 50 values found were 0·95 kr. for *Pseudomonas fluorescens*, 1·84 kr. for *Rhodopseudomonas spheroides* and 1·9 kr. for a haploid strain of *Saccharomyces cerevisiae*. For all three of these genera, the percentage survival was a simple exponential function of dose. The survival curve of *Escherichia coli*, strain W-1485, was found to have two exponential components, about 66 % of the cells were more sensitive, with an LD 50 value of 1·2 kr. and about 34 % of the cells were more resistant with an LD 50 value of 3·5 kr. Subculture of either the sensitive or the resistant cells always gave cell populations with the same mixture of sensitivities. The basis for the differences in radiosensitivity of these two groups of cells remains unknown. The survival curves of *Azotobacter agile* and a diploid strain of *S. cerevisiae* were convex in shape and LD 50 values of 8·5 and 24 kr. were found respectively. Although the range in sensitivities observed in these experiments is not very large, strain differences in radiosensitivity seems firmly established.

An organism with extreme radioresistance has been studied by Kilburn, Bellamy & Terni (1958). This organism, a pigmented *Sarcina* species, identified as *Micrococcus rubens*, has an LD 37 up to 300,000 r. depending on the conditions of growth and irradiation. An attempt was made in these experiments to determine whether the extreme radioresistance was due to the presence of the pigment. The approach used was to try to alter the level of pigmentation, or to inhibit pigment production completely with variation in growth conditions or by the addition of certain respiration inhibitors, and to compare the sensitivity of such cells with the normally pigmented cells. The only change in the level of pigmentation obtained was that found under conditions of anaerobic growth, and a decrease in the radioresistance of such cells was obtained as compared with cells grown under aerobic conditions. The decrease in radioresistance was not large and no greater than might be accounted for on the basis of the differences obtained with *Escherichia coli* with anaerobic growth alone (see previous section, Hollaender *et al.* 1951). The nature of the radioresistance of this organism remains unknown. However, in this connexion, it would be of interest to compare the radiosensitivity of a variety of induced non-pigmented mutant strains of this organism with that of the original strain so that the experiment could be performed in the complete absence of pigment and some unknown number of possibly non-pigmented precursors.

GENERAL OBSERVATIONS AND CONCLUSIONS

The studies on the radiobiology of various micro-organisms have shown that marked changes in sensitivity can be obtained by any one of a number of different environmental or biological factors. One is particularly impressed by the extreme variability of the radiation response, whether it is measured in terms of the shape of survival curves or the fraction of cells surviving treatment. Equally impressive is the complete lack of experimental evidence for any organism that shows the same effect after irradiation with some constant dose regardless of how the treatment is given. The variability in response is often so great that in many cases comparable results can only be obtained by repeating experiments under *exactly* the same conditions. Such differences as changes in the age of the culture; the culture medium; the temperature before, during and after treatment; growth and plating media; pH, oxygen tension; and the timing of various operations are all important factors in determining the ultimate response.

A number of hypotheses have been suggested to account for the marked variation observed in radiosensitivity under certain experimental conditions. It seems most probable that one or a combination of these hypotheses will be able to bring these observations into an ordered picture in the future. However, at the present time a complete generalization of all of the facts is not possible.

REFERENCES

ADLER, H. I. & HASKINS, S. D. (1960). Heterogeneity of cultures of *Escherichia coli* B/r. *Nature, Lond.* **188**, 249.

ALPER, T. & GILLIES, N. E. (1958). 'Restoration' of *Escherichia coli* strain B after irradiation: its dependence on suboptimal growth conditions. *J. gen. Microbiol.* **18**, 461.

ALPER, T. & GILLIES, N. E. (1960). The relationship between growth and survival after irradiation of *Escherichia coli* strain B and two resistant mutants. *J. gen. Microbiol.* **22**, 113.

ALPER, T. & HOWARD-FLANDERS, P. (1956). Role of oxygen in modifying the radiosensitivity of *E. coli* B. *Nature, Lond.* **178**, 978.

ANDERSON, R. S. & TURKOWITZ, H. (1941). The experimental modification of the sensitivity of yeast to Roentgen rays. *Amer. J. Roentgenol.* **46**, 537.

BEAM, C. A. (1955). The influence of ploidy and division stage on the anoxic protection of *Saccharomyces cerevisiae* against X-ray inactivation. *Proc. nat. Acad. Sci., Wash.* **41**, 857.

BEAM, C. A. (1959). The influence of metabolism on some radiation effects. *Rad. Res., Suppl.* **1**, 372.

BEAM, C. A., MORTIMER, R. K., WOLFE, R. G. & TOBIAS, C. A. (1954). The relation of radioresistance to budding in *Saccharomyces cerevisiae*. *Arch. Biochem. Biophys.* **49**, 110.

BIAGINI, C. (1955). Studies on the mode of action of ionizing radiations. Influence of cell concentration on lethal effects of X-rays on *Escherichia coli*. *Arch. Biochem. Biophys.* **56**, 38.

BIRGE, A. C. & TOBIAS, C. A. (1954). Radiation sensitivity of yeast cells grown in aerobic and anaerobic environments. *Arch. Biochem. Biophys.* **52**, 388.

BURNETT, W. T., Jr., MORSE, M. L., BURKE, A. W., Jr. & HOLLAENDER, A. (1952). Reduction of X-ray sensitivity of *Escherichia coli* by sodium hydrosulfite and certain other inorganic sulfer compounds. *J. Bact.* **63**, 591.

CONGER, A. D. (1960). Genetic effects. In *Radiation Protection and Recovery*. Edited by A. Hollaender. New York: Pergamon Press.

CRABTREE, H. G. & CRAMER, W. (1933). The action of radium on cancer cells. II. Some factors determining the susceptibility of cancer cells to radium. *Proc. Roy. Soc.* B, **113**, 238.

DOUDNEY, C. O. & HAAS, F. L. (1958). Modification of ultraviolet-induced mutation frequency and survival in bacteria by post-irradiation treatment. *Proc. nat. Acad. Sci., Wash.* **44**, 390.

DOUDNEY, C. O. & HAAS, F. L. (1959). Mutation induction and macromolecular synthesis in bacteria. *Proc. nat. Acad. Sci., Wash.* **45**, 709.

GRAY, L. H. (1959). Cellular radiobiology. *Rad. Res., Suppl.* **1**, 73.

GUNTER, S. E. & KOHN, H. I. (1956*a*). The effect of X-rays on the survival of bacteria and yeast. I. A comparative study of the dose–survival curves of *Azotobacter agile, Escherichia coli, Pseudomonas fluorescens, Rhodopseudomonas spheroides* and *Saccharomyces cerevisiae* irradiated in the resting state. *J. Bact.* **71**, 571.

GUNTER, S. E. & KOHN, H. I. (1956*b*). The effect of X-rays on the survival of bacteria and yeast. II. Relation of cell concentration and endogenous respiration to sensitivity. *J. Bact.* **72**, 422.

GUNTER, S. E. & KOHN, H. I. (1958). Post-irradiation incubation temperature and X-ray sensitivity in micro-organisms. *Bact. Proc.* p. 34.

HENSHAW, P. S. & FRANCIS, D. S. (1935). A consideration of the biological factors influencing the radiosensitivity of cells. *J. cell. comp. Physiol.* **7**, 173.

HENSHAW, P. S. & HENSHAW, C. T. (1933). Changes in susceptibility of *Drosophila* eggs to X-rays. I. A correlation of changes in radiosensitivity with stages in development. *Radiology*, **21**, 239.

HILL, R. F. (1958). A radiation-sensitive mutant of *Escherichia coli*. *Biochim. biophys. Acta*, **30**, 636.

HOLLAENDER, A., BILLEN, D. & DOUDNEY, C. O. (1955). The modification of X-ray induced mutations in *Escherichia coli* by pre- and post-treatment. *Rad. Res.* **3**, 235.

HOLLAENDER, A. & STAPLETON, G. E. (1953). Fundamental aspects of radiation protection from a microbiological point of view. *Physiol. Rev.* **33**, 77.

HOLLAENDER, A. & STAPLETON, G. E. (1956). Studies on protection by treatment before and after exposure by X- and gamma radiation. In *Peaceful Uses of Atomic Energy. Proc. Intern. Conf. Geneva*, August 1955, Vol. 2, p. 106. New York: United Nations.

HOLLAENDER, A., STAPLETON, G. E. & MARTIN, F. L. (1951). X-ray sensitivity of *E. coli* as modified by oxygen tension. *Nature, Lond.* **167**, 103.

HOWARD-FLANDERS, P. & ALPER, T. (1957). The sensitivity of microorganisms to irradiation under controlled gas conditions. *Rad. Res.* **7**, 518.

HOWARD-FLANDERS, P. & MOORE, D. (1958). The time interval after pulsed irradiation within which injury to bacteria can be modified by dissolved oxygen. I. A search for an effect of oxygen 0·02 second after pulsed irradiation. *Rad. Res.* **9**, 422.

KILBURN, R. E., BELLAMY, W. D. & TERNI, S. A. (1958). Studies on a radiation-resistant pigmented *Sarcina* sp. *Rad. Res.* **9**, 207.

KIMBALL, R. F. (1957). Nongenetic effects of radiation on microorganisms. *Annu. Rev. Microbiol.* **11**, 199.

KIMBALL, R. F., GAITHER, N. & PERDUE, S. W. (1960). Metabolic repair of premutational damage in *Paramecium*. *Intern. J. Radiation Biol.* (in the Press.)

KØLMARK, G. (1961*a*). Protection with 2-mercaptoethylamine against the mutagenic and lethal effects of X-rays in *Neurospora crassa*. *J. gen. Microbiol.* (in the Press).

KØLMARK, G. (1961*b*). Protection with endogenous anoxia against the mutagenic and lethal effects of X-rays in *Neurospora crassa*. *J. gen. Microbiol.* (in the Press.)

LATARJET, R. & EPHRUSSI, B. (1949). Courbes de survie de levures haploides et diploides soumises aux rayons X. *C.R. Acad. Sci., Paris*, **229**, 306.

LEA, D. E. (1947). *Actions of Radiations on Living Cells*, 1st ed. Cambridge University Press.

LURIA, S. E. (1955). Radiation and viruses. In *Radiation Biology*. Edited by A. Hollaender, Vol. II, p. 333. New York: McGraw-Hill Book Co. Inc.

MAGNI, G. E. (1959). Genetic effects of radiation on yeast cells and genetic control of radiosensitivity. *Rad. Res., Suppl.* **1**, 347.

MORTIMER, R. K. (1955). Evidence for two types of X-ray-induced lethal damage in *Saccharomyces cerevisiae*. *Rad. Res.* **2**, 361.

MORTIMER, R. K. (1958). Radiobiological and genetic studies on a polyploid series (haploid to hexaploid) of *Saccharomyces cerevisiae*. *Rad. Res.* **9**, 312.

MORTIMER, R. K. & BEAM, C. A. (1955). Cellular radiobiology. *Annu. Rev. Nuclear Sci.* **5**, 327.

MOTTRAM, J. C. (1935). On the alteration in the sensitivity of cells towards radiation produced by cold and by anaerobiosis. *Brit. J. Radiol.*, N.S. **8**, 32.

OWEN, M. E. & MORTIMER, R. K. (1956). Dominant lethality induced by X-rays in haploid and diploid *Saccharomyces cerevisiae*. *Nature, Lond.* **177**, 625.

POMPER, S. & ATWOOD, K. C. (1955). Radiation studies on fungi. In *Radiation Biology*. Edited by A. Hollaender. Vol. II, p. 431. New York: McGraw-Hill Book Co., Inc.

POWERS, E. L., WEBB, R. B. & EHRET, C. F. (1959). Modification of sensitivity to radiation in single cells by physical means. *Progr. in Nuclear Energy*. Series VI, Vol. 2, 189. New York: Pergamon Press.

SCHWINGHAMER, E. A. (1958). The relation of survival to radiation dose in rust fungi. *Rad. Res.* **8**, 329.

STADLER, L. J. (1928). The rate of induced mutations in relation to dormancy, temperature, and dosage. *Anat. Rec.* **41**, 97.

STAPLETON, G. E. (1955). Variations in the sensitivity of *Escherichia coli* to ionizing radiations during the growth cycle. *J. Bact.* **70**, 357.

STAPLETON, G. E. (1960). Protection and recovery in bacteria and fungi. In *Radiation Protection and Recovery*. Edited by A. Hollaender. New York: Pergamon Press.

STAPLETON, G. E., BILLEN, D. & HOLLAENDER, A. (1952). The role of enzymatic oxygen removal in chemical protection against X-ray inactivation of bacteria. *J. Bact.* **63**, 805.

STAPLETON, G. E., BILLEN, D. & HOLLAENDER, A. (1953). Recovery of X-irradiated bacteria at suboptimal incubation temperatures. *J. cell. comp. Physiol.* **41**, 345.

STAPLETON, G. E. & EDINGTON, C. W. (1956). Temperature dependence of bacterial inactivation by X-rays. *Rad. Res.* **5**, 39.

STAPLETON, G. E. & ENGEL, M. S. (1960). Cultural conditions as determinants of sensitivity of *Escherichia coli* to damaging agents. *J. Bact.* **80**, 544.

STAPLETON, G. E. & HOLLAENDER, A. (1952). Mechanism of lethal and mutagenic action of ionizing radiation on *Aspergillus terreus*. II. Use of modifying agents and conditions. *J. cell. comp. Physiol.* **39** (Suppl. 1), 101.

STAPLETON, G. E., SBARRA, A. J. & HOLLAENDER, A. (1955). Some nutritional aspects of bacterial recovery from ionizing radiations. *J. Bact.* **70**, 7.

THOMPSON, T. L., MEFFERD, R. B., Jr. & WYSS, O. (1951). The protection of bacteria by pyruvate against radiation effects. *J. Bact.* **62**, 39.

WITKIN, E. M. (1946). Inherited difference in sensitivity to radiation in *Escherichia coli*. *Proc. nat. Acad. Sci., Wash.* **32**, 59.

WITKIN, E. M. (1959). Post-irradiation metabolism and the timing of ultraviolet-induced mutations in bacteria. *Proc. 10th Int. Conf. Genet.* **1**, 280.

WOOD, T. H. (1954). Influence of temperature and phase state on X-ray sensitivity of yeast. *Arch. Biochem. Biophys.* **52**, 157.

WOOD, T. H. (1958). Cellular radiobiology. *Annu. Rev. Nuclear Sci.* **8**, 343.

WOOD, T. H. (1959). Inhibition of cell division. *Rad. Res., Suppl.* **1**, 332.

WOOD, T. H. & TAYLOR, A. L. (1957*a*). Dependence of X-ray sensitivity of yeast on phase state and anoxia. *Rad. Res.* **6**, 611.

WOOD, T. H. & TAYLOR, A. L. (1957*b*). X-ray inactivation of yeast at freezing temperatures. *Rad. Res.* **7**, 99.

ZELLE, M. R. & HOLLAENDER, A. (1955). Effects of radiation of bacteria. In *Radiation Biology*. Edited by A. Hollaender, vol. II, p. 365. New York: McGraw-Hill Book Co. Inc.

ZIRKLE, R. E. (1936). Modification of radiosensitivity by means of readily penetrating acids and bases. *Amer. J. Roentgenol.* **35**, 230.

SYNCHRONIZATION OF DIVISION IN CULTURES OF *SACCHAROMYCES CEREVISIAE* BY CONTROL OF THE ENVIRONMENT

D. H. WILLIAMSON AND A. W. SCOPES

Brewing Industry Research Foundation, Nutfield, Surrey, England

In recent years a bewildering variety of processes has been devised for inducing synchronous division in cultures of micro-organisms. Although information about the mode of action of most of these processes is scanty, it is apparent that they may be subdivided into two distinct categories. The first of these includes methods whereby cells at the same stage of division are mechanically selected from randomly dividing populations and pooled to provide cultures in which division occurs in synchrony. The selection may be accomplished by filtration (Maruyama & Yanagita, 1956; Abbo & Pardee, 1960; Anderson & Pettijohn, 1960), centrifugation (Maruyama & Yanagita, 1956), or micromanipulation (Prescott, 1955).

The second group comprises several widely used methods in which entire populations are synchronized to varying degrees by manipulation of the organisms' environment. Examples of processes in this group include the use of shifts of temperature between two levels (Hotchkiss, 1954; Scherbaum & Zeuthen, 1954; Lark & Maaløe, 1954; Scott & Chu, 1958), treatment with a sublethal dose of radiation (Spoerl & Looney, 1959), alternation of light and darkness in the case of certain algae (Nihei *et al.* 1954), exposure to inhibitory substances (Scott, 1956), and changes in the medium usually involving starvation with respect to one or more nutrients (Barner & Cohen, 1956; Burns, 1959; Hayashi & Shichiji, 1959; Sylvén *et al.* 1959; Williamson & Scopes, 1960*b*).

Synchronously dividing cultures are usually employed as a means of following biochemical and cytological changes involved in the growth and division of the individual cell. However, it has frequently been suggested (Barner & Cohen, 1956; Campbell, 1957; Abbo & Pardee, 1960) that in contrast to the 'natural' synchrony resulting from the use of selective methods, synchronization induced by environmental changes may entail a disturbance of the normal balance between different enzymic processes in the cell, so that the subsequent behaviour of the synchronized organisms perhaps differs considerably from that of the untreated individual.

This contention appears to be well founded at least in regard to certain aspects of the growth of Gram-negative bacteria synchronized by environmental means. The evidence for this view comes from several sources. First, Lark & Maaløe (1956) found that deoxyribonucleic acid (DNA) synthesis in cultures of *Salmonella typhimurium* synchronized by repeated temperature shifts was a stepwise process occupying only a small fraction of the cell cycle. Similar results were obtained in the same year by Barner & Cohen (1956) using a strain of *Escherichia coli* synchronized by starvation for thymine, and by Maruyama (1956) making the first use of filtration synchronization of the same organism. At that time these observations were readily accepted, for there is abundant evidence that DNA synthesis follows a similar course in many other types of cell (Brachet, 1957; Campbell, 1957; Swann, 1957). However, examination of autoradiographs of *E. coli* cells from randomly dividing cultures exposed to tritiated thymidine led Schaechter, Bentzon & Maaløe (1959) to conclude that the unsynchronized individual cell synthesizes DNA throughout its division cycle, and McFall & Stent (1959) reached the same conclusion from a study of the death rates of cells from randomly dividing populations which had been exposed to radioactive phosphorus compounds for varying periods. These observations indicated that the stepwise synthesis seen in the synchronized cultures was an experimental artefact, though this seemed unlikely to apply to Maruyama's results, which were obtained using what was thought to be a selective and not an environmental method. However, Abbo & Pardee (1960) and Anderson & Pettijohn (1960) have now pointed out that Maruyama's technique involved inadvertent changes of temperature which could have distorted the behaviour of the synchronized cells. As proof of this contention, Abbo & Pardee (1960) showed that by carefully avoiding temperature changes, the filtration method provided highly synchronous cultures of *E. coli* which exhibited the continuous synthesis of DNA previously detected in randomly dividing cultures.

Thus, there can be little doubt that the earlier results obtained from environmentally synchronized cultures were not indicative of the true state of affairs in the untreated cell. This immediately raises doubts about the relevance to the latter of other biochemical measurements in these and similar systems.

In this connexion it is worth pointing out that the stepwise synthesis of DNA reported by Barner & Cohen (1956) only occurred when starvation for thymine was carried out in a synthetic medium lacking amino acids. When synchronized in the presence of the latter, the cells synthesized DNA throughout their division cycle at an apparently exponential rate. Here is clear evidence that 'normal' behaviour may be observed in at least some environmentally synchronized cultures. However, the marked variation in behaviour thus obtained by seemingly minor changes in technique emphasizes the need for better understanding of the mechanisms underlying environmental synchronization in order that the results obtained can be evaluated in terms of the behaviour of the untreated individual cell.

Apart from these considerations, environmental synchronization is clearly worthy of study in its own right, for it provides an interesting example of a biological system in which the net result of applying stimuli to a population depends on the fact that the response of the component individuals varies according to their age. Thus, such studies may help in elucidating some of the lesser known features of cellular growth and division, and may further our understanding of the complexities involved in the response of randomly growing cultures to environmental changes.

A GENERAL BASIS FOR ENVIRONMENTAL SYNCHRONIZATION

The procedures so far available are extremely varied, and few of them have been examined in detail. Accordingly, no attempt will be made here to survey them all or to seek a common rationale. Nevertheless, certain processes appear to embody a principle suggested by Hotchkiss (1954) that exposure of cells to an environmental change (notably of temperature) might selectively slow down or halt certain enzymic processes involved in cellular division, thus allowing individual cells to come into phase with respect to their progress through the division cycle.

Mechanisms of this type may underlie several techniques in which conditions are first applied which permit mass growth to proceed but selectively inhibit cell division; at the end of some predetermined period, the inhibitory condition is removed and the ensuing divisions show a marked degree of synchrony. It is thought that the mass growth that takes place during the period of inhibition furnishes each cell with nearly all it requires for division, the synthesis of perhaps one key compound only being blocked. On releasing the inhibition, synthesis of the missing compound is resumed at once, and division is triggered off as soon as a sufficient quantity has been prepared. An essential feature of this argument is that the conditions prevailing during the period of inhibition must in some way endow every cell with an equal capacity to meet its requirements for the blocked compounds once the inhibition is lifted, though this might not be necessary if the latter were normally capable of being made in a period which was short compared with the generation time of the cell.

Processes which may embody this type of mechanism include synchronization of protozoa by repeated prior temperature shifts (Scherbaum & Zeuthen, 1954), of yeasts by treatment with a sublethal dose of radiation (Spoerl & Looney, 1959), and of bacteria by deprivation of either thymine or thymidine (Barner & Cohen, 1956; Burns, 1959). Of these, the only process to have received more than passing attention is that involving synchronization of *Tetrahymena pyriformis* by means of multiple temperature shifts. Scherbaum (1957) has put forward a theoretical model for this system which accounts for many of the experimentally observed facts. Certain aspects of his scheme may be applicable in principle to other systems, so its main features will now be summarized.

The technique itself involves subjecting a randomly dividing culture growing exponentially at 28·5° to a series of half-hour 'heat shocks' at 33·9°, each shock being alternated with an equal period at the lower temperature. During the entire period of temperature shifts, which lasts about 7·5 hr., the size and DNA content of the cells increases considerably, while division is almost completely blocked. On finally returning the temperature to 28·5° there is a lag of about 75 min., whereupon all the cells divide simultaneously, and further synchronous divisions occur at intervals thereafter. Scherbaum proposed that the generation period of the cells is determined by the rate of synthesis of a substance which specifically triggers off division when it reaches a critical concentration. The enzyme responsible for synthesis of this compound is inactivated by thermal denaturation during the periods of heat shock and is partially reactivated during the intervals at normal temperature. Further enzyme is synthesized during the latter periods, and the net result of the entire treatment is that the cells finally contain no active enzyme, but a supply of the inactive form in proportion to their mass. At the same time, the pool of 'division substance' is completely emptied, perhaps by being used for growth processes relatively unaffected by the heat shocks. Thus, the treated cells cannot divide until their enzyme has been reactivated and has catalysed the synthesis of an adequate supply of the division substance. All the cells are now equivalent in their capacity to meet this need, and the first few divisions take place with a marked degree of synchrony.

ENVIRONMENTAL SYNCHRONIZATION OF *SACCHAROMYCES CEREVISIAE*

It is now proposed to discuss some observations relating to a process we have devised for synchronizing division of *Saccharomyces cerevisiae* (Williamson & Scopes, 1960*b*). Most of the data described here is being prepared for publication elsewhere, and so far as possible, details of experimental technique will be reduced to a minimum.

The process as currently used does not differ in essentials from that originally described. It utilizes a mature culture of *Saccharomyces cerevisiae* (National Collection of Yeast Cultures strain no. 239) grown in a version of the malt extract–yeast extract–glucose–peptone medium of Wickerham (1951). The culture is treated with a preparation of the gut juice of the snail, *Helix pomatia*, which effectively eliminates the small proportion of paired and bud-bearing cells usually found in such cultures, but has no noticeable effect on the viability of the remaining organisms. After washing, thc population is subjected to a series of alternating periods of suspension, first for about 18 hr. in the malt-extract medium maintained at 4°, and then for 6 hr. with aeration in a mineral salts solution (referred to as 'starvation medium') at 25°. Usually two such periods in the starvation medium are applied on successive days between the periods in the cold malt extract medium, the suspension being stored overnight at 4°. Following each period in cold malt-extract medium, the supernatant, bearing mostly small cells and debris resulting from the action of the snail enzyme, is decanted. After not less than about four such cycles of treatment, the population is found to comprise individuals of a uniform size (Pl. 1, fig. 2) which, on inoculation into a suitable medium (that originally described now being generally supplemented with 0·1 %, w/v, 'Difco' yeast extract), undergo at least two divisions with what may be described as a useful degree of synchrony (Fig. 1). In order to clarify this general description, it may be noted that the 'synchronization index' (Scherbaum, 1959) relating to the first division in these cultures is usually between 0·4 and 0·6. (The synchronization index is defined as $(n/n_0-1)\ (1-t/g)$, where n_0 and n are the cell numbers just before and just after a burst of synchronous multiplication, t is its time span, and g is the generation time). This should be compared with the theoretically attainable maximum of 1·0 and the actual values for various other systems which range between 0·26 and 0·83 (Bernstein, 1960). For reasons that will become apparent later in this discussion, we shall be concerned primarily with the degree of synchrony displayed on the emergence of the first generation of buds after inoculation, and it is uncertain if Scherbaum's synchronization index is directly applicable in this case. As a rough guide to the degree of synchrony of budding, however, cultures which on their first division exhibit values of the index between 0·5 and 0·6 are usually such that

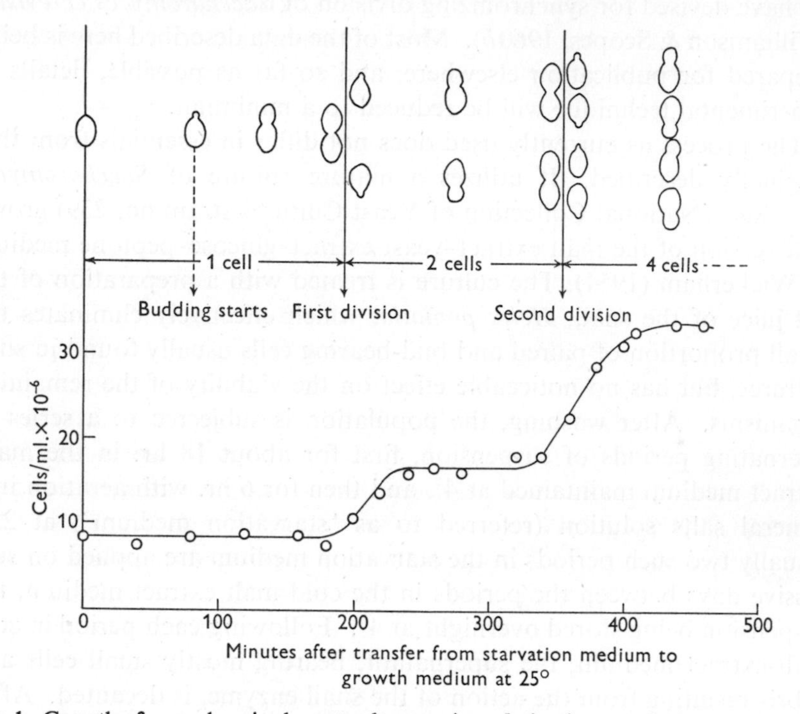

Fig. 1. Growth of a synchronized yeast culture on inoculation into synthetic medium. Cell numbers determined by haemocytometer counts. The diagram above the graph shows the corresponding stages in the cell's life cycle.

80–90 % of the cells produce their first buds in a burst lasting no more than 30 min. This 'useful' degree of synchrony may be compared with the course of appearance of buds in an unsynchronized population, which is usually distributed through a period of more than 100 min. (Fig. 3).

In common with most other strains of *Saccharomyces cerevisiae*, the organism used for these studies tends to give rise in mature cultures to cells of widely differing sizes. In fact, individuals in such cultures usually fall into one or other of two main size ranges, which will be referred to as 'large' and 'small' (Pl. 1, fig. 1). The actual mean size of each group varies with the medium on which they are grown but, typically, the 'large' cells are some 2–3 times bigger than the small ones (Williamson & Scopes, 1961 *a*).

In cultures grown in a liquid malt extract medium, the small cells usually account for about 40 % of the total number of individuals, and preliminary observations showed that on inoculation into a nutrient medium these cells were usually much slower than the large ones in giving rise to a bud. It was originally reasoned, therefore, that the degree of synchrony shown by the mass culture on inoculation might be improved by prior removal of the smaller cells. Initial attempts to achieve this by differential centrifugation were not very successful, and the synchronizing process now used, in which the smaller cells are selectively removed by successive decantation, was originally designed to overcome this difficulty. This being so, it might be thought that the process acts in a selective rather than an environmental manner. With regard to synchronizing the first appearance of buds, however, this is not the case, for it is possible to achieve this in whole cultures without resorting to selective decantation. Nevertheless, it is not yet certain if the synchrony displayed by the subsequent division of cultures so treated is as good as that shown by cultures in which decantation is used, and the latter procedure consequently still forms part of the current routine. It has now been found possible, however, to select the larger cells from the initial population by differential centrifugation in aqueous mannitol (15 %, w/v) (Williamson & Scopes, 1961*a*), and for certain purposes (e.g. in following changes in the composition of the cells during synchronization), this procedure is considered preferable. It has also been found that the initial synchrony of budding of the 'large' cell fraction isolated in this way is only slightly better than that of the whole population. This shows conclusively that selective removal of the small cells from the initial population can play only a minor role in inducing synchrony of budding.

Another feature of the process which could conceivably exert a selective effect is the initial treatment with snail enzyme, for by removing paired and bud-bearing cells from the original population, this treatment selects out cells which are not visibly in the process of division. This facilitates subsequent microscopic examinations, but the selection involved plays no significant role in synchronization, for paired and budding cells usually comprise only a very small proportion of the cultures used as starting material.

For these reasons there can be little doubt that the process may be correctly classed with those treatments whose effect results primarily from the response of the cells to alterations of the environment. The outstanding problem is how the change in the behaviour of the cells is brought about.

EFFECT OF SYNCHRONIZATION ON THE LAG PERIOD

A distinctive feature of the process is that it involves bringing into a potentially synchronous state, cells which are considered to be in a 'resting' condition. No buds or new cells are formed in the course of the treatment, and the improved synchrony of the population is manifest only after inoculation into a growth medium.

The possible significance of this will become apparent later in the discussion. First of all, in order to facilitate discussion, it is necessary to outline the sequence of events which follows inoculation of a resting yeast cell into a growth medium. The first bud appears only after a marked delay (referred to in this discussion as the 'lag period') during which no appreciable mass growth or macromolecular synthesis occurs. Such activities, however, start as soon as the bud appears and continue throughout its growth period (Williamson & Scopes, 1960*a*, *b*; 1961*b*). For this reason the first generation proper is assumed to start with the appearance of the first bud. The end of the generation period is taken as the time when the mature daughter cell, having attained a size approximating that of the parent, separates from the latter and starts an independent existence. Difficulty is sometimes encountered in accurately timing this stage in yeasts, but in the strain used for this work, separation of the parent and daughter cells is normally clear-cut, particularly if the culture is vigorously agitated, for instance, by aeration. It has also been found that under these conditions this separation coincides with, or perhaps slightly precedes, the appearance of the next generation of buds (Williamson & Scopes, 1960*b*). It is assumed therefore that the end-point of one generation and the start of the next may equally well be taken as the time of appearance of the new buds. The correctness of this view is further supported by evidence suggesting that the appearance of a bud on one or both of the partners signifies their attainment of physiological independence (Burns, 1956; Williamson & Scopes, 1960*b*). Advantage of this fact may be taken in situations where mechanical agitation is not possible and physical cleavage of the daughter and parent may not be detectable, as for instance in a slide culture.

With the above view of the early 'life history' of the yeast cell in mind (it is summarized diagrammatically in Fig. 2), one can begin to see the possible significance of the fact that the synchronizing process acts on 'resting' as opposed to growing cells. In a sense, the process maintains the cells in what may be regarded as an artificially extended lag phase. It therefore seems reasonable to suppose that it owes its effect, in part at least, to some modification of the normal behaviour of the cells during

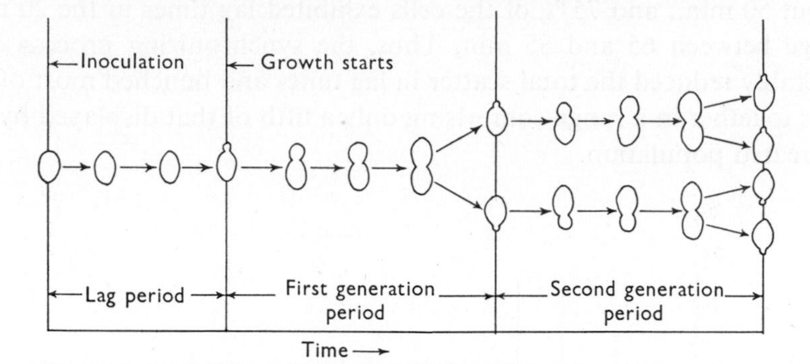

Fig. 2. Diagrammatic representation of the early life history of a yeast cell, illustrating lag and generation periods.

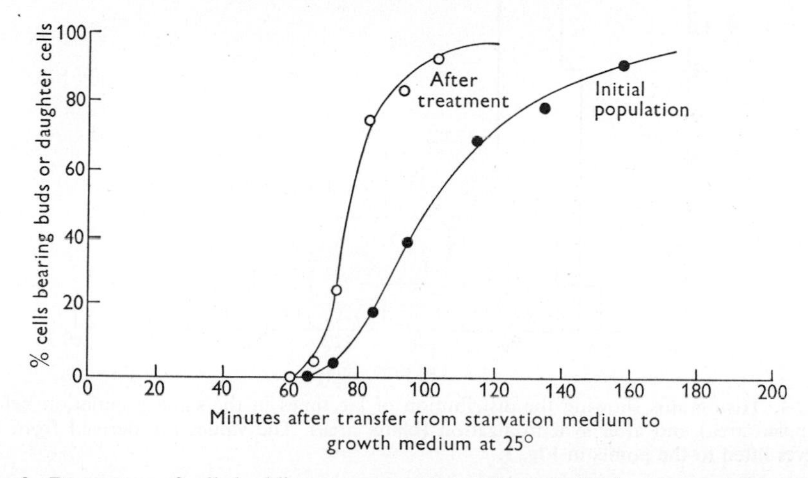

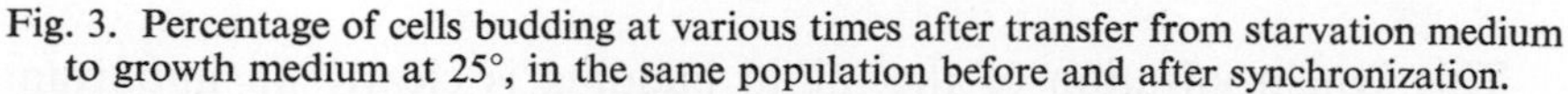

Fig. 3. Percentage of cells budding at various times after transfer from starvation medium to growth medium at 25°, in the same population before and after synchronization.

the lag period. The correctness of this view is supported by comparing the distribution of lag times in populations before and after synchronization. Figure 3 shows the course of budding in an unsynchronized 'large' cell fraction isolated by centrifugation, together with that displayed by the same population after synchronization. For ease of reference, histograms representing the distribution of individual lag times have been constructed from the curves fitted to these data, and are shown in Fig. 4.

It will be seen that, before treatment, the lag times of the cells were distributed fairly symmetrically about the mode (*c*. 90 min.), 95 % of the cells beginning to bud between 65 and 163 min., a total scatter of 98 min. After synchronization, on the other hand, this period was reduced to about 50 min., and 75 % of the cells exhibited lag times in the 20 min. range between 65 and 85 min. Thus, the synchronizing process considerably reduced the total scatter in lag times and bunched most of the cells together in a range comprising only a fifth of that displayed by the untreated population.

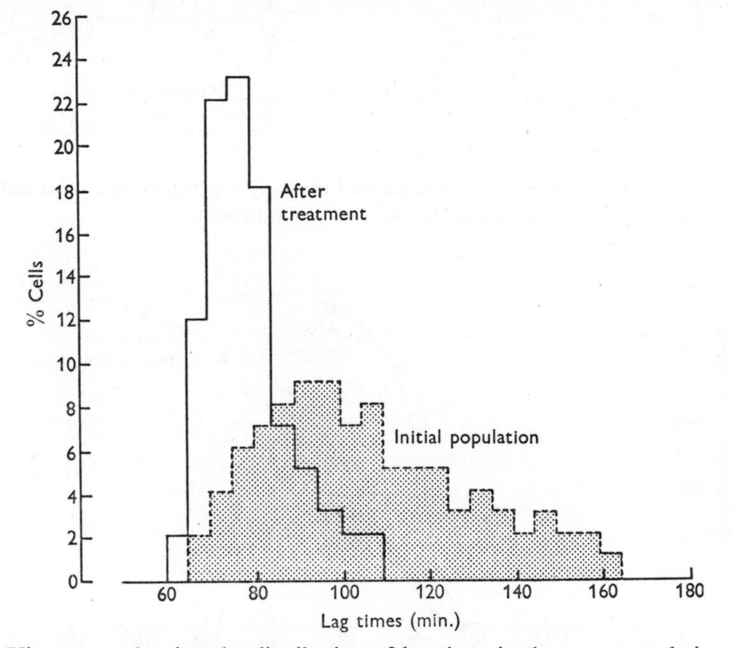

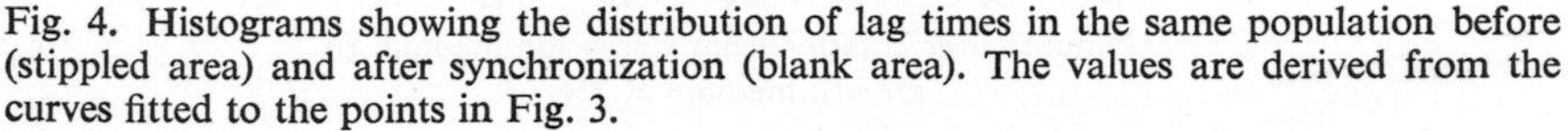

Fig. 4. Histograms showing the distribution of lag times in the same population before (stippled area) and after synchronization (blank area). The values are derived from the curves fitted to the points in Fig. 3.

This synchronization of the initial budding is clearly an important factor in bringing about the degree of synchrony displayed by the treated cells during their subsequent divisions. It is possible, in fact, that the division synchrony results entirely from the observed reduction of scatter in lag times. Nevertheless, a similar effect on the scatter of the generation times of the growing daughter cells might also play an important role. However, preliminary observations of individuals growing in slide cultures suggest that this is unlikely. These have indicated that the total scatter in the generation times of untreated cells is about the same as the duration of the burst of division which terminates the first generation in the synchronously dividing cultures. Consequently, it seems likely that the synchronizing process exerts its effect largely, if not entirely, by bringing into phase the emergence of the first buds; the generation times of the growing daughter cells may be comparatively unaffected.

Further work is needed to substantiate this view, but in any event, the synchronizing procedure clearly exerts a marked control on the lag periods of individual cells. Accordingly, in attempting to understand its action, attention will be primarily directed towards possible ways in which the activities of the cells during the lag period might be brought into phase in different individuals.

BIOCHEMICAL ACTIVITIES OF THE LAG PHASE YEAST CELL

Relatively little is known of the biochemical events preparatory to the initial appearance of the bud when a 'resting' yeast cell is inoculated into a growth medium. Evidence from various sources however suggests that prior to budding, the dry mass and volume of the cell normally remains constant, and in synchronized cultures there is no net uptake of nitrogen from the medium and no synthesis of ribonucleic acid or protein (Lindegren & Haddad, 1953; Sylvén *et al.* 1959; Williamson & Scopes, 1960*a*, *b*, 1961*b*). The deoxyribonucleic acid required for the daughter cell is apparently synthesized abruptly at the end of the lag, coincidentally with, or just after, the formation of the bud (Ogur, Minckler & McClary, 1953; Williamson & Scopes, 1960*b*). On the other hand, cytological observations suggest that nuclear material other than DNA may be replicated before budding commences (Swaminathan & Ganesan, 1958), and we have noted that the vacuole which is prominent in synchronized cells (Pl. 1, fig. 2) disappears, or at least becomes less evident, soon after inoculation.

The actual emergence of the bud is thought by Nickerson & Falcone (1959) to be a blow-out caused by the internal fluid pressure of the cell acting against a point on the wall weakened by the action of a reductase acting on protein disulphide bonds. The detection in yeasts of such an enzyme capable of reducing disulphide linkages in protein isolated from the cell wall (Nickerson & Falcone, 1956) certainly favours this hypothesis, but more direct evidence for its role in bud formation is desirable.

In any event, the processes culminating in emergence of the bud seem to require a supply of energy, for the cell respires throughout the lag period and budding may be completely prevented by omission of glucose from the medium (Scopes & Williamson, unpublished observations).

It has already been noted that the resting cell *prepares* for budding without any appreciable growth taking place. As will be discussed below, however, the converse is true of cells suspended at a high concentration in a nutrient medium maintained at 4°, for under these circumstances the cells increase in mass but do not produce buds. It has also been found that addition to synchronous cultures of Mitomycin C, an inhibitor of DNA synthesis (Shiba *et al.* 1959), prevents cell division and the appearance of second generation buds, although apparently permitting growth in mass to proceed, at any rate for a limited period (Pl. 2, figs. 3, 4). These observations imply that certain of the events essential to the appearance of the bud are dissociable from mass growth. Accordingly, it is suggested that the emergence of the bud may be initiated only in the presence of a critical concentration of one or more specific 'trigger' compounds, whose synthesis may proceed independently of cellular growth. It is tempting to speculate further that the protein disulphide reductase of Nickerson & Falcone (1956) might be such a compound, but there is at present no evidence justifying this contention.

The above proposition may be compared with the similar theories concerning cell division advanced by Hinshelwood (1946) and Scherbaum (1957), and with the energy reservoir hypothesis of Swann (1954). It must inevitably be a gross oversimplification of the true course of events leading up to budding, but it provides a basis for the working hypothesis to explain the action of the synchronizing process which is discussed below.

OBSERVATIONS RELATING TO THE MODE OF ACTION OF THE SYNCHRONIZING PROCESS

Any attempt to explain the action of the synchronizing process must take into account the following observations:

(1) Routine microscopic examination has shown that no new buds or cells are formed during any stage of the treatment.

(2) Analysis of synchronized populations by conventional methods has shown that the synchronizing process does not significantly alter the mass or gross composition of the cells (Table 1). Nevertheless, the mean dry mass per cell increases by about 25 % during each period in the cold malt extract medium, only to return to the initial value during subsequent starvation.

In contrast, the mean volume per cell is irreversibly increased by the treatment. In one example, using a 'large' cell fraction isolated by centrifugation, an increase of about 50 % was observed, the measurements being made by microscopic inspection. It seems that the reversible increase in mass brought about during the periods in the cold malt extract medium results in an irreversible stretching of the cell wall. As

Table 1. *The composition of yeast cells before and after synchronization*

	μg./cell × 10^6	
	Initial population	Synchronized population
Total nitrogen	4·1	4·3
Acid-insoluble nitrogen	2·4	2·5
Soluble nitrogen	0·8	1·4
Protein	16·6	17·6
Free amino acids	2·1	3·1
Free nucleotides	1·1	1·1
RNA	2·9	3·5
DNA	0·064	0·08
Total phosphorus	0·8	1·1
Total carbohydrate	13·8	10·2
Ash	2·9	3·9
Total dry mass	51·0	50·0

Nitrogen was determined by a micro-Kjeldahl method, phosphorus by the Fiske–Subbarow method and nucleotides by the absorption at 260 mμ of an ethanolic extract. Protein, amino acids, RNA, DNA and carbohydrate were estimated colorimetrically with the Folin-phenol, ninhydrin, orcinol, diphenylamine and anthrone reagents, respectively.

a corollary, the concentration of (dry) mass per unit volume in the synchronized cell is presumably less than that of the untreated individual.

(3) The synchronizing treatment reduces the mean lag time of the population to little more than the minimum displayed by any individual in the untreated culture (Figs. 3, 4). Moreover, the mean lag apparently cannot be further reduced by extending the series of cyclic treatments. It appears, therefore, that the treatment effectively reduces the lag periods of most individuals, but some factor prevents reduction below a certain minimum. Accordingly, the cells showing long lag periods prior to synchronization will be affected more than those which initially have lags nearer to the minimum.

(4) The degree of synchrony of budding improves during the periods of starvation. An example of this phenomenon is shown in Fig. 5, which compares the time course of budding observed on transferring a population directly from cold malt extract medium into a synthetic medium maintained at 25°, with that displayed by a portion of the same population after a subsequent period of starvation. It will be seen that the interval between the initial appearance of buds and the time at which 80 % of the cells had budded was reduced by starvation to about two-thirds of the initial value. The periods of starvation were originally envisaged merely as a device to prevent the cells from budding after their sojourn in the cold malt extract medium. It is now evident, however, that their use constitutes an integral part of the synchronizing

process, and for this reason the latter is always terminated by a period of starvation. It must be emphasized, however, that starvation is not the only requirement for the induction of synchrony, for if this were so it would be possible to synchronize the initial population merely by starving it for a sufficient period.

(5) Preliminary observations have been made relating to the possible role of temperature. As the process involves alternation of temperature between 4° and 25°, one is tempted to seek a comparison with temperature shift treatment such as that devised by Scherbaum & Zeuthen

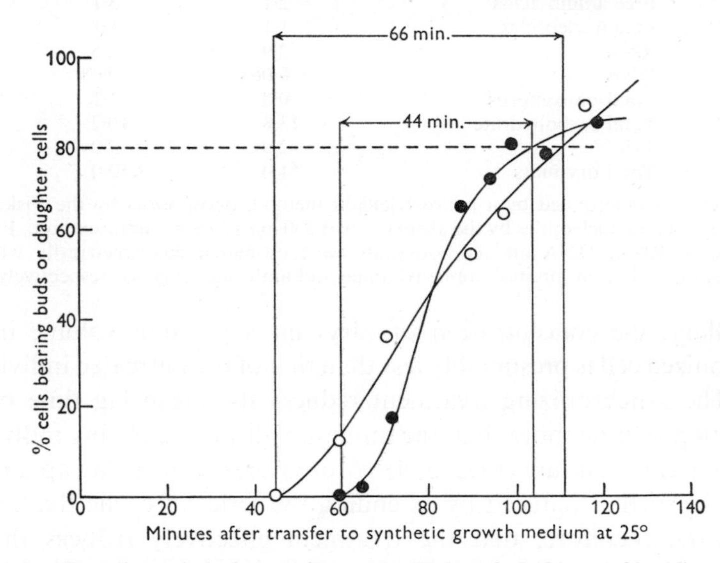

Fig. 5. The effect of a period of starvation on the degree of synchrony of budding. —○—: cells taken directly from the cold malt extract medium. —●—: cells from the same population after the subsequent period in starvation medium.

(1954) for *Tetrahymena*. However, the apparent similarity between the two processes is only superficial for, in the system under discussion, the yeast is prevented from growing during its periods at the normal temperature by the absence of nutrients, whereas *Tetrahymena* cells are continuously in contact with an adequate growth medium.

Originally, maintenance of the nutrient medium at a low temperature was employed as a device to prevent the cells from budding during their stay in the medium, which had to be long enough to permit preferential sedimentation of the larger cells. Little attention was accorded to the possible significance of biochemical changes that might be induced in the cells at the same time. In the light of developments outlined above, however, it became apparent that the latter changes were in fact all-important, selection of the larger cells playing a relatively insignificant role in the process. Accordingly it was desirable to test the possibility that the biochemical effects leading to synchronization might be brought about at a higher temperature, budding in this case being prevented by restricting the period of suspension in the warm medium.

A large cell fraction separated by centrifugation from a culture grown on solid medium was suspended at a high concentration in the liquid malt-extract medium held at 25°. After about 40 min., the population was centrifuged and transferred with washing into starvation medium previously warmed to 25°, care being taken to avoid fluctuations of temperature. After 6 hr. aeration, the suspension was cooled and stored at 4° overnight. The change of temperature involved here is

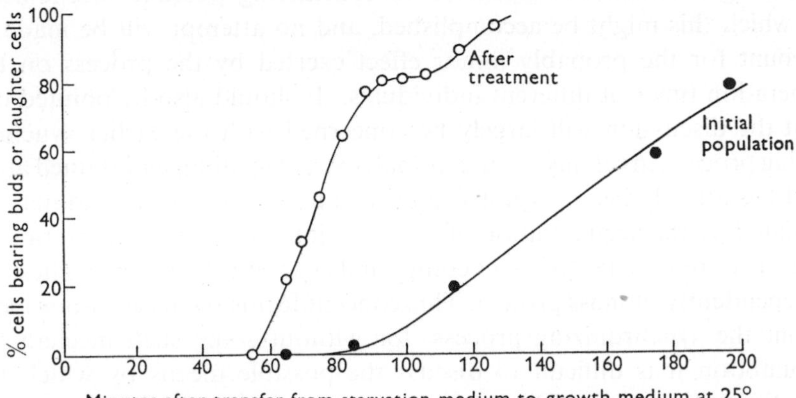

Fig. 6. Synchronization of budding achieved by subjecting a culture to 8 cycles of treatment, each consisting of 40 min. suspension in malt extract medium at 25° followed by 6 hr. starvation at the same temperature. The initial population was a 'large' cell fraction isolated from a culture grown on solid medium.

thought not to have had any significant effect relative to synchronization, for the starved cells are presumed to be comparatively inactive in the absence of nutrients. The following day the suspension was warmed to 25°, transferred to previously warmed malt extract medium, and the cycle of treatment repeated. After continuing this process for 8 days, a sample of the population was suspended in the synthetic growth medium enriched with yeast extract, and the time course of the appearance of buds determined. It will be seen from the results (Fig. 6) that the treatment had induced a marked degree of synchrony in the population.

The significance of this observation is twofold. First, it has revealed an apparently new (though related) way of synchronizing budding of yeast, and it will be interesting to compare the behaviour of the resulting synchronized organisms with that of populations prepared by the earlier method. Secondly, it showed that the temperature changes in the earlier method may be entirely dispensed with.

An attempt will now be made to relate the above observations to a hypothetical model to account for the action of the synchronizing process.

HYPOTHETICAL SCHEMES TO ACCOUNT FOR SYNCHRONIZATION

It will be recalled that the synchronizing process appears to exert its effect primarily by bringing into phase in different organisms the processes leading up to the appearance of the first bud. For this reason the following discussion is restricted to considering possible mechanisms by which this might be accomplished, and no attempt will be made to account for the probably minor effect exerted by the process on the generation times of different individuals. It should also be pointed out that the discussion will largely be concerned with the earlier synchronizing process, involving the use of malt extract medium maintained at 4°.

It has already been suggested that the appearance of the bud may be initiated by the accumulation of a critical intracellular concentration of one or more specific 'trigger' compounds capable of being synthesized independently of mass growth. This concept forms the basis of our ideas about the synchronizing process, for without some such mechanistic assumption it is difficult to discuss the possible means by which the process endows every cell with an equal ability to progress through the lag period. It is important to realize, however, that there is at present no direct evidence that such compounds exist, and it is possible that the effects attributed to them might be brought about by a structural reorganization, or the attainment of a particular physical state, in some part of the cell. For ease of discussion, however, the relevant changes will be assumed to be of a chemical rather than a physical nature.

The number of hypothetical systems that might be proposed to account for the action of the synchronizing process is limited only by the scope of one's imagination. However, two main types of model deserve attention. The first of these is illustrated in Fig. 7. It involves the assumption that the cells make some progress towards budding during their 18 hr. periods of suspension in the cold malt extract medium, but that the conditions during these periods impose a specific block at a particular stage in the ordered sequence of events that constitutes the budding cycle. Given sufficiently long in contact with the medium, the cells would all be halted at this stage and, on releasing the inhibition, would proceed to bud in synchrony.

To test this possibility, a portion of an unsynchronized population was suspended in cold malt extract medium at the high concentration normally used in synchronizing (*c.* 8×10^7 cells/ml.). Frequent microscopic examination during the course of several weeks showed that no buds were formed under these conditions, and this was taken as evidence for the existence of a block to the ultimate appearance of buds. However, this inhibition could apparently be lifted by daily transference of the population (without changing the concentration of cells) into fresh

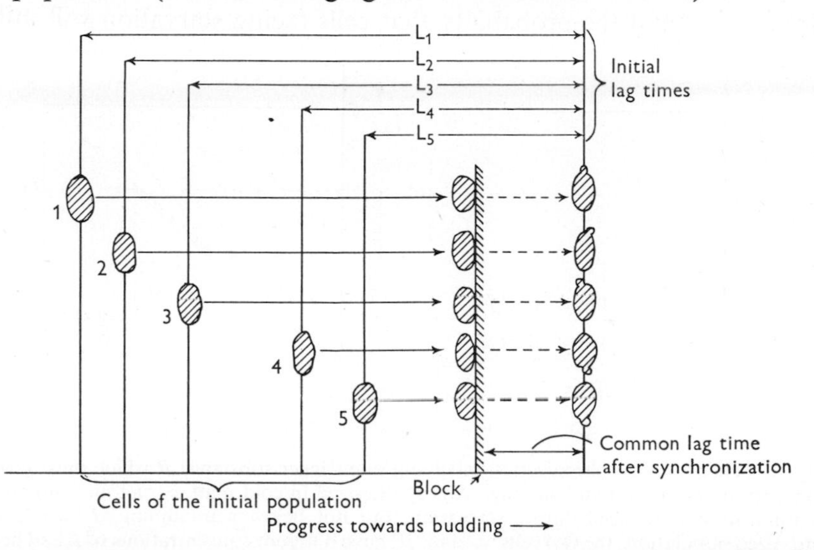

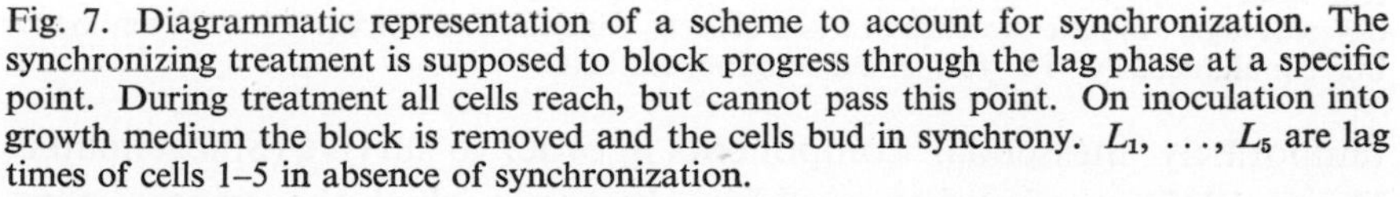

Fig. 7. Diagrammatic representation of a scheme to account for synchronization. The synchronizing treatment is supposed to block progress through the lag phase at a specific point. During treatment all cells reach, but cannot pass this point. On inoculation into growth medium the block is removed and the cells bud in synchrony. $L_1, \ldots, L_5$ are lag times of cells 1–5 in absence of synchronization.

cold medium, for in these circumstances buds appeared after about 138 hr. Coupled with the finding by chromatographic examination that the glucose content of the unchanged medium decreased to a negligible level in about 48 hr., it seemed that failure to bud when the medium was not renewed was ultimately due to exhaustion of the energy source. However, this condition could not have been operative while glucose was still present, and the absence of budding during the first 18 hr. of the cells' contact with the cold medium (i.e. during the normal period employed in the synchronizing process) seems to be attributable to a temperature-induced retardation of progress through the lag phase, rather than to the existence of an absolute block.

The above reasoning leads one to consider alternative schemes that do not entail the assumption of a block to budding imposed by the low temperature. As shown above, the cells form buds after about 140 hr. continuous contact with periodically renewed cold malt extract medium. During the synchronizing process, on the other hand, the cells may be in contact with the cold medium for a much longer total period without any buds being formed. This can only be attributed to the action of the intermittent periods of starvation. These must do more than merely halt the progress through the lag period made during contact with the malt extract medium; they must also in some way reverse this progress. Bearing in mind the probability that cells facing starvation will utilize

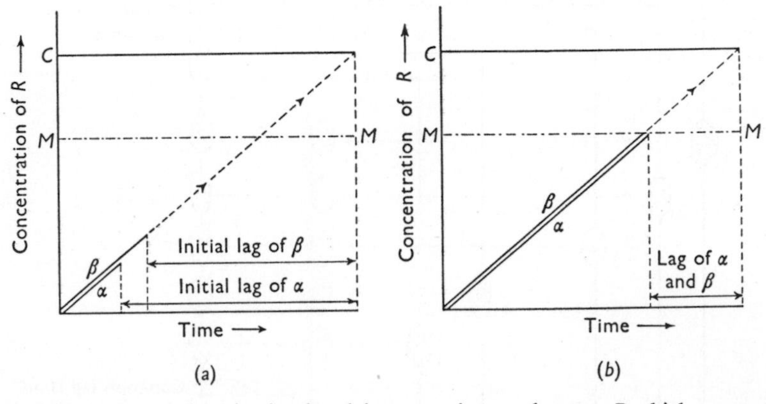

Fig. 8. Scheme for synchronization involving one trigger substance R which must reach a concentration C to initiate budding. R is synthesized in cold malt extract medium and its concentration is reduced during starvation, but not below a minimum M. (*a*) Unsynchronized population; the two cells 'α' and 'β' have different concentrations of R and hence, different lags. (*b*) After synchronization, all cells contain R at concentration M, and hence bud simultaneously. For further explanation, see text.

temporarily 'inessential' components in order to survive (Spiegelman & Dunn, 1947), one is led to consider schemes in which the 'trigger' compounds essential to bud formation are synthesized during each period of contact with the malt extract medium, and to some extent broken down again during the alternating periods of starvation.

A simple model along these lines is represented in Fig. 8, in which it is assumed that the initial emergence of the bud requires the presence at a critical intracellular concentration (C), of only one trigger compound, designated as substance R. It will be further assumed that the value of C and the rate of synthesis of R per unit cell mass is the same in all the cells. If this were so, the interval between inoculation and the appearance of the bud would be dependent only on the initial cellular concentration of R, and one could account for the action of the synchronizing process in terms of a mechanism which would adjust the latter to the same level in all the cells.

According to this model, the cell is assumed to synthesize *R* during its stay in the cold malt extract medium, and to break it down to some extent during the subsequent period of starvation. However, starvation could not completely empty the cell's pool of *R*, for this would imply that the initial population could be synchronized merely by subjecting it to an adequate period of starvation. It has therefore to be assumed that starvation cannot reduce the concentration of *R* below a certain level, indicated in Fig. 8 as the line at '*M*'. It will now be evident that the net effect of repeated alternate exposure to malt extract and starvation media would be to progressively level off the concentration of *R* in all the cells to the minimum designated as *M*. From what has already been said, it follows that on inoculation into a growth medium, each cell would take the same time to increase its pool of substance *R* from level *M* to the critical concentration *C*, and buds would appear in synchrony.

Since buds are not formed in the course of 18 hr. suspension in the cold malt extract medium, even in the later stages of synchronization when most of the cells would contain *R* at the concentration *M*, it is necessary to assume that at 4°, 18 hr. is insufficient to permit the raising of this concentration to the critical level, *C*. In this connexion it should be pointed out that all the reactions leading up to budding are assumed to be retarded to the same extent by the lowered temperature. While this may not in fact be the case, there is at present no evidence to the contrary, and any attempt to devise a more complex model in this respect is unnecessary and would be unjustified in the light of our present knowledge.

It is possible by this scheme to account for several of the experimental observations. Thus, the scatter of individual lag times prior to synchronization would be due to variation among different cells in the initial concentration of *R*. Moreover, if the minimum level, *M*, was not much less than the critical concentration *C*, the mean lag time after synchronization would be little greater than the minimum displayed by the untreated individuals. Further, the previously described synchronizing effect of a period of starvation on cells taken from the cold malt extract medium could be explained as a reduction of the concentration of *R* in different cells from varying levels between *M* and *C* to the common minimum, *M*.

An objection to the above model is that it entails the assumption of restricted breakdown of *R* during starvation. If the starving cell uses *R* for essential maintenance purposes, there is no clear reason why any of its available complement should be spared. To overcome this objection, the alternative scheme in Fig. 9 is presented.

In this model, the previously made assumptions about substance R are adhered to, but this time R is thought to be completely removed from the cell during starvation. It is further suggested that in addition to substance R, the emergence of the bud demands the presence of a

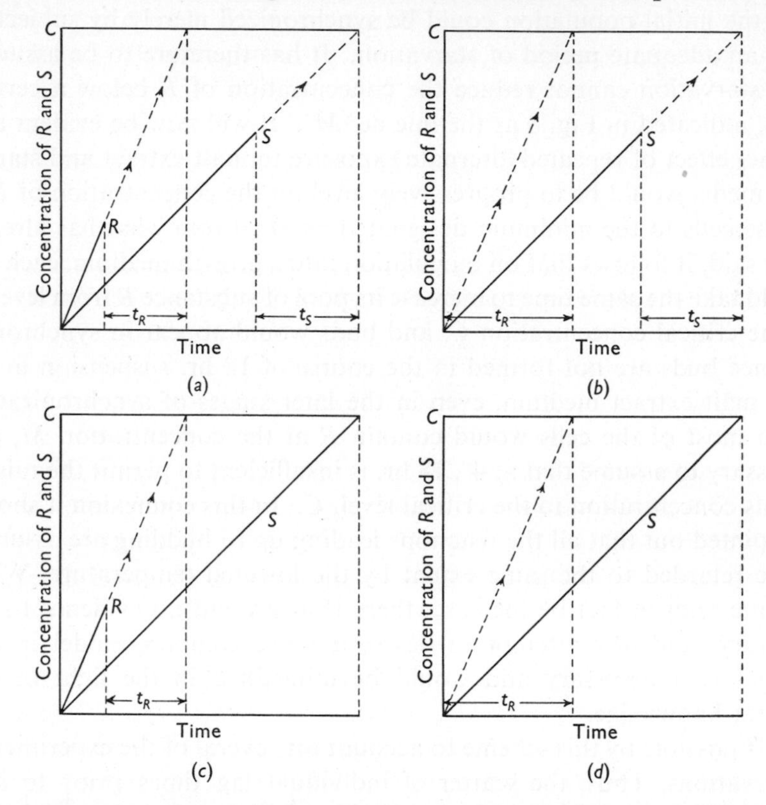

Fig. 9. Scheme for synchronization involving two trigger substances R and S which must both reach critical levels (C) to initiate budding. R and S are synthesized in cold malt extract medium. During starvation the concentration of R is reduced to zero, while that of S remains unchanged. (*a*) and (*b*) early in synchronization: (*a*) immediately after passage through malt extract medium; (*b*) after the subsequent period of starvation. S has not yet reached its critical level, R is reduced to zero by starvation. (*c*) and (*d*) fully synchronized cells immediately before and after the final period of starvation. S has now reached its critical level. In (*a*) and (*b*) the lag of a cell is equal to t_R or t_S whichever is the greater. In (*c*) and (*d*) the lag depends solely on the synthesis of R and hence equals t_R. For further explanation, see text.

second trigger substance, designated 'S', this substance normally being synthesized at a lower rate than R and, in contrast to the latter, being stable under starvation conditions. It will readily be appreciated that repeated alternate suspension in the malt extract and starvation media would lead to progressive accumulation of S, while the concentration of R in the fully starved cells would always be zero. Thus, after a sufficient number of cycles, the lag time displayed on inoculation would be that required for the synthesis of the critical amount of *R*. As in the previous model, this time would be the same for all the cells, and budding would consequently occur in a synchronous manner.

As in the first scheme, one would need to assume that 18 hr. in the cold malt extract medium is too short a time to permit synthesis of *R* to the critical level *C*. The scatter in lag times prior to synchronization would result from variation in the initial cellular concentrations of the more slowly synthesized (and hence limiting) substance, *S*. Moreover, the mean lag after synchronization, being the time required for the relatively rapid synthesis of *R*, would approximate to the minimum lag displayed by the untreated cells. This model could also account for the improved synchrony resulting from a period of starvation. Following certain of the passages through the cold malt extract medium, a proportion of the cells should show lag periods dependent on their varying levels of *S*, rather than on synthesis of *R* (t_S in Fig. 9*a*). After subsequent starvation, on the other hand (Fig. 9*b*), the lag times of many of these same individuals would be governed by synthesis of *R*, and they would therefore bud simultaneously. Hence the degree of synchrony displayed by the population would be enhanced by a period of starvation.

Thus it will be seen that both the above model schemes could account for the previously described experimental observations. Nevertheless, it is evident that they both depend on a chain of assumptions, and much more work will be needed to put either of them on a firm experimental basis. Future efforts should be mainly directed to detecting and identifying the hypothetical 'trigger' substances on which both schemes rest. This may prove difficult, for, as has already been noted, the gross composition of the cells is essentially unaltered by synchronization, and any chemical changes induced by the process are likely to be in quantitatively minor components.

THE BEHAVIOUR OF SYNCHRONIZED YEAST CELLS

Early in the course of this discussion it was pointed out that in contrast to the selective methods, the environmental procedures for synchronization have an intrinsic value inasmuch as they may yield evidence concerning cellular activities that the methods based on selection cannot supply. Nevertheless, environmental synchronization procedures, like their selective counterparts, are commonly used to provide a means whereby the processes constituting the cell division cycle can be ascertained by direct examination of the synchronously dividing organisms.

However, as has already been noted, certain methods of environmental synchronization lead to a distortion of the behaviour of the dividing cells as compared with that of untreated individuals, and there is a clear need to determine whether or not this is true of other such processes. Direct observation of the biochemical and cytological activities of untreated single individuals would provide the most straightforward approach to this problem, but present techniques for this purpose are of limited applicability. However, knowledge of the changes induced in the cells by the synchronizing procedure should permit a partial assessment of the degree of aberration of the synchronously dividing organisms, and such an assessment of the process under discussion will now be attempted in the light of the observations described in the preceding pages.

It has already been shown that there are grounds for believing that the synchronizing process exerts its effect largely, if not entirely, by synchronizing the appearance of the first generation of buds subsequent to inoculation of the cells into a growth medium. Though further work is needed to confirm this belief, it provides a general basis for the contention that the synchronizing process entails little or no alteration of the behaviour of the cells once growth is under way. Support for this view might be forthcoming if the detailed cellular responses underlying the process were fully understood. In the absence of such knowledge, however, it is perhaps permissible to speculate a little on the basis of the hypothetical models discussed above. If either one of them closely approximates the true course of events, it will be realized that at the time of appearance of buds, the synchronized cells would not differ greatly in composition from unsynchronized individuals at the same stage. For the emergence of buds, both treated and untreated organisms would have to be in possession of the critical amounts of hypothetical 'trigger' substances, and the theoretical models involved no assumptions about changes in the mode of synthesis of these, or about alterations in the proportions of other cell components.

To some extent this reasoning gains support from the finding that the synchronizing process did not significantly alter the mass or gross chemical composition of the cells, though the analytical techniques employed certainly would not permit detection of quantitatively minor changes. In more general terms, however, this finding clearly distinguishes the present process from treatments, such as that devised by Scherbaum & Zeuthen (1954) for *Tetrahymena*, in which the synchronized cells are considerably heavier and contain more DNA/cell than the untreated individuals (Scherbaum, Louderback & Jahn, 1959.)

Whereas the growth of such grossly 'unbalanced' organisms might be expected to deviate from the normal, there is no ground for believing this to be true of the comparatively unaltered yeast cells.

Turning to evidence based more directly on experimental observations, it is pertinent to compare the known behaviour of the synchronized yeast cells with earlier observations of untreated organisms of the same species. In this connexion, the pattern of synthesis of DNA is of prime interest. It has been shown (Williamson & Scopes, 1960*b*) that DNA synthesis in synchronous cultures is restricted to a small part of the division cycle, the DNA content of the cultures doubling abruptly with the appearance of the buds. There is abundant evidence that similar stepwise synthesis occurs in many other types of cell (Brachet, 1957; Swann, 1957), and the observations of Ogur *et al.* (1953) on yeast cells taken from old cultures show that such a pattern is displayed also by normal yeast. This is supported by the findings of Beam *et al.* (1954) that the resistance of yeast cells to X-irradiation increased with the appearance of buds, an observation they considered might be consistent with an increase in the DNA content of the budding organisms. Thus, there is good reason to believe that in respect of DNA synthesis, the synchronously dividing cells behave in the same way as untreated individuals.

In contrast to DNA, synchronized cells have been shown to synthesize ribonucleic acid (RNA) at a fairly steady rate throughout most of their division cycle (Williamson & Scopes, 1960*b*). The only relevant evidence with which this might be compared are the observations of Mitchison & Walker (1959), who found that individuals from randomly dividing cultures of the fission yeast, *Schizosaccharomyces pombe*, also synthesized RNA throughout their division cycle.

The situation with regard to protein is even less satisfactory, no direct observations of its course of synthesis in untreated individual cells being available. However, from observations of the growth in mass of isolated yeast cells, Mitchison (1957, 1958) inferred that protein synthesis continues at a constant rate throughout the cell cycle of these organisms, and we have found that synchronized cells do in fact exhibit such continuous synthesis (Williamson & Scopes, 1960*a*, 1961*b*). Although our analytical techniques were too insensitive to permit precise definition of the shape of the synthesis curve, there is at present no evidence that it was other than linear and, superficially at any rate, the synchronized cells do not seem to deviate in this respect from untreated organisms.

In conclusion, it will be seen that all the currently available information is consistent with the view that the environmental synchronizing process we have devised for *Saccharomyces cerevisiae* does not result in significant distortion of the activities of the growing cells. This carries the broader implication that the behaviour of organisms synchronized by environmental methods is not *ipso facto* 'pathological'. Examination of the cultures obtained using certain of these processes should therefore provide information of direct value in constructing a picture of the division cycle of the individual.

REFERENCES

ABBO, F. E. & PARDEE, A. B. (1960). Synthesis of macromolecules in synchronously dividing bacteria. *Biochim. biophys. Acta*, **39**, 478.

ANDERSON, P. A. & PETTIJOHN, D. E. (1960). Synchronization of division in *Escherichia coli*. *Science*, **131**, 1098.

BARNER, H. D. & COHEN, S. S. (1956). Synchronisation of division of a thymineless mutant of *Escherichia coli*. *J. Bact*. **72**, 115.

BEAM, C. A., MORTIMER, R. K., WOLFE, R. G. & TOBIAS, C. A. (1954). The relation of radio-resistance to budding in *Saccharomyces cerevisiae*. *Arch. Biochem. Biophys*. **49**, 110.

BERNSTEIN, E. (1960). Synchronous division in *Chlamydomonas moewusii*. *Science*, **131**, 1528.

BRACHET, J. (1957). *Biochemical Cytology*. New York: Academic Press Inc.

BURNS, V. W. (1956). Temporal studies of cell division. I. The influence of ploidy and temperature on cell division in *Saccharomyces cerevisiae*. *J. cell. comp. Physiol*. **47**, 357.

BURNS, V. W. (1959). Synchronised cell division and DNA synthesis in a *Lactobacillus acidophilus* mutant. *Science*, **129**, 566.

CAMPBELL, A. (1957). Synchronization of cell division. *Bact. Rev*. **21**, 263.

HAYASHI, M. & SHICHIJI, S. (1959). Effects of 2:4-dinitrophenol on endogenous respiration of yeast harvested during the first budding cycle. *Bull. agric. Chem. Soc. Japan*, **23**, 159.

HINSHELWOOD, C. N. (1946). *The Chemical Kinetics of the Bacterial Cell*. Oxford: Clarendon Press.

HOTCHKISS, R. D. (1954). Cyclical behaviour in pneumococcal growth and transformability occasioned by environmental changes. *Proc. nat. Acad. Sci., Wash*. **40**, 49.

LARK, K. G. & MAALØE, O. (1954). The induction of cellular and nuclear division in *Salmonella typhimurium* by means of temperature shifts. *Biochim. biophys. Acta*, **15**, 345.

LARK, K. G. & MAALØE, O. (1956). Nucleic acid synthesis and the division cycle of *Salmonella typhimurium*. *Biochim. biophys. Acta*, **21**, 448.

LINDEGREN, C. C. & HADDAD, S. A. (1953). The control of nuclear and cytoplasmic synthesis by the nucleocytoplasmic ratio in *Saccharomyces*. *Exp. Cell Res*. **5**, 549.

MARUYAMA, Y. (1956). Biochemical aspects of the cell growth of *Escherichia coli* as studied by the method of synchronous culture. *J. Bact*. **72**, 821.

MARUYAMA, Y. & YANAGITA, T. (1956). Physical methods for obtaining synchronous culture of *Escherichia coli*. *J. Bact*. **71**, 542.

MCFALL, E. & STENT, G. S. (1959). Continuous synthesis of deoxyribonucleic acid in *Escherichia coli*. *Biochim. biophys. Acta*, **34**, 580.

MITCHISON, J. M. (1957). The growth of single cells. I. *Schizosaccharomyces pombe*. *Exp. Cell Res.* **13**, 244.

MITCHISON, J. M. (1958). The growth of single cells. II. *Saccharomyces cerevisiae*. *Exp. Cell Res.* **15**, 214.

MITCHISON, J. M. & WALKER, P. M. B. (1959). RNA synthesis during the cell life cycle of a fission yeast *Schizosaccharomyces pombe*. *Exp. Cell Res.* **16**, 49.

NICKERSON, W. J. & FALCONE, G. (1956). Identification of protein disulphide reductase as a cellular division enzyme in yeasts. *Science*, **124**, 722.

NICKERSON, W. J. & FALCONE, G. (1959). Function of protein disulphide reductase in cellular division of yeasts. In *Sulfur in proteins, Proc. Symposium Falmouth, Mass.*, p. 409. Edited by R. Benesch *et al.* New York: Academic Press Inc.

NIHEI, T., SASA, T., MIYACHI, S., SUZUKI, K. & TAMIYA, H. (1954). Change of photosynthetic activity of *Chlorella* cells during the course of their normal life cycle. *Arch. Mikrobiol.* **21**, 156.

OGUR, M., MINCKLER, S. & MCCLARY, D. O. (1953). Desoxyribonucleic acid and the budding cycle in the yeasts. *J. Bact.* **66**, 642.

PRESCOTT, D. M. (1955). Relations between cell growth and cell division. *Exp. Cell Res.* **9**, 328.

SCHAECHTER, M., BENTZON, M. W. & MAALØE, O. (1959). Synthesis of deoxyribonucleic acid during the division cycle of bacteria. *Nature, Lond.* **183**, 1207.

SCHERBAUM, O. H. (1957). Studies on the mechanism of synchronous cell division in *Tetrahymena pyriformis*. *Exp. Cell Res.* **13**, 11.

SCHERBAUM, O. H. (1959). A comparison of the degree of synchronous multiplication in various microbial systems. *J. Protozool.* **6** (Suppl.), 17.

SCHERBAUM, O. H., LOUDERBACK, A. L. & JAHN, T. L. (1959). DNA synthesis, phosphate content and growth in mass and volume in synchronously dividing cells. *Exp. Cell Res.* **18**, 150.

SCHERBAUM, O. H. & ZEUTHEN, E. (1954). Induction of synchronous cell division in mass cultures of *Tetrahymena pyriformis*. *Exp. Cell Res.* **6**, 221.

SCOTT, D. B. M. (1956). The oxidative pathway of carbohydrate metabolism in *Escherichia coli*. 4. Formation of enzymes induced by 2:4-dinitrophenol. *Biochem. J.* **63**, 593.

SCOTT, D. B. M. & CHU, E. (1958). Synchronized division of growing cultures of *Escherichia coli*. *Exp. Cell Res.* **14**, 166.

SHIBA, S., TERAWAKI, A., TAGUCHI, T. & KAWAMATA, J. (1959). Selective inhibition of formation of deoxyribonucleic acid in *Escherichia coli* by Mitomycin C. *Nature, Lond.* **183**, 1056.

SPIEGELMAN, S. & DUNN, R. (1947). Interactions between enzyme-forming systems during adaptation. *J. gen. Physiol.* **31**, 153.

SPOERL, E. & LOONEY, D. (1959). Synchronised budding of yeast cells following X-irradiation. *Exp. Cell Res.* **17**, 320.

SWAMINATHAN, M. S. & GANESAN, A. T. (1958). Kinetics of mitosis in yeasts. *Nature, Lond.* **182**, 610.

SWANN, M. M. (1954). The control of cell division. In *Recent Developments in Cell Physiology*, p. 185. Edited by J. A. Kitching. London: Butterworth's Scientific Publications.

SWANN, M. M. (1957). The control of cell division: a review. I. General mechanisms. *Cancer Res.* **17**, 727.

SYLVÉN, B., TOBIAS, C. A., MALMGREN, H., OTTOSON, R. & THORELL, B. (1959). Cyclic variations in the peptidase and catheptic activities of yeast cultures synchronised with respect to cell multiplication. *Exp. Cell Res.* **16**, 75.

WICKERHAM, L. J. (1951). Taxonomy of yeasts. *Tech. Bull. U.S. Dep. Agric.* no. 1029.

WILLIAMSON, D. H. & SCOPES, A. W. (1960*a*). Some aspects of the behaviour of synchronously dividing yeast cultures. Paper read before Soc. gen. Microbiol., April, 1960.

WILLIAMSON, D. H. & SCOPES, A. W. (1960*b*). The behaviour of nucleic acids in synchronously dividing cultures of *Saccharomyces cerevisiae*. *Exp. Cell Res.* **20**, 338.

WILLIAMSON, D. H. & SCOPES, A. W. (1961 *a*). The distribution of nucleic acids and protein between different sized yeast cells *Exp. Cell Res.* (in the Press).

WILLIAMSON, D. H. & SCOPES, A. W. (1961 *b*). Protein synthesis and nitrogen uptake in synchronously dividing cultures of *Saccharomyces cerevisiae*. *J. Inst. Brew.* (in the Press).

EXPLANATION OF PLATES

PLATE 1

Fig. 1. 7-day-old culture in liquid malt extract medium before synchronization, showing the typical variation in cell size. Phase contrast.

Fig. 2. Cells from a synchronized population. Note the uniformity of size and the prominent vacuoles. Phase contrast.

PLATE 2

Fig. 3. Synchronously dividing cells just after their first division. In order to restrict physical separation of parent and daughter cells, this culture was not agitated. Most pairs have produced 2nd generation buds. Phase contrast.

Fig. 4. Synchronous culture a little older than that shown in Fig. 3, grown in the presence of Mitomycin C (200 μg./ml.). This inhibitor has prevented cell division and the appearance of 2nd generation buds. However, the volume (and presumably mass) of the cells is greater than that of their non-inhibited counterparts (fig. 3), which implies that some growth has taken place despite the block to budding. Phase contrast.

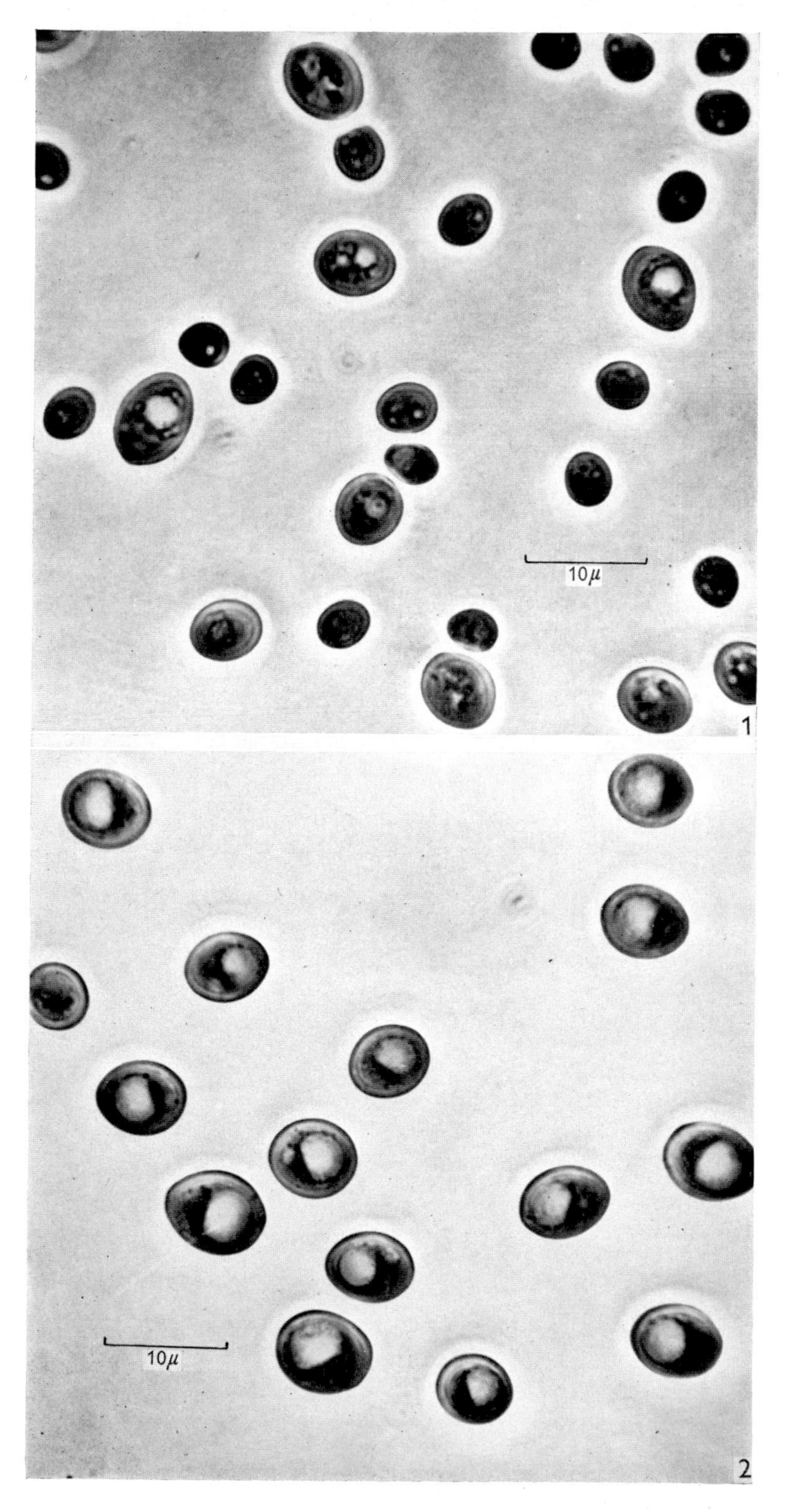

(*Facing page 242*)

PLATE 2

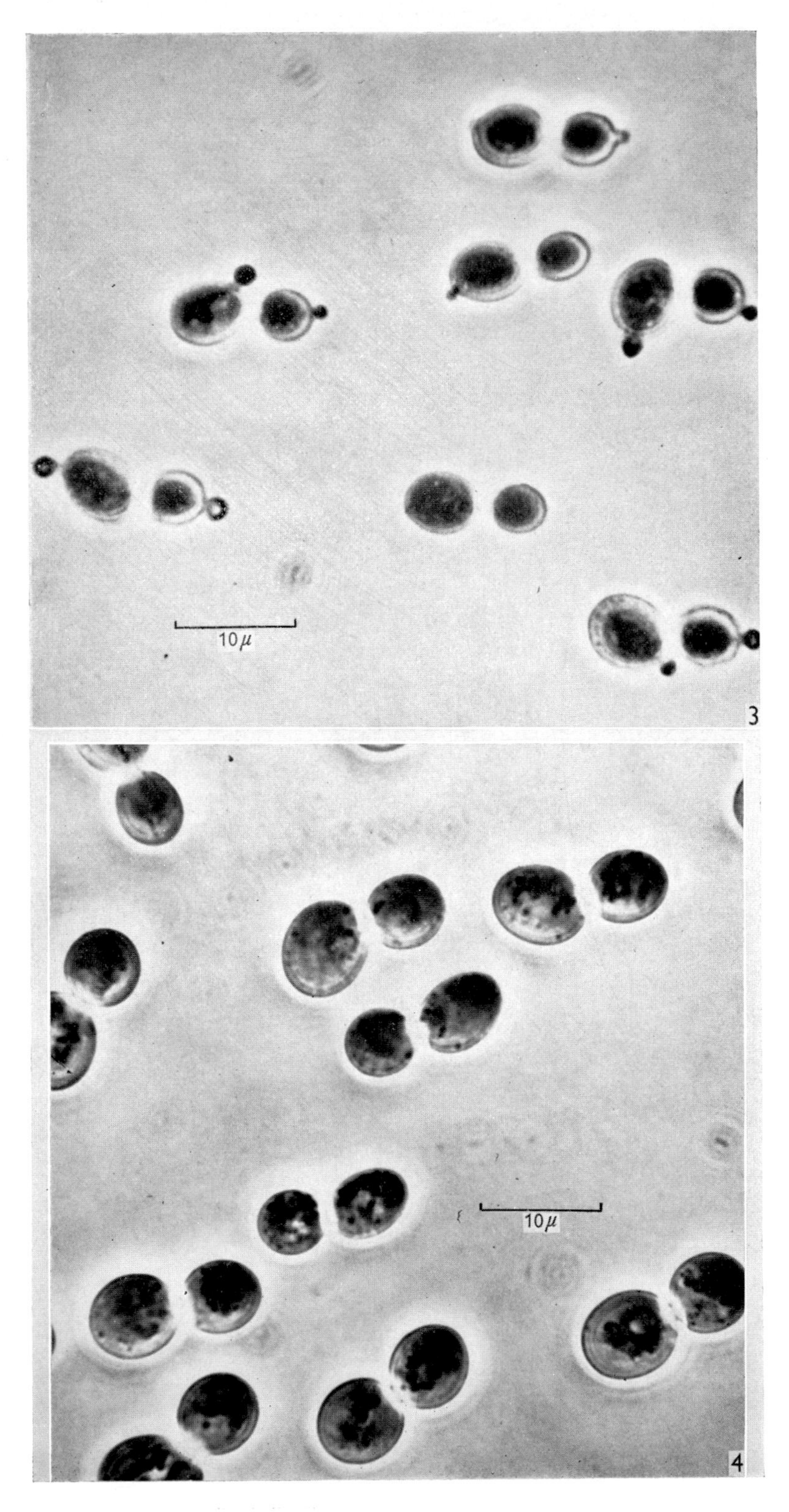

THE RELATIONSHIP BETWEEN BIOCHEMICAL AND MORPHOLOGICAL DIFFERENTIATION IN NON-FILAMENTOUS AQUATIC FUNGI

EDWARD C. CANTINO

Department of Botany and Plant Pathology, Michigan State University, East Lansing, Michigan

Almost twenty years ago, the writer was introduced by Professor Ralph Emerson to an amazingly versatile group of miniature organisms called the water moulds and to the rare combinations that these primitive Lilliputian creatures could provide for the developmental biologist interested in morphogenesis. Among them were the Chytridiales and Blastocladiales, most of which developed, at maturity, into non-filamentous, multinucleate, one- and two-celled plants; tiny thalli in their over-all dimensions, yet abundantly complex in their morphology and developmental pathways.

In 1948, we gazed for the first time at one such organism—which, with memories of him who introduced us to the antics of these creatures, we named *Blastocladiella emersonii*—and we have been looking at it ever since. It seems appropriate, therefore, before the reader becomes absorbed in scientific thought, to let him catch a hyperbolic glimpse of our pet as we have learned to view her (Pl. 1)—and so, perhaps, to lure him to the fold. For, among the vast assemblage of aquatic Phycomycetes, almost all the non-filamentous ones remain to be exploited for experimental studies of morphogenesis. But lest, in our enthusiasm, we convert *too* many to our point of view, we leave one parting thought. In the affectionate words of Fred Sparrow, a fellow mycologist whose lifetime has been devoted to aquatic fungi: 'I need not repeat the harsh words uttered in the past about these dazzling and elusive creatures—chytrids are minute, here today and gone tomorrow, recalcitrant wherever and whenever the opportunity offers, and all in all, thoroughly uncooperative. They have left generations of skilled mycologists wondering in their final hours if they *had* seen such and such a chytrid, forty years ago, for it had never afterwards been seen.'

EARLY STUDIES OF DEVELOPMENTAL PHENOMENA IN *BLASTOCLADIELLA EMERSONII*

The genus *Blastocladiella* was discovered by Matthews (1937) with her description of *B. simplex*; almost simultaneously, but independently, Harder & Sörgel (1938) described a second species of the genus, now known (Couch & Whiffen, 1942) as *B. variabilis*. *B. emersonii* (Cantino & Hyatt, 1953), not unlike either of these in size and general appearance (Pl. 1), does differ from both of them in certain essential respects; it has occupied our attention for the past 13 years.*†

The main developmental pathway

When the motile spore of *Blastocladiella emersonii* is transferred to the surface of a suitable agar medium, it retracts its flagellum and subsequently sends out a germ tube; the latter develops (Fig. 1) into an extensive system of tapering, non-septate rhizoids, while the spore itself gradually enlarges several hundred-fold and becomes the main body of the plant. Later on, a cell wall is laid down which separates the thallus into a basal, rhizoid-bearing stalk and an apical, multinucleate, fertile cell. Finally the terminal protoplast is partitioned into hundreds of uninucleate, motile spores which are released through pores formed by the dissolution of several papillae on the surface of the plant. In any population of *B. emersonii*, four different phenotypes may be produced, their proportion in the population depending upon the nature of the medium and other environmental conditions: (1, 2), ordinary colourless (OC) and late colourless (LC) plants, which do not produce an orange pigment and can be distinguished from one another by their different generation times; (3), orange (O) plants which are brightly coloured by virtue of their content of gamma carotene; and (4), resistant sporangial (RS) plants, which possess a brown, thick, pitted wall impregnated with melanin.

Under ordinary circumstances, the vast majority of the thalli in any population are thin-walled, colourless (OC) plants. Let us look at the picture of their ontogeny in greater detail.

* Our work has been supported by research grants from the Eli Lilly Company (Indianapolis, Indiana), The National Science Foundation and The National Institutes of Health, U.S.A.

† Our present picture of the biology of *Blastocladiella* would not have been possible without the earnest co-operation and skilled investigations of my more recent colleagues—Evelyn Horenstein, James Lovett, Howard McCurdy and Gilbert Turian. The results of their work, many of which are scattered throughout this report, are documented in detail in the references, to which the reader is referred.

The spore itself (Fig. 1) is organized in crafty and intriguing fashion, the largest and most obvious compartment being the nuclear cap overlying the nucleus and nucleolus. (The fine structure of these organelles is revealed in great detail by the ultrathin sections of motile cells of a close relative, *Allomyces*: Turian & Kellenberger, 1956). The spore of *Blastocladiella* contains, on the average, 15 μg. of deoxyribose nucleic acid (DNA) and 200 μg. of ribose nucleic acid (RNA)/mg. dry weight;

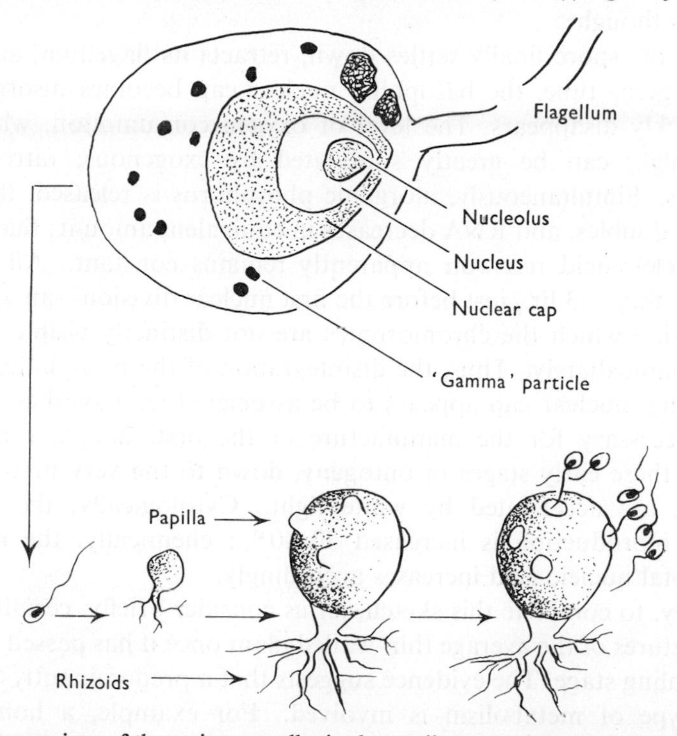

Fig. 1. Comparison of the main organelles in the motile spore of *Blastocladiella emersonii*, and the ontogeny of a thin-walled ordinary colourless (OC) plant derived therefrom.

judging from studies of *Allomyces*, and the basophilic nature of the nuclear cap in *Blastocladiella*, much of the RNA must be concentrated in this extra-nuclear organelle. In addition, a number of minute (0·5 μ dia.) bodies, which have been named 'gamma' particles, are visible in the cytoplasm of the spore. The flagellum of *B. emersonii* is of the whiplash type. Electron photomicrographs of ruptured flagella, which Dr A. L. Houwink kindly made for us during our stay in Professor Kluyver's laboratory at the University of Delft in 1950, revealed that the tail consisted of numerous, fibre-like strands surrounded by a sheath. Although we have done nothing further along these lines, it has been shown (Koch, 1956) that chytrid flagella, and recently, *Allomyces* flagella (Blondel & Turian, 1960), like those of many other motile cells (Hoffmann-Berling, 1959), consist of a pair of central fibrils and nine outer ones, all within a flagellar sheath. The tail of *B. emersonii* may well possess a similar structure. The speed with which the motile spore briskly darts about—and, so to speak, its state of apparent restlessness—suggests that its metabolic machinery must be in third gear. The extraordinarily high, endogenous Q_{O_2} of the spore (*c*. 100) is in keeping with this thought.

When the spore finally settles down, retracts its flagellum, and produces a germ tube, the basophilic nuclear cap becomes disorganized and quickly disappears. The level of oxygen consumption, which remains high, can be greatly stimulated by exogenous, nitrogenous materials. Simultaneously, inorganic phosphorus is released, the pool of DNA doubles, and RNA decreases in equivalent amount; that is, the total nucleic acid reservoir apparently remains constant. All of this occurs within 2–3 hr., just before the first nuclear division—an aberrant type during which the chromosomes are not distinctly visible—which follows immediately. Thus, the disintegration of the basophilic, RNA-containing, nuclear cap appears to be associated with synthesis of the DNA necessary for the manufacture of the first, daughter nucleus. Finally, these early stages in ontogeny, down to the very first nuclear division, are accelerated by white light. Cytologically, the rate of nuclear reproduction is increased 30–40%; chemically, the ratio of DNA/total nucleic acid increases accordingly.

Finally, to complete this sketch, let us consider briefly, certain metabolic features of the average thin-walled plant once it has passed beyond the germling stage. The evidence suggests that a predominantly fermentative type of metabolism is involved. For example, a homolactic fermentation is demonstrable under certain conditions during growth. In the absence of proliferation, lactate is the only labelled acid (other than amino acids) in the soluble pool which results from dissimilation of ^{14}C-glucose by pre-formed plants; furthermore, glucose does not induce an increase in O_2 consumption or CO_2 production over endogenous levels. And lastly, the following enzymic activities are readily detectable in cell-free preparations of these plants: (*a*) rapid conversion of fructose-1-6-diphosphate to phosphoglyceric acid via a diphosphopyridine nucleotide (DPN)-specific, iodoacetate-sensitive, triose phosphate dehydrogenase; (*b*) metabolism of triose via a slower, rate-limiting, fluoride-sensitive step between phosphoglyceric acid and pyruvic acid, and its conversion to lactic acid by a strong, DPN-specific, lactic dehydrogenase; (*c*) a very strong, DPN-specific, α-glycerophosphate dehydrogenase; and (*d*) a rate-limiting hexokinase yielding glucose-6-phosphate which, in turn, can feed either the hexose diphosphate reaction above or a strong, triphosphopyridine nucleotide (TPN)-specific, glucose-6-phosphate dehydrogenase found in OC plants. Finally, it is worth noting that acetate, at concentrations as low as 10^{-3} M, is toxic and prevents growth of *Blastocladiella*.

On the other hand, almost all the enzymic activities associated with the tricarboxylic acid cycle are demonstrable in cell-free preparations, as are the pools of the more stable intermediates of the cycle when labelling techniques are used. And this, then, leads to a final point of emphasis regarding OC plants. The stimulatory effect of light upon nuclear reproduction is by no means limited to the young germling stage. Suffice it to say for this discussion that, provided a minimum amount of CO_2 is available, white light induces OC plants to alter their generation time and to grow more rapidly (increase in *mass*), to fix CO_2 and consume glucose more rapidly, to accumulate a greater pool of soluble organic phosphorus, to consume glycine at a different rate, to produce more nuclei, and to exhibit a greater ratio of DNA/total nucleic acid than they do in the dark. The tentative picture as we see it now is this: light, in some unknown fashion, brings about increased CO_2 fixation via reductive carboxylation of ketoglutarate to isocitrate by means of a TPN-specific isocitric dehydrogenase, cleavage of isocitrate to succinate and glyoxylate via isocitritase, transamination of glyoxylate with alanine to yield pyruvate and glycine, and finally, increased synthesis of thymine and/or thymidine, DNA, and nuclei as a result (at least in part) of the increased availability of glycine. This 'lumisynthesis' in *Blastocladiella* represents another theatre of activity. It is mentioned here briefly because it is helping to establish the existence of metabolic pathways in this organism which can operate in the absence of a (functionally intact) tricarboxylic acid cycle; thus, it will also help to provide a pertinent background for the discussions which follow (Cantino, 1959, 1960; Cantino & Horenstein, 1959; Cantino & Lovett, 1960; Cantino & Turian, 1959*a*, 1961; McCurdy & Cantino, 1960; Turian & Cantino, 1959, 1960).

The carotenogenic developmental pathway

While some 99% of the spores of *Blastocladiella emersonii* generally develop into OC plants as described above, a few of them give rise instead to the thin-walled, orange plants (O plants) mentioned previously (see Fig. 2). It is particularly interesting that visible quantities of the pigment become detectable only at the very end of ontogeny (the last

2–3 hr.) at which time the full complement of pigment is laid down very quickly—usually in a matter of 10–15 min. Physiologically, O plants differ from OC plants in that they are more tolerant of reduced oxygen tensions; conversely, under ordinary aerobic conditions, the viability of their motile cells is much lower than that of OC plants. In all essential respects, however, the morphological sequence of events in ontogeny is like that described in Fig. 1 for OC plants. Because a convincing case can be made for considering O plants as non-functional males, they have played an important role in studies of sexual differentiation in the Bastocladiales. This again, however, represents a separate area of investigation which cannot be pursued further now (Cantino & Turian, 1959*a*, *b*).

The melanogenic developmental pathway

Actually, as we have seen, the uninucleate spores of *Blastocladiella emersonii* can give rise to four kinds of mature plants (Fig. 2). A first-generation population usually consists of three varieties of thin-walled plants, two of which have been described; the third is distinguishable from them by its longer generation time. The fourth, with its distinctive, brown, thick-pitted wall, seldom appears under these conditions.

In the presence of exogenous bicarbonate (*c*. 10^{-2}M), the nature of this population changes drastically; orange plants are produced as before, but the thin-walled, colourless plants are replaced by thick-walled, brown, resistant-sporangial (RS) plants. In effect, the bulk (*c*. 99 %) of the spores which ordinarily would have become OC plants are induced by bicarbonate to develop along a different developmental path. This dramatic influence of bicarbonate is, of course, circumscribed by time. Under any set of specific, environmental conditions, there exists a 'point of no return' in ontogeny beyond which the 'bicarbonate trigger mechanism', as we call it, can no longer be imposed upon a plant; on Emerson Y_pS_s agar (Difco Laboratories, Detroit) at 20°, for example, it is approximately three-fifths of the generation time of the OC plant. Introduction of bicarbonate at any stage in ontogeny up to this point induces formation of RS plants; beyond it, bicarbonate is ineffective. Thus, when this work began, we set out to examine the mechanism by which bicarbonate induced the *de novo* synthesis of carotene and melanin with increased deposition of chitin and fat, and the formation of the numerous other structural and functional components associated with a resistant sporangial plant. The lack of true, gametic copulation (thus giving a genotype which was stable, save for mutation) provided a favourable background for a study of this sort.

Very early in our work, it was learned that certain Krebs cycle intermediates (such as ketoglutarate) augment the effect of bicarbonate under nutritional conditions where the latter, by itself, cannot induce RS formation. Oxidative decarboxylation inhibitors (e.g. arsenite) and keto-reagents (e.g. semicarbazide) behave similarly; furthermore, once RS plants are formed, they are prevented from germinating by the

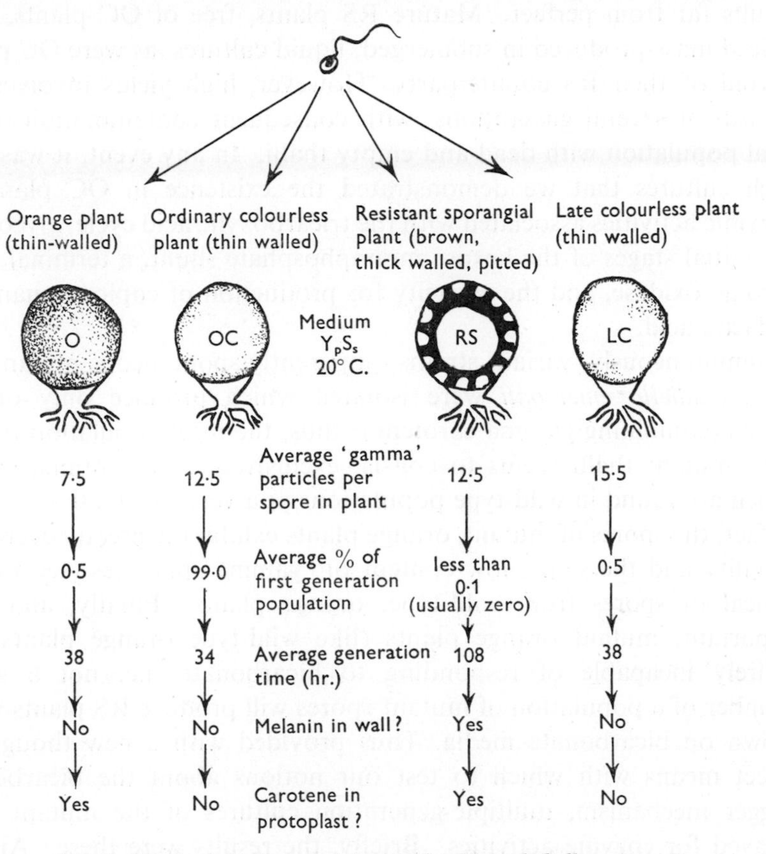

Fig. 2. The four developmental paths in *Blastocladiella emersonii*.

identical, narrow range of concentrations of inhibitors that induce their formation. Also, when OC plants are bathed in bicarbonate, the internal pool of ketoglutarate increases. These and other observations led to the hypothesis that bicarbonate interferes with the two successive oxidative decarboxylations in a weakly functional tricarboxylic acid cycle; that it gradually causes a reversal of this part of the cycle by way of a reductive carboxylation of ketoglutarate to isocitrate; and that the isocitrate formed is then further metabolized via shunt reactions which lead to the synthesis of the ingredients essential for the manufacture of a resistant sporangial plant.

The immediate hurdle was to learn to grow large quantities of mature OC and RS plants in liquid culture for further comparative analyses. This was accomplished although, as we look back upon it now with analytic hindsight, the approach was somewhat primitive and the results far from perfect. Mature RS plants, free of OC plants, were indeed mass-produced in submerged, liquid cultures, as were OC plants devoid of their RS counterparts. However, high yields involved the growth of several generations, with consequent contamination of the final population with dead and empty thalli. In any event, it was with such cultures that we demonstrated the existence in OC plants of enzymic activities associated with the tricarboxylic acid cycle, glycolysis, the initial stages of the hexose monophosphate shunt, a terminal cytochrome oxidase, and the capacity for production of copious quantities of lactic acid.

Simultaneously, variant strains (apparently spontaneous mutants) of *Blastocladiella emersonii* were isolated which produce only orange plants (containing gamma carotene); thus, the *total* population of mature, mutant thalli seems to consist exclusively of the orange plants which are found in wild-type populations at a very low (*c.* 0·5 %) level. In fact, the spores of mutant, orange plants exhibit the greatly decreased viability and the same, low content of 'gamma' particles (see Fig. 2) typical of spores from wild-type, orange plants. Finally, and most important, mutant orange plants (like wild-type orange plants) are entirely incapable of responding to bicarbonate, i.e. not a single member of a population of mutant spores will produce RS plants when grown on bicarbonate media. Thus provided with a new though indirect means with which to test our notions about the bicarbonate trigger mechanism, multiple-generation cultures of the mutant were assayed for enzyme activities. Briefly, the results were these. All but two of the terminal oxidase and Krebs cycle activities present in wild-type OC plants are demonstrable in the mutant, the exceptions being aconitase and ketoglutarate oxidase. Thus, the mutant appears to be incapable of carrying out the two, successive, oxidative decarboxylations of the tricarboxylic acid cycle as a functional unit—an observation consistent with our ideas about the mechanism of the bicarbonate effect.

Lastly, analyses of multiple-generation cultures of mature RS plants revealed the final, rather spectacular and manifold effect of bicarbonate (Fig. 3). In inducing formation of RS plants, bicarbonate brings about: (*a*) a *de novo* synthesis of carotene, melanin, a TPN-linked polyphenol oxidase, and an electrophoretically separable protein fraction, none of which is detectable in OC plants; (*b*) a large decrease in the free amino acid pool as compared to that of OC plants and, in particular, a very sharp drop in tyrosine—an established substrate for the polyphenol oxidase; (*c*) a loss of the terminal cytochrome oxidase activity and two electrophoretically separable protein fractions present in OC plants;

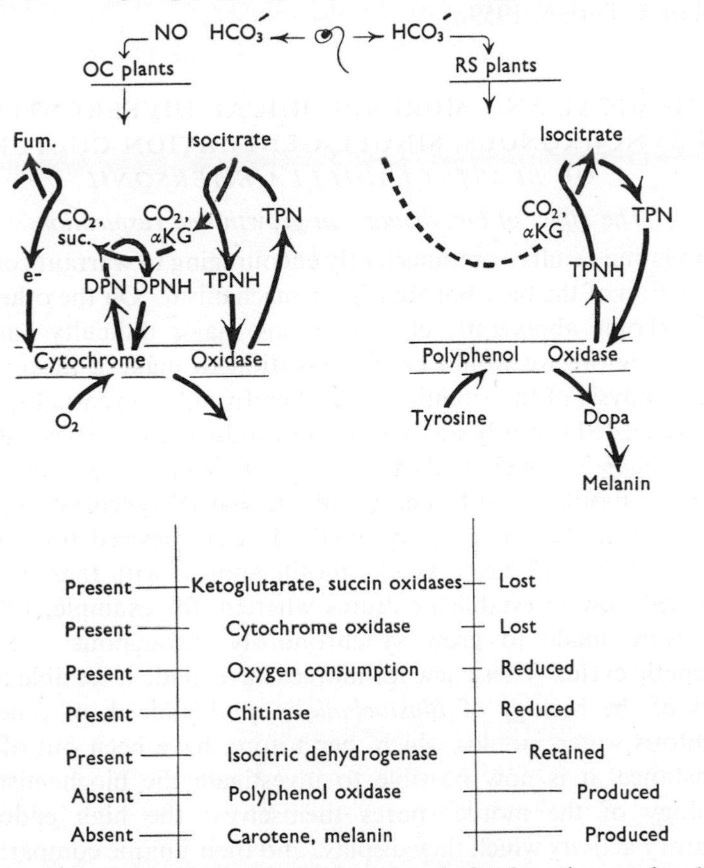

Fig. 3. Alterations in the activity of the Krebs cycle and related reactions produced in wild-type *Blastocladiella emersonii* by addition of bicarbonate to the medium.

(*d*) a very great decrease in the Q_{O_2} of RS plants relative to that of OC plants; and (*e*) a severe reduction, if not a total loss, in the ketoglutarate oxidase and the succinic dehydrogenase activities present in OC plants. However, in sharp contrast with these changes, the TPN-specific isocitric dehydrogenase present in OC plants is retained in RS plants; it mediates reductive carboxylation of ketoglutarate, and its activity in the forward direction is inhibited by bicarbonate. Parenthetically, it is noteworthy that: (*a*) bicarbonate induces a *de novo* synthesis of gamma carotene and multiple lesions in the tricarboxylic acid cycle, but it does not induce a loss of the isocitric dehydrogenase in RS plants; and (*b*) the orange mutant plants of *Blastocladiella*, like RS plants, possess the lesion—albeit a different one—and the carotene, but also retain the isocitric dehydrogenase. The losses and gains associated with differentiation of RS plants which these data emphasize are depicted in Fig. 3 (Cantino & Turian, 1959*a*).

BIOCHEMICAL AND MORPHOLOGICAL DIFFERENTIATION IN SYNCHRONOUS, SINGLE-GENERATION CULTURES OF *BLASTOCLADIELLA EMERSONII*

The effect of bicarbonate on growth and respiration

The foregoing results were sufficiently encouraging to warrant continued investigation of the bicarbonate trigger mechanism. On the other hand, it had become abundantly clear that one major difficulty had to be overcome before we would be in a position to make a thorough and definite analysis of the situation and, therefrom, an eventual synthesis. What we needed urgently was a means of producing very large quantities of *Blastocladiella* which could be studied at every single stage in ontogeny, from motile spore to mature plant, that is, synchronous, single-generation cultures. Eventually, methods were devised for preparing thick suspensions of clean, viable, motile spores; with these available, we learned how to establish cultures wherein, for example, 10^8 to 10^9 plants were made to grow synchronously throughout a complete ontogenetic cycle.* These new techniques have made it feasible to study aspects of the biology of *Blastocladiella*, and, indeed, of other non-filamentous water moulds which, heretofore, have been out of reach. For instance, it is now possible to investigate the biochemistry and physiology of the motile spores themselves; the high endogenous respiratory activity which they display, and their unique compartmenta-

* A suitable growth medium such as 'Cantino PYG broth' (Difco Laboratories Inc., Detroit) is heavily inoculated with viable motile spores. These can be collected by flooding agar plates bearing mature OC plants with water and by then concentrating the spore suspension (10^7–10^8 spores/ml.) by centrifugation in large (e.g. 100 ml.) cups at 400 *g* for exactly 5 min.; cooling the loosely-packed pellet of spores to 10° after transfer to a small centrifuge tube; followed by centrifugation at 1600 *g* for exactly 30 sec., the deposit being resuspended in 0·01 M phosphate buffer, pH 7·0. The last centrifugation is then repeated (McCurdy & Cantino, 1960).

The details are critical: when followed exactly, the concentrated suspension consists of spores showing 100 % motility and viability. Any deviation may lead to disintegration and/or loss of motility.

Spores prepared in this fashion develop almost synchronously when placed in vigorously aerated medium into mature, 2-celled plants without further treatment (McCurdy & Cantino, 1960; Lovett & Cantino, 1960*b*).

lization into packets of nucleic acids, suggest that this will be a most fruitful avenue for investigation. But more important for the immediate problem at hand, we can now analyse every single stage in the ontogeny of both RS plants and OC plants, and many aspects of our interpretation of the bicarbonate trigger mechanism can be put to a direct test.

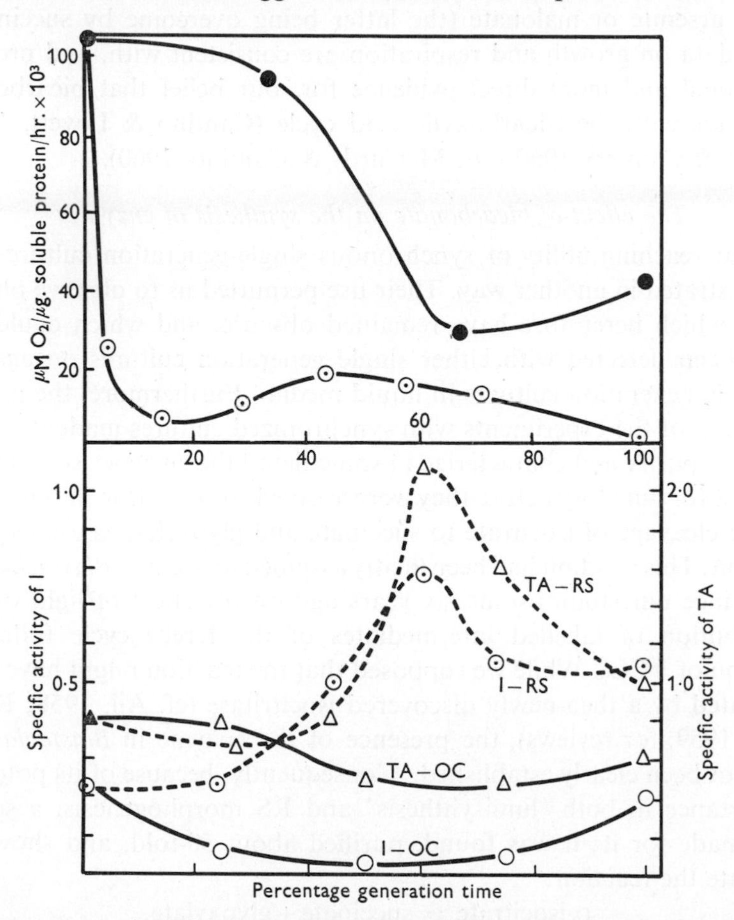

Fig. 4. *Top*. Comparison of the endogenous oxygen consumption of synchronously dividing OC (top curve) and RS (bottom curve) plants during ontogeny (derived from data of McCurdy & Cantino, 1960, and Cantino & Lovett, 1960). *Bottom*. Comparison of the specific activities of isocitritase (I) and glycine-alanine transaminase (TA) in synchronously dividing OC and RS plants during ontogeny (derived from data of McCurdy & Cantino, 1960).

For example, we now know that the course of oxygen consumption during ontogeny of an RS plant is strikingly different from that of a developing OC plant. Bicarbonate induces a precipitous drop in the respiration of spores as soon as they germinate and begin to develop along the path that leads to RS plants (Fig. 4, *top*). Furthermore, since the increase in dry weight was exponential in both cases, linear log plots of the dry weight/plant during the first 12 hr. of development reveal that the growth rate of a young RS plant in the bicarbonate medium is about half of that for a young OC plant in non-bicarbonate medium. Since the endogenous oxygen consumption of spores is sensitive to inhibition by either arsenite or malonate (the latter being overcome by succinate), these data on growth and respiration are consistent with, and provide additional and more direct evidence for, our belief that bicarbonate interferes with the tricarboxylic acid cycle (Cantino & Lovett, 1960; Lovett & Cantino, 1960*a*, *b*; McCurdy & Cantino, 1960).

The effect of bicarbonate on the synthesis of enzymes

The far-reaching utility of synchronous single-generation cultures can be illustrated in another way. Their use permitted us to observe phenomena which heretofore have remained obscure, and which could not have been detected with either single generation cultures on agar or multiple generation cultures in liquid media. Furthermore, the greater precision of the experiments with synchronized cultures made it worthwhile to purify and characterize in some detail the enzymes of potential interest for our story before they were assayed for their role in ontogeny.

The cleavage of isocitrate to succinate and glyoxylate is a good case in point. This reaction had been tacitly assumed to occur in *Blastocladiella* ever since our studies some six years ago on the effect of light on the distribution of labelled intermediates of the Krebs cycle following fixation of $^{14}CO_2$. While we supposed that the reaction might have been mediated by a then-newly discovered isocitritase (cf. Ajl, 1958; Kornberg, 1959, for reviews), the presence of the enzyme in *Blastocladiella* had not been clearly established. Consequently, because of its potential importance in both 'lumisynthesis' and RS morphogenesis, a search was made for it; it was found, purified about 50-fold, and shown to mediate the reaction:

$$\text{D-isocitrate} \rightleftharpoons \text{succinate} + \text{glyoxylate}.$$

Its reversibility, pH optimum (7·4), specificity for the *d*-isomer, requirement for SH groups and Mg^{++} (or Mn^{++}), and its Michaelis constant ($4 \cdot 8 \times 10^{-4}$ M), as well as the activation energy (10,700 cal./mole) and Q_{10} (1·8) for the enzymic reaction were established.

This, in turn, led to the investigation of a second enzyme involved in the chain of reactions leading away from isocitritase; namely, a glycine-alanine transaminase. This enzyme, never heretofore purified from any source as far as we can tell, mediates the reaction:

$$\text{glyoxylate} + \text{L-alanine} \rightleftharpoons \text{glycine} + \text{pyruvate}.$$

It was purified 80-fold, and its reversibility, pH optimum (8·5), probable requirement for pyridoxal phosphate, high degree of stability, and specificity for alanine were established.

With their properties characterized, the specific activities (total activity per unit soluble protein) of these enzymes were followed from spore stage to maturity for both OC and RS plants (Fig. 4, *bottom*). During development of thin-walled OC plants, the specific activity of isocitritase decreases during the first third of the generation time and does not rise again until the plants are nearing maturity; that is, as they approach the last stage in ontogeny when the protoplast is segmented into spores once again. It is precisely at this point (as shown at zero time for spores) that isocitritase activity is at its highest level in the life history of the thin-walled plant of *Blastocladiella*.

In sharp contrast to this state of affairs in OC plants, during the ontogeny of RS plants in bicarbonate media, isocitritase activity increases rapidly at *c.* 40–50 % of the generation time and then falls again; note that the rapid rise occurs in the general region of the 'point of no return' in the life history of *Blastocladiella*! Transaminase activity follows a similar course. Thus, there is a rapid increase in the specific activity of two sequential reactions which, at least potentially, constitute a device for removing the product of the reductive carboxylation of ketoglutarate; i.e. isocitrate. These results are consistent with, and supply further evidence for, our interpretation of the bicarbonate trigger mechanism for morphogenesis in *Blastocladiella*.

At this point, another aspect of the importance of synchronized, single-generation, cultures is best appreciated. It is a simple matter to set up parameters so that experimental data can be related to individual plants rather than the dry weight or soluble protein content of a population. Thus, when the *activity per plant* (as a log function) of the isocitritase is examined, it becomes abundantly clear (Fig. 5) that *no synthesis of isocitritase occurs until about half of the generation time of the OC plant has elapsed*! The large quantity of isocitritase originally present in the spore becomes quickly diluted as the plant increases in volume some 30- to 40-fold; only then does synthesis of isocitritase begin. However, when spores are placed in bicarbonate and induced to develop along the path which leads to RS plants, *synthesis of isocitritase apparently begins at once*; the increase in log enzyme activity per plant is linearly related to the increase in age (i.e. percentage-generation time) of the plant. Thus, directly or indirectly, bicarbonate induces immediate synthesis of isocitritase (an 'anaerobic' enzyme, and an adaptive enzyme in bacteria; cf. Reeves & Ajl, 1960). At the same time, bicarbonate causes the

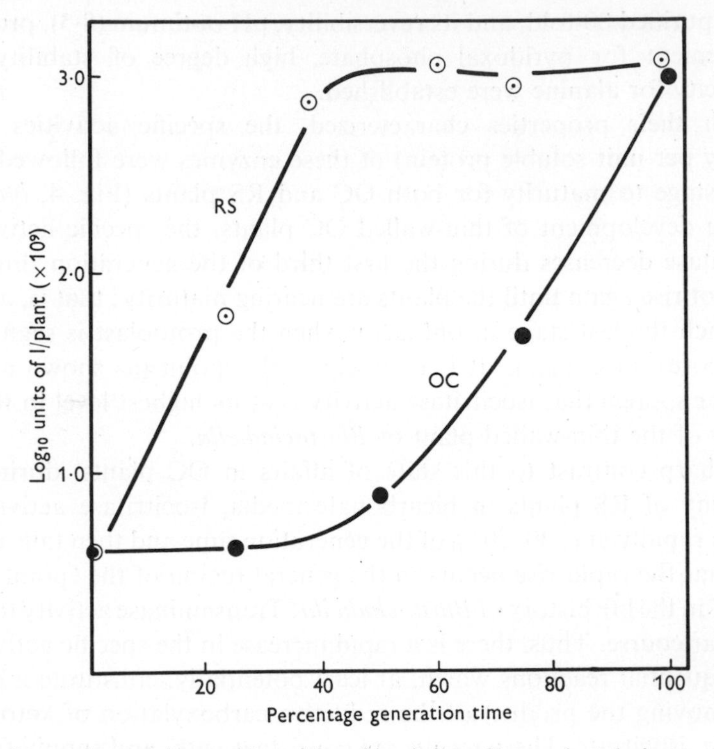

Fig. 5. Comparison of the specific activity of isocitritase (I) per plant during ontogeny of OC and RS thalli (derived by combination of data from McCurdy & Cantino, 1960, and Lovett & Cantino, unpublished).

precipitous drop in oxygen consumption and the sharp decrease in the growth rate; concomitantly, the young germling is on its way to an RS plant.*

* Parenthetically, it is of interest that in some bacteria (see references to Smith & Gunsalus, Kornberg *et al.* etc. in Kornberg, 1959), glucose seems to repress isocitritase formation. During growth of RS plants of *Blastocladiella* in bicarbonate media, there is no detectable glucose utilization from the medium throughout most of the period of active isocitritase synthesis. Subsequently, glucose uptake begins and rises very sharply, but only when synthesis of the enzyme has almost ceased. And yet, pre-formed RS plants at these same, early stages of development are perfectly capable of assimilating glucose if they are incubated under conditions (e.g. glucose and phosphate buffer) where they cannot grow (Cantino & Lovett, 1960). Curiously, this correlation is, on the surface at least, the reverse of the 'glucose effect' in *Escherichia coli* (e.g. Cohn & Horibata, 1959), where glucose inhibits production of β-galactosidase during growth in the presence of inducer until the sugar has been consumed, after which production of enzyme begins.

From another point of view, we apparently find here an interesting difference relative to higher plants. Carpenter & Beevers (1959) find that isocitritase is generally confined to regions of active fat breakdown. But, during growth of *Blastocladiella*, fat synthesis is exponential from zero time (as is isocitritase); in fact, maximum quantities (*c.* 200 μg. lipid/mg. dry weight) are deposited in the plant when the specific activity of isocitritase reaches its peak.

Without pursuing this story further, suffice it to say we now feel all the more certain that the sequence of reactions (Fig. 6) leading from the TPN-specific carboxylation of ketoglutarate to isocitrate, the cleavage of isocitrate to succinate and glyoxylate, and the further metabolism of glyoxylate via transamination are intimately involved in the bicarbonate trigger mechanism of morphogenesis in *Blastocladiella*. In the related,

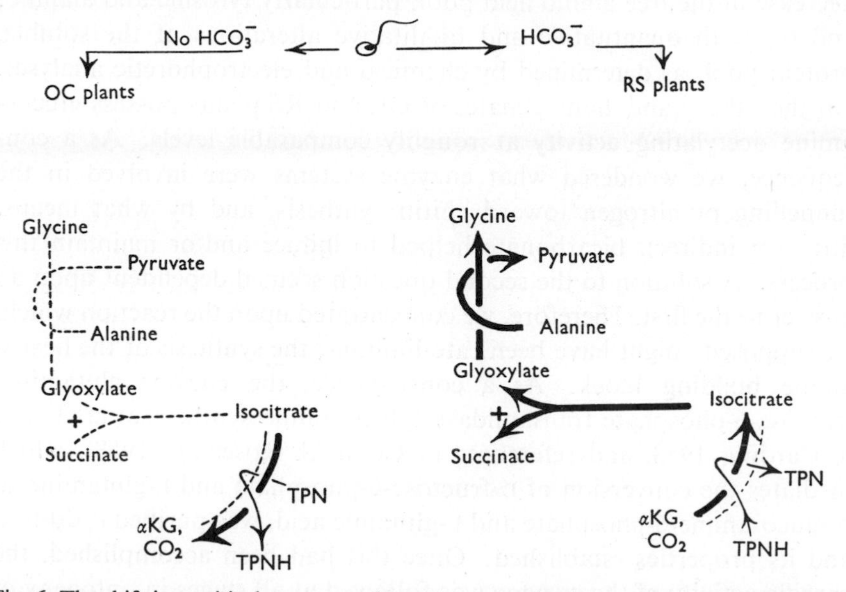

Fig. 6. The shift in a critical sequence of reactions associated with the bicarbonate trigger mechanism for morphogenesis, showing how bicarbonate alters the metabolism of isocitrate.

microaerophilic genus, *Blastocladia*, CO_2 and/or bicarbonate also induce RS formation and, simultaneously, a 100 % increase in production of succinic acid; a similar mechanism may well apply (Emerson & Cantino, 1948; Cantino, 1949). We feel confident that as our studies with synchronous cultures progress, the fate of glycine, in particular, will be shown to play an important role in the formation of RS plants and the nucleic acid transformations (see later) involved in their final maturation (Lovett & Cantino, 1960*b*; McCurdy & Cantino, 1960; Turian & Cantino, 1959).

The effect of bicarbonate on nitrogen metabolism

The transformations in the nitrogen pool which occur during morphogenesis deserve special attention for two reasons: (*a*) the bicarbonate trigger mechanism probably has a generalized effect on the machinery for nitrogen metabolism in *Blastocladiella*; and (*b*) our latest studies in this area further emphasize, in yet another way, the value of expressing metabolic data on a per-plant basis.

We had known from our early work that, relative to OC plants, the formation of RS plants involved: (*a*) a net increase in chitin synthesis and, simultaneously, a decrease in the activity of a chitinase capable of degrading chitin; (*b*) except for glutamate and aspartate, a drastic decrease in the free amino acid pool, particularly tyrosine and alanine; and (*c*) both quantitative and qualitative alterations of the soluble, protein pool, as determined by chemical and electrophoretic analyses. On the other hand, homogenates of OC and RS plants possess glucosamine acetylating activity at roughly comparable levels. As a consequence, we wondered what enzyme systems were involved in the funnelling of nitrogen toward chitin synthesis, and by what means, direct or indirect, bicarbonate helped to induce and/or maintain this process. A solution to the second question seemed dependent upon an answer to the first. Therefore, we concentrated upon the reaction which, we supposed, might have been rate-limiting; the synthesis of the hexosamine building block. As a consequence, the enzyme glutamine-fructose-6-phosphate transamidase (glucosamine synthetase; cf. Leloir & Cardini, 1953, and references in Comb & Roseman, 1958), which mediates the conversion of D-fructose-6-phosphate and L-glutamine to D-glucosamine-6-phosphate and L-glutamic acid, was purified *c*. 20-fold and its properties established. Once this had been accomplished, the specific activity of the enzyme was followed at all stages in ontogeny of RS plants.

Once again, it became apparent that the method of expressing enzyme activity, and a judicious interpretation thereof, is of critical importance in studies of morphogenesis. For example (Fig. 7), glucosamine synthetase activity per unit weight of organism reaches a summit at *c*. 29 % of the generation time of RS plants, specific activity (i.e. per unit of soluble protein) is greatest at 57 % of the generation time, and activity per plant reaches its peak at 43 % of the generation time. A log plot of the latter shows a linear relationship with age, thus revealing that glucosamine synthetase, like isocitritase, is synthesized exponentially from the time the spore germinates to about 43 % of the generation time, after which no further synthesis occurs. As it turns out, the 'premature' peak, obtained on a per-unit-weight basis, is due to more rapid synthesis of enzyme protein than soluble protein pool; thus, this peak could have been misleading. By the same token, the last peak obtained on a per-protein basis is perfectly real, but it is also misleading; it is due to a decrease in soluble protein which begins at 43 % of the generation time (see Fig. 8), rather than to an increase in net synthesis of enzyme after this point in ontogeny. Relative to the problem of morphogenesis, this last observation is important; it means that glucosamine synthetase protein is retained while other soluble proteins in the pool of the developing RS plant are being depleted. Thus, if we may speak teleo-

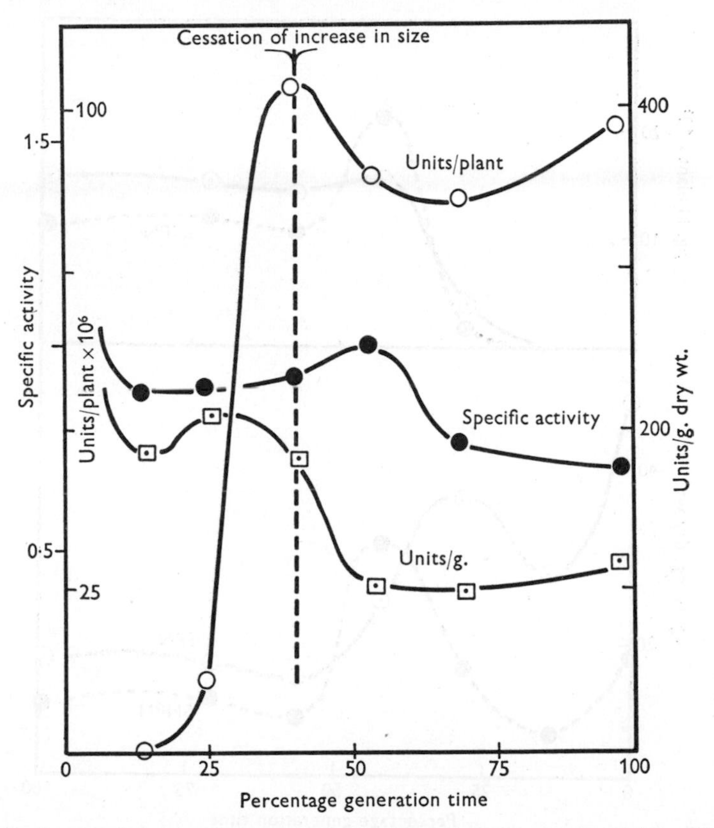

Fig. 7. The degree of glucosamine synthetase activity of synchronously dividing RS plants at various times during synchronous growth expressed in three different ways: (1) units of activity/plant; (2) specific activity (i.e. per unit soluble protein); and (3) units of activity/g. dry-weight plant.

logically for a moment, it leads to the conclusion that the enzyme is 'needed' and must be preserved by the plant for continued synthesis of chitin, half of which remains to be manufactured at this stage (43 % of the generation time) in ontogeny. The great utility of synchronous, single-generation cultures for morphogenetic studies, and the advantages of per-plant data for purposes of interpretation, are thus made clearly evident once again.

Finally, one more observation deserves particular attention (Fig. 8, bottom): on a per-weight basis, the soluble protein pool reaches a peak at about 29 % of the generation time; as it begins to drop again, the soluble non-protein pool starts to rise and subsequently reaches its apex later on in ontogeny, at *c.* 43 % of the generation time. Chromatographs of the free amino acid pool reveal that it, too—and in particular, alanine

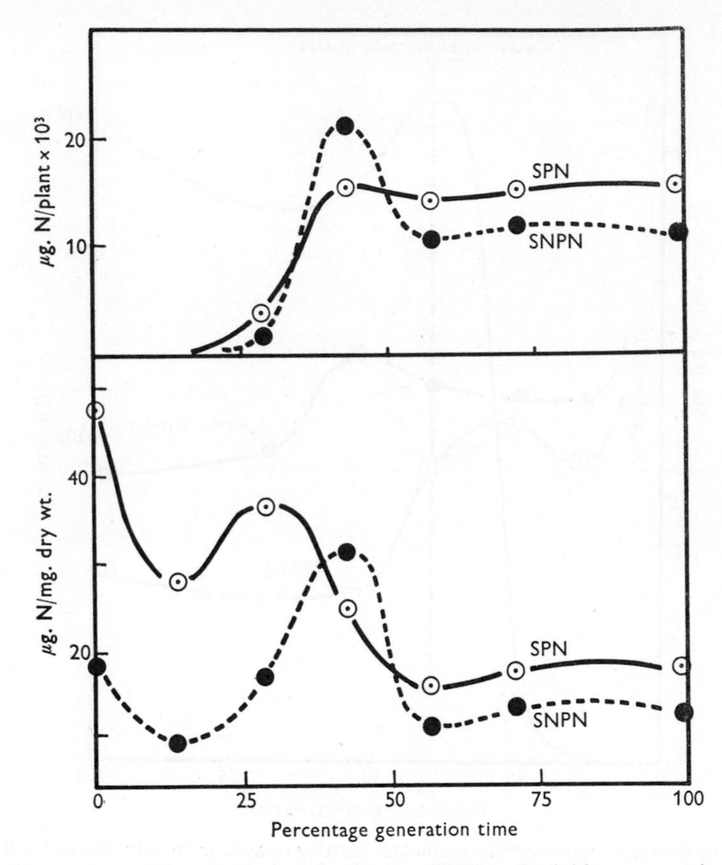

Fig. 8. Comparison of the soluble protein nitrogen (SPN) and soluble non-protein nitrogen (SNPN) pools during ontogeny of RS synchronously dividing plants, on a per-plant basis (*top*) and a per unit-weight basis (*bottom*). (Modified from Lovett & Cantino, 1960*b*.)

(Fig. 9)—reaches its highest value at this later time. It would be tempting to assume that this apparent sequence of events—i.e. a simultaneous decrease in the protein pool and an increase in the amino acid pool—reflects a conversion of soluble proteins into amino acids. That this is not the case, however, becomes evident from the per-plant data delineated in Fig. 8 (*top*). The fact is that both the soluble protein nitrogen and the soluble non-protein nitrogen reach their peaks simultaneously; however, there does occur a *differential* increase in the two!

At 43 % of the generation time, the soluble, non-protein nitrogen actually exceeds the soluble protein nitrogen by about 35 %, and the two together account for over 95 % of the total non-chitin nitrogen in the plant. Again, the advantage of using per-plant data is manifestly clear.

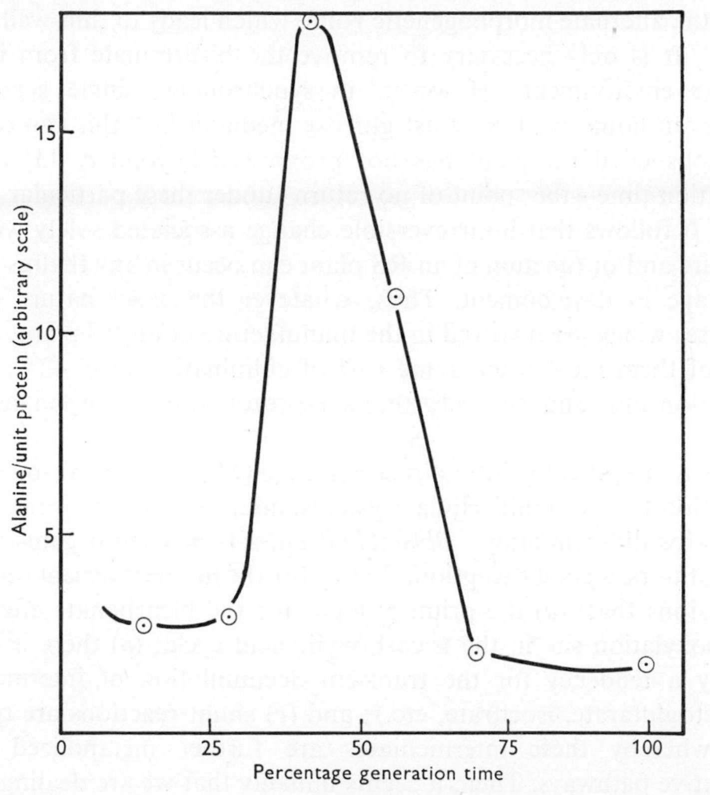

Fig. 9. The size of the alanine pool in synchronously dividing RS plants at various times during the growth cycle of a synchronized culture (in terms of an arbitrary scale, as determined from relative densities of ninhydrin spots after chromatography).

Without belabouring these details further, the over-all picture seems to be that during the bicarbonate-induced morphogenesis of RS plants, extensive rechannelling of nitrogenous components occurs; this involves the funnelling of nitrogen through the glucosamine synthetase reaction and, thence, to deposition of a chitinous cell wall. When we have analysed synchronous cultures of OC plants in similar fashion, we will be in a position to gauge better the causal role of bicarbonate in this sequence of events (Lovett & Cantino, 1960*a*, *b*; McCurdy & Cantino, 1960).

General conclusions about the bicarbonate trigger mechanism for morphogenesis in Blastocladiella emersonii

Even though a spore of *Blastocladiella* has been placed in bicarbonate media and started on its path to RS plants, it can still be made to proceed along the alternate morphogenetic route which leads to thin-walled OC plants. It is only necessary to remove the bicarbonate from its immediate environment. However, in synchronous, single generation cultures in liquid peptone-yeast-glucose media at 24°, this can only be accomplished if the plant has not progressed beyond *c.* 43% of its generation time—the 'point of no return' under these particular conditions. It follows that no irreversible change associated solely with the structure and/or function of an RS plant can occur in any thallus before this stage in development. Thus, whatever the exact nature of the processes which are involved in the manufacture of an RS plant, one or more of them must reach some sort of culmination at *c.* 43% of the generation time and, thenceforth, have profound effects upon development.

Apparently, there is little direct evidence (Markert, 1958) to support the notion that, in multicellular systems, adaptive enzyme formation is a basis for differentiation. *Blastocladiella*, a two-celled organism, may turn out to be a good exception. Let us for the moment accept our early conclusions that: (*a*) the primary locus for the bicarbonate effect is a decarboxylation site in the tricarboxylic acid cycle; (*b*) there is consequently a tendency for the transient accumulation of intermediates (i.e. ketoglutarate, isocitrate, etc.); and (*c*) shunt reactions are quickly used whereby these intermediates are further metabolized along alternative pathways. Then, it seems unlikely that we are dealing solely with a gradual, increased utilization of pre-existing, constitutive enzyme systems which lead to an over-all, quantitative shift in the intracellular balance within the plant. It is much more plausible to assume that bicarbonate brings about induction of new, enzymic systems and elimination of old ones. Three different sorts of evidence now support this point of view: (*a*) the *de novo* appearance of melanin and a polyphenol oxidase activity, a mechanism for synthesis of gamma carotene, and an electrophoretically-separable, soluble, protein fraction, and, at the same time (*b*) the apparent loss of an alpha-ketoglutarate oxidase, succinoxidase, and cytochrome oxidase system, and at least two electrophoretically-separable protein fractions; (*c*) the rapid synthesis of isocitritase following germination of spores in bicarbonate media, and the lack of isocitritase synthesis in the same media lacking bicarbonate; and finally (*d*) while no change in the amount of soluble protein occurs after 43 % of the generation time of the plant, the specific activity of certain enzymes (e.g. glucose-6-phosphate dehydrogenase) increases sharply, while the specific activity of others (e.g. isocitritase) decreases.

It seems highly likely that in *Blastocladiella emersonii*, we are dealing with a dynamic equilibrium wherein the influence of bicarbonate causes synthesis of some enzymes, the loss of others, and an abundant re-channelling of metabolic pathways. The increasingly altered metabolism finally becomes autocatalytic and irreversible—and the formation of an RS plant, inescapable.

MATURATION AND GERMINATION OF RESISTANT SPORANGIAL PLANTS

In the last analysis, the only absolute and incontestable criterion for establishing whether or not a resistant-sporangial (RS) plant is mature—or, for that matter, dead or alive—is its capacity to reproduce itself. This, in effect, means that the resistant sporangium must be capable of cutting up its large, multinucleate protoplast into lots of tiny protoplasts—the motile, uniflagellate, uninucleate spores which subsequently germinate and grow into a second generation of plants. In *Blastocladiella emersonii* (Fig. 10), as in other members of the family, this involves cracking of the thick, pitted wall, formation of papillae on the inner membrane which protrudes through the cracks, dissolution of the papillae to form pores, and escape of the motile spores through these openings.

We have not progressed very far in evaluating, at the biochemical

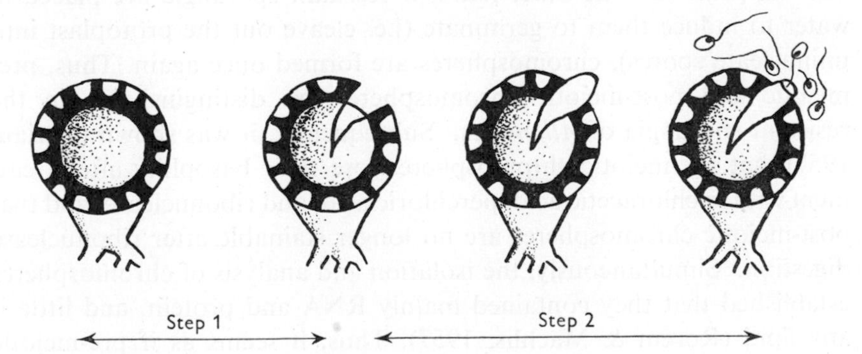

Fig. 10. Germination of RS plants of *Blastocladiella emersonii*.

level, the phenomenon of maturation in the RS plants of *Blastocladiella emersonii*. However, a great deal of progress along a different level of integration has been made on the close relative, *Allomyces*, by Emerson and his colleagues at the University of California. In many important respects, the resistant sporangia of *Allomyces* are much like those produced by *Blastocladiella*; it would be well to consider, first, what has been done with them. On the other hand, some fundamental differences do exist between the two, as we shall see.

Emerson & Wilson (1949) established that the resistant sporangium is the seat of meiosis in the life history of *Allomyces*, and described the interesting and unique nature of this reduction division. Meiosis begins as the resistant sporangium, with its dozen or so diploid nuclei, reaches maturity; shortly afterwards, it abruptly ceases when the nuclei reach advanced prophase I. In this condition, the resistant sporangium can be dried and stored away in a viable condition for many years. Then, when it is placed in water, meiosis commences from where it had left off, and goes on to completion; within several hours, some four dozen haploid spores are liberated.

As a result of these studies, an unusual series of events associated with maturation and germination of resistant sporangia was discovered (Wilson, 1952). Concurrent with the final thickening, pitting, and pigmentation of the sporangial wall, spherical bodies (1–4μ in diameter and staining with aceto-orcein) appear in the cytoplasm; they were named 'chromospheres'. Toward the end of the maturation period of resistant sporangia, when the diploid nuclei are entering prophase I of meiosis, the chromospheres rapidly disintegrate; provided sporangia are not placed under conditions where they will germinate, chromospheres do not reappear. On the other hand, if resistant sporangia are placed in water to induce them to germinate (i.e. cleave out the protoplast into uninucleate spores), chromospheres are formed once again. Thus, pre-meiotic and post-meiotic chromospheres are distinguishable in the resistant sporangia of *Allomyces*. Subsequently, it was shown (Turian, 1957) that pre-meiotic chromospheres lose their basophily after treatment with trichloracetic acid, perchloric acid, and ribonuclease, and that post-meiotic chromospheres are no longer stainable after ribonuclease digestion. Simultaneously, the isolation and analysis of chromospheres established that they contained mainly RNA and protein, and little if any lipid (Rorem & Machlis, 1957). Thus, it seems as if pre-meiotic chromospheres, as well as post-meiotic chromospheres (the latter are transformed into nuclear caps after cleavage of the protoplast into spores), are transitory organelles rich in RNA.

In view of these striking pre-meiotic transformations of RNA-containing organelles in *Allomyces*, it was tempting to hope that a similar phenomenon occurred in *Blastocladiella*. However, we have been unable to detect the presence of chromospheres in the RS of *B. emersonii*, nor have we found any clear-cut evidence for meiotic division of the sort described for *Allomyces*. In view of the enormously greater number of spores produced by RS plants of *Blastocladiella*, one might expect that, if chromospheres were formed, they would be much more numerous—albeit perhaps smaller—than those in the sporangia of *Allomyces*, and therefore readily detectable. Consequently, it seems probable from our cytological examinations that in *B. emersonii*, the final cleavage of an RS protoplast into spores, and the organization of RNA-rich nuclear caps and DNA-containing nuclei associated with it, does not involve the intermediate formation of chromospheres; at least, not the kind found in *Allomyces*.

On the other hand, the final, morphological division that occurs within a spore of *Blastocladiella emersonii*, and the presence of nucleic acids in the organelles thus formed, demand a fundamental reorganization of the various nitrogen and phosphorus pools within an RS plant during the last stages of ontogeny. But here, too, direct analyses had to await the development of synchronous, single generation cultures; therefore, we have just begun to look into this aspect of the story. As preliminary as they are, the following data are interesting here in that they bear directly upon the foregoing discussion. They suggest to us that perhaps we should still expect to find, after all, the existence of transient, cytologically-detectable, nucleic acid-rich organelles of some kind during maturation of RS plants of *B. emersonii*.

As a prelude to investigations of nucleic acid transformations during maturation, RS plants were analysed (Fig. 11) at different stages in development beyond the point where such plants cease to increase in size—but, as we have seen earlier, are far from fully formed. At about 50 % of the generation time, the insoluble, non-chitin nitrogen *per plant* (i.e. total N less the sum of chitin N + soluble protein N + soluble non-protein N) rises abruptly and then remains at this new level throughout maturation of an RS plant. Simultaneously, the percentage nucleic acid (as approximated by 260 $m\mu$ ratios; method of Warburg & Christian; cf. Colowick & Kaplan, 1955–7) in dialysed extracts of these same plants decreases rapidly and then, at about 50 % of the generation time, levels off during the last stages of ontogeny. If the likely assumption is made: (*a*) that this phenomenon results from a sudden, rapid conversion of soluble nucleic acid into an insoluble form; (*b*) that this insoluble nitrogen fraction is all nucleic acid; and (*c*) that it has a content of 16 % nitrogen, then the total insoluble nucleic acid fraction amounts to *c*. 200 μg of nucleic acid per mg. dry weight of protoplasm. *This is almost exactly equal to the nucleic acid content of the spores of* Blastocladiella, as determined by direct analysis—a remarkable correlation. Therefore,

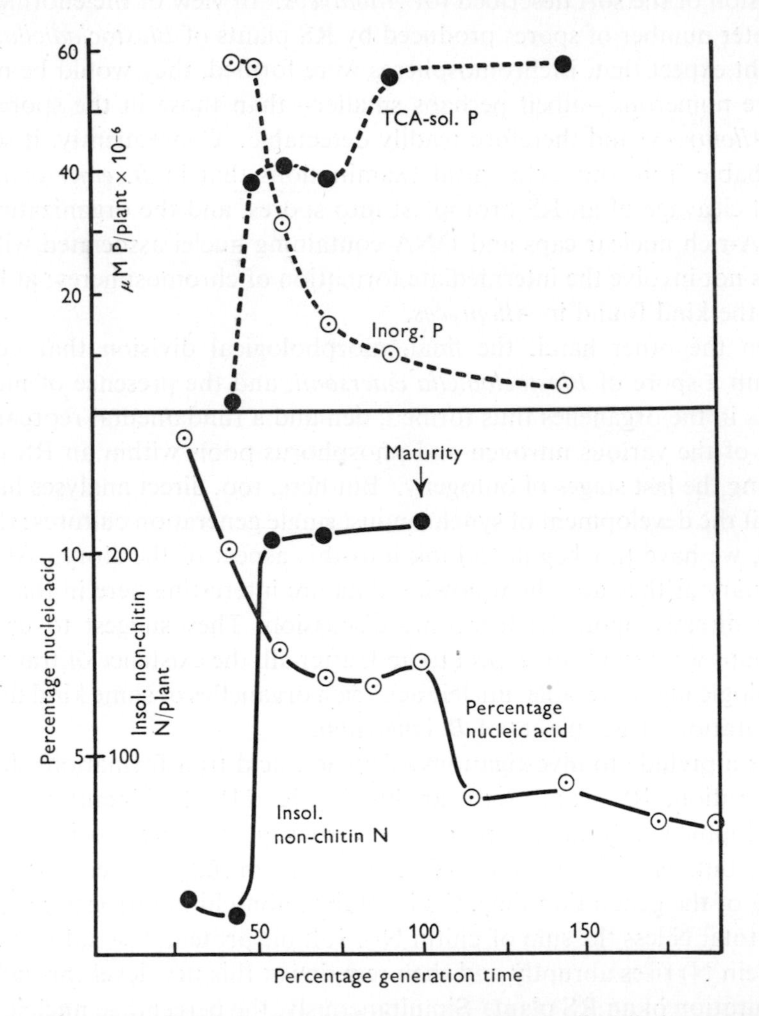

Fig. 11. Changes in nitrogen and phosphorus components per plant during ontogeny of synchronously dividing RS thalli. At 100 % of the generation time, plants are fully mature and will germinate (i.e. liberate spores) if placed in water. However, germination does not occur if plants are allowed to remain in the growth medium beyond this point (i.e. up to 150 %, 200 %, etc., of the generation time). The TCA-sol. P = the total 10 % TCA-soluble phosphorus less inorganic phosphorus. The Insol. non-chitin N = the total N less the sum of the chitin N + soluble protein N + soluble non-protein N. The % nucleic acid = an approximate value based upon the ratio of optical densities at 260 and 280 mμ obtained with dialysed extracts.

since the protoplast of a resistant sporangium of *Blastocladiella* is converted completely into spores at maturity, it seems to us an inescapable conclusion that at *c*. 50 % of the generation time, a considerable proportion of the soluble, non-protein nitrogen—presumably, the soluble nucleic acid and amino acid fraction—is mobilized, packaged into insoluble form, and made ready for its final incorporation into the nucleic acid-containing organelles of the spores.

The rather sudden changes in inorganic-phosphorus and TCA-soluble organic-phosphorus at 50 % of the generation time offer further evidence for an active reshuffling of internal pools at this stage in ontogeny. Since accumulation of soluble protein nitrogen per plant ceases abruptly just before this time, its relation to organic and inorganic phosphorus pools (cf. Hoagland's comments in Gale, 1958) deserve consideration. However, a more detailed interpretation of these alterations is unwarranted at this time, and must be postponed until more work is done.

Finally, attention should be focused on the last, precipitous drop in the curve for percentage nucleic acid (Fig. 11) which occurs almost precisely at the point when an RS plant becomes mature—i.e. the stage at which a resistant sporangium, if placed in water, liberates the whole of its protoplast in the form of thousands of individual, uninucleate spores. A final, second packaging of soluble nucleic acids into compartments of insoluble nucleic acids appears likely.

For these reasons, it seems likely that future studies will reveal cytological changes which will correspond to the chemical reorganizations now known to occur at 50 % of the generation time—and, once again thereafter, at maturity. When this has been accomplished, we should be in a position to clarify, at several different levels of organization, the relationship between nucleic acid metabolism and maturation of the resistant sporangial plants of *Blastocladiella emersonii* (Lovett & Cantino, unpublished data).

THE POINT OF NO RETURN IN ONTOGENY; A CRITICAL FULCRUM FOR THE BALANCE OF 'POWER' IN THE CELL

A casual glance at any, thin-walled, ordinary colourless plant (OC) of *Blastocladiella emersonii* shows immediately that it differs markedly from its brown, pitted, thick-walled RS counterpart, either of which may arise from the same, uninucleate spore. Furthermore, as has been seen, there are numerous other distinguishing characteristics, more fundamental and more subtle, that do not readily meet the eye. These manifold dissimilarities—some qualitative, others only quantitative—

are rendered all the more dramatic when it is recognized that they result from the presence or absence of such a simple agent as bicarbonate. But, as powerful and profound a voice as it may have in the intracellular affairs of *Blastocladiella*, the potential activity of bicarbonate becomes swiftly delimited by an invisible but impenetrable barrier, beyond which it can no longer exert its influence—namely, the 'point of no return' in ontogeny.

Considered simply as a point in developmental time—a sort of superficial, two-dimensional view—this critical period has no real meaning. But, if it could be viewed in terms of the numerous entities which, during this brief interlude, change, appear, or disappear—pools of metabolic intermediates, enzymes, end-products, and the transient steady states and over-all activities of the metabolic superhighways and their crossroads—then this place of no return would take on a kind of third dimension and, thereby, a tenor of significance.

But even so, no matter how exhaustive our analyses might eventually become, the descriptive taxonomy of these enzymes, pools, and products in the plant will not provide us, automatically, with the understanding we are searching for. As Markert (1958) points out, '...no description of structure or function, however sophisticated it may be, is truly developmental biology, for development is primarily concerned with the transformation of one condition into another'. A solution will be brought within reach only when we begin to comprehend the operational mechanics of the changing intracellular network which underlies the 'point of no return'—the hump over which the metabolic machine must pass, but beyond which the down-hill ride is smooth and its destination certain.

The capacity of bicarbonate to induce immediate synthesis of the enzyme, isocitritase, which otherwise does not occur, has provided us with a neat, focal point for continued exploration—and, hopefully, achievement of a partial understanding of this critical period in ontogeny. It is important to re-emphasize that, at the organismal level, the effect of bicarbonate is fully reversible before this developmental stage is reached. Therefore, with synchronous cultures now available, we can find out what will happen to isocitritase, and the capacity for its synthesis when bicarbonate, once added, *is again removed from the external environment of the plant before, during, and after the point of no return is reached* (Lovett & Cantino, 1961). In the process, we should be able to apply more elegant and modern techniques and concepts to achieve a better understanding of: (*a*) the fundamental nature of this enzyme-protein synthesis, and its probable relation to nucleic acids (Gale, 1958, and comments of M. B. Hoagland therein; Pardee & Prestidge, 1956, etc.); (*b*) the possible existence of different molecular forms ('isozymes') of catalytic proteins with the same enzymic specificity (Markert & Moller, 1959); and (*c*) the role of sequential inductions (Stanier, 1954) and permeases (Cohen & Monod, 1957) in morphogenetic phenomena in *Blastocladiella*. In similar fashion we hope to learn much more about the TPN-linked polyphenol oxidase which regenerates reduced nucleotide for ketoglutarate carboxylation, and the origin, activity, and fate of the other enzymes, pools, and products which seem to play a role in morphogenesis and help to establish final, irreversible differentiation in *B. emersonii*.

Some years ago, Stanier (1954) posed the question: 'To what extent do induced modifications of the enzymic machinery of the cell equip microorganisms to meet the needs of a changing environment?' It seems to us, now, that for *Blastocladiella*, the answer must be—'to a great extent'. The resistant sporangium—with its thick, resistant wall, its depressed state of metabolic activity, and its extended longevity—is the sole device possessed by *Blastocladiella* which, in nature, tides it over long unfavourable environmental conditions and ensures a continuity of the species.

REFERENCES

AJL, S. J. (1958). The evolution of a pattern of terminal respiration in bacteria. *Physiol. Rev.* **38**, 196.

BLONDEL, B. & TURIAN, G. (1960). Relation between basophilia and fine structure of cytoplasm in the fungus *Allomyces macrogynus* Em. *J. biophys. biochem. Cytol.* **7**, 127.

CANTINO, E. C. (1949). The physiology of the aquatic Phycomycete, *Blastocladia pringsheimii*, with emphasis on its nutrition and metabolism. *Amer. J. Bot.* **36**, 95.

CANTINO, E. C. (1959). Light-stimulated development and phosphorus metabolism in the mold *Blastocladiella emersonii*. *Developmental Biol.* **1**, 396.

CANTINO, E. C. (1960). Relations of metabolism to cell development. In *Handbk. Pflanzenphysiologie*, vol. 15. Edited by W. Ruhland. Heidelberg: Springer-Verlag. (in the Press.)

CANTINO, E. C. & HORENSTEIN, E. A. (1959). The stimulatory effect of light upon growth and carbon dioxide fixation in *Blastocladiella*. III. Further studies, *in vivo* and *in vitro*. *Physiol. Plant.* **12**, 251.

CANTINO, E. C. & HYATT, M. T. (1953). Phenotypic 'sex' determination in the life history of a new species of *Blastocladiella*, *B. emersonii*. *Leeuwenhoek ned. Tijdschr.* **19**, 25.

CANTINO, E. C. & LOVETT, J. S. (1960). Respiration of *Blastocladiella* during bicarbonate-induced morphogenesis in synchronous culture. *Physiol. Plant.* **13**, 450.

CANTINO, E. C. & TURIAN, G. F. (1959*a*). Physiology and development of lower fungi (Phycomycetes). *Annu. Rev. Microbiol.* **13**, 97.

CANTINO, E. C. & TURIAN, G. F. (1959*b*). Gamma particles, Nadi-positive mitochondria, and development in the water fungi *Blastocladiella* and *Allomyces*. *Nature, Lond.* **184**, 1889.

CANTINO, E. C. & TURIAN, G. F. (1961). A role for glycine in light stimulated nucleic acid synthesis by *Blastocladiella*. *Arch. Mikrobiol.* (in the Press).

CARPENTER, W. D. & BEEVERS, H. (1959). Distribution and properties of isocitritase in plants. *Plant Physiol.* **34**, 403.

COHEN, G. N. & MONOD, J. (1957). Bacterial permeases. *Bact. Rev.* **21**, 169.

COHN, M. & HORIBATA, K. (1959). Inhibition by glucose of the induced synthesis of the β-galactoside-enzyme system of *Escherichia coli*. Analysis of maintenance. *J. Bact.* **78**, 601.

COLOWICK, S. P. & KAPLAN, N. O. (1955–1957). *Methods in Enzymology*, Vols. I–IV. New York: Academic Press Inc.

COMB, D. G. & ROSEMAN, S. (1958). Glucosamine metabolism. IV. Glucosamine-6-phosphate deaminase. *J. biol. Chem.* **232**, 807.

COUCH, J. N. & WHIFFIN, A. J. (1942). Observations on the genus *Blastocladiella*. *Amer. J. Bot.* **29**, 582.

EMERSON, R. & CANTINO, E. C. (1948). The isolation, growth and metabolism of *Blastocladia* in pure culture. *Amer. J. Bot.* **35**, 157.

EMERSON, R. & WILSON, C. M. (1949). The significance of meiosis in *Allomyces*. *Science*, **110**, 86.

GALE, E. F. (1958). Protein synthesis in sub-cellular systems. In *Biochemistry of Morphogenesis* (*Symp. VI Proc. 4th Int. Congr. Biochem., Vienna*; p. 156). Edited by W. J. Nickerson. New York, London: Pergamon Press.

HARDER, R. & SORGEL, G. (1938). Über einen neuen planoisogamen Phycomyceten mit Generationswechsel und seine Phylogenetisch Bedeutung. *Nachr. Ges. Wiss. Göttingen, Fachgruppe VI* (*Biol.*) (N.F.), **3**, 119.

HOFFMAN-BERLING, H. (1959). The role of cell structures in cell movement. In *Cell, Organism, and Milieu*. Edited by D. Rudnick. New York: The Ronald Press.

KOCH, W. J. (1956). Studies of the motile cells of chytrids. I. Electron microscope observations of the flagellum, blepharoplast, and rhizoplast. *Amer. J. Bot.* **43**, 811.

KORNBERG, H. L. (1959). Aspects of terminal respiration in micro-organisms. *Annu. Rev. Microbiol.* **13**, 49.

LELOIR, L. F. & CARDINI, C. E. (1953). The biosynthesis of glucosamine. *Biochim. biophys. Acta*, **12**, 15.

LOVETT, J. S. & CANTINO, E. C. (1960*a*). The relation between biochemical and morphological differentiation in *Blastocladiella emersonii*. I. Enzymatic synthesis of glucosamine-6-phosphate. *Amer. J. Bot.* **47**, 499.

LOVETT, J. S. & CANTINO, E. C. (1960*b*). The relation between biochemical and morphological differentiation in *Blastocladiella emersonii*. II. Nitrogen metabolism in synchronous culture. *Amer. J. Bot.* **47**, 550.

LOVETT, J. S. & CANTINO, E. C. (1961). Reversible, bicarbonate-induced enzyme activity and the point of no return during morphogenesis in *Blastocladiella*. *J. gen. Microbiol.* (in the Press).

MARKERT, C. L. (1958). Chemical concepts of cellular differentiation. In *A Symposium on the Chemical Basis of Development*. Edited by W. D. McElroy and B. Glass. Baltimore, Md.: Johns Hopkins University Press.

MARKERT, C. L. & MOLLER, F. (1959). Multiple forms of enzymes: tissue, ontogenic, and species-specific patterns. *Proc. nat. Acad. Sci., Wash.* **45**, 753.

MATTHEWS, V. D. (1937). A new genus of the Blastocladiaceae. *J. Elisha Mitchell Sci. Soc.* **58**, 191.

PLATE 1

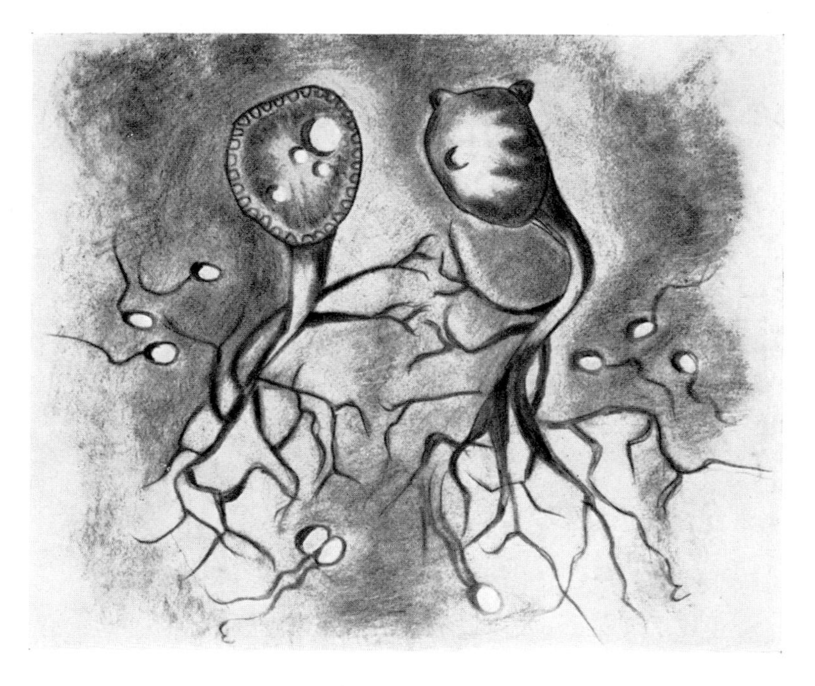

(*Facing page 270*)

McCurdy, H. D. & Cantino, E. C. (1960). Isocitritase, glycine-alanine transaminase, and development in *Blastocladiella emersonii*. *Plant Physiol.* **35**, 463.

Pardee, A. B. & Prestidge, L. S. (1956). The dependence of nucleic acid synthesis on the presence of amino acids in *Escherichia coli*. *J. Bact.* **71**, 677.

Reeves, H. C. & Ajl, S. (1960). Occurrence and function of isocitritase and malate synthetase in bacteria. *J. Bact.* **79**, 341.

Rorem, E. S. & Machlis, L. (1957). The ribonucleoprotein nature of large particles in the meiosporangia of *Allomyces*. *J. biophys. biochem. Cytol.* **3**, 879.

Stanier, R. Y. (1954). The plasticity of enzymatic patterns in microbial cells. In *Aspects of Synthesis and Order in Growth*. Edited by D. Rudnick. New Jersey: Princeton University Press.

Turian, G. (1957). Détection cytochimique de l'acide ribonucléique dans les chromosphères pre- et post-méiotiques des sporanges de résistance d'*Allomyces*. *Experientia*, **13**, 315.

Turian, G. T. & Cantino, E. C. (1959). The stimulatory effect of light on nucleic acid synthesis in the mould *Blastocladiella emersonii*. *J. gen. Microbiol.* **21**, 721.

Turian, G. T. & Cantino, E. C. (1960). A study of mitosis in the mould *Blastocladiella* with a ribonuclease-aceto orcein staining technique. *Cytologia*, **25**, 101.

Turian, G. & Kellenberger, E. (1956). Ultrastructure du corps paranucléaire, des mitochondries et de la membrane nucléaire des gamètes d'*Allomyces macrogynus*. *Exp. Cell Res.* **11**, 417.

Wilson, C. M. (1952). Meiosis in *Allomyces*. *Bull. Torrey bot. Cl.* **79**, 131.

EXPLANATION OF PLATE

Plate 1

A hyperbolic view of *Blastocladiella emersonii* (× *c.* 150).

ENVIRONMENTAL INFLUENCE ON GENETIC RECOMBINATION IN BACTERIA AND THEIR VIRUSES

K. W. FISHER

Medical Research Council, Microbial Genetics Research Unit, Hammersmith Hospital, Ducane Road, London, W. 12

Modern genetic studies of bacteria originated from the work of Griffiths (1928) when he recognized the conversion of rough avirulent *Streptococcus pneumoniae* to smooth virulent organisms by a factor derived from heat-killed, smooth virulent cells. From these observations stems the entire field, now known as transformation, which has finally established, with the experiment of Hershey & Chase (1952) on bacteriophage infection, that deoxyribonucleic acid (DNA) is the genetic material in a large number of species with the notable exception of those viruses which contain ribonucleic acid.

Since the middle of the 1940's, the study of genetic mechanisms in micro-organisms has become a major tool in the analysis of genetic material at all levels down to the molecular. Undoubtedly the most profitable studies for understanding the processes of recombination have been concerned with studies of recombination frequencies, mapping, etc., but the study of environmentally induced variations of recombination frequencies has aided in the elucidation of phenomena not encountered in the genetics of higher organisms.

One of the most important advantages of microbial genetics is the availability of selective methods which facilitate the screening of enormous populations of individuals. In the case of bacteriophage genetics, this enables recombination to be detected between two adjacent pairs of nucleotides of the DNA molecule (Benzer, 1957).

A number of distinct ways for transfer of genetic material between bacterial cells have been described. The transfer may be effected by means of free, chemically-purified DNA. This mechanism is referred to as 'transformation'. A second mechanism, 'transduction', involves a bacteriophage as the carrier of part of the donor genetic material which is introduced into the recipient cell during bacteriophage infection. A third system requires intercellular contact between donor (male) and recipient (female) cells; this is called 'conjugation'. More recently, two other systems have been described, viz. those in which genetic characters may be transferred along with colicinogeny (cited by Stocker, 1960), and those in which there is an association of genetic markers with the component which confers maleness on *Escherichia coli* K12 (Jacob & Adelberg, 1959). The two latter systems are at present outside the scope of this paper as no work has been done on modifications induced by changes of environment.

Of all bacterial mechanisms for genetic transfer, only in the case of conjugation, and there very rarely, is all the genetic material of the donor cell transferred to the recipient. Thus the recipient usually becomes a partially diploid cell, or 'merozygote' (Wollman, Jacob & Hayes, 1956). With conjugation, it is justifiable to refer to male and female strains as equivalent to donor and recipient, but in other systems the terms have no real meaning, the two strains being indistinguishable except for the purposes of the particular experiment.

The environment may interfere with the processes of genetic recombination at a number of levels, determining when genetic material may be transferred between cells, how much is transferred, and whether or not the genetic material newly acquired by the recipient cells will be integrated to form a recombinant.

Although it could be argued that the environment of genetic material includes everything surrounding it, we have chosen to consider the other components of the cell as not forming part of the environment and consider only factors which may be varied in a controlled manner. The discussion has been limited to genetic recombination in bacteria and their viruses.

TRANSFORMATION

Dawson & Sia (1931) established transformation as an *in vitro* process in *Streptococcus pneumoniae*. Subsequently, successful transformations of a number of other species have been reported, notably *Haemophilus influenzae* (Alexander & Leidy, 1951), *Bacillus subtilis* (Spizizen, 1958), *Salmonella typhimurium* (Demerec *et al.* 1958), *Escherichia coli* (Sinai & Yudkin, 1959; Chargaff, Schulman & Shapiro, 1957) and *Rhizobium* sp. (Balassa, 1958).

Chemical nature of the transforming agent

Avery, MacLeod & McCarty (1944) showed that extracts of *Streptococcus pneumoniae*, prepared from cells with a particular capsular type, can change the capsular type of recipient cells. The active factor of these extracts was chemically purified and shown to be DNA. The transforming activity was destroyed by the enzyme, deoxyribonuclease

(DNase), which destroys the DNA molecule. DNA is now generally accepted as the only active constituent of transforming principle (TP), but various suggestions that at least some of the activity of TP lies in very small contaminating amounts of protein cannot be absolutely excluded.

Preparation of transforming principle

The method of preparation of the DNA can have very profound effects on the activity of the TP produced. Sinai & Yudkin (1959) described the preparation of TP by three different techniques, only two of which gave preparations which showed any possible transforming activity. Dangers which might be foreseen include too harsh deproteinization resulting in some degradation of DNA, persistence of an active DNase, and physical rupture of the DNA molecules during sonic disintegration of the donor cells (Litt *et al.* 1958). Techniques most generally employed to prepare active TP have included lysis of cells by sodium deoxycholate (McCarty, Taylor & Avery, 1946), sodium dodecyl-sulphate (Sinai & Yudkin, 1959), physical disintegration (Spizizen, 1958), and osmotic shock of spheroplasts (Chargaff *et al.* 1957). The crude lysate has been deproteinized either with chloroform-octanol (Lerman & Tolmach, 1957) or with chloroform-amyl alcohol mixtures (McCarty *et al.* 1946), except where sodium dodecyl-sulphate has been used, for it is itself a deproteinizing agent. Protection against marauding DNase has in general been provided through the agency of sodium citrate or versene which chelate the Mg^{++} ions necessary to activate the DNase of many species. When sodium dodecylsulphate is used as lysing agent as well as for deproteinization, there is no apparent need for separate inhibition of DNase activity.

The possibility of alterations to the DNA occurring during its extraction have been underlined recently by the work of Demerec *et al.* (1958) who described a system of transformation in *Salmonella typhimurium* in which a large proportion of the transformants for streptomycin resistance (S^r) had a requirement for thiamine where both the donor and recipient cells had previously been thiamine-independent.

The necessity for protecting TP from DNase during preparation has been stressed. Equally important is the need for protection in the transforming mixture (recipient cells + TP), where the existence of an extracellular DNase could account for the apparent absence of transformation in some systems.

Under suitable conditions the addition of active TP to recipient cells results in the formation of cells with one or occasionally more of the characters of the donor strain (Hotchkiss & Marmur, 1954). The nature of the conditions and the state of the cells necessary for transformation to occur have been actively investigated by workers in a number of laboratories.

McCarty *et al.* (1946) discovered the necessity for bovine serum albumin in order that efficient transformation of pneumococcal cells might take place. The function of the albumin is not known, but it is thought that it may have a temporary role during the binding of the DNA to the recipient cell surface (Fox & Hotchkiss, 1957). Albumin is not required in all transformation systems yet investigated. In addition to the albumin fraction, Ca^{++} ions are necessary for transformation of pneumococci (Fox & Hotchkiss, 1957). The formation of transformed cells can be detected in a defined medium containing glucose, potassium phosphate, serum albumin, Ca^{++} ions (10^{-3} to 10^{-4} M) and TP. The limiting requirements for other systems have not been rigorously defined.

Competence

The ability of cells to be transformed has been found to vary greatly. A cell which can accept DNA is said to be 'competent'. The extent of competence of a culture varies from system to system in its time of appearance, duration and proportion of cells affected. In *Streptococcus pneumoniae*, competence affects only a small proportion of cells at one time and appears late in the logarithmic phase of growth. Hotchkiss (1954) and Thomas (1955) obtained evidence that an individual pneumococcal cell can maintain competence for 15–20 min. The prolonged period of competence in a culture is due to the appearance of successive waves of competent cells. Hotchkiss (1954) further showed that competence in a synchronized culture was cyclical, making its appearance at approximately the same stage of each division cycle, although waves of division did not exactly correspond to successive appearances of competent cells. If a competent culture is diluted into fresh medium, competence is lost exponentially at a rate which is directly affected by the temperature of incubation. Upon prolonged incubation of such a diluted culture at 37°, competent cells reappear. However, competent cells do not appear in a culture if protein synthesis is inhibited (Fox & Hotchkiss, 1957).

In the case of *Haemophilus influenzae*, cells show competence towards the end of the logarithmic phase of the growth cycle. Schaeffer (1957*c*), using streptomycin resistance as marker, followed the competitive inhibition produced by added DNA prepared from a streptomycin-sensitive strain, and was able to show that under optimal conditions every cell in a culture of *H. influenzae* is competent to accept TP.

Strains of increased competence may be selected by the technique described by Sinai & Yudkin (1959). The method is an adaptation of the velveteen replica-plating technique of Lederberg & Lederberg (1952). A plate was spread with a strain of *Escherichia coli* sensitive to proflavine, and after 6 hr. incubation at 37° was replica-plated to a nutrient agar plate containing TP extracted from a proflavine-resistant strain. Precautions were taken to prevent the action of DNase. After 24 hr. incubation, the colonies on this plate were replicated to agar containing proflavine at concentrations tolerated by the resistant but not by the sensitive strain. Resistant clones developed in a number of areas. By reference to corresponding areas on the original masterplate, clones were isolated and tested against TP prepared from resistant and sensitive donor strains. One of the ten clones tested appeared to be consistently more competent than the other nine as measured by a higher production of resistant cells when grown in the presence of TP isolated from a resistant donor strain.

Fox & Hotchkiss (1957) demonstrated that the competence of a culture can be maintained for several weeks by inducing synchronous division in a culture of *Streptococcus pneumoniae* by temperature shifts and by storing the competent culture which results after incubation, at −20° in 10% glycerol. The use of this technique will help to remove some of the day-to-day variations in the level of competence which has hitherto hindered work on the mechanism of transformation.

DNA uptake

The uptake of DNA by competent cells has been studied by several authors; some have used genetic markers and others DNA labelled with ^{32}P to determine changes in uptake with variation of conditions. Thomas (1955) described the reaction of competent cells of *Streptococcus pneumoniae* with TP derived from a streptomycin-resistant strain (S^r). The number of S^r transformations obtained was directly related to the concentration of TP at low concentrations, and reached a plateau at higher TP concentrations where the limiting factor was the number of competent cells. Using ^{32}P-labelled TP, Lerman & Tolmach (1957) showed that the number of cells becoming S^r and the amount of TP incorporated by the culture was directly related to the concentration of DNA in the medium. Uptake of labelled DNA and the number of cells transformed both showed the same temperature-dependence, with an optimum near 33°. The ratio, number of cells transformed to S^r/ amount of ^{32}P (i.e. DNA) incorporated, was constant. Thomas (1955) had previously reached a similar conclusion.

In common with other workers, notably Hotchkiss (1954) and Schaeffer (1957*a*), Lerman & Tolmach (1957) reported inhibition of transformation by calf thymus DNA. They showed that DNA bound to cells was either transiently bound and DNase-sensitive or permanently incorporated and insensitive to DNase. The amount of DNA bound in either of these states can be profoundly influenced by temperature, in that the change from the transiently bound form to the permanently bound condition is dependent on a higher temperature than is initial binding. If the initial step resulted from random collision between the cell and the DNA molecule, it might be expected to be relatively insensitive to temperature change over the narrow range employed. This is apparent from the data of Lerman & Tolmach (1957). Since both permanently and transiently bound DNA are affected to the same extent by the addition of calf thymus DNA, the latter probably competes at the stage of transient binding rather than permanent uptake of the DNA into the cell, suggesting that there are perhaps a limited number of specific sites available for binding the DNA. Competent pneumococci can incorporate DNA prepared from *Escherichia coli* to approximately the same extent as pneumococcal DNA (Lerman & Tolmach, 1957). Other high molecular weight substances, e.g. protein and ribonucleic acids, are not bound by competent cells (Lerman & Tolmach, 1957).

Transiently bound DNA which is not incorporated may be later released from the cells and can be shown not to have suffered significant degradation. DNA degraded by DNase is not incorporated and, as would be expected, does not produce transformants. Transformation, in terms of the ability to transmit a particular characteristic, is much more sensitive to DNase than is the ability to be permanently incorporated. The decline of this latter property is approximately proportional to the fall in molecular weight (Lerman & Tolmach, 1957). Litt *et al.* (1958) showed that initial binding of the DNA is dependent on its molecular weight. They used sonic disintegration to break DNA and, from their results, concluded that the smallest piece taken up by competent cells has a molecular weight of 4×10^5.

Transformation between species

As already mentioned, heterologous DNA reduces transformation by competition at the cell surface. Schaeffer (1956) has attempted transformation in *Haemophilus influenzae* by treating recipient strains with DNA from related species, such as *H. parainfluenzae*. Cultures of *H. influenzae* become completely competent near the end of the logarithmic phase of the growth cycle. Although such cells can bind DNA efficiently, they cannot express the newly-acquired characteristic unless further growth can occur. Although strains of *H. influenzae* can be transformed to S^r by DNA prepared from a streptomycin-resistant mutant of *H. parainfluenzae*, the frequency of this interspecific transformation is about 10^4 times lower than for intraspecific transformations. As in the case of pneumococci (Lerman & Tolmach, 1957), Schaeffer (1957*b*) showed with *H. influenzae* that competent cells are perfectly able to bind heterospecific DNA irreversibly. Therefore, the low rate of transformation cannot be ascribed to a lack of penetration by heterologous DNA. Strains successfully transformed by heterospecific DNA can be used as a source of DNA for further homospecific transformation. In such cases, the higher frequency characteristic of homospecific transformation is obtained. This might be expected if the lower rate of heterospecific transformations were due to poor pairing of the genetic material prior to genetic recombination, the poor pairing being due to fairly extensive differences of base sequence of the DNA from the two species. Once successful genetic integration of a marker is achieved, non-homologous regions are probably eliminated. The implication of this is that the two types of DNA have similarities confined to small regions, e.g. in Schaeffer's experiments, the S^r locus. Another implication is that the regions of DNA surrounding such loci are most important in deciding the occurrence of synapses necessary for the integration of a particular marker into the recombinant.

Schaeffer (1958) showed that transformation of *Haemophilus influenzae* was inhibited less by calf thymus DNA or DNA from bacterial sources not closely related to the recipient, than by DNA isolated from closely similar strains of *Haemophilus*. The DNA of the former group must have a lower affinity for the receptors of the recipient strain, suggesting that the receptors of the cell surface have some kind of specificity.

Effect of the environment on integration

Hotchkiss & Evans (1958) investigated the effect of the environment on the outcome of transformations involving three closely linked markers determining resistance to sulphanilamide in *Streptococcus pneumoniae*. Each marker confers resistance to a given level of the drug, their combined effect being considerably greater than additive. The minimum inhibitory concentrations of sulphanilamide (in μg./ml.) are 4 for wild-type organisms; 20 for type *a* mutants; 80 for type *d*; 15 for type *b*; 400 for type *ad*; 300 for type *db*; 70 for type *ab*; and 800 for type *adb*.

When DNA from the *adb* strain was added to a sensitive recipient strain at different times during the growth cycle, marked differences in the phenotypes of the transformants were found. When the culture first showed competence, addition of the triply-marked DNA was followed by the appearance of resistant clones, and amongst those carrying the *d* marker, approximately one-third inherited simultaneously either *a* or *b*. Addition of the same DNA somewhat later in the growth cycle caused more than 50 % of the transformants inheriting *d* to acquire also *a* and *b*. When DNA was added at still later times there was a fall in the amount of joint incorporation of the three markers.

Other factors influenced the incorporation of markers after and during exposure to transforming DNA. The work was performed with cells of *Streptococcus pneumoniae* stored at $-20°$ in 10 % glycerol to rule out variations in the level of competence, but this at the same time introduced the difficulty of not knowing what other effects the storage may induce. The results indicated, however, that exposure to the DNA at 30° produced more transformants having jointly integrated 2 or 3 of the sulphanilamide markers. Also, at the lower temperatures, a significantly higher number of transformants had also acquired S^r, which shows loose linkage to the group of sulphanilamide markers. The amount of ^{32}P-DNA fixed was directly related to the number of S^r transformants in this experiment. Hotchkiss & Evans (1958) were unable to influence significantly the number of cells becoming S^r by alteration of the environment after DNA is irreversibly bound. It was concluded that environmental influences largely affect uptake of the DNA rather than the subsequent steps producing a transformant (cf. Lerman & Tolmach, 1957).

Ephrussi-Taylor (1958) described the effect of chloramphenicol on the establishment of new characters within transformed cells. Cells which have reacted with DNA for a short time are exposed to chloramphenicol. If this is done within 10 min. after mixing the cells with DNA, most of the potential transformants are reversed by 100 min. exposure to 10 μg./ml. of chloramphenicol. Between 10 and 14 min. after treatment with DNA, a change begins in the cells so that the chloramphenicol treatment is less effective. The change is stepwise, which may be due to the use of synchronously dividing cultures in these experiments. It can be argued that stabilization of the genetic marker occurs in waves, the first just before the first division is completed, and the second wave preceding the second division. The proportion of stabilized cells differs, however, for different markers (cited by Ephrussi-Taylor, 1958).

Ephrussi-Taylor (1958) also showed that incorporated DNA does not increase in amount before the first cell division, but that a proportion of cells, approximately the same as that which achieved stabilization of the marker at the first division, replicates it before completion of the second division. Thus it appears that if stabilization of a marker occurs during one division it will be replicated at the next division of the genetic material of the cell.

Marmur & Lane (1960) followed the thermal denaturation of DNA isolated from *Streptococcus pneumoniae* by changes in absorption at 260 mμ as well as by changes in the biological activity of DNA preparations which carried three unlinked markers, viz. streptomycin, erythromycin and bryamycin resistance. They concluded that after heating there was reformation of the DNA base pairs, provided the temperature fell slowly. The three markers were found to have different thermal sensitivities which, from the work of Sueoka, Marmur & Doty (1959), is very likely a reflection of differing base composition of the markers.

The rate of cooling of DNA converted to a single-stranded form by heating (Meselson & Stahl, 1958) profoundly affected its transforming activity. Rapid cooling resulted in a great loss of transforming activity of the DNA, whereas if heated DNA was cooled over a period of about 2 hr. as much as 50 % of the original activity was restored. The activity can be restored to rapidly cooled preparations by reheating and subsequent slow cooling. Marmur & Lane (1960) also found that the DNA concentration affected the degree of restoration of transforming activity, presumably by increasing the probability of contact between different single-stranded DNA molecules. By increasing the ionic strength of the medium, the repulsive forces between the single strands were reduced and a higher restoration of transforming activity achieved.

An interesting effect was the ability of homologous DNA carrying a different allele, to pair and reform duplexes; i.e. if a preparation of DNA isolated from a streptomycin-resistant strain was mixed with DNA from a streptomycin-sensitive strain and then heated and cooled slowly, the transforming activity with respect to S^r was increased. Thus it appears that hybrid molecules are formed by cooling so that some duplexes may carry the S^r locus in one strand and the S^s locus in the other, the mixing being produced by random aggregation of the two types of single-stranded molecules during slow cooling. The fact that the S^r marker can be recovered by addition of unmarked homologous DNA strongly suggests that each strand of the DNA molecule can carry information independently. This conclusion agrees with results obtained by Pratt & Stent (1959) and Tessman (1959) working with bacteriophage.

CONJUGATION

Conjugation between cells of different genotype has been observed in a number of bacterial species, including *Escherichia coli* (Lederberg & Tatum, 1946), *Pseudomonas aeruginosa* (Holloway, 1955), *Serratia marcescens* (Belser & Bunting, 1956), *Shigella sonnei* (Luria & Burrous, 1957) and *Salmonella typhimurium* (Baron, Carey & Spilman, 1958). The greater part of the work has been carried out using strains of *E. coli* K12, and studies of the effects of environment on recombination are confined to this organism.

Hayes (1952) suggested that transfer of genetic material occurs solely from one parental strain to the other. This has been confirmed by Wollman & Jacob (1957) and by Jacob & Wollman (1954), working on zygotic induction, and by Skaar & Garen (1956) studying transfer of ^{32}P-labelled DNA. The fact that transfer is unidirectional facilitates the separate study of factors affecting transfer from those affecting genetic integration within the zygote.

Donor and recipient strains

Hayes (1952) made one of the initial observations of an environmental effect on the outcome of a cross when he observed the differential effect of streptomycin on strains hitherto unrecognized as donor and receptor. The explanation for this difference is that the receptor cell must be viable in order to support the processes of recombination, segregation and phenotypic expression. All that is required of the donor strain is that it should be viable long enough to pair with and transfer genetic material to the recipient cell.

The differences noted by Hayes (1952) were soon correlated with the mating types found by Cavalli-Sforza, Lederberg & Lederberg (1953) who showed, together with Hayes (1953*b*), that donor cells possessed a transmissible agent named the 'fertility factor' or F, which could convert recipient to donor strains. When mixed with suitable F− strains (recipients) about $1/10^5$ donor F+ cells can produce recombinants. Other donor strains called 'Hfr' (high frequency of recombination) were isolated from F+ cultures by Cavalli (1950) and by Hayes (1953*a*) in which about 10% of the Hfr cells produced recombinants for some markers. Old cultures of F+ cells behave phenotypically like F− cultures: these cells are called 'F− phenocopies'. Their F+ character is rapidly restored by reincubation of the culture in fresh broth. Matings between two cultures of F− phenocopies show a very much reduced fertility.

Merozygote formation

The development by Hayes (1957) and Wollman & Jacob (1955, 1958) of techniques for separating transfer of genetic material from the subsequent steps of genetic recombination within the recipient cell (merozygote) enabled detailed investigation of merozygote formation. Hayes (1953*a*) reported an increase in the number of recombinants after irradiation of the F+ donor cells with low doses of ultraviolet light (u.v.l.) before mating. Irradiation of F− or Hfr cells did not produce this effect. Jacob & Wollman (1955*a*) analysed zygotes formed at different times after mixing the parental strains and observed a characteristic time of entry of each marker into the merozygote. The best explanation for their observations is that the genetic material of the Hfr donor cell entered the F− cell in a definite order starting with an 'origine', O, succeeded by the other markers in order of their situation on the donor chromosome. The observations were confirmed by Hayes (1957) and Fisher (1957*b*) using different techniques to interrupt chromosomal transfer.

Jacob & Wollman (1955*a*) suggested that zygote formation involves the following steps: *collision* involving random pairings between cells; *effective contact*—subsequent to collision there must be a specific attachment between cells of opposite mating type which is presumably dependent on the surface properties of the two cells; and *genetic transfer* in which the donor genome (or part of it) is transferred to the recipient cells, which thus become merozygotes. Wollman & Jacob (1958) showed that the number of effective contacts between Hfr and F− cells increased directly with the period of mixing for approximately 25 min., after which no more occurred. The linear relationship suggests that the stage of collision, at least, is reversible.

Nutritional effects

Fisher (1957*a*) found that the recombination rate between Hfr and F− cells could be varied widely by the medium in which the cells were mixed. Broth, which was used for earlier work on conjugation in *Escherichia coli* K12, was effectively replaced by a defined medium consisting of 0·05 M-phosphate buffer (pH 7·2) supplemented with 20 μM./ml. of D-glucose and 80 μg./ml. of D- or L-aspartic acid (BGA medium). It was further found that dicarboxylic acids of the tricarboxylic acid cycle or glutamic acid could replace aspartate in supporting recombination at a rate closely approaching that observed in broth. Various metabolic inhibitors, with more or less specific modes of action, were tested for their effects on recombination. The results suggested that conjugation was an energy-requiring process, dependent on the tricarboxylic acid cycle, which did not require protein or nucleic acid synthesis for the transfer of genetic material. Mating mixtures in BGA medium oxidized glucose at the same rate as unmixed parental cells as measured by the rate of oxygen uptake in manometric experiments (Fisher, unpublished experiments). This may mean that genetic transfer takes place at the expense of other energy-requiring reactions.

The energy requirement during transfer of genetic material was further investigated (Fisher, 1957*b*). The female (F−) strain was found to be a passive partner during mating while the important energy-requiring step was the movement of genetic material from the donor to the recipient cell. If the initiation of chromosome transfer was prevented by mixing the parents in the absence of oxygen, and then restarted at intervals afterwards by admitting oxygen, it was found that entry of the first detectable marker into the F−cell occurred at a time equal to the duration of the anaerobic period plus the normal delay between mixing of parental cells and the appearance of the first marker. Furthermore, the curves for zygote formation were parallel irrespective of the duration of the anaerobic period, thus excluding the possibility that excess cell pairs had accumulated which were able to perform the next step in zygote formation. It is of significance for later discussion to mention that the mating mixtures were agitated during the experiment just described.

Effect of temperature

Hayes (1957) showed that the rate of chromosomal transfer is affected by temperature, whereas formation of pairs is independent of temperature over the range 32°–37°. Assuming that the physical distance between markers is not affected by temperature changes, Hayes showed that, for the chromosomal segment O-*Lac* (*Lac* is the locus controlling lactose fermentation) at least, the rate of transfer of the chromosome is halved by reducing the incubation temperature from 37° (optimal) to 32°. The temperature coefficient for a 10° change (Q_{10}) is therefore 4.

A value for the Q_{10} for pair formation (i.e. the stages of collision and effective contact) is now available for a wider range of temperatures than that studied by Hayes (1957). This was obtained by using unshaken cultures, so reducing dissociation of parents, and by adding 2:4-dinitrophenol (DNP) which permits pair formation but does prevent chromosomal transfer. Accordingly, the parental strains, 58–161 Hfr and W-1 F−, were mixed at 0°, 20°, 30° and 37° in BGA medium

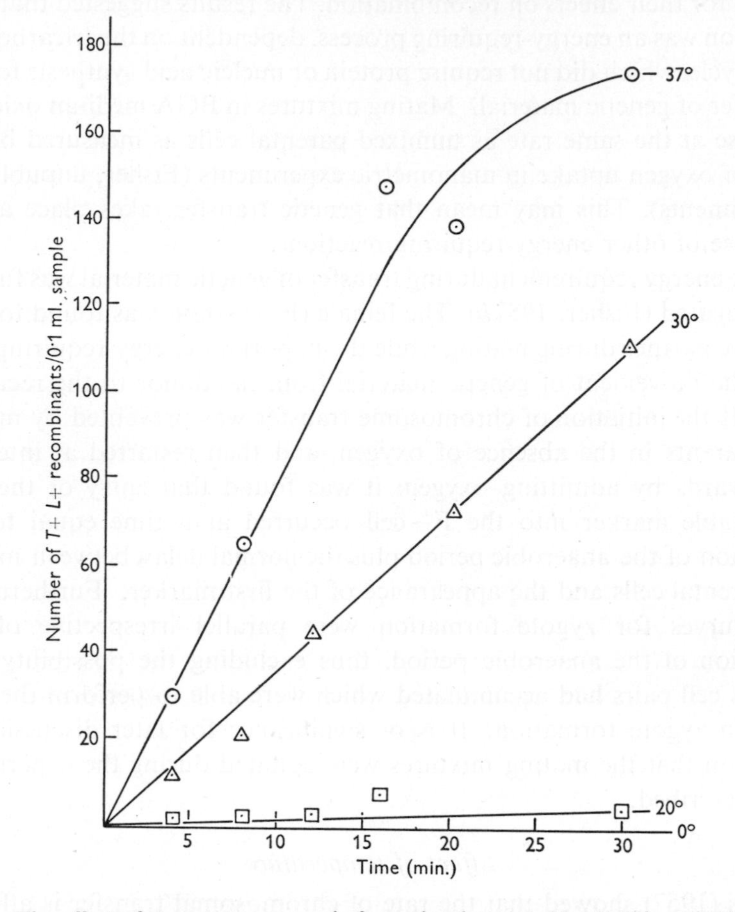

Fig. 1. The effect of temperature on pair formation between 58–161 Hfr and W-1 F−. Equal numbers (2×10^8 organisms/ml.) of each parent were mixed in BGA medium + 10^{-3} M-DNP at 0°, 20°, 30° and 37°. At the times indicated on the abscissa, 0·1 ml. was removed carefully to 100 ml. of BGA medium at 37°, gently dispersed and left undisturbed for 30 min. to allow transfer of the markers *T* and *L* (conferring independence of L-threonine and L-leucine respectively). 1 ml. of coliphage T6 was added to an equal volume of the diluted suspensions and 0·1 ml. of the phage-bacteria mixtures plated on minimal agar + vitamin B_1. (Since the recipient strain was $T- \; L- \; B_1-$). The number of colonies appearing after 18–24 hr. incubation at 37° is taken as an indication of the number of effective contacts (ordinate) formed at the time the sample was taken (abscissa).

(pH 6·5) + 10^{-3} M-DNP. This was taken as zero time. At intervals afterwards, aliquots were removed from each tube, and, with as little disturbance as possible (to avoid separating pairs) were diluted 1/1000 in BGA medium (pH 6·5) at 37° and left undisturbed for 30 min. At the end of this period, sufficient of the coliphage, T6, was added to destroy the Hfr cells (sensitive to T6) and 0·1 ml. aliquots plated on minimal agar + vitamin B_1. This allows the growth of recombinants which are taken as a measure of the number of pairs formed at the time the sample was diluted 1/1000. All the pairs present in the BGA medium + DNP at 37° after 30 min. can be separated either by vigorous pipetting or shaking before the second incubation in BGA medium, as shown by the fact that no recombinants appear when such samples are plated on minimal agar + vitamin B_1. Other controls show that when each parent is diluted 10^{-3} before mixing, no recombinants are formed during 30 min. at 37° in BGA.

The number of recombinants formed at various temperatures of incubation are shown in Fig. 1. All curves arise from the origin, showing that collision occurs without delay at all temperatures. Thus it appears that collision is independent of temperature over the range studied. Pair formation evidently involves a temperature-dependent process with a Q_{10} of 2·7; which, being at least partially resistant to DNP, is distinguishable from genetic transfer. There is no indication as to the nature of this step, e.g. it may represent the conversion of random collisions into effective contacts either by the alignment of partner cells in specific configurations or by something as complex as the formation of the intercellular bridge visible in electron-micrographs, or it may even represent the formation of a hole in the rigid part of the cell wall between the two cells which would allow the chromosomal material to pass across. This possibility was cursorily examined by mixing Hfr cells with unfractionated, unwashed F− cell walls (unpublished experiments performed jointly with Dr E. Work at the Institut Pasteur). Although pairings were microscopically visible between Hfr cells and fragments of cell wall, the Hfr cells remained viable suggesting that if a hole had been formed it could be 'plugged', possibly by the cytoplasmic membrane, to prevent death by leakage of cell constituents. Alternatively, under the conditions of these experiments holes were not formed.

Because the temperature coefficient for the second stage of pair formation is different from that of chromosomal transfer, attempts were made to induce synchronized injection of genetic material from Hfr to the F− cells. As yet there is no indication that this happens. The reason may be technical or perhaps because there is yet another step between collision and the start of genetic transfer.

The experiment shown in the figure is not consistent with that described by Hayes (1957), studying the effect of temperature on chromosome transfer, which gave no indication of the second step since his curves for different temperatures were parallel. One possible explanation for the discrepancy is that cultures in the earlier experiments were agitated so increasing the possibility that pairs were separated before genetic transfer could occur.

Fisher (1957*b*) determined the effect of pH of the medium on the rate of zygote formation. With the strains, 58–161 Hfr and W-1 F−, the optimal pH was 6·1. Hayes (1957) examined the pH effect in greater detail and concluded that pH changed the configuration of the cell surface so that a higher or lower proportion of random collisions became effective. Maccacaro (1955) and Maccacaro & Comoli (1956) had reached similar conclusions regarding the importance of the surface properties of the cells involved in mating.

Zygotic induction

Jacob & Wollman (1954) showed that in crosses between Hfr strains lysogenized by the temperate phage lambda, and non-lysogenic F− strains, zygotes which receive the prophage during conjugation will lyse and liberate normal lambda particles. This phenomenon, which occurs when the zygotes are formed in nutrient broth but not in BGA medium, was called 'zygotic induction'. Since the zygote is destroyed during zygotic induction, the mating medium clearly affects the outcome of a cross when the prophage is transferred to the zygote. Under conditions where zygotic induction occurs, the production of recombinants testifies to the existence of a class of zygotes which have not received the lambda prophage. This is evidence for incomplete transfer of the donor genetic material. The closer a marker is to lambda, the greater the probability that it will be eliminated by zygotic induction (Wollman & Jacob, 1957).

Integration

A further environmental effect on recombination was observed by Jacob & Wollman (1955*a*) and Wollman *et al.* (1956). If in a non-lysogenic system, the Hfr parent is irradiated with ultraviolet light (u.v.l.) before mixing with F− cells, the apparent linkage between markers is drastically reduced, although the inheritance of individual markers is only slightly reduced. Mating mixtures were also subjected to u.v.l. at intervals after mixing, to determine sensitivity of the zygotes. They were found to be relatively sensitive up to 70 min. after the parents were mixed and then to become more resistant and to reach a constant level at *c*. 120 min. The interpretation was that the formation of recombinant chromosomes begins to be complete at 70 min. and is completed by 120 min. Because the proportional inheritance of markers over a quarter of the chromosome was unaffected by u.v.l. and because a similar effect is obtained by irradiation of the zygotes it was suggested that u.v.l. affected the processes of recombinant formation rather than chromosomal transfer. If this is taken to mean that u.v.l. does not cause rupture of the chromosome, it can be suggested that recombination depends on a 'copy-choice' model (Levinthal, 1954), the switching from one strand to another being caused by some form of damage produced by the u.v.l. Similar results have been obtained by irradiation of the bacteriophage lambda: these are discussed below.

TRANSDUCTION

Bacteriophage-mediated transduction was first demonstrated by Zinder & Lederberg (1952) in *Salmonella typhimurium*. Subsequently, this mechanism of genetic transfer has been demonstrated in *Escherichia coli* (Lennox, 1955; Morse, Lederberg & Lederberg, 1956*a*, *b*; Lederberg, 1960), *Shigella* spp. (Adams & Luria, 1958), *Pseudomonas aeruginosa* (Loutit, 1958) and *Staphylococcus pyogenes* (Ritz & Baldwin, 1958; Morse, 1959).

In order to demonstrate transduction between genotypically different strains, a bacteriophage grown on the donor strain is allowed to infect a recipient strain of different genotype. With the exception of the high-frequency transduction systems (Morse *et al.* 1956*a*, *b*; Luria *et al.* 1958) the infected recipient cells must be plated on a selective medium to isolate recombinants from the recipient population. In this way, the number of recombinant clones which can be isolated is several hundred times greater than would be expected if the change in genotype had a mutational origin. A variety of characters has now been transduced, including nutritional markers, resistance to antibiotics, prophages, etc. (see Clowes, 1960).

When a bacteriophage particle infects a cell either of two sequences of events can occur. In one case, virtually every infected bacterium lyses at the end of a period characteristic for each bacteriophage-host cell system and liberates up to several hundred bacteriophage particles, each capable of repeating this lytic cycle. A bacteriophage giving this response is said to be 'virulent'. On the other hand, following infection by certain bacteriophages, a large proportion of the cells do not lyse. These cells, however, can be shown to carry the bacteriophage genetic material in a special manner (Lederberg & Lederberg, 1953) closely associated with the bacterial chromosome with which it divides so that almost all the cells carry the bacteriophage genetic material in the form now termed 'prophage' (Lwoff & Gutmann, 1950). Bacteriophages which can give this type of response are called 'temperate', and a cell carrying the prophage is said to be 'lysogenic'. With the exception of the temperate bacteriophage P 1, the prophages associated with *Escherichia coli* K 12 are always linked to specific chromosomal markers (Jacob & Wollman, 1958). Very rarely, the association between prophage and host genetic material is disturbed and the prophage enters the vegetative cycle of multiplication within the bacterium. The particles released upon lysis of such a cell are capable of giving either the lysogenic or the lytic response when they infect a sensitive cell.

It is obvious that bacteriophage infections which always lead to lysis will destroy the recipient cell so that one cannot tell if transduction had occurred. In the *Salmonella* system, mutants of temperate bacteriophages have been isolated which cannot lysogenize sensitive bacteria, i.e. infection with these bacteriophages always results in lysis. Such mutants, although virulent, can be shown to have transducing activity, since they cannot multiply vegetatively in and lyse lysogenic recipient strains and in such a system the occurrence of transduction can be detected (Zinder, 1953). Temperate bacteriophages can give a lytic response after infection. Obviously, this should be kept to a minimum to obtain the highest possible number of recombinants, especially if recombinants for extremely closely linked markers are required. The proportion of infections giving a lysogenic response can be varied within wide limits. Bertani (1957) studied lysogenization of *Escherichia coli* by temperate phage P2. She found that lysogeny is favoured in this system either by exposing infected cells to chloramphenicol, proflavine or amino acid analogues, or by starvation for a required amino acid. Pretreatment with u.v.l. had no effect. The establishment of lysogeny in *Salmonella typhimurium* after infection with phage P22 was affected in a like manner.

Fry (1959) has studied the environmental conditions which favour or reduce lysogenization of *Escherichia coli* K 12 by the temperate bacteriophage, lambda. Lysogeny is favoured either by high multiplicity of infection, by increasing the concentration of Mg^{++} ions or by lowering the temperature of incubation after infection.

Several authors (Zinder, 1955; Lennox, 1955) have shown that transduced clones inheriting a variety of characters frequently do not carry the transducing phage. There is obviously a need to reconcile this finding with the observation that higher transduction frequencies are obtained under conditions which favour lysogeny. One explanation assumes that transducing particles have lost the ability to lysogenize and are also unable to initiate the lytic cycle of phage multiplication, i.e. they are defective in a number of viral functions. In *Salmonella*-P22 transductions it was observed by Luria *et al.* (1958) that both sensitive and lysogenic cells can issue from the same infected cell. A second explanation then suggests itself. When infection occurs, the bacteriophage genetic material and the transducing fragment enter the multi-nucleate recipient bacterium. The two entities may perhaps associate with different bacterial nuclei, leading to transduced sensitive cells and lysogenic non-transduced cells. Such a mechanism suggests that the transduced fragment is not integrated with the bacteriophage genome as appears to occur with lambda and the galactose marker of *Escherichia coli* (Arber, 1958).

Garen & Zinder (1955) examined the effect of u.v.l. on the lytic, lysogenizing and transducing activities of the *Salmonella* phage, P22. Its lytic and lysogenic abilities were almost equally sensitive to u.v.l., whereas the transduction curves showed no decrease in activity with doses of u.v.l. sufficient to reduce the lytic activity to approximately 10^{-8} of its initial value. At low doses of u.v.l., a moderate increase in transducing activity of the phage was detected. The results can be interpreted in terms of the target affecting transduction being smaller than that needed for normal lytic or lysogenic functions. The moderate increase in transduction frequency may be equated with the stimulating effect of u.v.l. on recombination in the temperate bacteriophage, lambda, and the effect on *Escherichia coli* discussed elsewhere.

Kirchner & Eisenstark (1957) and Kirchner, Rouhandeh & Eisenstark (1958) reported the effects of treating recipient cells of *Salmonella typhimurium* with various agents and their effect on the transduction frequencies for various markers. They found that versene, manganous chloride, u.v.l. and higher multiplicity of adsorbed bacteriophage all increased the frequency of some but not all tryptophan markers. They suggested that some of their results were associated with requirements for divalent cation bridges in the genetic material.

RECOMBINATION IN BACTERIOPHAGES

The growth of bacteriophages and recombination between different bacteriophages are both obligatory intracellular events. Doerman (1953) first noticed, by using the technique of premature lysis, that the proportion of recombinants among the progeny of a mixed infection increased during the latent period. Levinthal & Visconti (1953) extended these studies, prolonging the latent period by making use of lysis inhibition. For closely linked markers, there was a close correlation between increase in frequency of recombination and the prolongation of the latent period. With less closely linked markers, the correlation was less perfect. As the latent period is extended under these conditions, the burst size (i.e. the average number of infective particles liberated/cell) increases. It would therefore seem that, as the burst size increases, markers show less linkage. Although the recombinant frequency may approach 0·5, the value observed for unlinked markers, there are factors operating to prevent this value from being reached. For instance, after part of the latent period has elapsed, bacteriophage particles begin to mature and in so doing are withdrawn from the genetic pool. Maturation is an irreversible process so that these particles are prevented from further mating. Because they are present in the vegetative pool for a shorter period of time, the early formed particles are much less likely to have recombined and consequently tend to be predominantly of the parental types. The equilibrium value of 0·5 for unlinked markers will not hold for bacteria infected with unequal numbers of the parent phages. Different rates of adsorption of the parents can grossly alter the recombinant frequencies, perhaps caused by lack of organic or inorganic co-factors and operation of exclusion phenomena whereby bacteria can only be superinfected up to a certain time after the primary infection.

Jacob & Wollman (1955*b*) studied the effect of u.v.l. on the temperate bacteriophage, lambda. During normal mixed infections by two mutants of this phage, there is a low probability that a vegetative bacteriophage will mate, and recombination is likely to occur relatively late in the growth cycle of the bacteriophage. Parental bacteriophages can be irradiated before infection, or the infected bacteria can be irradiated in cyanide to prevent bacteriophage development. With low doses of u.v.l., such that more than 70% of free bacteriophage would be unaffected, the recombination frequencies are markedly enhanced, the increase being directly related to the dose of u.v.l. The slope of the curve is more steep, the less closely linked are the markers. This effect is photo-reversible in the early part of the latent period. Recombination is a relatively rare event during infection by unirradiated lambda bacteriophage and occurs late in the growth cycle. After u.v.l. treatment, however, recombination occurs much earlier as well as to a greater extent. In crosses involving three markers in which one parental phage was irradiated, the centre of its three markers was present more frequently in the progeny than that of the unirradiated parent. These results suggest that after irradiation the markers are more easily separable. It has been argued from the effects of u.v.l. on recombination in *Escherichia coli*, K12, that u.v. irradiation does not cause breakage of the chromosome. If one takes this as true of phage lambda, the increased recombinant frequencies are perhaps due to production, by u.v.l., of more regions in which switches between parental strands can occur rather than to recombination of fragmented parental material.

CONCLUSIONS

The environment can play an important role in genetic recombination in bacteria and their viruses. In all the mechanisms studied, the two parental contributions cannot be brought together unless certain well-defined conditions prevail. The environment can influence the amount of the donor genetic material which can be transferred to the recipient and in some cases apparently alter linkage relationships between markers. Processes succeeding zygote formation can be influenced by various means. One can seek information about the recombination mechanism from the studies described but an unequivocal picture of the events involved cannot be derived.

There is no information regarding the importance of sexual mechanisms to micro-organisms in their natural environment. It can be argued that organisms which have a sexual mechanism have the advantage of being able to effect large changes of genotype more rapidly than those which do not. This faculty allows them to withstand a wider range of environmental stress.

ACKNOWLEDGEMENTS

I would like to express my thanks to Professor A. Lwoff for the facilities of the Service de Physiologie Microbienne made available to me whilst working at the Institut Pasteur, during which time some of these experiments were performed.

REFERENCES

Adams, J. N. & Luria, S. E. (1958). Transduction by bacteriophage P1: abnormal phage function of the transducing particles. *Proc. nat. Acad. Sci., Wash.* **44**, 590.

Alexander, H. E. & Leidy, G. (1951). Determination of inherited traits of *H. influenzae* by desoxyribonucleic acid fractions isolated from type-specific cells. *J. exp. Med.* **93**, 345.

Arber, W. (1958). Transduction des caractères *gal* par le bactériophage lambda. *Arch. Sci.* (*Geneva*), **11**, 259.

Avery, O. T., MacLeod, C. M. & McCarty, M. (1944). Studies on the chemical nature of the substance inducing transformation of pneumococcal types. Induction of transformation by a desoxyribonucleic acid fraction isolated from *Pneumococcus* type III. *J. exp. Med.* **89**, 137.

BALASSA, R. (1958). Transformationserscheinungen bei *Rhizobien*. *Abstr. 7th Int. Congr. Microbiol., Stockholm*, p. 49.

BARON, L. S., CAREY, W. F. & SPILMAN, W. M. (1958). Hybridisation of *Salmonella* species with *E. coli*. *Abstr. 7th Int. Congr. Microbiol., Stockholm*, p. 50.

BEILSER, W. C. & BUNTING, M. I. (1956). Studies on a mechanism for genetic transfer in *Serratia marcescens*. *J. Bact.* **72**, 582.

BENZER, S. (1957). The elementary units of heredity. In *The Chemical Basis of Heredity*. Edited by W. D. McElroy and B. Glass. Baltimore: The Johns Hopkins Press.

BERTANI, L. E. (1957). The effect of the inhibition of protein synthesis on the establishment of lysogeny. *Virology*, **4**, 53.

CAVALLI, L. L. (1950). La sessualita nei batteri. *Boll. Ist. sieroter, Milano*, **29**, 1.

CAVALLI-SFORZA, L. L., LEDERBERG, J. & LEDERBERG, E. M. (1953). An infective factor controlling sex compatibility in *Bacterium coli*. *J. gen. Microbiol.* **8**, 89.

CHARGAFF, E., SCHULMAN, H. M. & SHAPIRO, H. S. (1957). Protoplasts of *E. coli* as sources and acceptors of deoxypentose nucleic acid: rehabilitation of a deficient mutant. *Nature, Lond.* **180**, 851.

CLOWES, R. C. (1960). Fine structure as revealed by transduction. In *Microbial Genetics*. *Symp. Soc. gen. Microbiol.* **10**, 92.

DAWSON, M. H. & SIA, R. H. P. (1931). *In vitro* transformation of pneumococcal types: technique for inducing transformation of pneumococcal types *in vitro*. *J. exp. Med.* **54**, 681.

DEMEREC, M., LAHR, E. L., MIYAKE, T., GOIDMAN, I., BAILBINDER, E., BERNIC, S., HASHIMOTO, K., GLANVILLE, E. V. & GROSS, J. D. (1958). *Yearb. Carneg. Instn*, **57**, 390.

DOERMANN, A. H. (1953). The vegetative state in the life cycle of bacteriophage: Evidence for its occurrence and its genetic characteristics. *Cold Spr. Harb. Symp. quant. Biol.* **18**, 3.

EPHRUSSI-TAYLOR, H. (1958). The mechanism of desoxyribonucleic acid-induced transformations. In *Recent Progress in Microbiology, 7th Int. Congr. Microbiol., Stockholm*, p. 51.

FISHER, K. W. (1957*a*). The role of the Krebs cycle in conjugation in *Escherichia coli* K-12. *J. gen. Microbiol.* **16**, 120.

FISHER, K. W. (1957*b*). The nature of the endergonic processes in conjugation in *Escherichia coli* K-12. *J. gen. Microbiol.* **16**, 136.

FOX, M. S. & HOTCHKISS, R. D. (1957). Initiation of bacterial transformations. *Nature, Lond.* **179**, 1322.

FRY, B. A. (1959). Conditions for the infection of *Escherichia coli* with lambda phage and for the establishment of lysogeny. *J. gen. Microbiol.* **21**, 676.

GAREN, A. & ZINDER, N. D. (1955). Radiological evidence for partial genetic homology between bacteriophage and host bacteria. *Virology*, **1**, 347.

GRIFFITHS, F. (1928). The significance of pneumococcal types. *J. Hyg., Camb.* **27**, 113.

HAYES, W. (1952). Recombination in *Bact. coli* K12; unidirectional transfer of genetic material. *Nature, Lond.* **169**, 1017.

HAYES, W. (1953*a*). The mechanism of genetic recombination in *Escherichia coli*. *Cold Spr. Harb. Symp. quant. Biol.* **18**, 75.

HAYES, W. (1953*b*). Observations on a transmissible agent determining sexual differentiation in *Bacterium coli*. *J. gen. Microbiol.* **8**, 72.

HAYES, W. (1957). The kinetics of the mating process in *Escherichia coli*. *J. gen. Microbiol.* **16**, 97.

HERSHEY, A. D. & CHASE, M. (1952). Independent functions of viral protein and nucleic acid in growth of bacteriophage. *J. gen. Physiol.* **36**, 39.

HOLLOWAY, B. W. (1955). Genetic recombination in *Pseudomonas aeruginosa*. *J. gen. Microbiol.* **13**, 572.

HOTCHKISS, R. D. (1954). Cyclical behaviour in pneumococcal growth and transformability occasioned by environmental changes. *Proc. nat. Acad. Sci., Wash.* **40**, 49.

HOTCHKISS, R. D. & EVANS, A. H. (1958). Analysis of the complex sulphonamide resistance locus of *Pneumococcus*. *Cold Spr. Harb. Symp. quant. Biol.* **23**, 85.

HOTCHKISS, R. D. & MARMUR, J. (1954). Double marker transformations as evidence of linked factors in desoxyribonucleate transforming agents. *Proc. nat. Acad. Sci., Wash.* **40**, 55.

JACOB, F. & ADELBERG, E. A. (1959). Transfert de caractères génétiques par incorporation au facteur sexuel d'*E. coli*. *C.R. Acad. Sci., Paris*, **249**, 189.

JACOB, F. & WOLLMAN, E. L. (1954). Induction spontanée du dévelopment du bactériophage λ au cours de la recombinaison génétique chez *E. coli* K12. *C.R. Acad. Sci., Paris*, **239**, 317.

JACOB, F. & WOLLMAN, E. L. (1955*a*). Étapes de la recombinaison génétique chez *Escherichia coli* K12. *C.R. Acad. Sci., Paris*, **240**, 2566.

JACOB, F. & WOLLMAN, E. L. (1955*b*). Étude génétique d'un bactériophage tempéré d'*E. coli*. III. Effet du rayonnement ultraviolet sur la recombinaison génétique. *Ann. Inst. Pasteur*, **88**, 724.

JACOB, F. & WOLLMAN, E. L. (1958). Sur les processus de conjugaison et de recombinaison génétique chez *E. coli*. IV. Prophages inductibles et mesure des segments génétiques transférés au cours de la conjugaison. *Ann. Inst. Pasteur*, **95**, 497.

KIRCHNER, C. & EISENSTARK, A. (1957). The importance of divalent cations to genetic exchange in *E. coli* K12 recombination and *S. typhimurium* transduction. *Bact. Proc.* p. 50.

KIRCHNER, C., ROUHANDEH, H. & EISENSTARK, A. (1958). Factors influencing transduction frequencies in *S. typhimurium*. *Bact. Proc.* p. 46.

LEDERBERG, E. M. (1960). Genetic and functional aspects of galactose metabolism in *Escherichia coli* K12. In *Microbial Genetics*. *Symp. Soc. gen. Microbiol.* **10**, 115.

LEDERBERG, J. & LEDERBERG, E. M. (1952). Replica plating and indirect selection of bacterial mutants. *J. Bact.* **63**, 399.

LEDERBERG, E. M. & LEDERBERG, J. (1953). Genetic studies of lysogenicity in *E. coli*. *Genetics*, **38**, 51.

LEDERBERG, J. & TATUM, E. L. (1946). Novel genotypes in mixed cultures of biochemical mutants of bacteria. *Cold Spr. Harb. Symp. quant. Biol.* **11**, 113.

LENNOX, E. S. (1955). Transduction of linked genetic characters of the host by bacteriophage P1. *Virology*, **1**, 190.

LERMAN, L. S. & TOLMACH, L. J. (1957). Cellular incorporation of DNA accompanying transformation in *Pneumococcus*. *Biochim. biophys. Acta*, **26**, 68.

LEVINTHAL, C. (1954). Recombination in phage T2; its relationship to heterozygosis and growth. *Genetics*, **39**, 169.

LEVINTHAL, C. & VISCONTI, N. (1953). Growth and recombination in bacterial viruses. *Genetics*, **38**, 500.

LITT, M., MARMUR, J., EPHRUSSI-TAYLOR, H. & DOTY, P. (1958). The dependence of pneumococcal transformation on the molecular weight of deoxyribose nucleic acid. *Proc. nat. Acad. Sci., Wash.* **44**, 144.

LOUTIT, J. S. (1958). A transduction-like process within a single strain of *Pseudomonas aeruginosa*. *J. gen. Microbiol.* **18**, 315.

LURIA, S. E. & BURROUS, J. W. (1957). Hybridisation between *Escherichia coli* and *Shigella*. *J. Bact.* **74**, 461.

Luria, S. E., Fraser, D. K., Adams, J. N. & Burrous, J. W. (1958). Lysogenisation, transduction and genetic recombination in bacteria. *Cold. Spr. Harb. Symp. quant. Biol.* **23**, 71.

Lwoff, A. & Gutmann, A. (1950). Recherches sur un *Bacillus megatérium* lysogène. *Ann. Inst. Pasteur*, **78**, 711.

Maccacaro, G. A. (1955). Cell surface and fertility in *Escherichia coli. Nature, Lond.* **176**, 125.

Maccacaro, G. A. & Comolli, R. (1956). Surface properties correlated with sex compatibility in *Escherichia coli. J. gen. Microbiol.* **15**, 121.

Marmur, J. & Lane, D. (1960). Strand separation and specific recombination in deoxyribonucleic acids: Biological studies. *Proc. nat. Acad. Sci., Wash.* **46**, 453.

McCarty, M., Taylor, H. E. & Avery, O. T. (1946). Biochemical studies of environmental factors essential in transformation of pneumococcal types. *Cold Spr. Harb. Symp. quant. Biol.* **11**, 177.

Meselson, M. & Stahl, F. W. (1958). The replication of DNA in *Escherichia coli. Proc. nat. Acad. Sci., Wash.* **44**, 671.

Morse, M. L. (1959). Transduction by staphylococcal bacteriophage. *Proc. nat. Acad. Sci., Wash.* **45**, 722.

Morse, M. L., Lederberg, E. M. & Lederberg, J. (1956*a*). Transduction in *E. coli* K12. *Genetics*, **41**, 142.

Morse, M. L., Lederberg, E. M. & Lederberg, J. (1956*b*). Transductional heterogenotes in *E. coli. Genetics*, **41**, 758.

Pratt, D. & Stent, G. S. (1959). Mutational heterozygotes in bacteriophages. *Proc. nat. Acad. Sci., Wash.* **45**, 1507.

Ritz, H. L. & Baldwin, J. N. (1958). Induction of penicillinase production in staphylococci by bacteriophage. *Bact. Proc.* p. 40.

Schaeffer, P. (1956). Transformation interspécifique chez des bactéries du genre *Hemophilus. Ann. Inst. Pasteur*, **91**, 192.

Schaeffer, P. (1957*a*). Existence d'une compétition entre molecules d'acide desoxyribonucléique pour la pénétration dans les bactéries transformables. *C.R. Acad. Sci., Paris*, **245**, 230.

Schaeffer, P. (1957*b*). La pénétration de l'acide nucléique dans les bactéries réceptrices au cours des transformations interspécifiques. *C.R. Acad. Sci., Paris*, **245**, 375.

Schaeffer, P. (1957*c*). L'inhibition de la transformation comme moyen de mésure de la 'compétence' bactérienne. *C.R. Acad. Sci., Paris*, **245**, 451.

Schaeffer, P. (1958). Interspecific reaction in bacterial transformation. In *Biological replication of Macromolecules. Symp. Soc. exp. Biol.* **12**, 60.

Sinai, J. & Yudkin, J. (1959). Transformation to proflavine resistance in *E. coli. J. gen. Microbiol.* **20**, 400.

Skaar, P. D. & Garen, A. (1956). The orientation and extent of gene transfer in *Escherichia coli. Proc. nat. Acad. Sci., Wash.* **42**, 619.

Spizizen, J. (1958). Transformation of biochemically deficient strains of *Bacillus subtilis* by deoxyribonucleate. *Proc. nat. Acad. Sci., Wash.* **44**, 1072.

Stocker, B. A. D. (1960). Introduction: micro-organisms in genetics. In *Microbial Genetics. Symp. Soc. gen. Microbiol.* **10**, 1.

Sueoka, N. J., Marmur, J. & Doty, P. (1959). Dependence of the density of deoxyribonucleic acids on guanine-cytosine content. *Nature, Lond.* **183**, 1429.

Tessman, I. (1959). Mutagenesis in phages ϕX 174 and T4 and properties of the genetic material. *Virology*, **9**, 375.

Thomas, R. (1955). Recherches sur la cinétique des transformations bactériennes. *Biochim. biophys. Acta*, **18**, 467.

WOLLMAN, E. L. & JACOB, F. (1955). Sur le méchanisme du transfert de matériel génétique au cours de la recombinaison chez *E. coli* K12. *C.R. Acad. Sci., Paris*, **240**, 2449.

WOLLMAN, E. L. & JACOB, F. (1957). Sur les processus de conjugaison et de recombinaison chez *Escherichia coli.* II.—La localisation chromosomique du prophage λ et les conséquences génétiques de l'induction zygotique. *Ann. Inst. Pasteur*, **93**, 323.

WOLLMAN, E. L. & JACOB, F. (1958). Sur les processus de conjugaison et de recombinaison génétique chez *Escherichia coli.* V.—Le méchanism du transfert de matériel génétique. *Ann. Inst. Pasteur*, **95**, 641.

WOLLMAN, E. L., JACOB, F. & HAYES, W. (1956). Conjugation and genetic recombination in *E. coli* K12. *Cold Spr. Harb. Symp. quant. Biol.* **21**, 141.

ZINDER, N. D. (1953). Infective heredity in bacteria. *Cold Spr. Harb. Symp. quant. Biol.* **18**, 261.

ZINDER, N. D. (1955). Bacterial transduction. *J. cell. comp. Physiol.* **45** (Suppl. 2), 23.

ZINDER, N. D. & LEDERBERG, J. (1952). Genetic exchange in *Salmonella. J. Bact.* **64**, 679.

SOME EFFECTS OF CHANGING ENVIRONMENT ON THE BEHAVIOUR OF PLANT VIRUSES

F. C. BAWDEN

Rothamsted Experimental Station, Harpenden, Hertfordshire, England

When the organizers of this Symposium first invited me to contribute, they said its object was 'to survey the nature and significance of individual microbial responses to environmental changes' and they expressed the hope that contributors 'will review the causes and features of non-genetic variation, elucidate as far as possible the mechanisms involved and speculate on the role of such variation in survival and evolution of species'. This would be difficult to undertake with genuine microbes that have measurable metabolic activities of their own, but is impossible with plant viruses. In saying this politely when I refused the invitation, I rashly added that there were plenty of phenomena showing that environment affects the behaviour of plant viruses in many and various ways, but that it was impossible to fulfil their hopes because the mechanisms involved are largely unknown. Many of the changes most likely to be important for survival and evolution are probably genetic, and effects on virus multiplication or symptom production as often as not probably reflect changes in the host cells rather than in the virus particles. My reply brought a second invitation, which said that a major purpose of the Symposium was to indicate the range of effects produced by changing environments, that it would be incomplete without something about viruses and which asked me to review the phenomena even though I could not interpret them. This invitation I reluctantly accepted, so if, as I rather fear, my contribution will seem to have strayed into the wrong Symposium, the organizers rather than I must take the blame.

PROPERTIES THAT AFFECT SURVIVAL AND TRANSMISSION

An elementary start by defining a virus as a submicroscopic infective entity that multiplies only intracellularly will make the critical importance of the environment to viruses immediately obvious. With no independent metabolism, the most viruses can do outside susceptible cells is to retain their ability to infect. Once established in susceptible cells, plant viruses are perhaps more secure than animal viruses, because plants lack the antibody-forming mechanisms whereby higher animals combat their virus infections, and infected plant cells usually contain virus for as long as they remain alive. However, as plant cells are mortal, the continued existence of a virus demands that it infects a continuous series of susceptible cells. This demand is met in various ways, of which the most common are efficient methods of spread from organism to organism and the ability to spread from cell to cell within plants. It is worth stressing that not all hosts provide environments equally favourable to the survival of a virus. A host that is infected only locally or is rapidly killed provides only a brief period of multiplication compared with one that is invaded systemically and is little harmed by infection. In a perennial plant or one propagated vegetatively, a virus that is tolerated can be supplied with a continuing series of susceptible cells year after year without running the risks entailed in moving from one organism to another and, as often as not, such plants are the sources of virus for annual plants. Survival, too, can be helped by the ability to be perpetuated through the seed of infected plants, but most viruses seem not to occur in the pollen or egg mother cells of most of their plant hosts. What prevents a virus able to invade and multiply in vegetative cells from infecting these reproductive cells is unknown, but many viruses also seem not to occur in apical meristems, which may point to the possibility that the nucleic acid metabolism of actively dividing cells is often an environment unfavourable to virus multiplication.

Most plant viruses are poorly adapted to survive long outside susceptible cells, but a few, of which tobacco mosaic (TMV) is the prime example, are stable enough to remain infective for long periods in dead cells or extracellularly. The stability of plant viruses seems to be determined by the structure of their protein or by the way the protein is bonded to their nucleic acid. For a long time the nucleoprotein particles seemed to be the minimum replicating entities, but it is now clear that the protein is not essential for virus multiplication and that the reproductive parts are the nucleic acids (Gierer & Schramm, 1956; Fraenkel-Conrat, 1956). The nucleic acid of TMV seems no more stable when freed from its protein than are those from other viruses, and various treatments that inactivate some other viruses but are harmless to intact TMV rapidly inactivate its free nucleic acid (Bawden & Pirie, 1959). Some treatments have long been known to inactivate viruses without greatly affecting their physical properties or serological behaviour. Some of these, such as exposure to radiations or to formaldehyde, do so with all viruses, though different viruses differ in their resistance to inactivation, but others, such as exposure to gentle heat, act only with some. While the nucleoprotein particles were thought to be the minimal infective units, the phenomenon remained unexplained, but the discovery that infectivity and reproduction are primarily functions of the nucleic acid suggests that such inactivation treatments have little effect on the surface proteins, which mainly determine physical and serological behaviour, and inactivate only when they can reach and change the embedded nucleic acid. The extent to which potentially inactivating treatments affect the nucleic acid in virus particles clearly depends on its immediate environment, that is, on the way it is carried in or bonded to its protein. Thus, the nucleic acids of viruses such as the Rothamsted tobacco necrosis seem almost unprotected by their proteins against many inactivating treatments that are successfully kept at bay by the protein of TMV.

It is not only during the extracellular period when virus particles are moving from one organism to another that they are at risk. Establishing infection in a fresh host also seems a process beset by hazards. This is plausibly explained by the current idea that initiation of infection entails the nucleic acids of virus particles separating from their containing protein. Experiments in which either viruses or leaves are exposed to ultraviolet radiation provide evidence that virus particles change their state soon after inoculation to leaves, but unfortunately give no unequivocal evidence about the nature of the changes. However, as these experiments vividly illustrate the extent to which changing the environment in intrinsically susceptible cells can affect the ability of viruses to infect and multiply, they merit a brief description. Particles of many viruses, though not of TMV, after damage by radiation will not infect inoculated plants kept in the dark but will infect plants kept in the light after inoculation. (The lack of photo-reactivation with TMV again results from its protein protecting the nucleic acid from changes it readily undergoes when irradiated free from protein: Bawden & Kleczkowski, 1959). Potato virus X shows the phenomenon very strongly, and *c.* 30 min. elapse after inoculation before the damaged particles reach a state in which exposing the leaf to visible light affects them; the leaves then need illuminating for only a few minutes to allow infections to develop. If the leaves are not illuminated within the next hour or so, exposure to visible light has no effect, presumably because by then infectivity has been irreversibly destroyed (Bawden & Kleczkowski, 1955). This could mean that only freed nucleic acid is photoreactivated, that 30 min. is the time needed for the virus particles to undress and 1 hr. is the maximum time the naked nucleic acid can survive in the leaf cells. Equally, however, 30 min. may be the time it takes the virus particles to associate with some light-sensitive component of the cells and 1 hr. as long as the intact virus particles retain their infectivity *in vivo*.

Viruses can also be inactivated *in vivo* by ultraviolet radiation; after leaves are inoculated with virus particles, there is a period, which differs with different viruses and virus strains, during which the same dose of radiation will prevent infection from developing, but after this period, the dose needs to be increased. There is no such lag period with inocula of free nucleic acids (Siegel, Ginoza & Wildman, 1957; Kassanis, 1960); the immediate gain in resistance to inactivation clearly suggests that nucleic acid starts doing something sooner than do intact virus particles, but it is an assumption that the virus particles are slower off the mark because they have to undress. The nucleic acid is labile and if it does not quickly do something that protects it, it is likely to be inactivated, whereas the intact virus particles can be more leisurely, at least when they are as stable as TMV. Certainly there is no suggestion that TMV soon separates into protein and nucleic acid after it is inoculated to leaves that have been exposed to ultraviolet radiation. After appropriate irradiation, leaves that show no signs of damage are nevertheless changed so that they cannot support infection by viruses to which they are usually susceptible. This acquired resistance is only temporary provided the leaves are exposed to visible light, but whether they become infected by inocula applied while the leaves are regaining their susceptibility depends on the quality of the inocula. Inoculations with nucleic acid from TMV, or with relatively unstable viruses such as the Rothamsted culture of tobacco necrosis, do not produce infection, whereas inoculations with intact TMV do (Bawden & Kleczkowski, 1952, 1960). This hardly suggests that the nucleic acid of TMV becomes free, for, if it did, it might be expected to be inactivated, as it is when plants are inoculated with it already free. Consequently, the failure of the tobacco necrosis virus to survive through the period of acquired host resistance may well lie in the inability of its protein to protect the nucleic acid from inactivation, rather than because the particles disrupt and set the nucleic acid free.

The idea that plant viruses disrupt as an essential step in the infection process derives from work with coliphages, which leaves no doubt that most of the phage protein remains outside the infected bacterium and that little except nucleic acid penetrates. Plant viruses may undergo some similar process, but analogies between phages and viruses of higher plants should not be pushed too far. Not only are phages chemically different in containing deoxyribose, instead of ribose, nucleic acid, but they operate in very different environments and, as they are lethal pathogens of short-lived hosts, their survival problems are greater and seem to have been solved in different ways. For instance, there is no evidence that plant viruses ever enter into any intimate relationships with the genetic systems of their hosts comparable to the harmless and self-perpetuating state of prophage in lysogenic bacteria. Also, whereas plant viruses depend on other agencies for their spread from one organism to another, phages do this unaided, an activity for which their complex morphology fits them excellently. If the replication of ribose nucleic acids demands that they are free from proteins, there is an obvious reason for the particles disrupting, but this is uncertain and to assume that plant viruses, which are morphologically so much simpler, spread in such different ways and are less at risk of extinction, must follow a similar infection process to bacteriophages is, to say the least, premature.

The morphology and behaviour of bacteriophages suggest an efficient method of independent spread has been the main factor influencing their evolution. Plant viruses are less standardized, their methods of spread are more varied and seem more chancy. The one common feature is that they all depend on some other agency to make the wounds in plants through which they enter. Some, such as TMV, which are stable outside cells, can survive extracellularly almost indefinitely and enter a plant through any wound they chance to meet, but most must get rapidly from one susceptible organism to another and for this they depend on animals that feed on their plant hosts. For many, this entails the necessity of surviving in the environment of the mouthparts of their animal vectors, but some avoid this by using these vectors as alternative hosts. A few of this second type also increase their chances of survival by being able to invade the egg cells of their vectors and so be handed on from one generation of animals to the next without the risk of exposure to an extracellular environment. Little is known about the structure or properties of these viruses, but they are likely to be very labile outside their host cells. Perhaps the most surprising thing about them is that no change in their behaviour has been reported as a consequence of multiplying in such diverse environments as plant and animal cells, though it must be added that they have yet to be studied in detail.

Any change in a virus that increases its range of animal vectors is likely to be important to survival and evolution, for it not only increases its chances of spread between individual plants of a species the virus is already infecting, but it also brings the possibility of spread to other species from which it was previously debarred. The effective host range of many plant viruses is set by the feeding habits of the animals that transmit them, rather than by the number of plant species that they are intrinsically able to infect. An extension of host range in its turn may have far-reaching consequences, for in the environment of a new host a virus may multiply to much greater amounts than in an old one and be acquired and transmitted by animal vectors much more readily.

What confers on a virus the ability to remain infective in or on the mouthparts of a given species of animal is unknown, but one obvious possibility is some specific configuration of the surface protein; it is something that can be lost and gained, and it can be lost for one species while retained for another. It also seems to be something that may need continual selection to preserve, for there are several examples of transmissibility by a given insect being lost when a virus has either long been maintained in one plant or transmitted by some other agency.

The prevalence of individual viruses varies greatly from place to place and from season to season, reflecting mainly the numbers and activity of the specific animals that transmit them. Such effects of changes in environment are outside the scope of our present discussion, but need passing mention because their practical consequences are probably far greater than most of the changes with which we are concerned.

CHANGES IN THE PHYSIOLOGICAL CONDITIONS OF HOST PLANTS

Whether a given plant can be a host for a given virus is presumably determined by the genetic constitution of the plant and virus but, within this genetic limitation, phenotypic changes can have enormous effects. The physiology of plants responds rapidly and strikingly to changes in the environment and whereas a plant in one state may be readily infected, in another it may be highly resistant; should infection occur, the environment of the plant may determine whether it is localized or systemic and the kind of symptoms shown. Here is no place for details, but some effects of changing light intensity and temperature merit describing.

Keeping plants in darkness or at temperatures of around 37° for a day or two before they are inoculated increases their susceptibility to infection by mechanically transmitted viruses, in the sense that a standard inoculum will cause more infections than in comparable plants not given these treatments. Similarly, plants grown in shade are more susceptible than those grown under a high light intensity. This seems to be a general effect with all viruses and in part might mean only that the treatments make leaves more susceptible to injury when rubbed. However, that there is more to it than this is suggested by the fact that the size of the response to these conditioning treatments depends on the quality of the inoculum. Table 1 shows that, whereas with inoculum of intact TMV the number of lesions was approximately doubled by keeping *Nicotiana glutinosa* in darkness or at 37° before they were inoculated, with inoculum of nucleic acid the number was increased 10 or more times. This difference can hardly be explained by the two kinds of inocula giving different numbers of wounds and it strongly suggests that keeping leaves in darkness or at 37° decreases the amount of some inactivator that affects free nucleic acid more than intact TMV. Plants of different ages, or plants raised under widely differing light regimes, can show much greater differences in their susceptibility to infection by the two kinds of inocula than those shown in Table 1. With differently treated plants that differ in their susceptibility to infection by intact TMV by factors of *c.* 10, the more susceptible may give 100 lesions per leaf with a nucleic acid inoculum that fails to infect the more resistant plants. Extracts of leaves contain ribonuclease and various other components that will inactivate the nucleic acid *in vitro*, but although some extracts of resistant leaves are more powerful inactivators than extracts from susceptible ones, the quantitative differences seem inadequate to explain the differences in susceptibility (Bawden & Pirie, 1959). That some product of photosynthesis may be concerned in the changes in susceptibility is suggested by the fact that plants in full light have their susceptibility increased by freeing the air from CO_2, a treatment without effect when applied to plants in the dark (Wiltshire, 1956).

Infections by some other viruses depend very much more on the physiological state of their potential host plants than do those by TMV. During summer in the glasshouses at Rothamsted, for example, *Nicotiana* sp. that are readily infected by TMV, and so are presumably prone to wounding, resist infection by tobacco necrosis virus, again possibly because the leaves in summer are rich in systems that inactivate this less stable virus. One of the most striking phenotypic changes occurs with some French bean varieties, which during winter serve as good hosts for assaying cucumber mosaic virus, whereas during summer they seem to be immune (Bhargava, 1951).

Changes in susceptibility to infection produced by differences in the environment to which plants are exposed before they are inoculated are reasonably attributed to effects on the host only, but when environmental differences are applied after inoculation, effects on the host are less easily disentangled from those on the infecting virus. However, exposing plants to 36° after inoculation produces such different results from exposures before that they seem likely to be largely effects on the virus, especially as the results differ greatly with different viruses. Instead of increasing the number of infections, which is the general effect of high temperature before inoculation, the number is decreased, only slightly with intact TMV as inoculum, more so with its free nucleic acid, and so greatly with tobacco necrosis and some other viruses that inocula which give 100 lesions or more per leaf in plants kept at 20° give few or none at 36° (Kassanis, 1957, 1959). The viruses that fail to infect inoculated plants kept at 36° are ones whose concentration also falls when fully infected plants are maintained at 36°. The infectivity of leaf extracts falls much more rapidly than does their content of specific antigens, indicating that the inactivation is not caused primarily by protein denaturation but by changes in the nucleic acids, which again suggests that these are less well protected in such viruses than in TMV.

Table 1. *The effect of darkness and high temperature on the numbers of lesions caused by tobacco mosaic virus and its nucleic acid*

Inoculum		Treatment of plants			
		Control	2 days in darkness	Control	2 days at 37°
Virus	5 mg./l.	610	960	512	800
	1	168	270	213	395
	0·2	21	56	65	210
Nucleic acid	5 mg. P/l.	24	384	100	760
	1	9	90	15	175
	0·2	2	18	2	50

The control *Nicotiana glutinosa* plants were kept continuously in ordinary glasshouse conditions, whereas the others were in darkness or at 37° for the 2 days immediately before they were inoculated. The numbers are the total local lesions produced on twelve half leaves by each inoculum at the specified concentration; opposite halves of the same leaves were inoculated with virus and nucleic acid at the same relative concentration, i.e. virus at 5 mg./l. was opposite nucleic acid at 5 mg. P/l., and so on.

In cells that survive infection, the synthesis of virus seems not to be a once-for-all affair but a continuing process, with the amount of virus at any one time representing the balance between synthesis and degradation. The rate at which viruses multiply increases with increasing temperature, but so also does the rate of inactivation, and for each virus there is a critical temperature at which inactivation becomes faster than multiplication: many that increase fastest at about 25° cannot maintain themselves in plants at over 30°. Their inactivation *in vivo* may be a direct effect of heat, but as some are stable at 36° for long periods in plant extracts, it seems more likely to be indirect, caused by some inactivator in the host that works more effectively than at lower temperatures. Some effect of the host is also suggested by the fact that a given virus may multiply in one host species but not in another kept at 36°.

Many virus-infected plants show increasingly mild symptoms as the ambient temperature is raised; this is often accompanied by, and presumably a consequence of, a decreased virus content. With TMV in tobacco, however, the symptoms shown by many plants when grown at over 30° are not milder because their virus content is less than at 20°, but because the higher temperature favours the multiplication of avirulent strains. TMV is far from being the single entity that my remarks so far may have suggested. Every bulk culture comprises a multitude of types that differ from one another in almost every property for which tests have been made. Whenever the environment is changed, conditions may preferentially select what was previously a quantitatively minor component of the culture and rapidly turn it into the bulk of a culture. Sukhov (1956) has argued that exposing plants to high temperatures causes the virus to change, but for this there is no evidence. It does seem likely that the avirulent forms often arise in the plants kept at high temperatures, rather than that they are introduced as such in the inoculum (Kassanis, 1957), but there is no reason to consider them other than randomly produced variants that then romp ahead in the new environment. Whether they are 'mutations' is impossible to decide, because it is never sure that any starting source of TMV is genetically homogeneous. The nearest approximation to a single-cell culture of a microbe obtainable with a virus is a culture from a single local lesion. Many workers assume that such lesions are caused by infections with one virus particle, but there is little substantiating evidence. To produce a single lesion, a leaf has to be rubbed with inoculum containing hundreds of thousands of virus particles. No doubt most of these are wasted because they never meet an entry point, but a statistical analysis (Kleczkowski, 1950) of the dilution curve shows incompatibility with the hypothesis that infection depends on single virus particles encountering uniform sites and compatibility only with the hypothesis of a minimum infecting dose, the size of which varies at different infection sites. Not all strains of TMV throw variants with equal readiness and keeping plants inoculated with tomato aucuba mosaic virus at 36° did not produce variants better able to multiply at 36° than the parent culture (Kassanis, 1957). Similarly, over many years this virus has not produced a variant that invades tobacco systemically at 20° to give mosaic symptoms, whereas variants that produce only necrotic local lesions seem always to occur in cultures of type TMV, even those recently derived from a single lesion. The greater variability of the type strain suggests that it is genetically richer than most of its derivatives, which in turn suggests that these may be mostly derived by losing rather than acquiring some activity.

Changing host plants also often changes the behaviour of a virus culture, either by eliminating strains to which the new host is insusceptible or by encouraging ones that multiply faster in it than the main form. Such changes, as with those produced by keeping TMV-infected plants at high temperatures, are usually slight, showing mainly in a different virulence towards a given host, and the general properties or constitution of the particles are little altered. However, changes in a strain of TMV found in leguminous plants are much larger than this and are also reversible (Bawden, 1958). The form in which the strain occurs in systemically infected legumes differs greatly from type TMV; the two are only remotely related serologically, cause different types of lesions in *Nicotiana glutinosa*, have many different properties *in vitro*, and distinctive amino acid compositions. Comparable differences have been noted between type TMV and a few other strains, which mostly fail to infect tobacco and have been considered to have achieved their different constitution by a protracted series of changes while they evolved in different hosts. That a long period is not always needed to produce large changes, however, is shown by passing the strain from leguminous plants to tobacco. At 20° it causes only necrotic local lesions in most plants, but an occasional one becomes systemically infected by a form that closely resembles type TMV. Most bean plants inoculated with this tobacco form remain healthy, but occasional ones become systemically infected and these contain virus resembling the original leguminous form.

Again, it cannot be decided whether these very different forms arise as random mutants, by genetic recombination or by some other process. The one clear thing is that the two are separated by the environment of the host plant, because one infects leguminous plants but not tobacco systemically, and the other tobacco but not legumes. When tobacco plants inoculated with the legume form are kept above 30°, they show no local lesions, although the virus content of leaves becomes greater than at 20° because the legume form now spreads more readily from cell to cell. In these conditions, the tobacco form seems not be produced or, if it is, it no longer predominates over the bean form as it does in plants at 20°. The two forms also differ widely in their behaviour in *Nicotiana glutinosa* at different temperatures. At 20°, both produce necrotic local lesions, but whereas those from the legume form are small and appear slowly, those from the tobacco form, like those from type TMV, are larger and appear sooner. In plants kept at over 30°, neither form produces necrotic lesions, but whereas the tobacco form, again like type TMV, produces spreading yellow lesions, the legume form multiplies without causing external symptoms. At the higher temperature both forms invade more cells than at the lower, but the virus content per cell is lower, and when infected plants are first kept at over 30° for some days and then put at 20°, their virus content rapidly increases greatly and the invaded leaf tissue dies.

This strain from leguminous plants was the first to show such changes, but it is not unique. Type TMV and a few other strains can do similar things, for occasional French bean plants of many inoculated with type TMV became infected systemically and contained virus that differed greatly from type TMV and closely resembled the strain found naturally in leguminous plants. Inoculations with tomato aucuba mosaic virus, however, have so far failed to infect beans systemically, again suggesting that, although type TMV embraces all or most of the genetic possibilities shown by its many and varied relatives, other strains are more limited.

EFFECTS OF INFECTION BY OTHER VIRUSES

Interpretations of changes in the behaviour of plant viruses is difficult not only because the genetic homogeneity of the culture is always uncertain, but also because related strains of one virus interfere with the ability of one another to multiply and produce their characteristic effects. This phenomenon is shown most vividly by the resistance to the effects of virulent strains acquired by plants when already infected with an avirulent one, but the effects of mixing strains in inocula can also be striking. For example, whereas tomato aucuba mosaic on its own at about 1 mg./l. gives many of its characteristic, large, brown, necrotic, local lesions in Java tobacco, inocula that contain in addition 50–100 mg./l. TMV produce no such lesions but only a few small white necroses. Various explanations for such interference phenomena have been advanced, but there is no conclusive evidence for any. The most attractive is probably Best's (1954) suggestion that related strains exchange genetic determinants when multiplying: if they do, a specific determinant for some activity, such as virulence towards a given host, which was introduced as only a quantitatively minor component of the total genetic pool would be submerged unless it had unusual survival value in the particular environment of the host.

Whereas infections by pairs of related strains of one virus usually produce effects intermediate between those of either alone, simultaneous infection by two unrelated viruses often causes a more severe disease than is caused by either virus alone. With some pairs of viruses this enhanced virulence is accompanied by one of the pair multiplying to greater levels than when on its own; tobacco leaves simultaneously infected by potato viruses *X* and *Y*, for example, contain several times as much virus *X* as do comparable leaves infected with virus *X* alone (Rochow & Ross, 1955). Again the phenomenon is unexplained, but obviously infection with potato virus *Y* may either destroy some host system that normally inactivates virus *X* and prevents it reaching such high levels, or it may increase the amount of some substance essential for the synthesis of virus *X* that was previously limiting multiplication. A comparable effect occurs with dodder latent virus in plants which when first infected show severe symptoms and have high virus content, but whose virus content later falls and symptoms decrease until in time the plants look normal. Infecting such 'recovered' plants with TMV apparently takes some brake off a process that limits the accumulation of dodder latent virus, which now again increases to its initial amount and the severe symptoms are restored and retained indefinitely (Bennett, 1949).

An even more striking effect of one virus on the activities of another has been uncovered by recent work with the Rothamsted culture of tobacco necrosis virus. Extracts from leaves infected with this culture have long been known to contain specific particles of at least two sizes, but the relationship between these particles was obscure. Partially purified preparations containing both sizes were highly infective and crystalline preparations of the small ones had little or no infectivity, so the small ones seemed likely to be either inactivated derivatives of the larger or specific products of their multiplication, analogous to material antigenically related to the infecting virus that occurs in plants infected with some other viruses (Bawden & Pirie, 1950). However, Kassanis & Nixon (1960) have now shown that the particles of two sizes are not serologically related and that, although the small ones fail to multiply when inoculated alone to plants, they infect and multiply when inoculated together with the larger ones. Hence, the particles of different sizes seem to be the equivalent of unrelated viruses, one of which depends on the other for creating the conditions in which it can multiply. Infection with the large particles alone probably produces none of the small ones, but this has yet to be proved because of the difficulty of getting inocula entirely free from the small ones. It seems likely because the concentration of small particles in extracts from inoculated leaves depends on the ratio of small to large particles in the inoculum. This ratio also affects the type and number of lesions produced: the large particles produce large lesions in both French bean and tobacco leaves; the small particles produce no lesions in either, but when mixed in inocula with large particles, they decrease the size of the lesions and, when present in great excess, they also decrease the number of lesions formed (Table 2). Whether the large particles multiply less in the presence of the small ones than when on their own has yet to be determined, but the effect on lesion size suggests that the synthesis of the small particles may be at least in part at the expense of the larger ones.

Table 2. *The interaction between the large and small particles from preparations of the Rothamsted tobacco necrosis virus*

Composition of inoculum (mg./l.)		Number and size of lesion	
Large particles	Small particles	Large lesions	Small lesions
4	0	322	0
4	5	12	314
4	20	4	295
4	80	1	184
0	80	0	0

In striking contrast to the activation of one component in the culture of Rothamsted tobacco necrosis virus by the other is the interaction between potato virus *Y* and tobacco severe etch virus (Bawden & Kassanis, 1945). When the two are inoculated simultaneously to healthy plants, only severe etch virus becomes established, and when plants already systemically infected with virus *Y* are inoculated with severe etch virus, virus *Y* is supplanted and disappears. Again, no definite explanation can be offered, but the phenomenon may mean that the two occupy the same ecological niche and illustrates that in an intracellular environment, as in the larger world, evolution follows the general principle of competitive exclusion, which states that two organisms with the same ecological requirements but unable to interbreed cannot coexist; the one with the greater multiplication rate always ousts the other. Strains of one virus are, of course, also likely to occupy the same ecological niche in their host cells, but that they may interbreed has already been mentioned as a possible explanation for the way they interact; this also finds its analogy with organisms, with which introducing a new variety to an environment already containing others may add something to the gene pool, but will rarely oust those already established.

A further effect of one virus on another, which could greatly influence survival and evolution, is that one not usually transmitted by a given species of animal may become transmissible when multiplying together with another that is. For such an interaction between related strains (such as potato virus *C* which becomes transmissible by *Myzus persicae* when present in plants together with potato virus *Y*: Watson, 1960), there is the ready explanation, whether right or not, that genetic recombination between the two produces variants with the pathogenic peculiarities of virus *C* and the transmissibility of virus *Y*. However, the phenomenon also occurs with unrelated viruses, such as tobacco mottle and vein-distorting viruses (Smith, 1946), and here another explanation must presumably be sought. As some viruses increase the concentration reached by others in infected leaves, this could be a factor but alone it seems unlikely to explain the gain of transmissibility, unless the extra virus is produced at a different place from the usual and is more readily obtained by the potential vector. Two other possible explanations are obvious; the virus not normally transmitted may simply become attached to the one that is, and so be carried along with it by the vector; or phenotypic mixing may occur when the two viruses multiply in the same cells, so that the protein constitution of the particles becomes changed and the one not normally transmitted now contains enough protein from the other for it to be specifically acquired by the other's vector.

INHIBITORS OF INFECTION AND VIRUS MULTIPLICATION

Although failure to be transmitted by mechanical inoculation is often treated as though it were an intrinsic property of a virus and looms large in descriptions of virus properties, it sometimes means no more than that the environment of the virus precludes successful transmission. Infection of a susceptible host does not depend simply on the amount of active virus in the inoculum, but on what other materials are also present. Many substances, as chemically diverse as proteins, polysaccharides, tannins and analogues of purines and pyrimidines, although not inactivators of viruses *in vitro*, can by their presence in inocula prevent infection from occurring. Most of these produce their full effect immediately they are added to a virus; their ability to inhibit depends more on the identity of the host inoculated than of the virus and is proportionally much greater at high than at low concentrations, so that diluting non-infective mixtures of concentrated inhibitor and virus often allows them to infect.

Tannic acid behaves differently: its inhibitory effect is increased by increasing the time that viruses are exposed to it *in vitro*; its action is independent of the identity of the plant inoculated but differs with different viruses, and there is a fixed ratio between amount of virus and the amount of tannin that is needed to lower infectivity by fixed amount (Cadman, 1959). Hence, whereas tannin seems to act directly on the virus particles, perhaps because it forms links between, and so clumps, the virus particles, other inhibitors seem to act indirectly and prevent infection by altering the host cells. One that obviously may act directly is pancreatic ribonuclease for, on the hypothesis that nucleic acids become free as an initial step in the infection process, its action can be plausibly interpreted as hydrolysis of the nucleic acid (Casterman & Jeener, 1955). However, it behaves so like other inhibitors that have no ribonuclease activity (Bawden, 1954), whereas leaf ribonuclease which hydrolyses virus nucleic acids *in vitro* is not a powerful inhibitor of infection, that even its action is also probably best sought in the metabolic disturbances it causes in the host cells.

Extracts from many plants contain inhibitors of infection, but what role, if any, they play in determining the spread of viruses is unknown. It may well be slight, except in experimental transmissions by inoculation of sap from plants that contain them, for viruses multiply in plants that contain them, insects acquire viruses readily from such plants, and even with mechanical transmissions by sap, they often inhibit infection less in plants that contain them than in other species.

Most inhibitors are substances with large molecules and prevent infection only when they are in inocula or are inoculated to leaves within an hour or two of the virus being introduced. In contrast, the analogues of purines and pyrimidines are small molecules, which readily diffuse into leaves, and they not only inhibit infection but may also decrease the rate at which viruses multiply in already infected leaves. The two most studied are thiouracil and azaguanine; both these become incorporated in the nucleic acid of TMV when the virus is multiplying in leaves treated with them (Matthews, 1953; Jeener & Rosseels, 1953), so their action can be plausibly explained by postulating that particles containing these analogues instead of uracil or guanine are inactive. However, the explanation is not fully satisfying, for the TMV produced in the presence of thiouracil is not readily demonstrated to be less infective than usual. Also, the fact that thiouracil prevents further multiplication when it is applied to leaves some days after they have been inoculated and they already contain much infective virus, suggests that some other factor than 'sterility' of the first-formed particles is needed to explain the phenomenon.

The ability of such analogues of purines and pyrimidines to interfere with virus multiplication depends on the identity of both the infecting virus and the plant infected. Thus, the multiplication of tobacco necrosis viruses is affected by thiouracil in tobacco but not in French bean leaves, and whereas TMV is inhibited strongly in tobacco leaves by thiouracil but not by azaguanine, cucumber mosaic virus is inhibited strongly by azaguanine but not by thiouracil (Badami, 1959). The extent to which thiouracil inhibits TMV multiplication in tobacco, however, also depends greatly on the physiological state of the leaves. The virus normally reaches much higher concentrations in detached leaves floated in nutrient solution and kept in the light than in leaves floated in water and kept in the dark, but in the presence of thiouracil the usual difference is reversed, because thiouracil almost stops virus formation in illuminated leaves floated in nutrient solution but has little effect when leaves are in water and darkness. This may mean no more than that leaves floated in water and kept dark contain more free uracil than the others, for the inhibiting effect of thiouracil can be counteracted by an excess of uracil, but this possibility has not been tested. Uracil, but not cytosine and thymine, also counteracts the inhibition of TMV multiplication by thiocytosine and thiothymine (Commoner & Mercer, 1951, 1952), which perhaps suggests that the inhibiting substances all act by affecting some uracil-demanding host system rather than simply because they substitute for normal purines and pyrimidines in virus nucleic acids. However, although uracil counteracts effects of these analogues on virus multiplication, it does not counteract all their adverse effects on plants, and much more work will be needed before their behaviour is understood. Nevertheless, there is an obvious common meeting ground for most inhibitors of infection and virus multiplication, as well as for the action of ultraviolet radiation in making plants resist infection, in the idea that they all disturb nucleic acid metabolism and produce an intracellular environment inclement for virus survival and multiplication.

THE ACTION OF NITROUS ACID

Of the many treatments that can inactivate viruses without denaturing the protein or changing serological behaviour, exposure to nitrous acid calls for comment because some conclusions about its action on TMV have such great genetical significance. What is significant is not so much whether an environment of nitrous acid is mutagenic as well as inactivating, but whether the experimental results establish the prevailing idea that genetic specificity resides solely in the order in which nucleotides follow one another and is changed by changing their order. In doubting that they do, I seem to be in a minority of one, but I shall air my doubts in the hope that they may stimulate critical experiments to provide conclusive evidence one way or the other.

What are the facts? Exposing TMV nucleic acid to nitrous acid converts adenine, guanine and cytosine respectively to hypoxanthine, xanthine and uracil, and there is no need to doubt that it can change the nucleotide pattern (Schuster & Schramm, 1958). Also, there is no doubt that the treatment changes the behaviour of TMV cultures when inoculated to certain plants and that variants differing from the type strain are more readily isolated than previously (Mundry & Gierer, 1958; Bawden, 1959; Tsugita & Frankel-Conrat, 1960; Siegel, 1960). This fits together, so why question the conclusion? There would perhaps be little reason to do so if the starting culture was genetically homogeneous and stable, or if the treatment produced novel variants, but cultures of TMV are heterogeneous, and nitrous acid has so far produced nothing very different from forms already known to exist in cultures of TMV. For example, if treating tomato aucuba mosaic virus, which does not normally contain forms able to cause a systemic mosaic in Java tobacco, produced forms with this ability, it would be much more significant than is the production from TMV of forms that cause necrotic local lesions in Java tobacco, because all cultures of TMV already contain many particles able to do this. My attempts to get variants from aucuba mosaic virus, however, have so far failed, whereas every treatment of TMV with nitrous acid increases the numbers of local lesions produced in Java.

The failure to get novel variants by treatment with nitrous acid is, of course, not evidence against nitrous acid being mutagenic, for it can be argued plausibly that such a treatment is most likely to produce the kind of variants that arise most often spontaneously. However, the fact that similar variants already exist before treatment with nitrous acid means that the possibility of the treatment favouring their selection must be carefully considered. Mundry (1959) unhesitatingly rejects the idea that selection can explain the increased number of local lesions produced in Java tobacco when TMV is treated with nitrous acid. In doing so, he may be correct, and there is no doubt that the simplest explanation of the results is that nitrous acid is mutagenic. However, where I disagree with him is that this is necessarily the only possible conclusion, for knowledge about the infection process, what is entailed in the formation of a single lesion, and the meaning of infectivity assays made on different hosts or by counting different types of lesions, is not yet adequate to exclude the possibility that changes in the ability of strains to interfere with one another plays some part in selecting pre-existing variants. Quantitative assays in *Nicotiana glutinosa* or Xanthi tobacco may measure inactivation of all strains equally, but this is an assumption and one not easily justified when there are different strains that not only produce different kinds of lesions in these hosts, but also differ greatly in the amounts of them needed to cause a lesion (Bawden, 1958). Like Mundry & Gierer (1958), after treating TMV with nitrous acid I have obtained preparations that gave up to 10% as many necrotic local lesions in Java tobacco as in Xanthi, but I could find no evidence that 10% of the lesions in Xanthi contained virus that produced comparable necrotic local lesions in Java. Indeed, of 200 single-lesion isolates from such Xanthi leaves, all behaved like TMV and produced a systemic mosaic in Java. This proves nothing, but it does suggest that the two plants may preferentially select different strains or that infection of Xanthi reverses changes caused in TMV by nitrous acid. The suggestion of preferred hosts receives some support from work with a variant isolated from a local lesion in Java, for this infects Java more readily than Xanthi and reaches greater amounts in local lesions in Java than in Xanthi. Such a variant cannot be compared accurately with type TMV because type TMV does not cause necrotic local lesions in Java. However, when lesions caused by type TMV in Xanthi are macerated and used as inocula for other Xanthi, they often produce more infections than do lesions produced by variants that cause necrotic local lesions in Java. Until the interactions between these variants in different hosts, and the ways in which nitrous acid affects their infectivity, are better understood, it seems wise to postpone a decision about the origin of the variants that produce necrotic lesions in Java tobacco.

Should further work leave no doubt that the changes in the behaviour of TMV preparations caused by nitrous acid are mutations, in the sense that they reflect changes in the internal structure of individual virus particles, rather than changes in the manner and extent to which different particles interact, there still seems no reason to accept the conclusion that 'the replacement of one single NH_2-group by one OH-group *in vitro* can change the genetic character of the whole TMV-RNA molecule' (Mundry & Gierer, 1958). That nitrous acid can destroy infectivity is obvious enough and it may be able to destroy other specific activities individually. For example, while destroying whatever character of

TMV that allows it to invade Java tobacco systemically, it may leave unimpaired characters that allow it to infect this plant and multiply locally. Presumably any change in a particle that still leaves it infective is legitimately described as a mutation, but changes in a genetic character that confer new functions are obviously likely to differ from and be of greater evolutionary significance than changes that simply put it out of action. The changes in the behaviour of TMV produced by treatment with nitrous acid that have so far been studied quantitatively are all of the kind that would follow automatically from eliminating characters responsible for such dominating features as ability to invade Java tobacco systemically or to cause large lesions with a high virus content in *Nicotiana glutinosa* (Siegel, 1960). I think it has yet to be proved that changes in individual particles have changed their biological behaviour, but even if this were established and the cause shown unequivocally to be a single deamination, there would still be no reason to assume that 'the genetic character of the whole TMV-RNA molecule' had been altered; the deamination need have had no more than a local effect, destroying one genetic character while leaving the others unimpaired. Sequences of nucleotides may carry information, as do the dots, dashes and spaces of the Morse code, the analogy often used by those applying 'information theory' to genetics, but even on this analogy there is no reason to expect such far-reaching effects from single changes. The change of one dot to a dash in a long message might confuse or make nonsense of one part, but would rarely change the character of the whole message.

CONCLUSION

Of the many effects of environment on the behaviour of plant viruses, I have touched on only a few, selecting largely those that interest me most. That phenotypic changes in the host cells can be important in determining the survival of viruses is clear enough but of their occurrence in virus particles there is little evidence. The temporary acquisition of the ability to be transmitted by an animal when multiplying with a virus that is transmissible points to the possibility, and the possible importance of phenotypic changes is suggested by the report that a strain of TMV poorly infective towards *Nicotiana glutinosa* has its infectivity increased by combining its nucleic acid with the protein from type TMV (Frankel-Conrat & Singer, 1957). However, there is little doubt that the most important way environmental changes affect the behaviour of viruses is by selecting genetically differing variants that are favoured by the different environments.

Most of the phenomena that have been discovered remain unexplained. For some, plausible explanations can be advanced that fit with current ideas about the separation of protein from nucleic acid during the infection process or about the coding of genetical information by rhythms built up with patterns of four nucleotides. Some of these explanations may be right, and my treatment of them may have been too iconoclastic. However, it was deliberately so, because there is the danger of plausibility being treated as evidence and of deductions becoming accepted as facts. This way there can be no progress. Although to have no explanation for a phenomenon is an unsatisfactory state, it does not carry the same dangers, because the need for information is obvious. The need is no less, but only less obvious, with many phenomena for which explanations can be advanced, because however plausible and topical these explanations may be, they are largely based on unproven concepts.

REFERENCES

BADAMI, R. S. (1959). Some effects of changing temperature and of virus inhibitors on infection by cucumber mosaic virus. *Ann. appl. Biol.* **47**, 78.

BAWDEN, F. C. (1954). Inhibitors and plant viruses. *Advanc. Virus Res.* **2**, 31.

BAWDEN, F. C. (1958). Reversible changes in strains of tobacco mosaic virus from leguminous plants. *J. gen. Microbiol.* **18**, 751.

BAWDEN, F. C. (1959). Effect of nitrous acid on tobacco mosaic virus: Mutation or selection? *Nature, Lond.* **184**, 27.

BAWDEN, F. C. & KASSANIS, B. (1945). The suppression of one plant virus by another. *Ann. appl. Biol.* **32**, 52.

BAWDEN, F. C. & KLECZKOWSKI, A. (1952). Ultraviolet injury to higher plants counteracted by visible light. *Nature, Lond.* **169**, 90.

BAWDEN, F. C. & KLECZKOWSKI, A. (1955). Studies on the ability of light to counteract the inactivating action of ultraviolet radiation on plant viruses. *J. gen. Microbiol.* **13**, 370.

BAWDEN, F. C. & KLECZKOWSKI, A. (1959). Photoreactivation of nucleic acid from tobacco mosaic virus. *Nature, Lond.* **183**, 503.

BAWDEN, F. C. & KLECZKOWSKI, A. (1960). Some effects of ultraviolet radiation on the infection of *Nicotiana glutinosa* leaves by tobacco mosaic virus. *Virology*, **10**, 163.

BAWDEN, F. C. & PIRIE, N. W. (1950). Some factors affecting the activation of virus preparations made from tobacco leaves infected with a tobacco necrosis virus. *J. gen. Microbiol.* **4**, 464.

BAWDEN, F. C. & PIRIE, N. W. (1959). The infectivity and inactivation of nucleic acid preparations from tobacco mosaic virus. *J. gen. Microbiol.* **21**, 438.

BENNETT, C. W. (1949). Recovery of plants from dodder latent mosaic. *Phytopathology*, **39**, 637.

BEST, R. J. (1954). Cross-protection by strains of tomato spotted wilt virus and a new theory to explain it. *Aust. J. biol. Sci.* **7**, 415.

BHARGAVA, K. S. (1951). Some properties of four strains of cucumber mosaic virus. *Ann. appl. Biol.* **38**, 377.

CADMAN, C. H. (1959). Some properties of an inhibitor of virus infection from leaves of raspberry. *J. gen. Microbiol.* **20**, 113.

CASTERMAN, C. & JEENER, R. (1955). Sur le méchanisme de l'inhibition par la ribonucléase de la multiplication du virus de la mosaïque du tabac. *Biochim. biophys. Acta*, **16**, 433.

COMMONER, B. & MERCER, F. (1951). Inhibition of the biosynthesis of tobacco mosaic virus by thiouracil. *Nature, Lond.* **168**, 113.

COMMONER, B. & MERCER, F. (1952). The effect of thiouracil on the rate of tobacco mosaic virus biosynthesis. *Arch. Biochem. Biophys.* **35**, 278.

FRAENKEL-CONRAT, H. (1956). The role of nucleic acid in the reconstitution of active tobacco mosaic virus. *J. Amer. chem. Soc.* **78**, 882.

FRAENKEL-CONRAT, H. & SINGER, B. (1957). Virus reconstitution. II. Combination of protein and nucleic acid from different strains. *Biochim. biophys. Acta*, **24**, 540.

GIERER, A. & SCHRAMM, G. (1956). Infectivity of ribonucleic acid from tobacco mosaic virus. *Nature, Lond.* **177**, 702.

JEENER, R. & ROSSEELS, J. (1953). Incorporation of 2-thiouracil-^{35}S in the ribose nucleic acid of tobacco mosaic virus. *Biochim. biophys. Acta*, **11**, 438.

KASSANIS, B. (1957). Some effects of varying temperature on the quality and quantity of tobacco mosaic virus in infected plants. *Virology*, **4**, 187.

KASSANIS, B. (1959). Comparison of the early stages of infection by tobacco mosaic virus and its nucleic acid. *J. gen. Microbiol.* **20**, 704.

KASSANIS, B. (1960). Comparison of the early stages of infection by intact and phenol-disrupted tobacco necrosis virus. *Virology*, **10**, 353.

KASSANIS, B. & NIXON, H. L. (1960). The activation of one plant virus by another. *Nature, Lond.* **187**, 713.

KLECZKOWSKI, A. (1950). Interpreting relationships between the concentrations of plant viruses and numbers of local lesions. *J. gen. Microbiol.* **4**, 53.

MATTHEWS, R. E. F. (1953). Chemotherapy and plant viruses. *J. gen. Microbiol.* **8**, 277.

MUNDRY, K. W. (1959). The effect of nitrous acid on tobacco mosaic virus: mutation, not selection. *Virology*, **9**, 722.

MUNDRY, K. W. & GIERER, A. (1958). Die erzeugung von mutationen des tabakmosaikvirus durch chemische behandlung seiner nucleinsäure *in vitro*. *Z. Vererbungslehre*, **89**, 614.

ROCHOW, W. F. & ROSS, A. F. (1955). Virus multiplication in plants doubly infected by potato viruses *X* and *Y*. *Virology*, **1**, 10.

SCHUSTER, H. & SCHRAMM, G. (1958). Bestimmung der biologisch wichtigen Einheit in der Ribosenucleinsäure des TMV auf chemischem Wege. *Z. Naturf.* **13*b***, 697.

SIEGEL, A. (1960). Studies on the induction of tobacco mosaic virus mutants with nitrous acid. *Virology*, **11**, 156.

SIEGEL, A., GINOZA, W. & WILDMAN, S. G. (1957). The early events of infection with tobacco mosaic virus nucleic acid. *Virology*, **3**, 554.

SMITH, K. M. (1946). Transmission of a plant virus complex by aphids. *Parasitology*, **37**, 131.

SUKHOV, K. J. (1956). The problem of hereditary variation of phytopathogenic viruses. Reprint by Akad. Nauk SSSR, Moscow.

TSUGITA, A. & FRAENKEL-CONRAT, H. (1960). The amino acid composition and C-terminal sequence of a chemically evoked mutant of TMV. *Proc. nat. Acad. Sci., Wash.* **46**, 636.

WATSON, M. A. (1960). Evidence for interaction or genetic recombination between potato viruses *Y* and *C* in infected plants. *Virology*, **10**, 211.

WILTSHIRE, G. H. (1956). The effect of darkening on the susceptibility of plants to infection with viruses. 1. Relation to changes in some organic acids in the French bean. *Ann. appl. Biol.* **44**, 233.

ANTIBIOTIC PRODUCTION AS AN EXPRESSION OF ENVIRONMENT

H. B. WOODRUFF

Merck Sharp & Dohme Research Laboratories, Division of Merck & Co., Inc., Rahway, New Jersey, U.S.A.

There are few areas of microbiology in which the record of progress has been as great as in that of antibiotics. Today, the clinically useful antibiotics are obtained in yields of 5–10 g./l. Many of these antibiotic processes are derived from initial laboratory cultures which formed the antibiotics in amounts of only 50–100 mg./l. The record with penicillin is especially striking, with a 5000-fold increase realized in broth potency from laboratory to industrial-scale operation. The improved penicillin process and the yield increase of approximately 100-fold observed with many of the other antibiotics manufactured on a commercial scale have been accomplished in part by studies on the environment to which the micro-organisms are exposed, and in part by studies on the selection of new strains of micro-organisms.

Often credit for the amazing advances in antibiotic yield is assigned primarily to the genetic modification of the producing cultures. The responses of asexual micro-organisms to mutagenic agents have been well documented. Mutant cultures have been reported to give increased yields, to respond to new precursors, to suppress unwanted products, to grow faster, and to respond more efficiently to limited air supply. Searches have been made for more efficient mutagenic agents, means for specifically selecting mutant cultures, and methods of preserving mutants from culture degeneration. Recently, research has been undertaken on parasexual processes as a method of extending the advantages to be reaped from genetic modification.

If one calculates the yield of penicillin carbon on the basis of carbon available in the culture medium, the fermentation yield has advanced from 0·005 % to nearly 10 %. The trail of the culture in this advance has progressed from Fleming's chance air contaminant, *Penicillium notatum*, to *P. notatum* 832, a strain particularly adapted to submerged fermentation conditions; to *P. chrysogenum* 1951, selected from a spoiled cantaloupe in Peoria, Illinois; to 1951B25, a subisolate obtained by plating spores; to an X-ray-induced mutant X1612; to an ultraviolet-induced mutant, Q-176; then through divergent lines of natural selection and treatment with X-rays, nitrogen mustards, and other mutagenic agents applied in many industrial laboratories and at the University of Wisconsin Department of Botany (Raper & Alexander, 1945).

Through this trail of modified cultures, suggestions of the influence of environment are apparent but often overlooked. It is impossible to assess accurately the proportionate contribution of a genetic change and a slight environmental modification which together gave a significant increase in yield of antibiotic.

Evidence of the quantitative aspects of environment may be found in a report on the genealogy of a series of mutants of *Streptomyces griseus* which led to an eightfold increase in antibiotic production. The fourth mutant obtained produced streptomycin in an amount 3·2 times that of the original culture. The antibiotic yield was increased an additional 25% by modification of the culture medium. Furthermore, the final mutant culture of the series, found to produce 2000 μg. of streptomycin/ml., was raised to a productivity level of 2300 μg./ml. by minor changes in environment, as expressed by temperature, aeration, and agitation (Dulaney, 1953).

A survey of various review articles on antibiotic production shows that more than 100 pages are devoted to listing the influence of environmental factors on antibiotic yield. A tabulation could be made of antibiotic process, environmental factor, and response. Even such an abbreviated listing would be extensive and would contribute little to the interpretation of the reasons for the observed responses.

For the purposes of this report, it is preferable to select for emphasis a single antibiotic process, to discuss the response to a wide variety of environmental factors, and, when possible, to present explanations for the response in terms of biosynthetic and metabolic pathways. The conclusion may be drawn that there is a wide area for research remaining on the response of a micro-organism to its environment; however, the significance of environment in the antibiotic fermentation process will become obvious.

The penicillin fermentation has been chosen as the example for discussion because it is the antibiotic process least complicated by patents and industrial secrecy. However, examples will be drawn from other antibiotic fermentations for illustration of certain environmental responses.

In assessing the impact of environmental changes on production of an antibiotic, it is important to have knowledge of the biosynthetic pathway employed by the micro-organism in forming the antibiotic. This knowledge is valuable, especially if one is to study variations in the nutrition of the micro-organism following a logical sequence. Unfortunately, exact knowledge of biosynthetic pathways is not available until long after the immediate need for yield improvement is passed. To evaluate the influence of environment on antibiotic yield, therefore, one must design experiments by analogy, based on past experience of the formation of antibiotics of related structure. Knowledge of the antibiotic structure is essential if one is to proceed on any basis other than empirical trial and error.

ENVIRONMENTAL CONTROL BY PRECURSORS

We now know through studies with penicillin and novobiocin that moulds and streptomycetes find difficulty in synthesizing the phenolic group at rates commensurate with maximum antibiotic biosynthesis. The presence of such a ring compound in an antibiotic should be an immediate indication for supplementation of the culture medium with compounds of related structures, with fragments of the antibiotic obtained by degradation, or with natural metabolites, for example, shikimic acid, phenylalanine, or other substances which might fit logically into the biosynthetic pathway. Preferably, these studies should be undertaken with isotopically-labelled compounds. The detection of labelling in the antibiotic can be accepted as significant evidence for precursor action and indicates that the compound should be included in the fermentation medium, even though no immediate yield response is obtained.

In the early investigations with penicillin, minor responses in antibiotic potency were observed following addition of phenylalanine to the fermentation medium (Behrens, 1949). In attempts to explain the major yield increase brought about through use of the nutrient complex, corn steep liquor, β-phenylethylamine was detected in corn steep liquor. A significant yield response was obtained by supplementation of the medium with phenylacetic acid, provided care was taken to adjust the pH of the culture medium to overcome the toxic effects of the undissociated acid (Moyer & Coghill, 1947; Smith & Bide, 1948). Oxidative destruction of phenylacetic acid by the mould presented a problem which was overcome through use of a difficultly hydrolysable derivative, such as phenylacetylethanolamine, or by continuous addition of phenylacetic acid throughout the course of the fermentation. Between 71 and 94% of the added phenylacetic acid was metabolized in pathways other than penicillin biosynthesis, regardless of whether all of the phenylacetic acid was added as a single supplement at the 24th hr. of fermentation or at 12 hr. intervals in a series of nine additions throughout the fermentation (Singh & Johnson, 1948).

The potential of the phenolic precursor for the penicillin fermentation was made more obvious, however, by the discovery that 92·5 % of the benzylpenicillin isolated from a fermentation supplemented with deuterophenylacetyl DL-valine was derived from the added deuterated precursor (Behrens *et al.* 1948). Today, no one could operate successfully the commercial fermentation for benzylpenicillin with highly developed

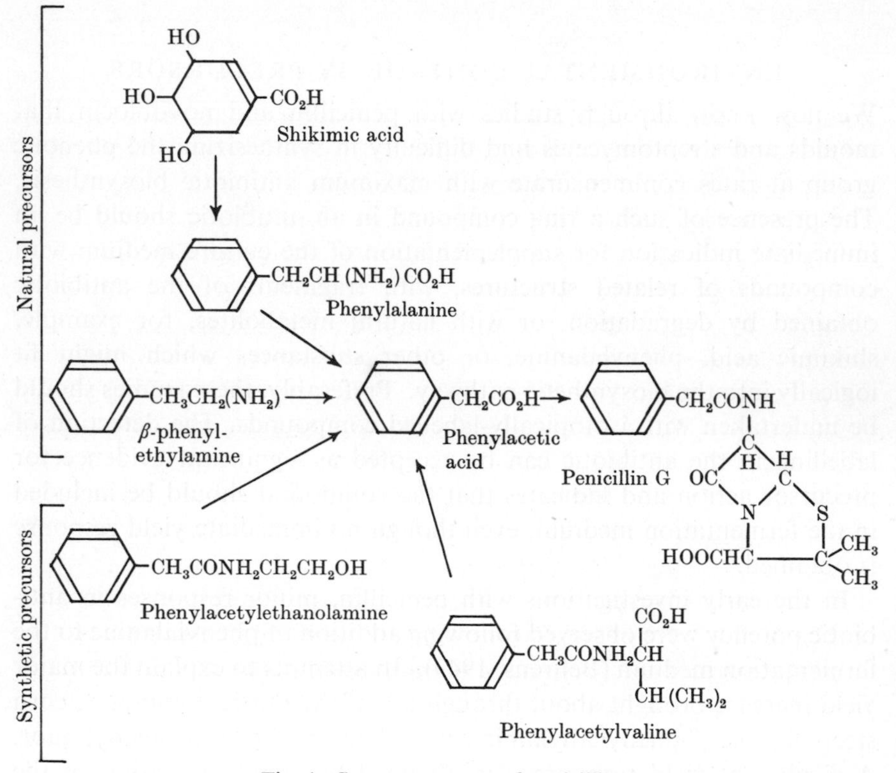

Fig. 1. Some precursors of penicillin.

mutant cultures without aiding the mutants through the addition to the culture medium of the difficultly biosynthesized phenylacetic group (Fig. 1). The significance of this addition is demonstrated by the 900 % increase in penicillin bioactivity observed with the mutant culture, Wisconsin 51–20, when grown in the presence of a precursor, as compared with a control fermentation. In the absence of precursor, a fourfold increase in the accumulation of desbenzylpenicillin (6-aminopenicillanic acid) occurred (Rolinson, 1960).

The full impact of precursors as an environmental factor for micro-organisms may be realized in connexion with genetic modification of the antibiotic-producing culture. The prime purpose of a mutation programme is to improve fermentation yield through successive release of the yield-limiting reactions in a biosynthetic pathway. Thus, if a compound is shown by labelling experiments to be utilized as a precursor for an antibiotic, but gives no increase in yield of that antibiotic, another rate-limiting step must be assumed to be controlling the fermentation with the specific culture under investigation. In subsequent mutants of the culture, synthesis of the isotopically labelled component may become rate-limiting. Mutation experiments, therefore, should be carried out in the presence of all possible precursors in order that no environmental limitation is placed on success of the mutation approach.

To summarize the research approaches based on precursors as a factor in fermentation environment, potential precursors of antibiotics should be ferreted out through the use of the sensitive isotopic-labelling approach, as well as by cruder observations of direct yield increase. Because of the interplay of genetic modifications and environmental factors, the selected nutrient medium should contain all possible precursors during the course of development and selection of mutant cultures. Precursors of little commercial significance in the first stages of a mutation programme are likely to become yield-limiting at the later stages. Of course, in the commercial application of any mutant culture, only those precursors need be included in the medium which give an economically significant response.

The interplay of genetic factors and of environmental factors in controlling yields of antibiotics will be a constantly recurring theme throughout this report. Excellent examples are found in the consideration of physical environmental factors which influence yield, i.e. temperature and oxygen-transfer coefficient.

NUTRITION AS AN ENVIRONMENTAL FACTOR

Simultaneously with studies on precursors, more general investigations on nutrition are required to define the influence of environmental factors on antibiotic yield. These studies consist of the determination of a fairly complete carbon and nitrogen balance for the nutrients added and products produced at various times during the fermentation, as well as measurements of the pH of the fermentation medium and the antibiotic yield. Such studies are generally conducted with a complex medium and allow one to differentiate roughly several phases in the fermentation. The three major phases often distinguished have been named the growth phase, the maturation phase, and the lytic phase, based on changes in mycelial or cell weight (Fig. 2; Table 1). A careful evaluation of antibiotic production with relation to the three phases, as the medium components are varied, may lead one to detect the controlling environmental factors in the fermentation. For example, from such studies it was observed that high streptomycin yields were correlated with very low soluble phosphate levels during the maturation phase. Initial growth is at the expense of proteins in the medium, whereas streptomycin production and glucose utilization occur during the maturation phase. Lysis of the mycelium begins before the maximum amount of streptomycin is liberated; therefore streptomycin accumulation overlaps both the maturation and lytic phases (Dulaney & Perlman, 1947).

Table 1. *Fermentation phases for production of four antibiotics*

	Phase 1, growth	Phase 2, maturation	Phase 3, lysis
Penicillin			
Mycelium	Rapid growth	Slow growth	Decrease
Lactic acid	Rapid use	—	—
Lactose	Lag in use	Steady use	—
pH	Sharp rise	Slow rise	Slow rise
NH_3-N	Released	Slow use	Released
Penicillin	Lag in formation	Steady production	Decrease
Streptomycin			
Mycelium	Rapid growth	Lysis starts	Decrease
Soluble N	Rapid use	Released	Released
Glucose	Little use	Rapid use	—
Inorganic P	Rapid use	Minimal	Released
Streptomycin	Lag in formation	Steady production	Levels-out
Chlortetracycline			
Mycelium	Rapid growth	Continued growth	—
Sucrose	Rapid use	Rapid use	—
Protein N	Rapid increase	Levels-out, then drops	—
RNA	Rapid increase	Levels-out	—
Chlortetracycline	Steady production	Steady production	—
Bacitracin			
Cells	Rapid growth	—	Decrease
Glucose	Rapid use	—	—
pH	Little change	—	Rises
Bacitracin	Steady production	—	Levels-out

Formation of penicillin is also associated with the maturation phase of the mycelium and occurs simultaneously with slow utilization of NH_3-N, a gradual rise in pH, and slow utilization of lactose. Growth occurs at the expense either of the lactic acid present in corn steep liquor or of glucose if this is provided to the mould. Loss of penicillin occurs in the lytic phase (Koffler *et al.* 1945).

The maturation phase is missing in the bacitracin fermentation. Cell formation, glucose utilization, and bacitracin production occur simultaneously in the growth phase. With exhaustion of the carbohydrate,

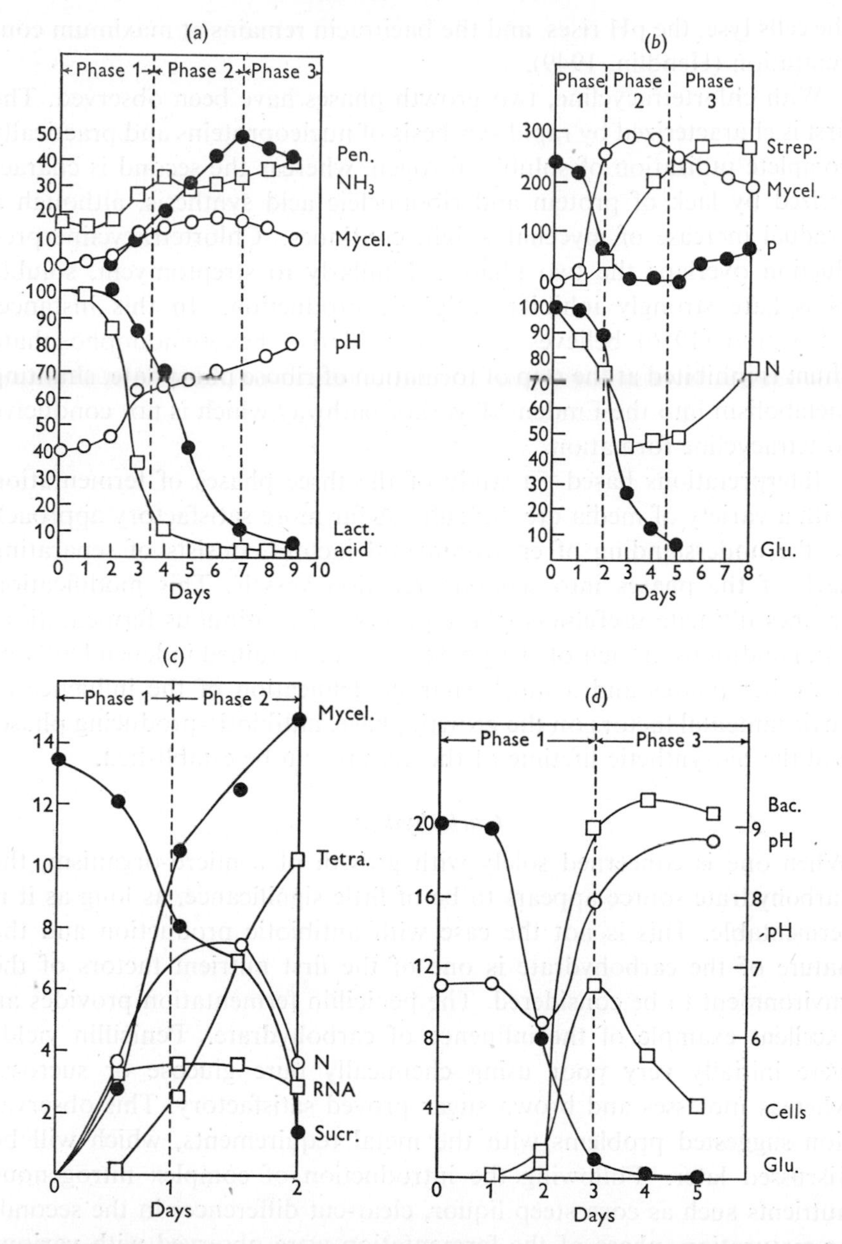

Fig. 2. Relation between concentration of antibiotic, some media constituents, and mass of organisms during the various growth phases in four fermentations. (*a*) Penicillin fermentation: pen., units penicillin/ml.; NH_3, mg. NH_3-N/100 ml.; mycel., dry weight of mycelium in mg./ml.; pH, ×10; lact., % of initial lactose remaining; acid, % of initial lactic acid remaining. (*b*) Streptomycin fermentation: strep., μg. streptomycin/ml.; mycel., dry weight of mycelium in mg./50 ml.; P, inorganic phosphorus in mg./2 ml.; N, % of initial soluble N remaining; glu., % of initial glucose remaining. (*c*) Chlortetracycline fermentation: mycel., dry weight of mycelium in mg./50 ml.; tetra., mg. chlortetracycline/10 ml.; N, mg. protein N/10 ml.; RNA, mg. RNA/10 ml.; sucr., gm. sucrose/400 ml. (*d*) Bacitracin fermentation: bac., units bacitracin/ml.; cells, dry weight of cells in mg./10 ml.; glu., mg. glucose/4 ml. No distinct maturation phase is apparent in the bacitracin fermentation.

the cells lyse, the pH rises, and the bacitracin remains at maximum concentration (Hendlin, 1949).

With chlortetracycline, two growth phases have been observed. The first is characterized by rapid synthesis of nucleoproteins and practically complete utilization of soluble nitrogen, whereas the second is characterized by lack of protein and ribonucleic acid synthesis, although a gradual increase of mycelial weight continues. Chlortetracycline production overlaps the two phases. Similarly to streptomycin, soluble phosphate strongly inhibits antibiotic production. In this instance, Di Marco (1956) believes that the oxidative hexosemonophosphate shunt is inhibited at the step of formation of ribose phosphate, shunting metabolism into the Emden-Meyerhof pathway which is not conducive to tetracycline formation.

Interpretations based on study of the three phases of fermentation with a variety of media are difficult. A far more satisfactory approach to the understanding of environmental factors consists of separating each of the phases into separate reaction vessels. This modification reaches ultimate usefulness in the process of continuous fermentation. The conditions in each of the phases may be modified independently of the other phases and a much sharper delineation of the influence of environmental factors on the growth phase, antibiotic-producing phase, and the biosynthetic lifetime of the culture can be established.

Carbohydrate

When one is concerned solely with growth of a micro-organism, the carbohydrate source appears to be of little significance, as long as it is fermentable. This is not the case with antibiotic production and the nature of the carbohydrate is one of the first nutrient factors of the environment to be considered. The penicillin fermentation provides an excellent example of the influence of carbohydrate. Penicillin yields were initially very poor using chemically pure glucose or sucrose, whereas molasses and brown sugar proved satisfactory. This observation suggested problems with the metal requirements, which will be discussed later. Following the introduction of complex nitrogenous nutrients such as corn steep liquor, clear-cut differences in the second, or maturation, phase of the fermentation were observed with various carbon and energy sources. Polysaccharides, such as soluble starch or alkali-treated gum, proved more favourable for penicillin production than glucose or sucrose (Moyer, 1949). This was also true of lactose. A comprehensive survey of 18 carbon sources showed lactose to be best and the high yield to correlate with low respiration rate (Nielsen, 1952).

Lactose provides an optimum slow respiration rate in the penicillin-producing phase, probably due to the low levels of β-galactosidase produced by the culture. However, lactose should be present from the beginning of the fermentation, even though glucose is present also, as there is evidence that β-galactosidase is an adaptive enzyme and it is desirable that the mycelium progress without delay from the growth phase to the maturation, or penicillin-producing, phase. A low respiration level can be established also by feeding glucose to the culture at less than the optimum rate. Under this condition, the penicillin yield was found to be as high with glucose as with lactose. The best yield of penicillin was obtained by feeding glucose during the maturation phase at a rate of 0·042 % (w/v)/hr. (Davey & Johnson, 1953). The importance of considering the phases of the fermentation separately became obvious through the observation that unlimited glucose is far more desirable than lactose for *Penicillium chrysogenum* during the growth phase.

Lipids

Lipids play a special role in antibiotic fermentations, and the investigation of environmental factors should include a thorough study of these compounds. For many years, lipids have been an essential component of industrial fermentations because they have been required as defoaming agents. Silicones are now preferred as defoaming agents in many fermentations and the influence of lipids on antibiotic yield may be overlooked. Studies for the selection of non-toxic lipid defoaming agents often show increased antibiotic activity as compared with the non-lipid-containing controls. Early explanations of this effect as due to a greater oxygen solution rate with lowered surface tension have been disproved by clear-cut evidence that the oxygen-transfer coefficient is lowered by as much as 50 % in the presence of lipids, probably due to increased film resistance to oxygen transfer from the gaseous to the liquid phase (Deindoerfer & Gaden, 1955).

Studies by Birch and his associates (Birch, Massy-Westropp, Rickards & Smith, 1958*a*; Birch, English, Massey-Westropp & Smith, 1958*b*; Birch, Ryan & Smith, 1958*c*) on the importance of acetate in the biosynthesis of a great variety of antibiotic molecules support the suggestion that lipids are especially valuable as a source of active acetate, serving as a direct precursor for antibiotics. With penicillin, labelled acetate enters the penicillanic acid fragment of the molecule in the components derived from both valine and cysteine, but the degree of incorporation is low (Stone & Farrell, 1946). Acetate utilization by the mould varies inversely with the pH. The optimum growth with acetate can be obtained only at pH values well above the optimum on a glucose substrate (Jarvis & Johnson, 1947). This response has been traced to the toxicity of the un-ionized acetic acid molecule. Although acetate cannot be thought to have the direct precursor significance in penicillin production which it has in the formation of tetracyclines, griseofulvin, mycophenolic acid, and helminthosporin, it is known to participate directly in valine biosynthesis. The suggestion that lipids serve as a source of acetate or of acetyl coenzyme A for penicillin biosynthesis is strengthened by the observation that a lipid of a chain length of more than 14 carbon atoms is required for optimum antibiotic yield (Goldschmidt & Koffler, 1950). β-Oxidation of lipids, leading to acetate, also occurs at the greatest rate with fatty acids having a chain length of 14 to 18 carbon atoms.

Oils are not only important candidates among environmental factors to be tested for the stimulation of antibiotic yield, but should also be considered as primary sources of energy. In the penicillin fermentation, the fatty acids are preferred energy sources under conditions of high oxygenation during the maturation phase of fermentation (Rolinson & Lumb, 1953). Failure of oils in the growth stage of fermentation may be due to the toxicity of saturated acids for young cultures (Ishida, Isono & Wakita, 1952). Oils also have been used successfully as the prime nutrient for synthesis of streptomycin (Perlman & Langlykke, 1950).

When considering acetate as an environmental factor, the potentialities of mevalonic acid should not be overlooked. Direct incorporation into mycophenolic acid has been proved (Birch *et al.* 1958*b*); and mevalonic acid, with calcium ion, plays a determinate role in the formation of the polyene antifungal agent antimycoin A, which is produced by a streptomycete (Schaffner *et al.* 1958).

As with many nutrients, fatty acids can serve more than one function but the requirements may differ. Caproic acid and β,γ-hexenoic acid have been shown to be precursors for penicillin dihydro F and penicillin F (Thorn & Johnson, 1950). A configuration capable of blocking β-oxidation is conducive to precursor utilization with the fatty acids, and several new aliphatic penicillins can be produced with appropriate precursors. In contrast, as noted previously, a configuration conducive to β-oxidation is desirable in lipids to cause increases in the yield of benzylpenicillin in the presence of a specific benzyl precursor.

Nitrogen compounds

The importance of the nitrogenous nutrient as an environmental factor controlling antibiotic production cannot be overlooked, although complex materials such as corn steep liquor, cottonseed meal, soybean meal, or distillers solubles are commonly used in industrial fermentations. Occasionally, empirical surveys point to a nutrient with almost specific antibiotic-promoting properties, such as corn steep liquor for penicillin, cotton endosperm meal for chlortetracycline, and distillers solubles for streptomycin production. After long periods of research, the explanation for the favourable effects of these crude complexes has been unravelled. The presence of β-phenylethylamine in corn steep liquor has been mentioned previously. In addition, corn steep liquor provides *Penicillium chrysogenum* with an exceptionally favourable balance of organic acids and free and complex forms of nitrogen. Penicillin fermentations conducted in its presence are maintained at the optimum pH by utilization of organic acids, release of free NH_3, followed by slow utilization of NH_3. Under these conditions, antibiotic production continues for an exceptionally long time during the maturation phase of the fermentation pattern.

In surveying complex nitrogenous nutrients on an empirical basis, the requirement for supplemental nutrients must not be overlooked. Peptone and tryptone can supply the essential nitrogen needs for penicillin formation, but utilization of free ammonia is far too slow unless a mixture of the salts of various major and trace elements is added. If ammonia accumulates in the medium, the pH becomes too high for optimum yield of penicillin. Yeast extract and fish solubles can also provide the essential requirements for penicillin production, but release of ammonia is so rapid that glucose supplementation is necessary to control the pH. Growth then becomes so heavy that a higher-than-normal lactose concentration is required to promote penicillin formation in the maturation phase of the fermentation. Autolysis occurs if less than 5% lactose is present in the medium (Bhuyan & Johnson, 1957). The empirical survey of nitrogenous nutrients, therefore, must be undertaken with knowledge of the requirements of the various phases in the fermentation pattern and with careful control to maintain optimum conditions for antibiotic production.

Amino acids

As mentioned previously, commercial antibiotic fermentations are seldom based on pure amino acids or on inorganic nitrogen sources. However, studies with these compounds can have significance in industrial fermentations. To isolate the effects of nitrogenous nutrients on the antibiotic-producing phase of the fermentation, resting cells are often utilized. Phenylacetic acid, inorganic salts (including phosphate), valine, and cystine (as a source of cysteine) supply the total requirements for penicillin biosynthesis with resting cells of *Penicillium chrysogenum*. Individual amino acids were tested for availability for antibiotic synthesis at a concentration of 2×10^{-2} M. Of 26 amino acids investigated, all supported penicillin formation at levels above that of resting cells in a basal medium, with the exception of L-lysine which not only failed as a substrate for penicillin biosynthesis but blocked the endogenous biosynthesis as well (Demain, 1957). Furthermore, lysine decreased the rate of penicillin biosynthesis from 4·3 units/ml./hr. to 0·8 unit/ml./hr. in the presence of cystine and valine. Lysine is a significant constituent of corn steep liquor, the favoured nutrient for penicillin production. Additional increments of lysine depressed penicillin biosynthesis in the commercial production medium and dictated a search for approaches to neutralize the toxicity.

In resting-cell systems, DL-α-aminoadipic acid, which is a precursor of lysine, exerts a stimulatory rather than an inhibitory effect on antibiotic production. Adipic acid also is stimulatory at high concentrations. Furthermore, these acids increase penicillin production when added to a corn steep liquor medium and neutralize the unfavourable action of supplemental lysine (Somerson, Demain & Nunheimer, 1960). An apparent explanation for this action is provided by the discovery of Arnstein, Morris & Toms (1959) that a tripeptide consisting of α-aminoadipic acid, cysteine, and valine is present in the mycelium of *Penicillium chrysogenum*. The tripeptide has been postulated as an intermediate in benzylpenicillin biosynthesis. The lysine inhibition may be explained by its structural similarity to α-aminoadipic acid and lysine is thought to serve as an inhibitor of the formation of the tripeptide.

Minor elements

Important environmental factors which have a controlling influence on antibiotic biosynthesis are the various minor elements and minerals available to the mould. In the case of penicillin, sulphur plays an essential role because of the presence of this element in the antibiotic molecule. Sulphate, cysteine, methionine, glutathione, cysteic acid, taurocholic acid, and choline sulphate can all provide sulphur as a nutrient (Stevens *et al.* 1953). The commercial nutrient—corn steep liquor—is derived from steeping of corn in the presence of sulphite and, therefore, provides a source of sulphur not found in many complex nutrients. To avoid discarding otherwise suitable nutrients during the initial determination of optimum environmental conditions, one must be certain to supplement the nutrient complex with all elements found to be contained in an antibiotic molecule.

A similar situation exists in the provision of chloride ion for chlortetracycline, chloramphenicol, and griseofulvin; iron for grisein; or sulphur for the many other antibiotics which contain this element. Thiosulphate serves a special function in the case of penicillin and will be discussed later.

In a sufficiently purified synthetic medium, every element previously reported to influence growth of a micro-organism could equally well be claimed to influence antibiotic production. Many quantitative experiments have shown that antibiotic formation is even more demanding than growth for certain mineral elements. Manganese at a concentration many times greater than the optimum level for growth appears to play a special part in promoting biosynthesis of polypeptide antibiotics (Jansen & Hirschmann, 1944). Phosphorus and iron are important for penicillin formation. With phosphorus, the final weight of the mould is linear at P concentrations from 0 to 65 μg./ml. Penicillin production is initiated at 65 μg./ml. and its increase is linear up to a P level of 200 μg./ml. With iron, a linear yield of mycelium was shown with levels of 0·07–0·25 μg./ml., and penicillin yields showed a linear response in the range of 0·2–3·5 μg. of iron/ml. (Jarvis & Johnson, 1950).

Quantities required for growth and for penicillin production were found to be the same with copper, magnesium, potassium, and sulphur, if the concentration of the latter element is corrected to reflect the amount of sulphur entering the penicillin molecule. The response of growth and antibiotic production was not linearly related to the concentration of potassium. Under conditions of potassium deficiency, the growth rate and antibiotic production rate were decreased by as much as 3·5-fold, although maximum yields were eventually attained.

Zinc has a special significance in penicillin production as a deficiency of zinc leads to an incomplete oxidation of sugar, with gluconic acid accumulating in the medium. An unfavourably low pH is produced, preventing penicillin biosynthesis, again emphasizing the significance of the interplay of environmental factors in controlling antibiotic production (Foster, Woodruff & McDaniel, 1943).

Another mineral which has a specific place in the penicillin fermentation is calcium. At 2·5 % (w/v) in a synthetic medium, $CaCl_2.2H_2O$ promoted spore formation in cultures grown under aerated submerged conditions (Foster *et al.* 1945). Many of the highly productive mutant cultures of *Penicillium chrysogenum* are genetically unstable, and submerged spores have proved satisfactory as a method of maintaining the inoculum in a stable form for subsequent factory utilization.

Vitamins and growth factors

Most of the micro-organisms isolated from nature for antibiotic production do not show requirements for growth factors. However, because of the mutation approach employed in attaining the best commercial processes, the possibility of the induction of a nutritionally exacting mutant as part of the mutation programme should always be considered.

Cytidine has been identified as a factor present in yeast extract which stimulates the growth of *Penicillium chrysogenum* and also the production of penicillin when added to a purified culture medium (Kaplan, 1956). Commercial antibiotic operations almost invariably involve complex media and, therefore, the influence of vitamin factor requirements can generally be disregarded as a significant environmental variable.

POISONS AND INHIBITORS

Among other environmental factors of significance in the nutrition of an antibiotic-producing micro-organism, one must consider the possible application of poisons or inhibitors. Chlorination inhibitors have played an important part in the development of commercial processes for tetracycline. Of a variety of compounds studied, the compounds which were moderately active and very active in blocking chlortetracycline biosynthesis and allowing increased levels of tetracycline, all shared the grouping

[- - - -N=C(—SH)—X- - - -] or the tautomer [- - - -N(H)—C(=S)—X- - - -]

in a cyclic system, where X was a nitrogen, oxygen, or sulphur atom (Lein, Sawmiller & Cheney, 1959). In the case of penicillin, the reported inhibitors are varied and the mechanism of their effects not easily understood. A patent has been issued in which the alkaloids quinine, heroin, and berberine have been reported to increase yields (Fujimasa *et al.* 1948). Increased penicillin synthesis also has been reported with ethylamine (Calam & Hockenhull, 1949), 2:4-dinitrophenol (Rolinson, 1954), methylene blue (Takida & Tawara, 1955), and phthalate (Darkin, 1954) added to the fermentation medium. Depression of penicillin synthesis by cyanide (Rolinson, 1954) is more understandable since penicillin formation occurs best with mycelia having high Q_{O_2} capabilities. Such mycelia are produced under highly aerated conditions and contain the cyanide-sensitive cytochrome system as observed by spectroscopic examination of mycelial homogenates (Rolinson, 1955).

Thiosulphate is reported to play a special role in penicillin biosynthesis somewhat similar to that frequently ascribed to methylene blue. Thiosulphate acts as a primary hydrogen acceptor in the oxidation of pyruvate with certain bacteria, and the suggestion has been made that with *Penicillium chrysogenum* it serves specifically to block the oxidation of pyruvate by irreversibly combining with the enzyme system, thus promoting biosynthesis of the valine and cysteine components of the penicillin molecule (Hockenhull, 1959). While there is little experimental evidence for this proposed mechanism, it seems clear that thiosulphate does not exert its favourable influence through provision of more readily available sulphur, but that it behaves in a fashion analogous to poisons.

In considering toxins as environmental factors, the effects of excessive heating of the fermentation medium during sterilization should be considered. In a study of penicillin production with a chemically defined medium, increases in sterilization time from 15 to 55 min. gave a progressive fall in yield and an increase in the time required for the fermentation. With 55 min. sterilization, the penicillin yield was only 75 % of that in a medium with separate sterilization of glucose (Jarvis & Johnson, 1947). A practical solution to the sterilization problem may be found either by separate sterilization of the medium ingredients or through the use of continuous sterilization at a high temperature for very short times. $Cobalt^{60}$ has been employed for sterilization of corn steep liquor but had an unfavourable effect on lactic acid production by *Lactobacillus delbrueckii* (Gillies & Kempe, 1957). The utility of corn steep liquor sterilized in this fashion for antibiotic synthesis has not been evaluated.

ENZYMES

Certain enzymes produced during the course of antibiotic formation may cause unfavourable responses by the micro-organism. The addition of unsaturated fatty acids during the final phase of the penicillin fermentation is unfavourable because autolysis of the mycelium is enhanced. Stimulation of lipase formation by the fatty acids apparently plays a part in the autolysis, since the addition of various lipase inhibitors to the fermentation broth, for example boric acid, will prevent autolysis (Yasuda, Yamasaki & Mizoguchi, 1952).

A slow loss of penicillin activity has been observed in the third phase of the penicillin fermentation and this loss is accentuated greatly with a mutant culture made resistant to sulphathiazole. This loss has been ascribed to the formation of penicillin amidase, an enzyme capable of removing the phenylacetyl side chain (Murao, 1955). In selecting optimum environmental conditions, care should be taken to minimize formation of enzymes destructive to the products produced.

PHYSICAL FACTORS

Although the nutritional environment of a micro-organism probably has greater influence on antibiotic production, the physical factors of the environment have been established with greater precision, probably because they can be controlled more exactly and are economically important in factory-scale operations.

pH

The dependence of pH on other environmental factors influencing penicillin production has already been mentioned. In fact, an unfavourable pH response may well be the most important factor influencing the choice of concentration of medium ingredients, proportions of ingredients, and the nature of the ingredients. In addition to the different pH optima observed for growth and for penicillin production, unfavourable influences, such as autolysis, appear to be pH-controlled. The toxicity of acetate or of precursors, such as phenylacetic acid, under acidic conditions where ionization is repressed, has already been mentioned.

In studies at the University of Wisconsin (Bautz, Hosler & Johnson, 1950), the optimum rate of sugar utilization, rate of mycelial nitrogen production, and ratio of mycelial nitrogen produced per gram of sugar consumed were found to occur at a pH of approximately 4·7. The rates were decreased by as much as 50% as the pH approached 7·0. Exact studies such as these became possible only after the development of sterilizable electrodes, allowing one to control the pH of the fermentation by introduction of alkali or acid.

A separate study of the optimum pH for maximum rates of penicillin formation showed this to be 7·1 (Brown & Peterson, 1950). The controlling influence of pH as an environmental factor was demonstrated very clearly by fermentations in which the nutrition of the micro-organism was out of balance. In one, an excess amount of lipids led to an acidic fermentation; in the second, inability of the mould to adapt to a normal fermentation rate of lactose led to an alkaline-type fermentation. Each of these types of fermentation in factory-scale operation would be expected to be failures. However, 200 m-equiv. of sodium hydroxide added at 26 hr. served to adjust the medium of the high lipid fermentation from pH 6·1 to 6·9. The alkaline fermentation required 800 m-equiv. of H_2SO_4 added in increments at the 35th, 61st, and 100th hr. of fermentation to maintain the pH between 7 and 8. The fermentation with well-balanced nutrients was supplemented with acid at the 71st hr., at a time when the pH reached 7·5, in order to prolong the fermentation. The 100 hr. yields in these pH-adjusted fermentations were 1275, 1300, and 1560 units/ml., respectively.

Because of the impossibility of maintaining accurate pH control on an industrial scale by nutritional balance alone, wide application has been made of the mechanical adjustment of pH in fermentors by addition of acid or alkali. The degree of control required varies from fermentation to fermentation. Because complex and variable nutrients are employed on an industrial scale for reasons of cost, a variable pH response is to be expected and a need for external control is evident.

It is to be anticipated that a much more exact determination of optimum pH in the growth phase and penicillin production phase will develop from studies now in progress with two-stage continuous fermentations. Optimum pH in the growth phase as determined in the continuous units will have much more meaning than the values reported above because the effect on the subsequent ability of the mycelium to produce penicillin can be evaluated quantitatively. From preliminary reports on penicillin formation, it is evident that a pH nearer 7·0 for the growth phase is satisfactory in the continuous situation (Pirt & Callow, 1960).

Temperature

Temperature is an easily controllable environmental factor in industrial-scale fermentations. Laboratory studies have shown that antibiotic biosynthesis is frequently sharply influenced by the fermentation temperature. For example, with streptomycin, a gradually increasing yield is obtained with increasing temperature, until a critical point is reached beyond which streptomycin production may fall by as much as 80% with a temperature change of as little as one degree (Dulaney, 1951). The optimum temperature level differs with different strains of *Streptomyces griseus*.

Penicillin production, likewise, has a sharp optimum with respect to temperature. The temperature optimum for the various phases of fermentation has been investigated. With Wisconsin culture W49-133, a temperature of about 30° proved best for the mycelium-producing phase, whereas a temperature of approximately 20° proved best for the penicillin-producing phase. Exchange of cultures from one temperature to the other at the time of change of phase of the fermentation proved desirable. Yields were increased by 50 % by a preliminary incubation at 30° for 42 hr. followed by incubation at 20° until 90 hr., compared with the constant 25° control fermentation (Owen & Johnson, 1955). It was shown that the maximum rate of penicillin biosynthesis was correlated with the minimum rate of lactose utilization, and that this minimum rate during the penicillin-producing phase was favoured by the low temperature.

The effect of temperature, therefore, may be indirect in providing a favourable sugar utilization rate. A more direct approach to this problem would consist of establishing the optimum sugar fermentation rate by continuous feed of dextrose, followed by increases in temperature to reach the maximum rate of penicillin synthesis.

As with all environmental factors in antibiotic production, the optimum must be re-established with each mutant culture that is developed. *Penicillium chrysogenum* var. *brevisterigma*, which has been shown to give yields more than double that of its parent at the normal 24° fermentation temperature, shows a markedly unfavourable response in penicillin-producing activity at 29°. The yield was reduced by 81 % as compared to a yield loss of 27·5 % with the parent (Foster, 1949). Studies with the two cultures at the two temperatures showed an increase from Q_{O_2} 2·6 to Q_{O_2} 5·7 with *P. chrysogenum* var. *brevisterigma* following the five degree temperature rise; whereas, with the parent culture, the increase was from 2·4 to 3·0, indicating that the unfavourable influence of temperature was associated with an enhanced utilization of carbohydrate during the penicillin production phase.

Moisture and light

Moisture and light are two environmental factors which are seldom considered in industrial antibiotic fermentation because of the use of enclosed tanks for sterility and the use of liquid fermentation media. Antibiotics were produced on solid substrates for a short time prior to the development of the more efficient submerged fermentation technique. Penicillin was produced commercially on a solid grain substrate in a rotating drum fermentor. Pilot plant studies with streptomycin gave yields equivalent to those observed in liquid submerged culture, but a prolonged incubation period was required (Woodruff, 1947). At the present time, in Czechoslovakia, chlortetracycline is manufactured by a composting procedure for the preparation of animal feed-grade antibiotic supplements (Herold & Müller, 1959). Under these conditions, moisture content does become important and it is necessary to establish the optimum moisture for each type of substrate in order to obtain maximum yields.

While light has not been found to influence favourably yields of clinically significant antibiotics, the yield of the antibiotic chlorellin, produced by an alga grown in the light, is increased when cultures are exposed to intense illumination immediately prior to harvest (Pratt *et al.* 1944). Several pure antibiotics are known to be light-sensitive but the influence of light on fermentations during their formation has not been reported.

Oxygen, carbon dioxide, pressure, and agitation complex

The remaining physical factors of the environment are an important complex which must be considered in association. These are oxygen and carbon dioxide concentrations, pressure, and agitation. A number of antibiotic fermentations have been investigated in which oxygen supply and power supply have been increased with correlating increase in antibiotic potency. Finally, a level is reached whereupon increases in oxygen availability result in no further change in antibiotic production. Studies have shown that the critical level of power and oxygenation is usually associated with the maintenance of a definite dissolved oxygen concentration throughout the fermentation. Availability of dissolved oxygen to the cell is considered to be the critical factor in controlling the antibiotic formation, although this variable cannot be dissociated from power input, CO_2 concentration, or mechanical pressure imposed upon the cell.

The low solubility of oxygen in water presents a special technical problem in that the mechanical means of introducing oxygen into the solution must be continuous and is generally a costly aspect of the fermentation. Theoretical considerations of oxygen supply and demand have resulted in the defining of seven stages of specific transfer resistances for the oxygen supplied in an air bubble. These are the resistances imposed by gas film, interface, liquid film, liquid path, cell liquid film, intracellular and intraclump resistances, and the reaction resistance (Bartholomew *et al.* 1950*a*; Arnold & Steel, 1958).

The optimum oxygen concentration in solution may be determined for each fermentation in any specific type of laboratory fermentor. However, for commercial operation it is necessary to have methods available for predicting the dissolved oxygen levels which will be obtained in large-scale equipment. A number of techniques have been proposed for such predictions and scale-up in design has been accomplished with variable success. Mould and streptomycete fermentations present the greatest difficulty because of the problem of measuring and controlling the intraclump resistance to oxygen transfer. Power must be applied to overcome this resistance through mixing or stirring.

Other than considerations of the mechanical aspects, which are overcome by engineering approaches, one of the most important studies on the influence of oxygen concentration on antibiotic production has been put forward by Calam, Driver & Bowers (1951) who showed that growth rate, respiration rate, and penicillin production rate are independent although related processes. When aeration rates were altered solely by changing the rate of agitation in a given fermentor, the metabolic processes were altered in a similar but independent manner. Furthermore, by studying the influence of temperature within the range of 12–32° on the three aspects of the fermentation, it was possible to calculate the energies of activation for growth rate, respiration rate, and penicillin production rate. These values were 8230 gram calories for growth rate, 17,800 gram calories for respiration rate, and 26,800 gram calories for penicillin production rate (Arnold & Steel, 1958). This means that the controlling reaction in each of these three processes was independent, although obviously the over-all reactions were linked.

It is possible to determine with great accuracy the level of dissolved oxygen that it is necessary to maintain in order to have optimum antibiotic yield with any given process, but one must remember that this determination will apply only with the specific mutant strain under investigation. Different mutants may vary greatly in their efficiency of utilization of dissolved oxygen. A mutation programme directed towards selection of improved mutant cultures under fermentation conditions of limited oxygen tension may lead to superior cultures adapted to the specific limited condition, but which show no superiority when grown under adequate aeration conditions.

Power has been mentioned as an environmental factor, expressed through agitation, which aids the transfer of oxygen across the resistances associated with the air bubble and liquid path. It also has the specific advantage of aiding transfer through the intraclump resistance and the cell liquid film. In addition, power input will aid the transfer of dissolved nutrients other than oxygen, such as the minor mineral elements, if they are in limited supply. Under conditions of severe clumping, antibiotic yield can show a direct correlation with horsepower input (Wegrich & Shurter, 1953).

Power problems are intensified because antibiotic fermentation broths increase in rigidity as much as 100-fold throughout the course of the fermentation, the observed rigidity being as high as 148 centipoises. A reduction of as much as 85% in oxygen absorption rate can be observed in fermentations containing a mycelial concentration of 13·4 g./l. As the fermentation progresses, supplemental power is required to overcome these resistances (Deindoerfer & Gaden, 1955).

Power introduced into a fermentor through mechanical agitation can have an unfavourable influence on a filamentous micro-organism, independent of any influence on gas transfer. The effect of mechanical agitation on morphology of *Penicillium chrysogenum* has been studied and, with increased agitation, the type of hyphae associated with high penicillin production was obtained (Dion *et al.* 1954). Although the point of damage to *P. chrysogenum* mycelium has not been reached in fermentors with conventional types of agitators, such damage has been observed at high power inputs with *Streptomyces griseus* (Bartholomew *et al.* 1950*b*). That high speeds of agitation can fragment penicillium mycelium has been shown through the use of a Waring Blendor to produce inocula of much higher growth potential (Vander Brook & Savage, 1949). The inocula became unsatisfactory with extended blending.

Pure oxygen has in some instances been reported to be unfavourable for antibiotic production. Among explanations which have been offered is the loss of essential volatile material, interaction of oxygen with constituents of the mash (forming toxic materials), loss of carbon dioxide which may be essential to the fermentation, and inactivation of certain oxygen-sensitive enzyme systems. The carbon dioxide concentration does not appear to be critical for penicillin production. This fermentation may be unusually insensitive since carbon dioxide does not enter into the penicillin molecule, as demonstrated with radioactive experiments (Martin *et al.* 1953). As anticipated, carbon dioxide enters the mycelial protein and is found primarily in the aspartic acid, glutamic acid, and arginine of the mycelium (Gitterman & Knight, 1952).

Respiration rates of micro-organisms grown in complex media are usually adequate so that no degree of aeration can deplete the carbon dioxide level below that required for growth. The situation may be different for antibiotic production with a product such as streptomycin in which carbon dioxide is fixed in the guanidine groups as an integral portion of the antibiotic molecule (Hunter, Herbert & Hockenhull, 1954).

A factor which influences oxygen supply, as well as carbon dioxide retention, is pressure. Air under pressure is generally applied to commercial fermentations as a means of preventing contamination from external sources. Within normal ranges, the effect of pressure may be related directly to its influence on oxygen solubility and the antibiotic response to pressure will be equivalent to the antibiotic response to oxygen in solution. Other factors related to pressure may be observed. Increased pressure lowers the pH of a fermentation and the pH change is equivalent to that calculated to be due to increased carbon dioxide in solution. Because of the controlling effect of pH on many antibiotic fermentations, a change in pressure may favourably or unfavourably influence antibiotic production unless compensated for by external control of pH. In a fermentation with *Penicillium chrysogenum* X1612, variations in pressure up to 20 lb./sq. in. had no significant effect, but yields were consistently low at 40 lb./sq. in. gauge pressure (Stefaniak *et al.* 1946). At this level, growth was somewhat heavier. A slow pH rise had an adverse effect early in the fermentation, with a low ammonia level having an adverse effect later in the fermentation. That the effect was not due entirely to increased carbon dioxide concentration was shown by inclusion of 2·5 % (v/v) carbon dioxide in the in-coming air with no unfavourable response.

As can be seen from the previous discussions, the environmental factors which influence a fermentation are many and are interrelated one with the other. Most of the studies in determining optimum environment must proceed on an empirical basis, although knowledge of the antibiotic structure and of basic fermentation mechanisms can provide leads toward important environmental factors. To test each and every environmental factor in association with variations in the interrelated environmental factors, at several concentration levels, can require an impossible number of laboratory experiments and can be an insurmountable burden when the variations are conducted in pilot plant or production-scale fermentor equipment.

A factorial design has been proposed as a special approach for evaluating environmental factors in factory fermentors. The design permits greater accuracy for observation of small changes, permits the discovery of interaction of variables, and allows the collection of data without serious detriment to production schedules. In an investigation of six variables—consisting of variations in concentration of three ingredients; and a study of the initial pH, of the age of the inoculum, and of the amount of air introduced into the fermentor—application of a factorial design provided an accuracy of measurements in the ratio of 4:1 over a classical design. A half-replicate type of arrangement allowed collection of data from 32 fermentations, rather than 64 fermentations, and the use of confounding in four blocks of eight gave an accuracy in the ratio of 18:13 over an unconfounded arrangement (Brownlee, 1950).

While microbiologists must take advantage of such simplified schemes of experimentation, there is no question that success in the field of antibiotic fermentations will come to those who can develop a rational approach to the antibiotic fermentation based on broad biochemical knowledge. There is no field of microbiological investigation which can give greater responses to environmental variables than that of the production of antibiotics. Also, there are few aspects of microbiology which are more complicated.

REFERENCES

ARNOLD, B. H. & STEEL, R. (1958). Oxygen supply and demand in aerobic fermentations. In *Biochemical Engineering*, p. 149. Edited by R. Steel. London: Heywood.

ARNSTEIN, H. R. V., MORRIS, D. & TOMS, E. J. (1959). Isolation of a tripeptide containing α-aminoadipic acid from the mycelium of *Penicillium chrysogenum* and its possible significance in penicillin biosynthesis. *Biochim. biophys. Acta*, **35**, 561.

BARTHOLOMEW, W. H., KAROW, E. O., SFAT, M. R. & WILHELM, R. H. (1950*a*). Oxygen transfer and agitation in submerged fermentations. Mass transfer of oxygen in submerged fermentation of *Streptomyces griseus*. *Industr. Engng Chem.* **42**, 1801.

BARTHOLOMEW, W. H., KAROW, E. O., SFAT, M. R. & WILHELM, R. H. (1950*b*). Oxygen transfer and agitation in submerged fermentations. Effect of air flow and agitation rates upon fermentation of *Penicillium chrysogenum* and *Streptomyces griseus*. *Industr. Engng Chem.* **42**, 1810.

BAUTZ, M. W., HOSLER, P. & JOHNSON, M. J. (1950). Effect of control of nutrient level and pH on penicillin production. Abstracts of Papers. *118th Meeting of the American Chemical Society, Chicago, Illinois*, p. 17A.

BEHRENS, O. K. (1949). Biosynthesis of penicillins. In *The Chemistry of Penicillin*, p. 657. Edited by H. T. Clarke, J. R. Johnson and R. Robinson. Princeton, New Jersey: Princeton University Press.

BEHRENS, O. K., CORSE, J., JONES, R. G., KLEIDERER, E. C., SOPER, Q. F., VAN ABEELE, F. R., LARSON, L. M., SYLVESTER, J. C., HAINES, W. J. & CARTER, H. E. (1948). Biosynthesis of penicillins. II. Utilization of deuterophenylacetyl-N^{15}-DL-valine in penicillin biosynthesis. *J. biol. Chem.* **175**, 765.

BHUYAN, B. K. & JOHNSON, M. J. (1957). The effect of medium constituents on penicillin production from natural materials. *Appl. Microbiol.* **5**, 262.

BIRCH, A. J., ENGLISH, R. J., MASSY-WESTROPP, R. A. & SMITH, H. (1958*a*). Studies in relation to biosynthesis. Part XV. Origin of the terpenoid structures in mycelianamide and mycophenolic acid. *J. chem. Soc.* **1**, 369.

BIRCH, A. J., MASSY-WESTROPP, R. A., RICKARDS, R. W. & SMITH, H. (1958*b*). Studies in relation to biosynthesis. Part XIII. Griseofulvin. *J. chem. Soc.* **1**, 360.

BIRCH, A. J., RYAN, A. J. & SMITH, H. (1958*c*). Studies in relation to biosynthesis. Part XIX. The biosynthesis of helminthosporin. *J. chem. Soc.* **4**, 4773.

BROWN, W. E. & PETERSON, W. H. (1950). Factors affecting production of penicillin in semi-pilot plant equipment. *Industr. Engng Chem.* **42**, 1769.

Brownlee, K. A. (1950). A plant-scale planned experiment in penicillin production. *Ann. N.Y. Acad. Sci.* **52**, 820.

Calam, C. T., Driver, N. & Bowers, R. H. (1951). Studies in the production of penicillin. Respiration and growth of *Penicillium chrysogenum* in submerged culture, in relation to agitation and oxygen transfer. *J. Appl. Chem., Lond.* **1**, 209.

Calam, C. T. & Hockenhull, D. J. D. (1949). The production of penicillin in surface culture, using chemically defined media. *J. gen. Microbiol.* **3**, 19.

Darkin, M. A. (1954). Penicillin production with potassium acid phthalate as a fermentation buffer. United States Patent 2,685,554.

Davey, V. F. & Johnson, M. J. (1953). Penicillin production in corn steep media with continuous carbohydrate addition. *Appl. Microbiol.* **1**, 208.

Deindoerfer, F. H. & Gaden, E. L., Jr. (1955). Effects of liquid physical properties on oxygen transfer in penicillin fermentation. *App. Microbiol.* **3**, 253.

Demain, A. L. (1957). Inhibition of penicillin formation by lysine. *Arch. Biochem. Biophys.* **67**, 244.

Di Marco, A. (1956). Metabolism of *Streptomyces aureofaciens* and biosynthesis of chlortetracycline. *Giorn. Microbiol.* **2**, 285.

Dion, W. M., Carilli, A., Sermonti, G. & Chain, E. B. (1954). L'effetto della agitazione meccanica sulla morfologia del *Penicillium chrysogenum* Thom in fermentatori agitati. *R.C. Ist. sup. Sanit.* **17**, 1304.

Dulaney, E. L. (1951). Process for production of streptomycin. United States Patent 2,571,693.

Dulaney, E. L. (1953). Observations on *Streptomyces griseus*. VI. Further studies on strain selection for improved streptomycin production. *Mycologia*, **45**, 481.

Dulaney, E. L. & Perlman, D. (1947). Observations on *Streptomyces griseus*. I. Chemical changes occurring during submerged streptomycin fermentations. *Bull. Torrey bot. Cl.* **74**, 504.

Foster, J. W. (1949). Processes of fermentation. United States Patent 2,458,495.

Foster, J. W., McDaniel, L. E., Woodruff, H. B. & Stokes, J. L. (1945). Microbiological aspects of penicillin. V. Conidiospore formation in submerged cultures of *Penicillium notatum*. *J. Bact.* **50**, 365.

Foster, J. W., Woodruff, H. B. & McDaniel, L. E. (1943). Microbiological aspects of penicillin. III. Production of penicillin in surface cultures of *Penicillium notatum*. *J. Bact.* **46**, 421.

Fujimasa, S., Koyama, Y., Tsubota, M. & Kurosawa, A. (1948). Process for preparing penicillin. Japanese Patent 174,816.

Gillies, R. A. & Kempe, L. L. (1957). Comparison of gamma radiation and heat for sterilization of fermentation mashes. *J. Agric. Fd Chem.* **5**, 706.

Gitterman, C. O. & Knight, S. G. (1952). Carbon dioxide fixation into amino acids of *Penicillium chrysogenum*. *J. Bact.* **64**, 223.

Goldschmidt, M. C. & Koffler, H. (1950). Effect of surface-active agents on penicillin yields. *Industr. Engng Chem.* **42**, 1819.

Hendlin, D. (1949). The nutritional requirements of a bacitracin-producing strain of *Bacillus subtilis*. *Arch. Biochem.* **24**, 435.

Herold, M. & Müller, Z. (1959). *Int. Congr. Biochem., 4th Congr., Vienna*, 1958, p. 196. (Discussion of paper 'Antibiotikazusatz zu Futtermittein' by J. Brüggemann.)

Hockenhull, D. J. D. (1959). The influence of medium constituents on the biosynthesis of penicillin. In *Progress in Industrial Microbiology*, Vol. I, p. 1. Edited by D. J. D. Hockenhull. New York: Interscience Publishers.

Hunter, G. D., Herbert, M. & Hockenhull, D. J. D. (1954). Actinomycete metabolism. Origin of the guanidine group in streptomycin. *Biochem. J.* **58**, 249.

ISHIDA, Y., ISONO, M. & WAKITA, K. J. (1952). Effect of antifoaming oils on penicillin fermentation. VI. Individual effect of various kinds of fatty acid, and general consideration of oil behavior. *J. Antibiot.* **5**, 492.

JANSEN, E. F. & HIRSCHMANN, D. J. (1944). Subtilin—an antibacterial product of *Bacillus subtilis*. Culturing conditions and properties. *Arch. Biochem.* **4**, 297.

JARVIS, F. G. & JOHNSON, M. J. (1947). The role of the constituents of synthetic media for penicillin production. *J. Amer. chem. Soc.* **69**, 3010.

JARVIS, F. G. & JOHNSON, M. J. (1950). The mineral nutrition of *Penicillium chrysogenum* Q176. *J. Bact.* **59**, 51.

KAPLAN, M. A. (1956). Production of penicillin employing media containing cytidine and cysteine. United States Patent 2,768,117.

KOFFLER, H., EMERSON, R. L., PERLMAN, D. & BURRIS, R. H. (1945). Chemical changes in submerged penicillin fermentations. *J. Bact.* **50**, 517.

LEIN, J., SAWMILLER, L. F. & CHENEY, L. C. (1959). Chlorination inhibitors affecting the biosynthesis of tetracycline. *Appl. Microbiol.* **7**, 149.

MARTIN, E., BERKY, J., GODEZSKY, C., MILLER, P., TOME, J. & STONE, R. W. (1953). Biosynthesis of penicillin in the presence of C^{14}. *J. biol. Chem.* **203**, 239.

MOYER, A. J. (1949). Improvements in or relating to methods for producing penicillin. British Patent Spec. 618,416.

MOYER, A. J. & COHGILL, R. D. (1947). Penicillin. X. The effect of phenylacetic acid on penicillin production. *J. Bact.* **53**, 329.

MURAO, S. (1955). Studies on penicillin-amidase. Part II. Research on conditions of producing penicillin-amidase. *Nippon Nogei-Kagaku Kaishi*, **29**, 400.

NIELSEN, N. (1952). The influence of different carbohydrates on gross respiration and penicillin formation of *Penicillium chrysogenum*. *Int. Congr. Biochem., 2nd Congr., Paris*, 1952, p. 109.

OWEN, S. P. & JOHNSON, M. J. (1955). The effect of temperature changes on the production of penicillin by *Penicillium chrysogenum* W49–133. *Appl. Microbiol.* **3**, 375.

PERLMAN, D. & LANGLYKKE, A. F. (1950). The utilization of glycerides by *Streptomyces griseus*. *Abstracts of Papers, 117th Meeting of the American Chemical Society, Philadelphia, Pennsylvania*, p. 18A.

PIRT, J. S. & CALLOW, D. S. (1960). The production of penicillin by continuous flow fermentation. *1st International Fermentation Symposium, Rome.*

PRATT, R., DANIELS, T. C., EILER, J. J., GUNNISON, J. B., KUMLER, W. D., ONETO, J. F., STRAIT, L. A., SPOEHR, H. A., HARDIN, G. J., MILNER, H. W., SMITH, J. H. C. & STRAIN, H. H. (1944). Chlorellin, an antibacterial substance from *Chlorella*. *Science*, **99**, 351.

RAPER, K. B. & ALEXANDER, D. F. (1945). Preservation of molds by the lyophil process. *Mycologia*, **37**, 499.

ROLINSON, G. N. (1954). The effect of certain enzyme inhibitors on respiration and on penicillin formation by *Penicillium chrysogenum*. *J. gen. Microbiol.* **11**, 412.

ROLINSON, G. N. (1955). The effects of aeration on fungal metabolism. *Chem. and Ind.* (*Rev.*), p. 194.

ROLINSON, G. N. (1960). 6-Amino-penicillanic acid. *1st International Fermentation Symposium, Rome.*

ROLINSON, G. N. & LUMB, M. (1953). The effect of aeration on the utilization of respiratory substrates by *Penicillium chrysogenum* in submerged culture. *J. gen. Microbiol.* **8**, 265.

SCHAFFNER, C. P., STEINMAN, I. D., SAFERMAN, R. S. & LECHEVALIER, H. (1958). Role of inorganic salts and mevalonic acid in the production of a tetraenic antifungal antibiotic. In *Antibiotics Annual*, 1957–1958, p. 869.

Singh, K. & Johnson, M. J. (1948). Evaluation of precursors for penicillin G. *J. Bact.* **56**, 339.

Smith, E. L. & Bide, A. E. (1948). Biosynthesis of the penicillins. *Biochem. J.* **42**, xvii.

Somerson, N. L., Demain, A. L. & Nunheimer, T. D. (1960). Reversal of lysine inhibition of penicillin production by α-aminoadipic or adipic acid. *Bact. Proc.* p. 14.

Stefaniak, J. J., Gailey, F. B., Jarvis, F. G. & Johnson, M. J. (1946). The effect of environmental conditions on penicillin fermentations with *Penicillium chrysogenum* X-1612. *J. Bact.* **52**, 119.

Stevens, C. M., Vohra, F., Inamine, E. & Roholt, O. A. (1953). Utilization of sulfur compounds by *Penicillium chrysogenum. Fed. Proc.* **12**, 275.

Stone, R. W. & Farrell, M. A. (1946). Synthetic media for penicillin production. *Science*, **104**, 445.

Takida, T. & Tawara, K. (1955). Japanese Patent 443,455.

Thorn, J. A. & Johnson, M. J. (1950). Precursors for aliphatic penicillins. *J. Amer. chem. Soc.* **72**, 2052.

Vander Brook, M. J. & Savage, G. M. (1949). Seed cultures of filamentous organisms. United States Patent 2,488,248.

Wegrich, O. G. & Shurter, R. A., Jr. (1953). Development of a typical aerobic fermentation. *Industr. Engng Chem.* **45**, 1153.

Woodruff, H. B. (1947). The production of streptomycin in stationary culture on liquid and solid substrates. *J. Bact.* **54**, 42.

Yasuda, S., Yamasaki, K. & Mizoguchi, S. (1952). Effect of soybean oil on shaking cultures with *Penicillium chrysogenum* Q176. Part 4. Studies on counter-measures against toxicity of soybean oil. *J. agric. chem. Soc. Japan*, **25**, 361.

NON-GENETIC VARIATION OF SURFACE ANTIGENS IN *BORDETELLA* AND OTHER MICRO-ORGANISMS

B. W. LACEY

Department of Bacteriology, Westminster Medical School, London

Non-genetic variation of antigens, even more than non-genetic variation generally, has received little attention by writers of text-books on microbiology. Indeed, by most, the influence of environment, apart from its relation to enzyme induction, is considered only as a promoter of life or death. Partly as a result and partly as a cause it has become customary to regard antigens as having a one-to-one correspondence with genes that is maintained in all environments. Antigenic variation in consequence is often discussed as though it were exclusively a manifestation of genomic variation, occurring spontaneously, induced by mutagens, or following transduction, transformation or phage infection. How far from the truth this is can be seen from Table 1, in which all the published instances which could be found of variation of surface antigens with environment have been grouped according to the environmental change. In all these the antigenic change has been shown or can reasonably be presumed. Their variety and number suggest that many hitherto unexplored antigenic variations among viruses and rickettsias, as well as bacteria, will prove to be environmentally induced modifications (or 'modulations': Lacey, 1960) of this kind, and that a deliberate search in the laboratory would reveal many more.

From Table 1 it can be seen that no surface antigen appears invariable, since examples can be found of variation of mucoid, capsular, somatic, cell wall, flagellar, haemagglutinating and immobilizing antigens. Only in three genera, however—*Bordetella*, *Paramecium* and *Tetrahymena*—has a range of antigens of the same anatomical kind been found. It is of interest, and perhaps of significance, that most if not all species in these genera possess a predominant repertoire of three alternatives evoked at different temperatures, even though in some stocks of *P. aurelia* twelve alternatives have been found (Margolin, 1956) and inbreeding in *Tetrahymena* reveals the potential existence of more than three.

The inducing conditions include a wide variety of physical and chemical agents, many of which, such as temperature, atmosphere, pH, salts, carbohydrates and proteins, are known to influence also the

Table 1. *Influence of environment on development of microbial surface antigens (excluding transductional, transformational and prophage-induced antigenic changes)*

Factor	Antigen (or antigenic organelle)	Organism (not always the original name)	Observation	Reference
Temperature	Flagella	Typhoid-like bacillus	Present 23°; absent 38°	(Mironesco, 1899)
,,	,,	'Bacille TC'	Present 18°; absent 35°	(Nicolle & Trenel, 1902)
,,	,,	*Salmonella typhi*	Present 36°; absent 42°	(Nicolle & Trenel, 1902; Felix, Bhatnager & Pitt, 1934)
,,	,,	'*Bacillus inconstans*'	Present 22°; absent 37°	(Braun & Löwenstein, 1923)
,,	,,	*Pasteurella pseudotuberculosis*	Present 18°; absent 37°	(Arkwright, 1927; Preston & Maitland, 1952)
,,	,,	*Salmonella paratyphi* B	Present 22°; absent 37°	(Jordan, Caldwell & Reiter, 1934)
,,	,,	Psychrophils	Present 18°; absent 37°	(Schubert, 1952; Schubert & Schubert, 1953)
,,	,,	*Bordetella bronchiseptica*	Present 25°; absent 37°	(Lacey, 1953*b*)
,,	,,	Psychrophils	Present 37°; absent 18°	(Schubert & Schubert, 1953)
,,	,,	*Salmonella paratyphi* C	Present 37°; absent 18°	(Quadling, 1958)
,,	Flagellum	*Phytophthera infestans*	Present in cold; absent in warm	(Ferris, 1954)
,,	Flagellum	*Leishmania infantum*	Present < 28°; absent > 28°	(Lacey, unpublished)
,,	'Immobilizing'	*Paramecium aurelia*	Any one of 3–12 alternatives	(Beale, 1958; Margolin, 1956)
,,	'Immobilizing'	*Tetrahymena pyriformis*	Any one of 3 alternatives	(Margolin, Loefer & Owen, 1959)
,,	Vi	*S. typhi*	Optimum 37°	(Felix *et al.* 1934; Jude & Nicolle, 1952*a*)
,,	Vi	*S. paratyphi* C	Optimum 37°	(Jude & Nicolle, 1952*b*)
,,	Vi	*Escherichia coli*	Optimum *c.* 20°	(Nicolle, Jude & Diverneau, 1953)
,,	Vi	*Ballerup* sp.	Decreasing from 18 to 41·5°	(Nicolle & Jude, 1952*a*, *b*)
,,	B	*E. coli* (some strains)	More at 18° than 37°	(Ørskov, 1956)
,,	M	Group A streptococci	Optimum 20°; absent 37°	(Elliott, 1945)
,,	X and O	(*B. pertussis*, *B. parapertussis*, *B. bronchiseptica*)	X: present 37°; absent 25°. O: spectrum of 3	(Lacey, 1953*b*, 1960)

,,	Capsule	*Pasteurella pestis*	Present 37°; absent 26°	(Shibayama, 1905; Schütze, 1932)
,,	Capsule	*Pasteurella septica*	Present 20–40°; absent < 20° > 40°	(Priestley, 1936)
,,	Capsule	*Klebsiella pneumoniae*	Optimum 37°; absent 42°	(Hoogerheide, 1939)
,,	Mucoid	*S. paratyphi* B	Present < 27°; absent > 27°	(Brandenburg, 1952)
,,	Polysaccharide	*E. coli*	More at lower temperatures	(Wilkinson, Duguid & Edmunds, 1954)
,,	Polysaccharide	*Aerobacter cloacae*	More at higher temperatures	
,,	O	*S. typhi*	Minimal at 37°	(Jude *et al.* 1952)
,,	O	*Ballerup* sp.	Increasing from 18 to 41·5°	
,,	Surface	'*B. prodigiosus*'	Present 18°; absent 37°	(Kirstein, 1904)
,,	Surface	*Shigella shigae*	More O agglutinable at 20° than 37°	(Schütze, 1944)
,,	Labile surface	*Alcalescens dispar*	Present 37°; absent 20°	(Brandes, 1956)
,,	Phage sensitivity inhibitor	*E. coli*	Present 37°; absent 20°	(Bordet & Beumer, 1942)
,,	Absorbing Inaba serum	*Vibrio cholerae* (Ogawa)	Present 20°; absent 37°	(Kauffmann, 1950)
Light	Flagellum	*Chlamydomonas moewusii*	Increased length	(Lewin, 1953)
Solid media	Flagella	*Chromobacterium violaceum*	Lateral present	(Sneath, 1956)
Surface-active compounds	Flagella	*Proteus vulgaris*	Absent	(Lominski & Lendrum,1942)
Drying of agar	Flagella	*S. typhi*	Absent	(Felix, 1924)
CO_2	Mucoid	*Bacillus anthracis*	Increased	(Nungester, 1929)
CO_2	Capsule	*B. anthracis*	Increased	(Nungester, 1929; Ivánovics, 1937; Sterne, 1937*a*, *b*; Chu, 1952)
CO_2	Polypeptide	*B. anthracis*	Increased	(Thorne, Gómez & Housewright, 1952)
Anaerobiosis	M	Group A streptococci	Decreased	(Elliott & Dole, 1947)
Anaerobiosis	Vi	*S. typhi*	Absent	(Kozinski, Opara & Kraft, 1957)
Acid	Flagella	*Proteus* X19	Decreased	(Schiff & Nathorff, 1920)
Acid	X	*S. typhimurium*	Decreased	(Happold, 1929)
Acid	Capsule	*Streptococcus pneumoniae* type III	Increased	(Bernheimer, 1953)
Alkali	Surface	*S. typhi*, *S. paratyphi* A, B	Decreased	(Capone, 1919)
Alkali	Mucoid	*B. anthracis*	Increased	(Nungester, 1929)
Alkali	Capsule	*E. coli mutabile*	Increased	(Zamenhof, 1946)
Alkali	Flagella	*S. abortus-equi*	Absent pH > 7·8	(Kato, 1954)
Hypertonicity	Flagellum	*C. moewusii*	Decreased size	(Lewin, 1953)
Salt balance	X and O	*B. pertussis*, *B. parapertussis*, *B. bronchiseptica*	X: variable amount or absent. O: Spectrum of 3.	(Lacey, 1954; Lacey, 1960)

Table 1 (*cont.*)

Factor	Antigen (or antigenic organelle)	Organism (not always the original name)	Observation	Reference
Salt balance	Flagella	*B. bronchiseptica*	Present or absent	(Lacey, 1954*a*)
Salt balance	Flagella *	*N. gruberi*	Developed or lost	(Willmer, 1960).
Manganese	Polypeptide	*B. subtilis*	Increased and altered	(Leonard, Housewright & Thorne, 1958)
Potassium	Polysaccharide	*Aerobacter aerogenes*	Increased with less	(Duguid & Wilkinson, 1954)
Salts	Capsule	*K. pneumoniae*	Decreased	(Hoogerheide, 1940)
Salts	'Immobilizing'	*P. aurelia*	Changed	(Beale, 1954)
Sodium bisulphite / Sodium thiosulphate	Vi and H	*S. typhi* (Zwerg form)	Present	(Kauffmann, 1935)
Glucose	Mucoid	*B. anthracis*	Decreased	(Nungester, 1929)
Glucose	Vi	*S. typhi*	Increased	(Gladstone, 1937)
Glucose	Capsule	*K. pneumoniae*	Increased	(Hoogerheide, 1939)
Sucrose	Dextran	Group H streptococci and *Lactobacillus mesenteroides*	Increased	(Neill *et al.* 1941)
Sucrose	2:2-Glucopyranoside*	*Phytomonas tumefaciens*	Increased	(McIntire, Peterson & Riker, 1942)
Sucrose	Amylopectin*	*Neisseria perflava*	Increased	(Hehre & Hamilton, 1946; Hehre, 1949)
Sucrose and raffinose	Levan*	*Aerobacter levanicum*	Increased	(Hestrin & Avineri-Shapiro, 1944)
Sucrose and raffinose	Levan	*B. subtilis* and *Streptococus salivarius*	Increased	(Genghof, Hehre & Neill, 1946)
Maltose	Polysaccharide*	*E. coli*	Increased	(Torriani & Monod, 1949; Doudoroff *et al.* 1949)
Glucose-1-phosphate	Starch*	*Corynebacterium diphtheriae*	Increased	(Hehre, Carlson & Neill, 1947)
Dextrin	Dextran	*Aerobacter viscosum*	Increased	(Hehre & Hamilton, 1949)
Various carbohydrates	Mucoid	*E. coli* and salmonellae	Increased	(Morgan & Beckwith, 1939)
Various carbohydrates	Capsules	*V. comma, P. vulgaris, E. coli*	Increased	(Warren & Gray, 1954)
Lactose : nitrogen ratio	Polysaccharides	*A. aerogenes, A. cloacae, E. coli*	Varies directly	(Wilkinson *et al.* 1954)
33 % blood	—	*B. pertussis*	Altered agglutinability	(Bordet & Sleeswyk, 1910)
10 % blood	? Vi	*S. typhi*	Decreased agglutinability	(Gay & Claypole, 1913)
Rabbit blood	Complement fixing	*Histoplasma capsulatum*	Present (with yeast phase)	(Lindberg, 1951)

Rabbit or mouse serum	M	Group A streptococci	Present	(Elliott, 1945)
Ascitic fluid	Vi and H	*S. typhi* (Zwerg form)	Present	(Kauffmann, 1935)
Serum	Capsule	*B. anthracis*	Present	(Sterne, 1937*b*; Chu, 1952)
Serum	Capsule	*K. pneumoniae*	Present	(Knoll, 1953)
Heart extract	Vi	*S. typhi*	Decreased	(Horgan, 1936)
Ammonium as sole nitrogen source	Flagella	*S. typhi*	Absent	(Gladstone, 1937)
Phenol 0·1 %	Flagella	*Proteus* X19	Absent	(Braun & Schaeffer, 1919)
Phenol 0·1 %	Flagella	*S. typhi*	Absent	(Craigie, 1931)
Phenol 0·1 %	Flagella	*S. enteritidis*	Absent	(White, 1925)
Phenol 0·1 %	Vi and H	*S. typhi*	Absent	(Felix & Pitt, 1934)
Phenol 0·1 %	I	*Shigella dysenteriae*	Absent	(Shelubsky & Olitzki, 1948)
Phenol 0·1 %	Flagella	*E. coli*	Absent	(Smith, 1954)
Penicillin	L-form	*Proteus* sp.; *S. typhi* and *S. typhimurium*	Present	(Dienes, Weinberger & Madoff, 1950)
Penicillin	Flagella	*Proteus* sp.	Absent	(Tulasne, 1951)
Penicillin	Smooth surface	*Brucella abortus* (mutant)	Absent	(Braun *et al.* 1952)
Penicillin	Mucoid	*B. abortus* (mutant)	Absent	
Penicillin	Somatic	*E. coli* K12	Lost with appearance of protoplast	(Lederberg, 1956)
Streptomycin	Capsule	*K. pneumoniae*	Absent	(Lambkin & Desvignes, 1952)
L(+)-erythroisomer of chloramphenicol	Polypeptide†	*B. subtilis*	Decreased	(Hahn, Wissemen & Hopps, 1954)
Proflavine	Neutralizing	T_2 phage in *E. coli* B	Absent	(Lanni & Lanni, 1953)
Cyanide; malonic acid; dinitrophenol	Vi	*S. typhi*	Absent	(Kozinski *et al.* 1957)
Trypsin or chymotrypsin	'Immobilizing'	*P. aurelia*	Changed	(van Wagtendonk, 1951)
Growth time	X	*S. typhimurium*	Increased with long	(Topley & Ayrton, 1924)
Growth time	Vi	*S. typhi*	Optimum 4 hr.	(Craigie & Brandon, 1936)
Growth time	Vi and O	*S. typhi*	Present 48 hr.; absent 18 hr.	(Gladstone, 1937)
Growth time	Capsule	*P. septica*	Present 24 hr.; absent 72 hr.	(Priestley, 1936)
Growth time	Capsule	*K. pneumoniae*	Increased with long	(Hoogerheide, 1939)
Growth time	Polypeptide†	*B. subtilis*	Optimum 7 days	(Bovarnick, 1942)
Growth time	Haemagglutinin	*B. pertussis*	Optimum exists	(Fisher, 1948; Thiele, 1950; Masry, 1952)

* Antigenicity not demonstrated.

† Estimated in medium.

Table 1 (*cont.*)

Factor	Antigen (or antigenic organelle)	Organism (not always the original name)	Observation	Reference
Growth time	Tail	T_2, T_5 or T_6 phage in *E. coli* B	Absent in early stages	(de Mars *et al.* 1953)
Growth time	Neutralizing	T_3 or T_4 phage in *E. coli* B	Absent in early stages	(Barry, 1954)
Presence of phage: T_2 (or T_4)	—	T_4 (or T_2) phage in *E. coli* B	Phenotypic mixing	(Novick & Szilard, 1951; Streisinger, 1956)
Presence of virus:				
Influenza virus W	—	Influenza virus M	Phenotypic mixing	(Hirst & Gotlieb, 1953)
Influenza virus A	—	NDV	Phenotypic mixing	(Granoff & Hirst, 1954)
NDV	—	Influenza virus A	Phenotypic mixing	(Granoff & Hirst, 1954)
Unanalysed cause in medium	—	*S. typhi*	Changed agglutinability	(Kirstein, 1904)
Unanalysed cause in medium	—	*B. pertussis*	Changed agglutinability	(Povitzky & Worth, 1916)
Unanalysed cause in medium	—	Coliform	Changed agglutinability	(Dawson, 1919)
Unanalysed cause in medium	—	*B. pertussis*	Changed agglutinability	(Krumwiede, Mishulow & Oldenbusch, 1923)
Unanalysed cause in medium	—	*B. pertussis*	Changed agglutinability	(Cruickshank & Freeman, 1937)
Unanalysed cause in medium	—	*Vibrio* sp.	Changed agglutinability	(Linton, Shrivastava & Seal, 1937)
Unanalysed cause in medium	—	*K. pneumoniae*	Changed agglutinability	(Hoogerheide, 1939; Adler & Humphries, 1954)
Unanalysed cause in medium	—	*Staphylococcus pyogenes*	Changed agglutinability	(Pillet, Mercier & Orta, 1951)
Unanalysed cause in medium	—	*B. pertussis*	Changed agglutinability	(Andersen, 1952)
Unanalysed cause in medium	—	*Clostridium parabotulinum*	Changed agglutinability	(Hill & Weaver, 1955)
Unanalysed cause in medium	B	*E. coli*	More at room temperature on Worfel-Ferguson medium than at 37° on broth agar	(Ørskov, 1956)
Unanalysed cause in medium	O	*S. typhi*	More on liver-extract agar	(Czerniawski, Sedlaczek & Zabłocki, 1958)

production of enzymes and toxins and the production and germination of spores. They also include many of the factors known to affect the activity of purified enzymes with the notable exceptions of pressure, ion balance (apart from *Bordetella*), chelating agents and metals (apart from manganese on *Bacillus subtilis*).

In general, three types of change can be discerned: (i) in which the amount of antigen clearly varies continuously with a continuous environmental change; (ii) in which the surface antigenic change apparently occurs abruptly, in an all-or-none way, in spite of gradual change in the environment; and (iii) in which the antigenic and environmental changes are both patently discontinuous. The first type probably includes all modulations of O, capsular and slime antigens and the exceptional quantitative changes, in surface antigens D and M, observed in one stock of *Paramecium aurelia* (Margolin, 1956). That Vi antigen production varies continuously with temperature, and so belongs here, could have been inferred from the original paper of Felix, Bhatnager & Pitt (1934) but has been amply confirmed by the recent findings of Nicolle & Jude (1952*a*, *b*), Jude & Nicolle (1952*a*, *b*) and Nicolle, Jude & Diverneau (1953).

The second type comprises modulations of (*a*) flagellar antigens, (*b*) immobilizing antigens of *Paramecium* or *Tetrahymena* (when these appear as qualitative, mutually exclusive changes) and (*c*) cell wall antigens, when these appear as reversible *bacillary* ↔ *L-form* or *whole cell* ↔ *protoplast* changes. The differences between this type and the first may be more apparent than real and may reflect a situation in the second group where, in spite of the rate of antigen synthesis varying continuously with environmental change, the antigen appears (or disappears) suddenly from the surface because either (i) a threshold concentration is required for newly formed antigen to be retained at the cell surface (e.g. for maintaining the mechanical integrity of a surface layer), or (ii) the environmental conditions have not been held sufficiently close to those needed to detect a transitional or intermediate state. The third type includes at present only the influence of phage on phage or virus on virus. Examples of host influence on phage or virus, such as host-induced alteration of host-range (Luria, 1953; Hoskins, 1958), have not been included in Table 1 because no associated antigenic change has so far been shown. That this is likely to be present is suggested by the most interesting finding (Lieb, Weigle & Kellenberger, 1955) that the host-range of the phage produced by a given bacterium and the phage sensitivity of the same bacterium may be completely associated and inheritable as a single independent factor.

The precise role of the environment in antigenic variation has seldom been investigated and very rarely clarified. However, in theory at least, one can easily define a group of variations in which a more or less elaborate precursor is used directly as a building block. Most variations occurring in this group appear to arise from the polymerization of a hexose, derived from a di- or polysaccharide, into a polysaccharide antigen through the activity of a constitutive or induced transglycosidase (see Barker & Hassid, 1951). Among such polymers only levans and dextrans derived from sucrose, raffinose or dextrin have so far been proved antigenic, but there seems no reason to suppose that the dextrin-like compound produced from maltose by certain strains of *Escherichia coli* (Torriani & Monod, 1949), and all other bacterial polysaccharides, are not comparably antigenic for some animal. An analogous polymerization of an environmentally supplied unit almost certainly occurs in polypeptide formation from D-glutamic acid by *Bacillus subtilis*; and a formal resemblance to this kind of variation can be seen in the phenotypic mixing of vegetatively multiplying viruses and phages. Some polysaccharide polymerizations are subject to competitive inhibition; for example, maltose or isomaltose can almost abolish dextran synthesis by *Leuconostoc mesenteroides* (Koepsell *et al.* 1953) and the formation of dextrin-like polysaccharide by *E. coli* depends on the ratio of glucose to maltose (Barker & Hassid, 1951). It may well be therefore that the effect of glucose and other carbohydrates on the development of the Vi antigen of *Salmonella typhi*, or the capsules of *Aerobacter* or other bacteria (see Wilkinson, Duguid & Edmunds, 1954), stems from a direct action on a polymerizing system of this kind rather than from its well-known action on enzyme synthesis.

Although several examples are now known of the incorporation of amino acid analogues or abnormal proportions of normal amino acids into enzymes or flagella (see Kerridge in this Symposium), I have found no report of an alteration of immunological specificity being directly induced in this way. But since proteins vary considerably in amino acid composition, there seems little doubt that such changes will be demonstrated, especially in inessential products, such as flagella, capsules or inducible enzymes, whose absence, abnormal function or lack of function could not be incompatible with growth.

Among antigenic modulations where there can be no question of incorporation, the one whose mechanism is most understood is probably still that of the M protein of group A streptococci. Elliott (1945) showed that this antigen is absent from cultures of some strains of group A streptococci grown at 37° under normal (reducing) conditions because a proteinase precursor is then best formed and becomes activated to give an enzyme destroying the M protein. Conversely, the antigen may appear on cells grown at 20°, because proteinase production is then reduced or absent, or on cells grown at 37° in the presence of rabbit or mouse serum because both of these inhibit proteinase activity. But although these findings entirely account for the effect of the sera, they only partially explain the effect of temperature since one still has to ask how it is that temperature affects the production of the enzyme precursor. At present there is no satisfactory answer, and a similarly unanswerable question is at the heart of virtually all the other changes in this group.

With a view to gaining some insight into these problems I have for some time been exploring the reaction of the *Bordetella* group to its environment (Lacey, 1953*a*, *b*, 1960). For this purpose these three species appear peculiarly suitable. All three have a variety of responses involving at least three antigens; all three react to both temperature and chemical environment and only one (*B. pertussis*) will not grow within a fairly wide range of conditions. They are therefore complementary to *Paramecium* and *Tetrahymena* as subjects for investigating the expression of potential antigenic specificity. They probably lack the ability to conjugate which, in the ciliates, allows the interaction of genes and non-genic material to be explored, yet they have the advantage in speed and precision of response and in permitting the exploration of the effects of mutation and of greater differences of chemical environment.

ANTIGENIC MODULATION OF *BORDETELLA* SPECIES

(*a*) *Comparison of the three modes*

An overall view of antigenic variation in this genus is given in Table 2. This slightly oversimplifies the picture since the antigens are not mutually exclusive and over narrow ranges of intermediate conditions the antigenic state changes without a sharp break from one mode to another. Except under these conditions, any clone is readily assigned to one or other mode by its agglutinability on the slide with suitable sera. On ordinary bacteriological media, containing sodium chloride as the principal salt, 'X', 'I' and 'C' correspond to high, intermediate and low-temperature modes respectively, i.e. in *Paramecium* to D, G and S serotypes (Beale, 1954) and in *Tetrahymena* to H, I and L serotypes (Margolin, Loefer & Owen, 1959).

Growth is considered to be in X-mode as long as cells have agglutinable X-antigen on the surface even when, under conditions approximating to those inducing I-mode, the O antigen has also appeared on the surface (it is completely covered and inagglutinable under highly pro-X conditions) or when, in the case of *Bordetella bronchiseptica*, the bacteria are feebly motile by poorly developed flagella.

Table 2. *Agglutinable antigens of* Bordetella *species*

Species	Type of antigen	X-mode: Not near I-mode	X-mode: Near I-mode	I-mode	C-mode
pertussis	X	X*	X*	—	—
	O	[6, 1]	5, 1	5, 1	1
parapertussis	X	X	X	—	—
	O	[7, (2)]	7, 2	7, 2	2
bronchiseptica: O group 2	X	X*	X*	—	—
	O	[(8), 2]	(8), 2	(8), 2	2
	H	—	a, b or c	a, b or c	a, b or c
bronchiseptica: O group 3	X	X*	X*	—	—
	O	[(9), 3]	(9), 3	(9), 3	3
	H	—	a, b or c	a, b or c	a, b or c

[ ] = revealed by removing X-antigen with formamide. ( ) = not detectable in all strains. * = several serotypes recognizable with absorbed sera.

The X-antigen appears to be a protein or protein-like microcapsular antigen because it is partially destroyed by heating to 100° at pH 7·5, becomes much more heat-resistant after formalinization, and dissolves (in the unformalinized state) in 50 % formamide. It is also soluble in other protein solvents such as urea, thiourea, acetamide, guanidine, formamidine or acetamidine, but not in diethylene glycol, which dissolves the O antigens of *Shigella* (Morgan, 1937), or in potassium thiocyanate which dissolves the capsule of *Pasteurella pestis* (Amies, 1951). It can be precipitated from urea solution by dialysis and redissolved in 5 % urea without loss of blocking power, precipitability and antibody-inducing ability.

The X-antigen clearly corresponds roughly to the 'K' antigen first described by Andersen (1953), but for two reasons, heat has not been used in defining it. In the first place, all antibodies other than anti-X may be efficiently removed by absorption with a mixture of (i) cells in other modes of the same species, and (ii) either mutants lacking the ability to produce X-antigen under pro-X conditions or X-mode cells from which the X-antigen has been dissolved. In the second place, the heating to 120°, needed to abolish the reactivity of the X-antigen, also destroys the specificity of the underlying O antigens and at the same time appears to create a 'common O antigen' for the whole genus. The O antigens in this scheme therefore do not correspond with Andersen's generic O of whose existence in unheated organisms I can find no evidence.

All species overlap antigenically in X-mode although, as Andersen (1953) showed with *Bordetella pertussis*, serotypes can be distinguished within the *pertussis* and *bronchiseptica* species by suitably absorbed sera. In contrast, in C-mode, only *parapertussis* and some strains of *bronchiseptica* overlap by possessing an indistinguishable O antigen (O_2). Only one C-mode serotype of *pertussis* and one of *parapertussis* have been found, but freshly isolated strains of *bronchiseptica* may possess any of the six possible combinations of: one of two O antigens with one of three flagellar antigens.

When grown in X-mode all species produce a haemolysin, are highly sensitive to agglutination by mercuric or gold chlorides and produce a non-fimbrial agglutinin for the red blood cells of all species tested except the goat (human, monkey, dog, cat, sheep, pig, rabbit, ferret, mouse, rat, hamster, dormouse, cotton rat, guinea-pig, galago, horse, mule, cow, bat, goose, duck, pigeon, hen, turkey, tortoise, frog and eel). In C-mode no species produces an active haemolysin or is sensitive to heavy metal salts. *Bronchiseptica* strains alone produce a non-fimbrial agglutinin for the red blood cells of all the species listed above from the human to guinea-pig inclusive; i.e. not for those of birds, cold-blooded animals and a few others. The receptor on the surface of the red cell for C-mode (but not X-mode) agglutination is removed by digestion with papain. In pro-C-mode conditions, *bronchiseptica* strains also differ from those of the other species in producing a mucoid O antigen, of apparently normal specificity, under the influence of sub-bacteriostatic concentrations of cycloserine.

(*b*) *Prediction of the equilibrium state from ion balance, temperature and species*

At any given temperature some salts will favour production of X-mode (i.e. appear pro-X), some of C-mode, and others will appear neutral. These influences are a function of the ion species and their ratios and not of their absolute concentrations or ionic strengths. Certain ions at least can be arranged in a pattern (Fig. 1) in which the influence of any salt is represented at the centre of the line joining its ions and the influence of a mixture is represented at an imaginary centre of inertia. The ions appear to have the same relative activity at all temperatures and for all species. Hence the combined influence of temperature and salt balance for any species is represented by the sum of their separate influences, even though the value of this needed to induce the corresponding antigenic state differs for each species and for different strains of the same species. Thus, for example, it can be seen from Fig. 1 that with a given salt balance the change from X to I modes occurs at a higher temperature in *bronchiseptica* than *pertussis* and higher in *pertussis* than *parapertussis*.

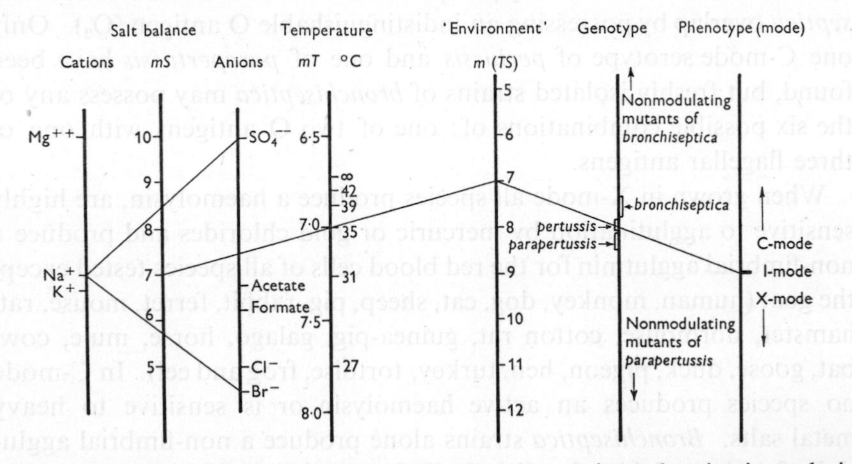

Fig. 1. Nomogram relating ion balance, temperature, species and antigenic mode in *Bordetella*. A salt is represented at the mid-point of the line joining its component ions. The influence of a salt mixture is represented at its imaginary centre of inertia. For calculating this, each salt is assumed to exert a second moment equal to the product of its concentration (in equivalents), conductance ratio (at the total salt concentration of the mixture) and square of the distance (along the centre (*mS*) line from salt to point of balance). The mode of growth is then found from this salt balance, temperature and species in the way illustated.

$$mT = 6{\cdot}7234 + \text{antilog}_{10}\left(\frac{27{\cdot}03 - t}{13{\cdot}93}\right) \; (t \text{ in °C})$$

$$m(TS) = mS + mT - 7.$$

(c) *Immediate cause of the antigenic differences*

All differences between antigenic states or modes at equilibrium, including such obvious 'qualitative' differences as the presence or absence of flagella, appear to stem from altered rates of synthesis and competition for space on the surface. Certainly in the case of the X and O antigens there is no question of any mutual exclusion, such as may occur in *Paramecium*, and the quantity of both X and O antigens almost certainly varies continuously with change in the conditions of growth. Hypothetical antigenic constitutions of a series of antigenic states of *Bordetella pertussis* are shown in Fig. 2. These interpret the observed changes of agglutinability and absorbing capacity of the antigens in terms of altered amounts as well as altered positions.

(*d*) *The influence of factors other than temperature and ion balance*

By a simple streak and disk method, the influence of some 2300 substances on the growth of a non-flagellated mutant strain (BR19) of *B. bronchiseptica* has been examined. Each substance, applied as a saturated or 10% neutral, aqueous or alcoholic solution on a blotting paper disk, has been tested for pro-X and pro-C activity on a standard Lemco-horse blood agar by incubating duplicate plates at 27° and 35°, i.e. temperatures at which, on this medium, growth is normally in

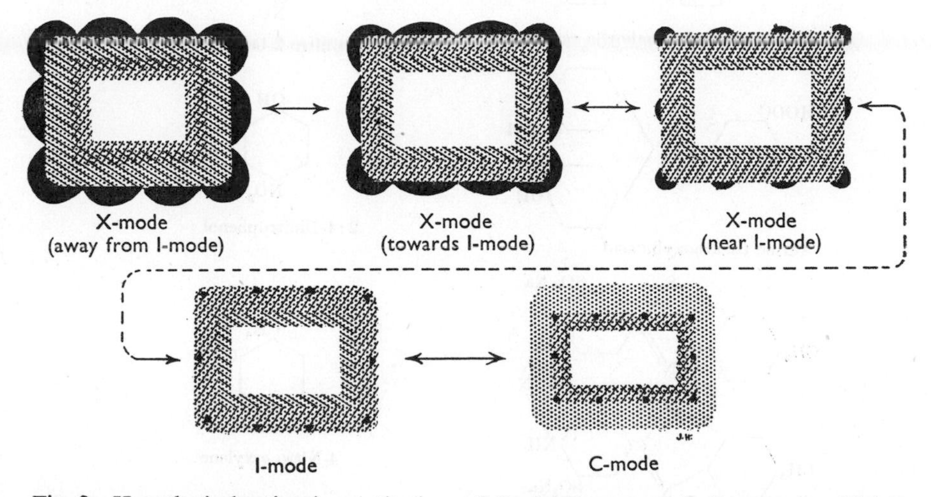

Fig. 2. Hypothetical antigenic constitutions of *Bordetella pertussis* in three modes. *Major antigens* are shown forming part of the surface, *minor antigens* touching the surface and *minimal antigens* submerged.

■, X antigen; ▧, 0_6; ▨, 0_5; ▒, 0_1.

C-mode and X-mode respectively. Diffusible pro-X substances, by inducing X-mode at 27°, lead to the formation of haemolysin and hence a zone of lysis near the disk, whereas, conversely, diffusible pro-C substances indirectly inhibit lysis. This method is fairly sensitive and works well with all substances that are not themselves markedly haemolytic and do not chelate the calcium necessary for the action (but not the formation) of the lysin. With such substances, however, the mode can readily be determined by agglutination on the slide.

About half the substances included were known to have an activity of biological interest, e.g. being a protein precipitant, denaturant or other reagent, enzyme activator or inhibitor, stain, dye, lake, mutagen, carcinogen, plant growth substance, allergen, pharmaceutical, poison, antibiotic, disinfectant, insecticide, vitamin, hormone, metabolite,

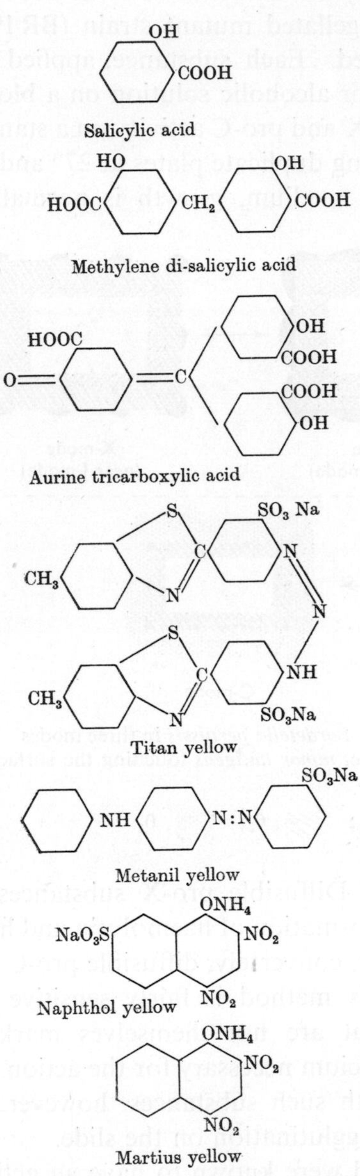

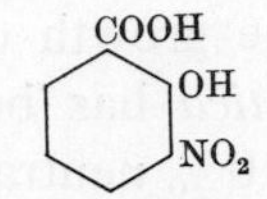

2-Hydroxy 3-nitrobenzoic acid

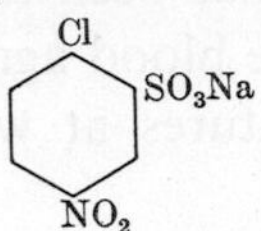

1-Chloro 4-nitro 2-benzenesulphonic acid

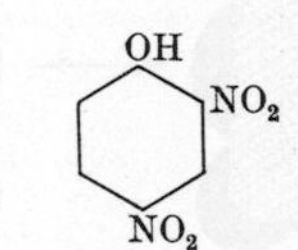

2:4-Dinitrophenol

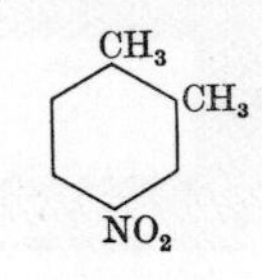

4-Nitro *o*-xylene

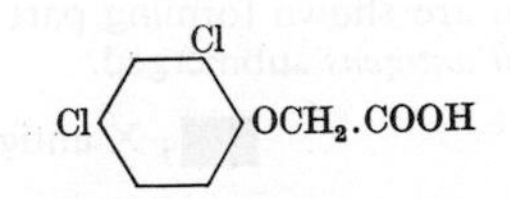

2:4-Dichlorophenoxyacetic acid

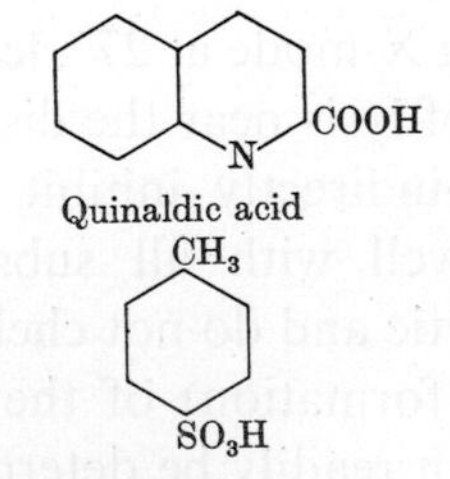

p-Toluenesulphonic acid

Fig. 3 *a*. Potent pro-C mode substances.

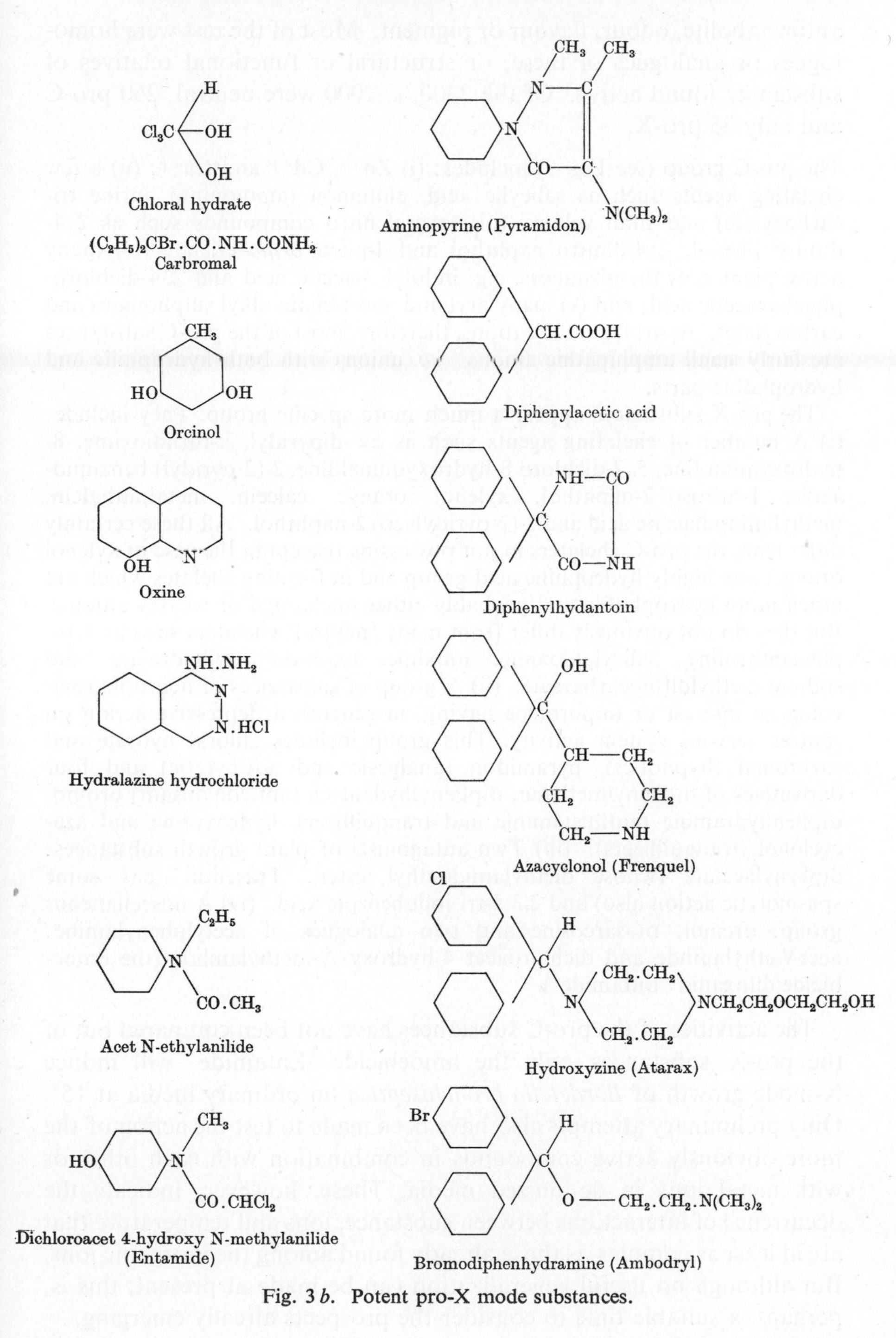

Fig. 3 *b*. Potent pro-X mode substances.

antimetabolite, odour, flavour or pigment. Most of the rest were homologues or analogues of these, or structural or functional relatives of substances found active. Of the 2300, *c.* 2000 were neutral, 250 pro-C and only 25 pro-X.

The pro-C group (see Fig. 3) includes: (i) Zn^{++}, Cd^{++} and Ca^{++}; (ii) a few chelating agents such as salicylic acid, aluminon (ammonium aurine tricarboxylate) and titan yellow; (iii) several nitro compounds such as 2:4-dinitro phenol, 2:4-dinitro naphthol and 4-nitro *ortho*-xylene; (iv) many active plant growth substances, e.g. indolyl 3-acetic acid and 2, 4-dichlorophenoxyacetic acid; and (v) many aryl and short-chain alkyl sulphonates and carboxylates. Apart from the cations, therefore, most of the pro-C substances are fairly small amphipathic anions—i.e. anions with both hydrophilic and hydrophobic parts.

The pro-X substances appear a much more specific group. They include: (i) A number of chelating agents such as $\alpha\alpha'$-dipyridyl, α-furildioxime, 8-hydroxyquinoline, 5, 7-dichloro 8-hydroxyquinaldine, 2-(2-pyridyl) benzimidazole, 1-nitroso 2-naphthol, xylenol orange, calcein, metalphthalein, methyliminodiacetic acid and 1-(2-pyridylazo) 2-naphthol. All these certainly differ from the pro-C chelaters in not possessing (except in the case of xylenol orange) any highly hydrophilic acid group and in forming chelates which are much more hydrophobic and probably either uncharged or weakly anionic. But they do not obviously differ from many 'neutral' chelaters such as 1:10-phenanthroline, salicylaldoxime, nioxime, *asym*-diphenylhydrazine and sodium diethyldithiocarbamate. (ii) A group of substances of neuropharmacological interest or importance having, in general, a depressive action on central nervous system activity. This group includes chloral hydrate and carbromal (hypnotics), pyramidon (analgesic and antipyretic) and four derivatives of diphenylmethane: diphenylhydantoin (anticonvulsant) bromodiphenhydramine (antihistaminic and tranquillizer), hydroxyzine and azacyclonol (tranquillizers). (iii) Two antagonists of plant growth substances: diphenylacetate (whose diethylaminoethyl ester, 'Trasentin' has some spasmolytic action also) and 2:3:5-tri-iodobenzoic acid. (iv) A miscellaneous group: orcinol, DL-sarcosine and two analogues of acetylphenylamine: acet*N*-ethylanilide and dichloroacet 4-hydroxy *N*-methylanilide (the amoebicide diloxanid 'Entamide').

The activities of the pro-C substances have not been compared but of the pro-X substances only the amoebicide 'Entamide' will induce X-mode growth of *Bordetella bronchiseptica* on ordinary media at 15°. Only preliminary attempts also have been made to test the action of the more obviously active compounds in combination with each other or with metal ions in de-ionized media. These, however, indicate the occurrence of interactions between substance, ions and temperature that are at least as complex as those already found among the inorganic ions. But although no useful generalization can be made at present, this is, perhaps, a suitable time to consider the prospects already emerging.

RELATIONSHIPS BETWEEN MODULATING AND OTHER BIOLOGICAL ACTIVITY

The modulating activity for *Bordetella bronchiseptica* of substances promoting or inhibiting a number of biological phenomena is shown in Table 3. Except in the development of the sea-urchin egg, this reveals a remarkable degree of association between modulating and other activity. Apart from Mg^{++}, whose role is considered below, the most important influence apparently out of place is probably that of temperature on fibrillation of the isolated rabbit heart. In this preparation a higher temperature predisposes to fibrillation (Milton, 1959), and hence counteracts a rise in the K^+/Ca^{++} ratio. This is the converse of the relationship found with *Bordetella* and, as shown in Table 4, the converse of that in three of four other situations (the only ones found in a prolonged search) where such interactions have been examined. No explanation of these differences can be offered, but *a priori* the usual relationship might be expected for two reasons. First, one or more pro-C substances, such as Ca^{++}, SO_4^{--} and amphipathic anions, may mitigate or protect against the effect of heat in several circumstances, for example the inactivation of phages (Burnet & McKie, 1930) or enzymes (Northrop, Kunitz & Herriott, 1948), precipitation of proteins (Boyer *et al.* 1946) and temperature shock on *Limnaea stagnalis* (Raven & Erkel, 1955). Secondly, a raised K^+ level has often been noted to mimic or aggravate the effect of a raised temperature, as, for example, on the activity of choline acetylase (Nachmansohn & John, 1945), the frequency of contraction of the isolated frog heart (Carmeliet & Lacquet, 1956) or the size of melanophores in isolated trout skin (Robertson, 1951). It seems likely, moreover, that other factors affecting cardiac fibrillation have yet to be found since, as Burn (1960) points out, in human surgery a lowered temperature appears to predispose to fibrillation.

In Table 3, Zn^{++} and amphipathic anions appear seven times and seven other factors more than twice; but their overlapping relationships are not readily seen in perspective. The more commonly occurring substances and a few others of interest have therefore been re-arranged in two-dimensional fashion in Fig. 4. This emphasizes the high degree of association between: (i) pro-C activity, (ii) antagonism in all eleven biological situations to neutral or pro-X substances, and (iii) except in the case of magnesium, general inhibitory activity for enzymes. At first sight Mg^{++} appears greatly out of place but although it activates, either alone or in combination with K^+, far more enzymes than any other ion, it nevertheless antagonizes K^+ in several instances other than antigenic

Table 3. *Modulating activity for* Bordetella bronchiseptica (*BR* 19) *of substances promoting or inhibiting other phenomena*

(Pro-X substances in bold type; pro-C substances in italics; neutral substances in roman type; untested substances in brackets.)

Phenomenon	Promoting phenomenon		Inhibiting phenomenon or promoter	
Activity of isolated enzymes in general (see Dixon & Webb, 1958)	**NH_4^+**	(0:0)*	*Zn^{++}*	(2:7)*
	K^+	(9:0)	*Cu^{++}*	(1:8)
	Cl^-	(1:0)	*Ca^{++}*	(6:1)
	Mn^{++}	(14:0)	*SO_4^{--}*	(0:0)
	Fe^{++}	(4:0)	*Li^+*	(0:0)
	Co^{++}	(3:0)	Na^+	(0:0)
	Mo^{++}	(2:3)	*Lower temperature*	
	Mg^{++}	(79:0)		
	Higher temperature			
Enzyme activity:				
Glutaminase of *Cl. welchii* (Hughes & Williamson, 1952)	**KBr**		*Sulphonphthaleins*	
Glyoxalate reductase (Zelitch, 1955)	**KI**		2:4-*Dichlorophenoxy acetic acid*	
Uncoupling of oxidative phosphorylation in mitochondria (see Schneider, 1959; Judah, 1960)	*Zn^{++}*	Arsenate	Mn^{++}	
	Cd^{++}	Azide	Co^{++}	
	Ca^{++}	Atabrine	*Mg^{++}*	
	Nitrite	Thyroxin	**EDTA**	
	Nitrate	(Gramicidin)	Serum	
	Dicoumarol		Diphenhydramine (Benadryl)	
	α, α′-dipyridyl			
	2:4-*Dinitrophenol*			
	Salicylate			
	Lauryl sulphate			
	Deoxycholate			
	o-Phenanthroline			
	2:4-*Dichlorphenoxy acetic acid*			
Sporulation of bacteria (see Murrell, 1955)	**K^+**		*Zn^{++}*	D-α-Alanine
	Mn^{++}		Fe^{++}	β-Alanine
	Mg^{++}		**NO_3^-**	Asparagine
	Cl^-		*SO_4^{--}*	Pyruvate
	Leucine		HPO_4^{--}	Lactate
	Isoleucine		*Fatty acids*	
			Glycerophosphate	
Germination of bacterial spores (see Murrell, 1955)	*Zn^{++}*	L-α-Alanine	**Oxine**	
	Cu^{++}	Adenosine	D-α-Alanine	
	Mg^{++}	Valine	Unsaturated fatty acids	
	Mn^{++}	Tyrosine		
	Fe^{++}	*Furfural*		
	Low pH			

* Number of enzymes for which ion is a specific activator: number of enzymes of which ion is an essential part.

Table 3 (*cont.*)

Phenomenon	Promoting phenomenon	Inhibiting phenomenon or promoter
Growth of shoots (see Audus, 1959)	*Zn^{++}*	Arsenite
	Ca^{++}	**Diphenylacetic acid**
	Coumarin	2:6-Dichlorophenoxyacetic acid
	Gibberellic acid	Naphthylmethylsulphide propionic acid
	2:4-*Dichlorophenoxyacetic acid*	**2:3:5-Tri-iodobenzoic acid**
	†*Low temperature*	
Fibrillation of rabbit heart (see Burn, 1960)	*Ca^{++}*	**K^+**
	NaF	*Mg^{++}*
	Acetylcholine	**EDTA**
	2:4-*Dinitrophenol*	ATP
	Sodium fluoride	Quinidine
	High temperature	*Low temperature*
Vegetalization of sea-urchin eggs (see Gustafson & Hörstadius, 1955; Lallier, 1959*a*, *b*)	*Li^+*	*Zn^{++}*
	α-Picolinic acid	**I^+**
	3-Acetylpyridine	*SCN^-*
	Phenyl lactic acid	Evans blue
	(8-Chloroxanthine)	Sulphocyanic acid
	(Deoxypyridoxine)	*Pyridine* 3-*sulphonate*
	Pyramidon	*Anionic detergents*
		Thiomalate
		(2-Thiomethyl 5-cytosine)
		(Iodosobenzoic acid)
		(Trypsin)
Convulsions and other activity of the central nervous system of animals (see McIlwain, 1957)	Strychnine	**K^+**
	Camphor	**Br**
	(Pentamethylene tetrazole)	**Chloral hydrate**
		Aminopyrin (Pyramidon)
		Diphenylhydantoin (Epanutin)
		Diphenylacetic acid (acid of Trasentin)
		Bromodiphenhydramine (Ambodryl)
		Bromodiethylacetyl Urea (Carbromal)
		γ-Aminobutyric acid
		Hydroxyzine (Atarax)
		Phenobarbitone
		Azacyclonol (Frenquel)
Diabetes (see Chenoweth, 1956; Krahl, 1957; Reid, 1958)	(Glucagon)	Insulin (*Zn*)
	Oxine	2:4-*Dinitrophenol*
	Albumin	*Salicyclic acid*
	Diphenylthiosemicarbazone	Tolbutamide
Rheumatic state (see Lansbury & Rogers, 1955)	**Hydralazine**	*Salicylic acid*
		Butazolidin

† Synergic with auxins in favouring development of female flowers (Heslop-Harrison, 1957).

Table 4. *Interaction of temperature and ion balance*

Rise in temperature equivalent to a rise in K^+/Ca^{++} or Na^+/Ca^{++} ratio	
Character of heart beat of frog	(Ringer, 1883)
Viscosity of gel of *Amoeba proteus*	(Thornton, 1935)
Inactivation of complement of guinea-pig	(Laporte, Hardre de Looze & Sillard, 1955)
Antigenic structure of *Bordetella pertussis*	(Lacey, 1960)
Fall in temperature equivalent to a rise in K^+/Ca^{++} or Na^+/Ca^{++} ratio	
After-potential, membrance potential etc. of 'A' fibres of motor roots of cat	(Lundberg & Laget, 1949)
Fibrillation of heart of rabbit	(Burn, 1960)

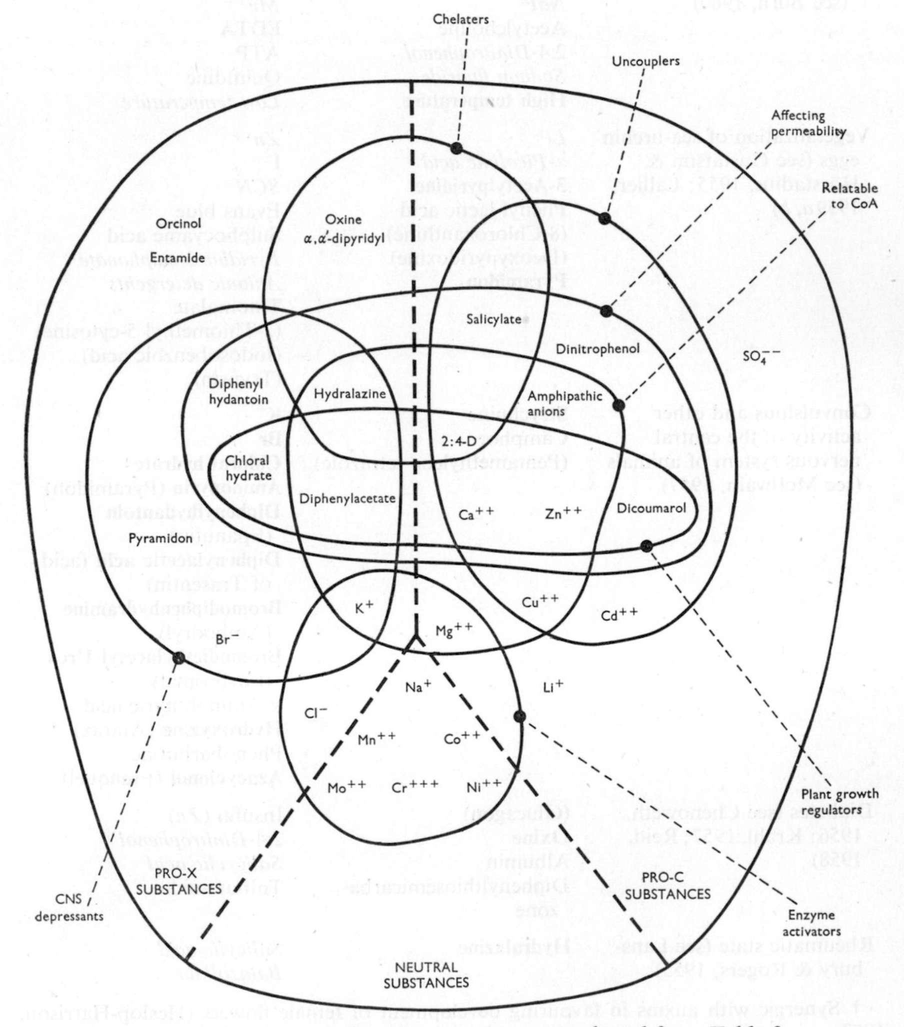

Fig. 4. Biological relationships of substances selected from Table 3.

development of *Bordetella*. Thus, for example, a balance of K^+ and Mg^{++} is needed for maximum activity of pantothenate synthetase of *Escherichia coli* (Maas, 1952) and ketohexokinase (Hers, 1952), and Mg^{++} antagonizes the activation by K^+ of choline acetylase (Mann, Tennenbaum & Quastel, 1939) and potassium-induced acceleration of glycolysis by extracts of *Lactobacillus arabinosus* (Clark & MacLeod, 1954). Indeed in review K^+ appears unique in never being an inhibitory antagonist of any other ion. In this context it is also of interest that the ions principally associated with ion antagonisms in biology may be arranged in a sequence of increasing inhibitory dominance such that for any ion, only ions to its right appear as inhibitor antagonists: K–Na–Li–Mg–Mn–Zn–Ca. Calcium activation and magnesium inhibition of myosin ATP-ase is, of course, a notable exception to this rule.

The justification for the position of some of the substances in Fig. 4 may not be evident. Diphenylacetate has been shown to antagonize the growth of shoots without promoting root growth (Veldstra & Åberg, 1953), unlike other antagonists—e.g. naphthylmethylsulphide propionic acid. It is relatable to substances which affect permeability by being an analogue of SKF 525A (the diethylaminoethyl ester of α-propyldiphenylacetic acid) which, as the ester *in vivo* or acid *in vitro*, prolongs the action of many drugs, probably by reducing the permeability of liver microsomes, and so delaying their conversion by the microsomes to more water-soluble and more easily excreted substances (see Brodie, 1956). Like many analogues of phenylacetic acid, it is a potent inhibitor of coenzyme A acetylation *in vitro* (Garattini, Morpurgo & Passerini, 1958) although in this respect less active than some alkyldiphenylylacetic acids and alkyloxydiphenylacetic acids which have a pro-C influence on *Bordetella*.

Diphenylhydantoin is probably the best suppressant for major epilepsy. In the rat it raises the electroshock seizure threshold and promotes the passage of sodium ions from inside to outside the cell and, because the electroshock seizure threshold varies with the intra/extra cellular brain Na^{++} ratio, it seems reasonable to attribute its anticonvulsant action to its effect on this ratio (Woodbury, 1955).

Hydralazine is a blood-pressure reducing agent believed to act through an effect on the midbrain (Lansbury & Rogers, 1955). It has an affinity for metals, inhibits the acetylation of co-enzyme A (Douglass *et al.* 1957) and inhibits histaminase. It can induce a rheumatic state in man which resembles disseminated lupus erythematosus and is controllable by antihistaminics and salicylates.

The central depressive action of ambodryl is probably attributable to a stabilizing action on permeability, not because of its anti-histaminic action but because its non-brominated analogue 'benadryl' (diphenhydramine) has been found a potent inhibitor of mitochondrial swelling induced by thioacetamide, calcium or thyroxine (Judah, 1960).

Both salicylate and dinitrophenol are well known as uncouplers of oxidative phosphorylation. Both have an antidiabetic action *in vivo* (Reid, 1958) and also accelerate loss of potassium from rat diaphragm (Hicklin, 1960).

The modes of action of plant growth substances are still unknown. Current speculations however (see Audus, 1959) centre round alteration of permeability, chelation (but see Fawcett, 1959), and interference with synthesis or utilization of acetyl co-enzyme A. The last possibility was suggested by Leopold & Guernsey (1953) and later withdrawn by Leopold & Price (1956). Nevertheless, one can certainly connect auxin action with Co A acetylation since diphenylacetate inhibits both.

Zinc is probably necessary for auxin production (Skoog, 1948; Tsui, 1948) and is synergic with gibberellin in promoting growth of the dwarf bean (Dancer, 1959). Zinc, copper and fatty acids can all affect Co A metabolism because both cations can accelerate the incorporation of acetyl Co A into fatty acids by extracts of rat liver (Dituri *et al.* 1957). *A priori* it seems likely that zinc would interfere, by chelation with imidazole, with any acetylation involving the energy-rich intermediate *N*-acetylimidazole, discovered by Stadtman & White (1953), as well as the imidazole-catalysed synthesis of acetyl Co A reported by Jencks (1957). There are also good reasons to associate zinc with alterations of membrane permeability: in efficiency as an uncoupling agent of some mitochondria it is second only to cadmium (Jacobs *et al.* 1956); it markedly promotes the attachment of some amphipathic anions to albumin (Klotz & Loh Ming, 1954) and conceivably, through imidazole chelation, plays a role in the interaction of insulin and glucagon in modifying membrane permeability (see Krahl, 1957; Weitzel, Schneider & Fretzdorf, 1957).

The effect of calcium on permeability needs no comment. It may, of course, also affect acetylation (involved in fatty acid synthesis or oxidation) through its activation of lipases and phospholipases to release fatty acids, or lysolecithin and fatty acids, from their substrates.

In sum, pro-C agents can be associated with anionic amphipathy, relative insolubility in fats, inability to pass the blood brain barrier, increase of permeability with letting-in of extracellular ions, inhibition or weak activation of enzymes, uncoupling of oxidative phosphorylation, promotion of Co A activity, promotion of insulin action, promotion of plant growth and promotion of cardiac fibrillation. Pro-X agents, in contrast, can be associated with fat solubility, chelation, depression of nervous activity, stabilization of membranes and intracellular ionic environment, activation of enzymes, promotion of oxidative phosphorylation, antagonism of plant growth, antagonism of insulin and antagonism of cardiac fibrillation.

Some of these associations are only partial or even fragmentary; but when considered together they strongly suggest that a similar control or trigger mechanism will be found to underlie many cyclical or modulatable phenomena in widely different forms of life.

THE SITE OF ACTION OF ANTAGONISTIC IONS AND OTHER SUBSTANCES

It seems reasonable to exclude from first consideration any question of modulating substances acting as building blocks or 'anti-building blocks' (Lwoff, 1957). The problem thus reduces to one of understanding how as modifiers they may induce the reversible changes observed, i.e. to the central and recurrent problem of pharmacology. Ideally, one wants to know four things: (i) the chemical nature, function and position of the molecules at the primary site of attachment; (ii) the nature of the change effected; (iii) the way in which this change triggers off or switches enzyme activity; and (iv) the way in which altered enzyme activity leads to the end response. In general, of course, one has to consider the possibility of action at or on any molecule or ion whether this be a fuel, a building block or part of an end-product such as DNA, enzyme-forming system, co-enzyme or other carrier, semi-permeable membrane, enzyme protein, or enzyme activator or inhibitor. But of these only the last three appear of immediate relevance.

Enzyme protein. All reversible alterations of enzyme activity, other than those caused by changes in concentrations of reactants or their competitors, appear to stem from one of three kinds of change of the protein molecule: an altered degree or pattern of folding, an altered distribution of charges, or a reversible inactivation of an essential or influential part such as a thiol group by mercury, a ferric ion by carbon monoxide, zinc by 1:10 phenanthroline (Vallee, Williams & Hoch, 1959), or calcium by EDTA (Vallee *et al.* 1959).

The first is well exemplified by the increased folding with change in optical rotation accompanying the activation of chymotrypsinogen (Rupley, Dreyer & Neurath, 1955). It almost certainly underlies many temperature- and pressure-sensitive responses such as luminescence of *Achromobacter fischeri* (Johnson, 1957), the interaction of calcium, pH, temperature and pressure on myosin ATP-ase (Brown *et al.* 1958), and the unique temperature activation of glutamic dehydrogenase in a mutant of *Neurospora crassa* (Fincham, 1957). In all these cases, ions probably play an important role. They are well known to predispose to unfolding or hyperfolding (Linderström-Lang & Schellman, 1959) and the differential effect of ion species as stabilizers to heat is clearly revealed by the marked difference in response of enolase to temperature when activated by zinc instead of magnesium (Malmström, 1955).

Both anions and cations may affect the total charge and hence electrophoretic mobility of a protein. At a given pH the charge of myosin may vary with the ratio of cations as well as with the nature of the anions in the medium (Erdős, 1955); and the activity and mobility of both fumarase (Massey, 1953*a*) and amylase (Myrbäck, 1926) vary with species of anions present. In this respect the interaction of amylase with chloride and acetate is of particular interest because its activity, mobility, stability and solubility all change in parallel with change in the anion (Muus, 1953). There is reason, therefore, to suppose that many ion antagonisms, particularly, of course, those demonstrable *in vitro* with isolated enzymes, reflect competitive effects on the three-dimensional pattern of charged and uncharged atoms. Certainly a most strikingly similar dependence on ion balance can be traced through a number of phenomena: growth of the bacterial cell (Macleod & Snell, 1948), glycolysis of the whole bacterial cell (Tsuyuki & Macleod, 1951), glycolysis of bacterial extract (Clark & Macleod, 1954), fructokinase activity (Hers, 1952), myosin ATP-ase activity (Mommaerts & Seraidarian, 1947), contraction and relaxation of actomyosin gel (Kuschinsky & Turba, 1950), contraction and relaxation of a synthetic polyelectrolyte model (Kuhn & Hargitay, 1951), charge of myosin (Erdős, 1955), phototaxis of *Platymonas subcordiformis* (Halldal, 1957), protein synthesis by calf thymus nuclei (Allfrey, Mirsky & Osawa, 1957) and antigen synthesis in *Bordetella pertussis*. According to Kirkwood (1956), H^+ and other ions may influence both the Michaelis-Menten (K_m) and intrinsic rate (K_3) constants by changing the charge configuration on the enzyme protein; and it seems at least possible that all eleven phenomena have in common a polyampholyte whose electrical charge, physical state, enzyme activity (if any) and substrate specificity (if an enzyme) are all aspects of the same thing: and all equally liable to reversible modification by change in the ionic environment.

Semi-permeable membrane. If the microbial cell were only a bag of enzymes, there would be only one membrane to consider. In reality, however, it is almost certainly organized structurally as well as by 'specificity' in the sense used by Dixon (1949), and there must exist many opportunities of regulation or modification of function by the opening up or closing down of various pieces of biochemical machinery with more or less specific agents. With such complexity of structure, indeed, external agents or hormones would seem *a priori* as likely to react primarily with membranes as with enzymes.

Some recent experimental findings strongly indicate the possibility of interaction between ions at a subcellular membrane or particle. For example, Ca^{++} can promote and Mn^{++} inhibit both uncoupling of oxidative phosphorylation in (Lindberg & Ernster, 1954), and physical swelling (Lehninger, 1956) of, isolated mitochondria. These effects almost certainly result from a primary action on mitochondrial structure, because the isolated enzymes behave quite differently. Thus Ca^{++} and thyroxin have no action on the enzyme activity of broken mitochondria although they uncouple oxidative phosphorylation in intact mitochondria. Conversely, Mg^{++} is an uncoupler of broken mitochondria, yet an antagonist of the uncoupling action of thyroid, and an essential activator of intact mitochondria (see Lehninger, 1956). Other studies with mitochondria also point to their structure as the primary site of uncoupling action by 1:10-phenanthroline and other chelating agents (Christie, Judah & Rees, 1953) and of the antagonism by diphenhydramine ('Benadryl') and other antihistamines of calcium, thyroxine and hypotonic solutions (Judah, 1960). The contrary influence of Ca^{++} and Mg^{++} on the amount of acetylcholine liberated from pre-ganglionic nerve endings (Hutter & Kostial, 1954), or motor-nerve endings (Castillo & Engbaek, 1954), is similarly attributable to antagonistic actions on the permeability of either the presynaptic membrane or the microsomes in which the acetylcholine may be stored (Castillo & Katz, 1956).

Changes in permeability almost inevitably lead to ion fluxes, and efflux of potassium is a fairly constant feature of mitochondrial uncoupling (see Spector, 1953; Berger, 1957). At times, many or all of the subsequent effects of changed permeability may represent the action of such ion movements on enzyme activity. But a change in permeability could obviously almost as well lead to a change in enzyme activity through a change in the access or release of any factor influencing the action of enzymes in solution. In addition, however, the presence of a membrane introduces new possibilities for enzyme modification since changes in the membrane may secondarily affect the area occupied by each molecule and the orientation and mobility of its active groups. Changes of this kind may well underlie: (i) the increased catalase activity of yeast that follows exposure to heat, chloroform, anionic detergents or ultraviolet light which Kaplan (1956) calls 'revelation'; (ii) the failure of magnesium to activate the ATP-ase activity of free myosin when it clearly activates the ATP-ase activity of actomyosin fibril (Perry, 1951); and (iii) some changes in sensitivity to activators or inhibitors that occur during the course of purification, and are not attributable to removal of an inhibitor or activator as, for example, change in sensitivity of phosphorylase phosphokinase to Mg^{++} and Ca^{++} (Krebs & Fischer, 1956).

HYPOTHETICAL MECHANISMS OF ANTIGENIC MODULATION

To define the site of ion antagonism or primary action of other modulating agents would solve only part of the problem. We should still want to know how altered enzyme activity could determine the proportions in which three or more antigens were formed. In essence this second question is only one aspect of the general problem, recently discussed at a CIBA Symposium (Wolstenholme & O'Connor, 1959) of how metabolic processes are normally controlled or subject to modulation by drugs, hormones, etc. Five conceivable mechanisms are represented in Fig. 5. The first three are based on a change of enzyme rate, the others

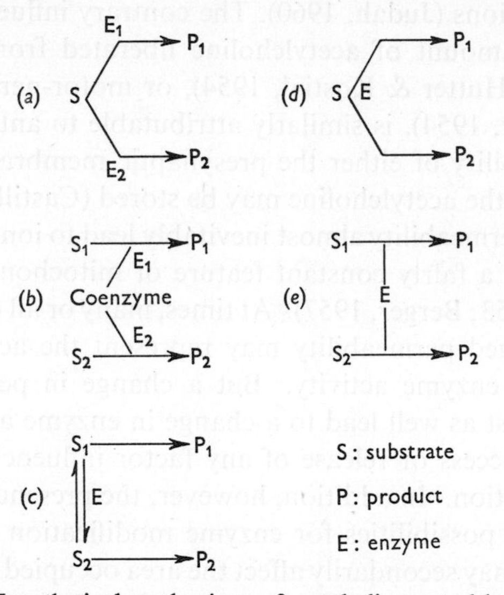

Fig. 5. Hypothetical mechanisms of metabolic control based on alterations of enzyme activity.

on a change of specificity. In scheme (*a*) two enzymes have the same substrate but are active under different conditions. This represents the way that, as Gale & Epps (1942) showed, pH may determine whether an amino acid is deaminated or decarboxylated. Scheme (*b*) illustrates the situation recently considered by Krebs (1959) in which two enzymes compete for the same co-enzyme, and (*c*) one in which the direction of the reaction is influenced by the environment: in the way salts may direct the catalysis by aspartase of *Escherichia coli* (Pereira & Soares, 1936), pH, anions and temperature the equilibrium constant of fumarase from pig's heart (Massey, 1953*a*, *b*), or anions the velocity constants of liver alcohol dehydrogenase (Theorell, Nygaard & Bonnichsen, 1955).

In (*d*) the environment determines which of two potential activities an enzyme shall show, in the way: (i) the redox potential governs the activities of the pyruvic oxidase of pigeon breast (Jagannathan & Schweet, 1952) and the α-ketoglutaric dehydrogenase of pig's heart (see Sanadi & Littlefield, 1953); (ii) pH markedly influences the ratio of hydrolysis to transamidation shown by cathepsin C (Fruton, 1954); and (iii) manganese ions and Co II determine whether the isocitric enzyme of pig's heart shall decarboxylate oxalosuccinate to α-oxoglutarate, or reduce it to D-*iso*citrate (Moyle, 1956).

Finally, (*e*) illustrates the effect of environment on substrate specificity. The liability of this to vary with change in environment has become well known and there are now over twelve instances of variation of specificity with pH (see, for example, Thayer & Horowitz, 1951; Olson & Anfinsen, 1953; Morton, 1955). Since highly purified or crystalline enzymes and synthetic, non-ionizable substrates were used in some of these studies there seems little doubt that pH change effectively alters the enzyme protein itself and it is of interest that with chymotrypsinogen, reversible transient changes of this kind (which might be called enzyme *modulations*) can be mimicked by oxidation or acetylation, i.e. by *modification* of the enzyme (Jansen, Curl & Balls, 1951). Change of substrate specificity has also been produced in proteinases by cysteine or ascorbic acid (Bergmann & Fruton, 1941), cholinesterase by inhibitors (Wilson & Bergmann, 1950; Myers, 1953), β-galactosidase by temperature (Kuby & Lardy, 1953) and in a number of enzymes by change in the species of activating ion: bacterial peptidases with Fe^{++}, Mg^{++}, Mn^{++} and Co^{++} (Maschmann, 1943); β-galactosidase with K^{+} and Na^{+} (Cohn & Monod, 1951); glutamine synthetase of pigeon liver with Co^{++} and Mg^{++} (Dénes, 1954); myosin ATP-ase with Ca^{++} and Mg^{++} (Blum, 1955) or with Ca^{++} and K^{+} (Greville, 1956); arylsulphatase with monovalent anions (Webb & Morrow, 1959) and leucine aminopeptidase with Co^{++} and Mn^{++} (Vescia, 1956). Although the last example has not been demonstrable after greater purification (Hill & Smith, 1957), it seems most improbable that all could ultimately prove to reflect the presence of two enzymes. The behaviour of glutamine synthetase is of particular interest because the change is one of stereospecificity. When activated by Mg^{++}, it will utilize either L(+) or D(−) glutamic acid whereas when activated by Co^{++}, its action is confined to the L(+) isomer (Dénes, 1954). A comparable mechanism might well be responsible for the influence, discovered by (Leonard *et al.* 1958), of Mn^{++} on the proportion of D- and L-glutamic acid polymerized by *Bacillus subtilis*.

In review it is obvious that no conclusion as to the mechanism of modulation in *Bordetella* is possible. Nevertheless, there is a little oblique evidence to support the view that all influences converge on an enzyme associated with CoA metabolism, whose rate of action or specificity is sensitive to temperature and ion balance. This would imply that all agents without direct access to this enzyme, i.e. all except temperature, many ions (including some charged chelates) and some un-ionized chelates, would act by hindering or promoting the access of agents capable of access, presumably either by chelation or by altering the permeability of a membrane.

SIMILARITIES AND DIFFERENCES BETWEEN *BORDETELLA* AND *PARAMECIUM*

No comparison of antigenic variation in *Paramecium* and *Bordetella* is possible without an appreciation of the confusing differences in the following list of equivalents:

Bordetella	*Paramecium*
Genus	*Species*
Species	*Variety*
Serotype	*Stock* or *subtype*
Strain	*Line*
Clone	*Clone*
Mode	*Serotype* or antigenic type
Modulation	*Transformation*

With this, however, the resemblances become striking and most of the differences comprehensible. In both kinds of organism: (i) growth is essential for change from one mode (*serotype*) to another; (ii) a principal repertoire of three equilibrium phenotypes may be evoked in a suitable medium by changing the temperature; (iii) serological overlap occurs between members of different species (*varieties*) and between members of different serotypes (*stocks*) of the same species (*variety*); (iv) antigens or mixtures of antigens not normally found at equilibrium may appear transiently during the change from one mode (*serotype*) to another; (v) when antigens of adjacent modes appear together on the surface of individual organisms, their proportions are negatively correlated, more of one being accompanied by less of the other. In both, also, the genotype determines: (i) the serological specificity of any antigen; (ii) the number of alternative phenotypes (modes or *serotypes*) that may be expressed; (iii) the ranges of environmental conditions over which the phenotypes are expressible or stable; and (iv) whether environmental conditions near the borderline between those characteristic of two modes (*sero-*

types) will lead to mutual exclusion or the simultaneous partial expression of both.

In contrast, the two organisms differ significantly in only three respects: (i) in practice the effects of recombination and hybridization can be studied only in the ciliates and the effects of mutation only in *Bordetella*; (ii) the antigenic differences in ciliates stem from serological variation of the same kind of substance in the same place, whereas in *Bordetella* they reflect differences between substances of markedly different chemical structure and function as well as differences of the first kind; (iii) mutual exclusion of antigens on the surface is a prominent feature in ciliates but never observed in *Bordetella*.

From the genetic analysis possible in *Paramecium*, Beale (1952) has shown that in heterozygotes each allele determines the synthesis of its corresponding antigen so that, under conditions when both alleles can be expressed, each individual will possess two antigens (of the same *serotype* but different specificities) at the same time. Similarly, Beale (1952) has clearly proven the existence of environmentally induced 'cytoplasmic states' which, once induced, will suppress or evoke, independently of the environment, the action of the macronuclear genes. Thus the sequence of events in ciliate transformation can be represented as follows: (→ = induces)

Environmental change → altered cytoplasmic state → altered pattern of gene suppression → altered channel of antigen synthesis → transformation to new *serotype*.

Because all alternative antigens in any ciliate are of the same chemical nature, it is feasible to consider (see Beale, 1954) that a transformation represents the addition of one or two different groups to an already elaborate precursor. The mutual exclusion (and negative correlation) would then result from a mutually exclusive competition for the precursor. In *Bordetella* and other organisms such a view is evidently untenable and the observed negative correlations must be attributed to another cause such as competititon for space on the surface.

Several important exceptions to the rule of mutual exclusion are now known. Antigens whose specificities are controlled by non-allelic genes have been found in apparently variable amounts (negatively-correlated) on the surface of stable *serotypes* of both *Paramecium aurelia* (Margolin, 1956) and *Tetrahymena pyriformis* (Margolin *et al.* 1959). With neither animal, however, is there yet sufficient evidence to decide whether this observed failure of mutual exclusion results from an oscillation between two exclusive syntheses or a mechanism of the kind which has to be postulated for the observed changes in other micro-organisms. Whether or not mutual exclusion is more than skin deep also remains to be determined. Antisera may contain antibodies of low titre to adjacent serotypes but their significance is debatable (see Sonneborn, 1947; Beale, 1954; Finger, 1957; Balbinder & Preer, 1959). In variety I of *Paramecium aurelia* the immobilizing antigens formally resemble both phage antigens and the flagellar antigens of diphasic salmonellae (Lederberg & Iino, 1956) in being either fully expressed or not detectable. But most other modulatable antigens vary in amount and sometimes appear in an inagglutinable, non-absorbing (minimal) form or in an inagglutinable absorbing (minor) form.

MINIMAL ANTIGENS

Apart from the X, O and H antigens of *Bordetella* species (see Table 2), records have been found of major↔minimal modulation of only three antigens: the C antigen of *Vibrio cholerae* Ogawa, which is major at 20° and minimal at 37° (Kauffmann, 1950), and the H and X antigens of *Salmonella typhimurium*, which are normally major but minimal either with phenol in the medium (White, 1925) or at low E_h (Happold, 1929) respectively.

In several instances, a minimal antigen has been revealed in a major form by rupture of the cell or other treatment removing an overlying antigen. Thus the nucleoprotein of pneumococci (Avery & Neill, 1925) and the A antigen of *Shigella shigae* (Shelubsky & Olitzki, 1948) are released in a major form by autolysis, the protoplast membrane of *Bacillus megaterium* (Vennes & Gerhardt, 1959) and of *Escherichia coli* B (Holme, Malmborg & Cota-Robles, 1960) by lysozyme, phage proteins by natural lysis (Barry, 1954), the nuclear membrane of hen red blood cells by saponin (Jossa & Beaumariage, 1955), antigen AF of sea-urchin egg by chymotrypsin (Tyler, 1946), antigen II of bull spermatozoa by ultrasonic rupture (Henle, Henle & Chambers, 1938), the O antigen of *E. coli* 'A' forms by heating to 100° (Knipschmildt, 1945), and the protein (PPD) of *Mycobacterium tuberculosis* (Meynell, 1954) and the cell wall polysaccharide of *B. megaterium* (Vennes & Gerhardt, 1959) by extraction.

Most minimal antigens, however, have been recognized by their appearance in a major form on a cell of another genotype. Thus, the R antigen of *Shigella shigae* appears major in the R form, minimal in the S form (Arkwright, 1920); the C antigen which is major in *Vibrio cholerae* Inaba appears minimal in the Ogawa strains (Kauffmann, 1950); the B antigen of *Corynebacterium insidiosum* is major in strain CIN53 and minimal in strain CIN (Rosenthal & Cox, 1953); the polysaccharide of certain Mycobacteria is major in *Mycobacterium tuberculosis* H37Rv and minimal in *M. phlei* (Meynell, 1954); the Z antigen of brucellae is major in the New Zealand strain and minimal in *Brucella abortus* and *B. melitensis* (Renoux & Mahaffey, 1955); the H antigen of some strains of *Bordetella bronchiseptica* is major in the C-mode of *bronchiseptica* and may be minimal in the C-mode of *B. parapertussis* (Lacey, unpublished); and the AF antigen of the sea-urchin is major in spermatozoa and minimal in ova (Tyler, 1946). In the same way, antigens that are normally in a major form on human red blood cells may appear minimal in exceptional genotypes or in the red cells of other animals: for example, antigen D may be minimal in some human *Du* cells (Argall, Ball & Trentleman, 1953) or in rhesus monkey cells (Murray, 1952), antigen B appears minimal in human *Bx* cells (Yokoyama, Stacey & Dunsford, 1957) and antigen A minimal in *Am* cells (Weiner *et al.* 1957). The last example is of particular interest since *Am* cells appear to be the expression of *A*, *B* genes in a person homozygous for the suppressor gene *y*.

In all these instances, the minimal behaviour is attributable solely to inaccessibility to antibody because of submersion. They almost certainly differ therefore from those in which minimal behaviour follows chemical maltreatment *in vitro* and hence presumably a molecular change making the determinant groups inaccessible to antibody. Changes of this kind have been produced in flagella of *Salmonella* (Arkwright, 1931) and *Bordetella* (Lacey, unpublished), and in bushy stunt virus with albumin (Bawden & Kleczkowski, 1941, 1942) by heating to 95–100°, and probably in flagella of enterobacteria by alcohol (Kauffmann, 1943).

MINOR ANTIGENS

Like minimal antigens, minor antigens may become agglutinable by modulation or by chemical modification of the whole organism; or they may appear in a major form on cells of another genotype or be derived *in vitro* by chemical means from major antigens.

Surface antigens of several Gram-negative bacteria show a minor↔ major modulation. For example, the O antigen of *Serratia marcescens* (Kirstein, 1904) or *Pasteurella pestis* (Schütze, 1932) may be minor at 37° and major at 20° or 28°; conversely, the O of a *Ballerup* species is minor at 18° and major at 37° (Nicolle & Jude, 1952*b*), the O of *Klebsiella pneumoniae* is minor on a rich medium and major on a poor one (Adler & Humphries, 1954); and the O of *Salmonella typhi* is minor when grown at 37° (McIntosh & Queen, 1914) and major on phenol agar (Felix & Pitt, 1934), in older liquid cultures (Craigie & Brandon, 1936) and at 18° or 45° (Jude & Nicolle, 1952*a*). In each of these cases the O is minor when covered by a capsule. In contrast, the Vi antigen of *S. typhi* may itself be minor on cells grown anaerobically with lactate as carbon source (Gladstone, 1937).

A hint that human red blood cell antigens may also be subject to modulation can be seen in the appearance of the antigen A in a minor form in a patient with 'hypoplastic anaemia' who was previously known to have normal A cells (Stratton, Renton & Hancock, 1958); for it seems most unlikely that this could have arisen from a somatic mutation.

The serological behaviour of minor antigens indicates that they are less submerged than minimal antigens; and it is not surprising that comparatively mild and non-disruptive treatments may often reveal them in a major form. Heating to 100° is of course well known as a means of removing 'O inagglutinability' and has proved effective with species of at least eight genera: *Shigella shigae* (Olitzki, Shelubsky & Koch, 1946), *Pasteurella pestis* (Schütze, 1944; Seal, 1951), *Salmonella typhi* (Felix, Bhatnager & Pitt, 1934), OL strains of *Escherichia coli* (Kauffmann, 1943), mucoid strains of *E. coli* (Henriksen, 1949), one strain of *Brucella abortus* (Wolff & Dinger, 1951), *Corynebacterium insidiosum* (Rosenthal & Cox, 1953) and *Bordetella* species in X-mode (Lacey, unpublished). Occasionally the overlying material responsible for the inagglutinability can be removed in solution: the capsule of *Pasteurella pestis* by solution in potassium thiocyanate (Amies, 1951), the X-antigen of *Bordetella* species in formamide (see page 352 above), and a polypeptide on pig red blood cells (to reveal a human A-like substance in major form) by digestion with trypsin (Saison, Goodwin & Coombs, 1955). The last example is, of course, strictly comparable with the routine use, in blood transfusion serology, of papain and trypsin to make ordinary human red blood cells agglutinable by 'incomplete antibodies'. Only one instance is known of a minor antigen appearing major on release from the cell: the group cell wall antigen of *C. diphtheriae* (Cummins, 1954).

Many antigens appear minor in one genotype or tissue and major in another. Bacterial mutants lacking the capsule of the wild-type usually possess the O antigen in a major form. Thus most of the O antigens already considered have a major form in a mutant or 'naked strain'. The Q antigen of *Salmonella typhimurium*, which is minor in 'smooth' forms, becomes major in mutant 'ρ' forms (White, 1925). Similarly, a β antigen appears minor on *Escherichia coli* $O_{111}B_4H_2$ and major on a paracolon strain '30' (Mushin, 1955). Among somatic antigens, however, the relationship of most interest is that of A and M antigens of brucellae discovered by Miles (1939). In this, it will be remembered, the A antigen appears major in *abortus* and minor in *melitensis*, whereas the M antigen shows the converse relationship.

Of great interest also is the occurrence of flagellar antigens in an apparently minor form in some strains of *Vibrio cholerae* (Balteanu, 1926) and *Bacillus polymyxa* (Davies, 1951). In view of the discovery of two independent antigens in preparations of flagella of one strain of *Proteus* X19H (Gard, Heller & Weibull, 1955), it seems possible that this reflects the presence of a second encapsulating antigen.

Several antigens of human red blood cells have been found in a minor form on red blood cells of other genotypes or on cells of other tissues—(see Race & Sanger, 1958). Thus, for example, the A antigen appears minor on human AO cells in persons homozygous for the suppressor gene *y* (Weiner *et al.* 1957), on appropriate human spermatozoa (Landsteiner & Levine, 1926) and on pig red blood cells (Saison *et al.* 1955). Similarly, in the rat, an antigen present in a minor form on rat Flexner-Jobling carcinoma cells, appears in a major form on rat red blood cells (Nungester & Halsema, 1953).

The conversion *in vitro* of antigens to an inagglutinable form has often been observed: the H antigen of *Salmonella typhi* by heating to 65° (Yokota, 1928) or by alcohol (Jenkins, 1946), the Vi antigen of *S. typhi* by heating to 60° (Felix *et al.* 1934) or by alcohol (Kauffmann, 1943), the B antigen of *Shigella shigae* by heating to 100° (Shelubsky & Olitzki, 1948), and the A antigen of human red blood cells by 20% formalin (Moskovitz & Carb, 1957).

In two instances, lysable cells with minor antigens have been found to have the interesting property of failing to lyse when exposed to complement and lytic serum against the minor antigen, i.e. to a system rapidly lytic for cells of the same kind possessing the antigen in a major form. This happens with Vi+O strains of *Salmonella typhi* in which the O antigen is minor (Nagington, 1956) and human red blood cells in which the A antigen is minor (Weiner *et al.* 1957). It might therefore well prove to be a trait of all lysable cells with minor antigens.

CONCLUSIONS

On a few occasions the synthesis of antigens normally present on the surface has been shown to be depressed below the level needed to elicit antibody formation. For example, rise in temperature of incubation may reduce production (i) of flagellar antigen in *Pasteurella pseudotuberculosis* (Preston & Maitland, 1952), *Bacillus megaterium* (Vennes & Gerhardt, 1959) and *Bordetella bronchiseptica* (Lacey, unpublished) and (ii) of immobilizing antigens (of low-temperature serotypes) in *Paramecium* (Beale, 1951) to this 'subminimal' or 'potential' form. In such circumstances, the activity of the corresponding gene is in effect completely suppressed even though, of course, synthesis of a large number of complete molecules of antigen may still be occurring. Antigens of differentiated tissues probably exist in this potential form in the early stages of embryonic development (see Ebert, 1954; Schectman, 1955; Woerdemann, 1955) and hence to some extent the process of differentiation can be viewed as a progressive antigenic modulation. As Tyler (1946) suggested: 'One might simply assume that all the genes are active in producing antigens but at different rates in different tissues so that certain of the antigens do not reach the surface of the cell. We may, then, conceive of differentiation involving differences in the rate of production of different antigens, at first in different regions of the uncleared egg, later in different blastomeres and embryonic tissues.' Nevertheless, antigenic change and differentiation clearly cannot be equated and, as Løvtrup (1959) points out: '...the presence of organ-specific antigens must be considered mainly as an indication of early differentiation, as a product of morphogenesis rather than a part of the morphogenetic mechanism itself'. At present the slime moulds appear to hold most promise as a means of investigating the relationships between differentiation and surface antigens. For Sussman (1958) has already brilliantly elucidated many of the factors concerned in or affecting their development, and Gregg (1956) has shown that antigenic changes accompany their cyclical changes of aggregation and differentiation.

The immediate practical importance in microbiology of non-genetic antigenic variation needs no emphasis. It extends to every situation in which surface antigens are themselves of everyday importance: in the identification of organisms, antibodies or allergies, the preparation of vaccines and antisera, and in taxonomy. In all these pursuits non-genetic variation and mutation have a complementary ability to hinder or help. In theory, at least, the occurrence of inducible antigens also provides opportunities for chemotherapy aimed at modification of virulence rather than stasis or death. Two alternative approaches of this kind can be conceived: (i) to induce an antigenic surface susceptible to the defence mechanisms of the host, for example, to sensitization by antibody and subsequent lysis or phagocytosis, even though this surface is toxic and contributes to invasiveness; or (ii) to induce a non-toxic surface even though the organism thereby becomes relatively insusceptible to the normal defence mechanisms. The first would contribute to immunity and elimination of the microbe, the second to suppression of disease with or without elimination of the infection. The action of diethylcarbamazine on microfilariae (see Bangham, 1955) may be an example of the first but not even a presumptive instance of the second is known.

For the microbe, antigenic plasticity will almost certainly have had considerable survival value. Many, if not most, pathogenic organisms have to survive in two or more environments and *a priori* it seems inevitable that surface properties most advantageous in one environment will, in general, be relatively disadvantageous in another. The advantage of a second antigenic form to *Bordetella bronchiseptica* is readily imagined. In a comparatively cold place, such as the anterior nares, the organism will grow in C-mode and be relatively immune to antibodies developed normally against the more toxic X-mode. Here, therefore, it will tend to persist and hence, and because of its motility in C-mode, will have an increased chance of dissemination. *In vitro* at 37°, mutants, which are motile and lack the ability to form X-antigen on the surface, tend to outgrow the wild-type (which are non-motile at 37°). Such mutants have not been isolated in the acute stages of infection and presumably therefore lack virulence. They might, however, still be as highly communicable; and it would be of interest to compare the communicability and virulence of such mutants, the wild-type and mutants that are non-flagellated in C-mode.

The sensitivity of *Bordetella* species to depressants of the central nervous system is not unique. Sanders & Nathan (1959) have drawn attention to the effects of antihistamines on the motility of *Ochromonas malhamensis* and *Tetrahymena pyriformis* and the possible value of these organisms as pharmacological models. Some strains of *Proteus vulgaris* are similarly sensitive to phenothiazine derivatives such as chlorpromazine and diethazine (Lacey, unpublished); and it would appear that micro-organisms might well prove of help in elucidating the mode of action of these drugs, in discovering new substances of interest among groups of known activity, or as indicators of activity in previously unexplored types of compound.

But whatever the outcome is in this direction there seems no reason to doubt that the following words will be as true in 1974 as they were in 1946 when Dubos wrote them:

'The extraordinary plasticity of bacteria, the ease with which they adapt themselves to the environment, either by reversible modification, or by hereditary variation, has not only determined their importance in the economy of nature; it also makes them ideal objects for the study of that organization and integration of independent characters which define and characterize life.'

REFERENCES

ADLER, L. T. & HUMPHRIES, J. C. (1954). The modifying effect of phenolized saline on the agglutination of a strain of *Klebsiella pneumoniae*. *J. Bact.* **67**, 126.

ALLFREY, V. G., MIRSKY, A. E. & OSAWA, S. (1957). Protein synthesis in isolated cell nuclei. *J. gen. Physiol.* **40**, 451.

AMIES, C. R. (1951). The envelope substance of *Pasteurella pestis*. *Brit. J. exp. Path.* **32**, 259.

ANDERSEN, E. K. (1952). Some observations made during experiments on mice inoculated with *H. pertussis*. *Acta path. microbiol. scand.* **31**, 546.

ANDERSEN, E. K. (1953). Serological studies on *H. pertussis*, *H. parapertussis*, and *H. bronchisepticus*. *Acta path. microbiol. scand.* **33**, 202.

ARGALL, C. I., BALL, J. M. & TRENTLEMAN, E. (1953). Presence of anti-D antibody in the serum of a D^u patient. *J. Lab. clin. Med.* **41**, 895.

ARKWRIGHT, J. A. (1920). Variation in bacteria in relation to agglutination by salts and by specific sera. *J. Path. Bact.* **23**, 358.

ARKWRIGHT, J. A. (1927). The importance of motility of bacteria in classification and diagnosis, with special reference to *B. pseudotuberculosis rodentium*. *Lancet*, i, 13.

ARKWRIGHT, J. A. (1931). Agglutination. In *A System of Bacteriology in Relation to Medicine*, vol. VI, p. 381. Med. Res. Council, London: H.M. Stationery Office.

AUDUS, L. J. (1959). *Plant Growth Substances*, 2nd ed. London: Leonard Hill (Books).

AVERY, O. T. & NEILL, J. M. (1925). The antigenic properties of solutions of pneumococcus. *J. exp. Med.* **42**, 355.

BALBINDER, E. & PREER, J. R. Jun. (1959). Gel diffusion studies on serotype and serotype transformation in *Paramecium*. *J. gen. Microbiol.* **21**, 156.

BALTEANU, I. (1926). The receptor structure of *Vibrio cholerae* (*V. comma*) with observations on variations in cholera and cholera-like organisms. *J. Path. Bact.* **29**, 251.

BANGHAM, D. R. (1955). The mode of action of diethylcarbamazine investigated with ^{14}C-labelled drug. *Brit. J. Pharmacol.* **10**, 406.

BARKER, H. A. & HASSID, W. Z. (1951). Degradation and synthesis of complex carbohydrates. In *Bacterial Physiology*, p. 548. Edited by C. H. Werkman and P. W. Wilson. New York: Academic Press.

BARRY, G. T. (1954). A study of the antigenicity of T_3 and T_4 coli-dysentery bacteriophages during the vegetative stage of development. *J. exp. Med.* **100**, 163.

BAWDEN, F. C. & KLECZKOWSKI, A. (1941). Some properties of complexes formed when antigens are treated in the presence of serologically unspecific proteins. *Brit. J. exp. Path.* **22**, 208.

BAWDEN, F. C. & KLECZKOWSKI, A. (1942). The antigenicity of non-precipitating complexes. *Brit. J. exp. Path.* **23**, 169.

BEALE, G. H. (1951). Antigen variation in *Paramecium aurelia*, variety 1. *Genetics*, **37**, 62.

BEALE, H. G. (1954). *The Genetics of* Paramecium aurelia. Cambridge University Press.

BEALE, G. H. (1958). The role of the cytoplasm in antigen determination in *Paramecium aurelia. Proc. roy. Soc.* B, **148**, 308.

BERGER, M. (1957). Studies on the distribution of potassium in the rat liver cell and the mechanism of potassium accumulation. *Biochim. biophys. Acta*, **23**, 504.

BERGMANN, M. & FRUTON, J. S. (1941). The specificity of proteinases. *Advanc. Enzymol.* **1**, 63.

BERNHEIMER, A. W. (1953). Synthesis of Type III pneumococcal polysaccharide by suspensions of resting cells. *J. exp. Med.* **97**, 591.

BLUM, J. J. (1955). The enzymatic interaction between myosin and nucleotides. *Arch. Biochem.* **55**, 486.

BORDET, J. & SLEESWYK (1910). Sérodiagnostic et variabilité des microbes suivant le milieu de culture. *Ann. Inst. Pasteur*, **24**, 476.

BORDET, P. & BEUMER, J. (1942). Corrélation entre la reproduction du bactériophage et la formation de substance inhibitrice par la bactérie sensible. *Schweiz. Z. Path.* **5**, 265.

BOVARNICK, M. (1942). The formation of extracellular D-(−)-glutamic acid polypeptide by *Bacillus subtilis. J. biol. Chem.* **145**, 415.

BOYER, P. D., LUM, F. G., BALLOU, G. A., LUCK, J. M. & RICE, R. G. (1946). The combination of fatty acids and related compounds with serum albumin. 1. Stabilization against heat denaturation. *J. biol. Chem.* **162**, 181.

BRANDENBURG, H. (1952). Beitrag zur Frage der Schleimwallbildung bei Paratyphus-B-Bacillen. *Z. Hyg. InfektKr.* **135**, 100.

BRANDES, S. (1956). Właściwości szczepów grupy *Alkalescens-dispar* wyhodowanych na terenie warszawy 1. Wpływ temperatury na wytwarzanie antygenu powierzchniowego. *Med. dośw. mikrob.* **8**, 307.

BRAUN, H. & LÖWENSTEIN, P. (1923). Über den *Bacillus inconstans*. Zugleich ein Beitrag zur Bedeutung der Züchtungstemperatur für die Entwicklung der Geisseln und des antigenen Apparates. *Zbl. Bakt.* (1. *Abt. Orig.*), **91**, 1.

BRAUN, H. & SCHAEFFER, H. (1919). Zur Biologie der Fleckfieber proteus bazillen. Ein Beitrag zur Frage der Wirkungsweise der Desinfektionsmittel und des Hungers auf Bakterien. *Z. Hyg. InfektKr.* **89**, 339.

BRAUN, W., KRAFT, M., MEAN, D. D. & GOODLOW, R. J. (1952). The effect of penicillin on genetic changes and temporary modifications in populations of Brucellae. *J. Bact.* **64**, 41.

BRODIE, B. B. (1956). Pathways of drug metabolism. *J. Pharm., Lond.* **8**, 1.

BROWN, D. E. S., GUTHE, K. F., LAWLER, H. C. & CARPENTER, M. P. (1958). The pressure, temperature and ion relations of myosin ATP-ase. *J. cell comp. Physiol.* **52**, 59.

BURN, J. H. (1960). The cause of fibrillation. *Brit. med. J.* i, 1379.

BURNET, F. M. & MCKIE, M. (1930). Balanced salt action as manifested in bacteriophage phenomena. *Aust. J. exp. Biol. med. Sci.* **7**, 183.

CAPONE, G. (1919). Ossevazione sull'agglutinabilità di alcuni microorganismi coltivati in terreni acidi ed alcalini. *Sperimentale*, **73**, 385.

CARMELIET, E. & LAQUET, L. (1956). L'influence de la température et des ions potassium et sodium sur la durée du potentiel d'action cardiaque en fonction de la fréquence. *Arch. int. physiol.* **64**, 513.

CASTILLO, J. DEL & ENGBAEK, L. (1954). The nature of the neuromuscular block produced by magnesium. *J. Physiol.* **124**, 370.

CASTILLO, J. DEL & KATZ, B. (1956). Biophysical aspects of neuro-muscular transmission. In *Progress in Biophysics and Biochemistry*, no. 6, p. 121. Edited by J. A. V. Butler. London: Pergamon Press.

CHENOWETH, M. B. (1956). Chelation as a mechanism of pharmacological action. *Pharmacol. Rev.* **8**, 57.
CHRISTIE, G. S., JUDAH, J. D. & REES, K. R. (1953). Cofactor and metal requirements of brain mitochondria. *Proc. roy. Soc.* B, **141**, 523.
CHU, H. P. (1952). Variation of *Bacillus anthracis* with special reference to the non-capsulated avirulent variant. *J. Hyg., Camb.* **50**, 433.
CLARK, J. A. & MACLEOD, R. A. (1954). Ion antagonism in glycolysis by a cell-free bacterial extract. *J. biol. Chem.* **211**, 531.
COHN, M. & MONOD, J. (1951). Purification et propriétés de la β-galactosidase (lactase) d'*Escherichia coli*. *Biochim. biophys. Acta*, **7**, 153.
CRAIGIE, J. (1931). Studies on the serological reactions of the flagella of *B. typhosus*. *J. Immunol.* **21**, 417.
CRAIGIE, J. & BRANDON, K. F. (1936). The identification of the V and W forms of *B. typhosus* and the occurrence of the V form in cases of typhoid fever and in carriers. *J. Path. Bact.* **43**, 249.
CRUICKSHANK, J. C. & FREEMAN, G. C. (1937). Immunising fractions isolated from *Haemophilus pertussis*. *Lancet*, ii, 567.
CUMMINS, C. S. (1954). Some observations on the nature of the antigens in the cell wall of *Corynebacterium diphtheriae*. *Brit. J. exp. Path.* **35**, 166.
CZERNIAWSKI, E., SEDLACZEK, L. & ZABŁOCKI, B. (1958). Properties of antigen O of *Salmonella typhi* in relation to culture media. *Bull. Acad. Polon. Sci.* **6**, 287.
DANCER, J. (1959). Synergistic effect of zinc and gibberellin. *Nature, Lond.* **183**, 901.
DAVIES, S. N. (1951). The serology of *Bacillus polymyxa*. *J. gen. Microbiol.* **5**, 807.
DAWSON, A. I. (1919). Bacterial variations induced by changes in the composition of culture media. *J. Bact.* **4**, 133.
DÉNES, G. (1954). Glutamine synthetase: its stereospecificity and changes induced by activating ions. *Biochim. biophys. Acta*, **15**, 296.
DIENES, L., WEINBERGER, H. J. & MADOFF, S. (1950). Serological reactions of L type cultures isolated from *Proteus*. *Proc. Soc. exp. Biol., N.Y.* **75**, 409.
DITURI, F., SHAW, W. N., WARMS, J. V. B. & GURIN, S. (1957). Lipogenesis in particle-free extracts of rat liver. 1. Substrates and cofactor requirements. *J. biol. Chem.* **226**, 407.
DIXON, M. (1949). *Multi-enzyme Systems*. Cambridge University Press.
DIXON, M. & WEBB, E. C. (1958). *Enzymes*. London: Longmans Green.
DOUDOROFF, M., HASSID, W. Z., PUTMAN, E. W., POTTER, A. L. & LEDERBERG, J. (1949). Direct utilization of maltose by *Escherichia coli*. *J. biol. Chem.* **179**, 921.
DOUGLASS, C. D., DILLAHA, C. J., DILLAHA, J. & KOUNTZ, S. L. (1957). Inhibition of biologic acetylation by 1-hydrazinonaphthalazine. *J. Lab. clin. Med.* **49**, 561.
DUBOS, R. J. (1946). *The Bacterial Cell*. Cambridge, Mass.: Harvard University Press.
DUGUID, J. P. & WILKINSON, J. F. (1954). Note on the influence of potassium deficiency upon the production of polysaccharide by *Aerobacter aerogenes*. *J. gen. Microbiol.* **11**, 71.
EBERT, J. D. (1954). Some aspects of protein biosynthesis in development. In *Aspects of Synthesis and Order in Growth*, p. 69. Edited by D. Rudnick. Princeton: Princeton University Press.
ELLIOTT, S. D. (1945). A proteolytic enzyme produced by group A streptococci with special reference to its effect on the type-specific M antigen. *J. exp. Med.* **81**, 573.
ELLIOTT, S. D. & DOLE, V. P. (1947). An inactive precursor of streptococcal proteinase. *J. exp. Med.* **85**, 305.

Erdős, T. (1955). The effect of ions on the charge of myosin (and of some other proteins). *Acta physiol. Acad. Sci. hung.* **7**, 1.

Fawcett, C. H. (1959). Plant-growth substances and the copper chelation theory of their mode of action. *Nature, Lond.* **184**, 796.

Felix, A. (1924). The qualitative receptor analysis in its application to typhoid fever. *J. Immunol.* **9**, 115.

Felix, A., Bhatnager, S. S. & Pitt, R. M. (1934). Observations on the properties of the Vi antigen of *B. typhosus*. *Brit. J. exp. Path.* **15**, 346.

Felix, A. & Pitt, R. M. (1934). Virulence of *B. typhosus* and resistance to O antibody. *J. Path. Bact.* **38**, 409.

Ferris, V. R. (1954). A note on the flagellation of *Phytophthora infestans* (mont) de Bary. *Science*, **120**, 71.

Fincham, J. R. S. (1957). A modified glutamic acid dehydrogenase as a result of gene mutation in *Neurospora crassa*. *Biochem. J.* **65**, 721.

Finger, I. (1957). Immunological studies of the immobilization antigens of *Paramecium aurelia* variety 2. *J. gen. Microbiol.* **16**, 350.

Fisher, S. (1948). The behaviour of *H. pertussis* in casein hydrolysate broth. *Aust. J. exp. Biol. med. Sci.* **26**, 299.

Fruton, J. S. (1954). The biosynthesis of protein and peptides. In *Aspects of Synthesis and Order in Growth*, p. 15. Edited by D. Rudnick. Princeton: Princeton University Press.

Gale, E. F. & Epps, H. M. R. (1942). The effect of the pH of the medium during growth on the enzymic activities of bacteria (*Escherichia coli* and *Micrococcus lysodeikticus*) and the biological significance of the changes produced. *Biochem. J.* **36**, 600.

Garattini, S., Morpurgo, C. & Passerini, N. (1958). A series of new compounds inhibiting the acetylation of choline *in vitro*. *Experientia*, **14**, 89.

Gard, S., Heller, L. & Weibull, C. (1955). Immunological studies on purified flagella from Proteus X19. *Acta path. microbiol. scand.* **36**, 30.

Gay, F. P. & Claypole, E. J. (1913). Agglutinability of blood and agar strains of the typhoid bacillus. Studies in typhoid immunization. II. *Arch. int. Med.* **12**, 621.

Genghof, D. S., Hehre, E. J. & Neill, J. M. (1946). Serological reactions of levans formed from sucrose and raffinose by certain bacilli. *Proc. Soc. exp. Biol., N.Y.* **61**, 339.

Gladstone, G. P. (1937). The antigenic composition and virulence of *Bact. typhosum* grown on a chemically defined medium. *Brit. J. exp. Path.* **18**, 67.

Granoff, A. & Hirst, G. K. (1954). Experimental production of combination forms of virus. IV. Mixed influenza A–Newcastle disease virus infections. *Proc. Soc. exp. Biol., N.Y.* **86**, 84.

Gregg, J. H. (1956). Serological investigations of cell adhesion in the slime molds: *Dictyostelium discoideum*, *Dictyostelium purpureum* and *Polysphondylium violaceum*. *J. gen. Physiol.* **39**, 813.

Greville, G. D. (1956). Effects of 2:4-dinitrophenol and other agents on the nucleosidetriphosphatase activities of L-myosin. *Biochim. biophys. Acta*, **20**, 441.

Gustafson, T. & Hörstadius, S. (1955). Vegetalization and animalization in the Sea Urchin egg induced by antimetabolites. *Exp. Cell Res.* (Suppl.), **3**, 170.

Hahn, F. E., Wisseman, C. L. Jr. & Hopps, H. E. (1954). Mode of action of chloramphenicol. II. Inhibition of bacterial D-polypeptide formation by an L-stereoisomer of chloramphenicol. *J. Bact.* **67**, 674.

Halldal, Per (1957). Importance of calcium and magnesium ions in phototaxis of motile green algae. *Nature, Lond.* **179**, 215.

HAPPOLD, F. C. (1929). The effect of cultural variation on the antigenic development of *B. aertrycke* Mutton. *Brit. J. exp. Path.* **10**, 263.
HEHRE, E. J. (1949). Synthesis of a polysaccharide of the starch-glycogen class from sucrose by a cell-free, bacterial enzyme system (amylosucrase). *J. biol. Chem.* **177**, 267.
HEHRE, E. J., CARLSON, A. S. & NEILL, J. M. (1947). Production of starch-like material from glucose-1-phosphate by diphtheria bacilli. *Science*, **106**, 523.
HEHRE, E. J. & HAMILTON, D. M. (1946). Bacterial synthesis of an amylopectin-like polysaccharide from sucrose. *J. biol. Chem.* **166**, 777.
HEHRE, E. J. & HAMILTON, D. M. (1949). Bacterial conversion of dextrin into a polysaccharide with the serological properties of dextran. *Proc. Soc. exp. Biol., N.Y.* **71**, 336.
HENLE, W., HENLE, H. & CHAMBERS, L. A. (1938). Studies on the antigenic structure of some mammalian spermatozoa. *J. exp. Med.* **68**, 335.
HENRIKSEN, S. D. (1949). Serological reactivity of mucoid strains of *Escherichia coli. J. Immunol.* **62**, 271.
HERS, H. G. (1952). Role du magnésium et du potassium dans la réaction fructokinasique. *Biochim. biophys. Acta*, **8**, 424.
HESLOP-HARRISON, J. (1957). The experimental modification of sex expression in flowering plants. *Biol. Rev.* **32**, 38.
HESTRIN, S. & AVINERI-SHAPIRO, S. (1944). The mechanism of polysaccharide production from sucrose. *Biochem. J.* **38**, 2.
HICKLIN, J. A. (1960). Metabolic inhibitors and membrane permeability. *Biochem. J.* **74**, 40P.
HILL, E. O. & WEAVER, R. H. (1955). The apparent antigenic structure of *Clostridium parabotulinum* as affected by the culture medium. *Bact. Proc., Baltimore*, p. 88.
HILL, R. L. & SMITH, E. L. (1957). Leucine aminopeptidase. VI. Inhibition by alcohols and other compounds. *J. biol. Chem.* **224**, 209.
HIRST, G. K. & GOTLIEB, T. (1953). The experimental production of combination forms of virus. I. Occurrence of combination forms after simultaneous inoculation of the allantoic sac with two distinct strains of influenza virus. *J. exp. Med.* **98**, 41.
HOLME, T., MALMBORG, A.-S. & COTA-ROBLES, E. (1960). Antigens of spheroplast membrane preparations from *Escherichia coli* B. *Nature, Lond.* **185**, 57.
HOOGERHEIDE, J. C. (1939). Studies on capsule formation. I. The conditions under which *Klebsiella pneumoniae* (Friedländer's bacterium) forms capsules. *J. Bact.* **38**, 367.
HOOGERHEIDE, J. C. (1940). Studies on capsule formation. II. The influence of electrolytes on capsule formation by *Klebsiella pneumoniae. J. Bact.* **39**, 649.
HORGAN, E. S. (1936). Notes on the Vi antigen of *Bacillus typhosus. J. Hyg., Camb.* **36**, 368.
HOSKINS, J. M. (1958). Host-controlled variation in animal viruses. *Symp. Soc. gen. Microbiol.* **9**, 122.
HUGHES, D. E. & WILLIAMSON, D. H. (1952). Some properties of the glutaminase of *Clostridium welchii. Biochem. J.* **51**, 45.
HUTTER, O. F. & KOSTIAL, K. (1954). The effect of magnesium and calcium ions on the release of acetylcholine. *J. Physiol.* **124**, 234.
IVÁNOVICS, G. (1937). Unter welchen Bedingungen werden bei der Nährbodenzüchtung der Milzbrandbazillen Kapseln gebildet? *Zbl. Bakt.* (1. *Abt. Orig.*), **138**, 449.
JACOBS, E. E., JACOB, M., SANADI, D. R. & BRADLEY, L. B. (1956). Uncoupling of oxidative phosphorylation by cadmium ion. *J. biol. Chem.* **223**, 147.

JAGANNATHAN, V. & SCHWEET, R. S. (1952). Pyruvic oxidase of pigeon breast muscle. I. Purification and properties of the enzyme. *J. biol. Chem.* **196**, 551.

JANSEN, E. F., CURL, A. L. & BALLS, A. K. (1951). A crystalline, active oxidation product of α chymotrypsin. *J. biol. Chem.* **189**, 671.

JENCKS, W. P. (1957). Non-enzymic reactions of acyl adenylate and imidazole. *Biochim. biophys. Acta*, **24**, 227.

JENKINS, C. E. (1946). The release of antibody by sensitized antigens. *Brit. J. exp. Path.* **27**, 111.

JOHNSON, F. H. (1957). The action of pressure and temperature. *Symp. Soc. gen. Microbiol.* **7**, 134.

JORDAN, E. O., CALDWELL, M. E. & REITER, D. (1934). Bacterial motility. *J. Bact.* **27**, 165.

JOSSA, P. & BEAUMARIAGE, M. L. (1955). Différenciation antigénique entre le noyau et le cytoplasme de l'hématie de Poule. *C.R. Soc. Biol., Paris*, **149**, 1515.

JUDAH, J. D. (1960). Antihistamines and mitochondrial swelling. *Exp. Cell Res.* **19**, 404.

JUDE, A. & NICOLLE, P. (1952*a*). Persistence, à l'état potentiel, de la capacité d'elaborer l'antigène Vi chez le bacille typhique cultivé en série à basse température. *C.R. Acad. Sci., Paris*, **234**, 1718.

JUDE, A. & NICOLLE, P. (1952*b*). Conditions thermiques différentes pour la production de l'antigène Vi par quelques entérobactériacées. *C.R. Acad. Sci., Paris*, **234**, 2028.

JUDE, A., SERVANT, J., NICOLLE, P. & SERVANT, P. (1952). Relation inverse entre les variations des teneurs en antigènes O et Vi de quelques entérobactériacées suivant la température d'incubation. *C.R. Acad. Sci., Paris*, **235**, 1443.

KAPLAN, J. G. (1956). The alteration of intracellular enzymes. 1. Yeast catalase and the Euler effect. *Exp. Cell Res.* **8**, 305.

KATO, E. (1954). The influence of the hydrogen ion concentration of culture media upon the development of *enx* factors of *Salmonella abortus-equi*. *Jap. J. vet. Res.* **2**, 189.

KAUFFMANN, F. (1935). Über einen neuen serologischen Formenwechsel der Typhusbacillen. *Z. Hyg. InfektKr.* **116**, 617.

KAUFFMANN, F. (1943). Über neue thermolabile Körperantigene der Coli-bakterien. *Acta path. microbiol. scand.* **20**, 21.

KAUFFMANN, F. (1950). On the serology of the *Vibrio cholerae*. *Acta path. microbiol. scand.* **27**, 283.

KIRKWOOD, J. G. (1956). The influence of fluctuations in protein charge and charge configurations on the rates of enzymatic reactions. In *The Physical Chemistry of Enzymes*. Discussions of the Faraday Society, no. 20. Aberdeen: Aberdeen University Press.

KIRSTEIN, F. (1904). Über Beeinflussung der Agglutinirbarkeit von Bakterien, insbesondere von Typhus bacillen. *Z. Hyg. InfektKr.* **46**, 229.

KLOTZ, I. M. & LOH MING, W. C. (1954). Mediation by metals of the binding of small molecules by proteins. *J. Amer. chem. Soc.* **76**, 805.

KNIPSCHILDT, H. E. (1945). Demonstration of capsular antigens in the colon group. *Acta path. microbiol. scand.* **22**, 44.

KNOLL, K. H. (1953). Beiträger zur Kapselsubstanz von *Klebsiella pneumoniae* (*Bacterium pneumoniae* Friedländer). I. Kulturelle physiologische und serologische Leistungen der Friedländerbakterien. *Zbl. Bakt.* (1. *Abt. Orig.*), **159**, 394.

KOEPSELL, H. J., TSUCHIYA, H. M., HELLMAN, N. N., KAZENKO, A., HOFFMAN, C. A., SHARPE, E. S. & JACKSON, R. W. (1953). Enzymatic synthesis of dextran. Acceptor specificity and chain initiation. *J. biol. Chem.* **200**, 793.

KOZIŃSKI, A. W., OPARA, Z. & KRAFT, Z. (1957). Effect of antimetabolites on antigen Vi synthesis and variability in *S. typhi*. *Nature, Lond.* **179**, 201.

KRAHL, M. E. (1957). Speculations on the action of insulin, with a note on other hypoglycemic agents. *Perspectives in Biol. and Med.* **1**, 69.

KREBS, E. G. & FISCHER, E. H. (1956). The phosphorylase b-to-a converting enzyme of rabbit skeletal muscle. *Biochim. biophys. Acta*, **20**, 150.

KREBS, H. (1959). Rate-limiting factors in cell respiration. In *Regulation of Cell Metabolism*, p. 1. Edited by G. E. W. Wolstenholme and C. M. O'Connor. London: Churchill.

KRUMWIEDE, C., MISHULOW, L. & OLDENBUSCH, C. (1923). The existence of more than one immunologic type of *B. pertussis*. *J. infect. Dis.* **32**, 22.

KUBY, S. A. & LARDY, H. A. (1953). Purification and kinetics of β-D-galactosidase from *Escherichia coli*, strain K-12. *J. Amer. chem. Soc.* **75**, 890.

KUHN, W. & HARGITAY, B. (1951). Muskelähnliche Kontraktion und Dehnung von Netzwerken polyvalenter Fadenmolekülionen. *Experientia*, **7**, 1.

KUSCHINSKY, G. & TURBA, F. (1950). Über den Chemismus von Zustandsänderungen des Aktomyosins. *Experientia*, **6**, 103.

LACEY, B. W. (1953*a*). Three dimensional patterns of antigenic modulation of *Haemophilus pertussis, H. parapertussis* and *H. bronchisepticus*. *J. gen. Microbiol.* **8**, iii.

LACEY, B. W. (1953*b*). The influence of growth conditions on the antigenic structure of *Haemophilus pertussis, parapertussis* and *bronchisepticus*. *Atti del VI Congresso Internazionale di Microbiologia*, 2, sez. vi–vii, 331.

LACEY, B. W. (1954). Variation of the antigenic structure of certain bacteria with temperature of incubation and ionic composition of the medium. *Biochem. J.* **56**, xiv.

LACEY, B. W. (1960). Antigenic modulation of *Bordetella pertussis*. *J. Hyg., Camb.* **58**, 57.

LALLIER, R. (1959*a*). Research on the animalization of the sea-urchin egg by zinc ions. *J. embryol. exp. Morph.* **7**, 540.

LALLIER, R. (1959*b*). Les effets de la phénazone sur la détermination embryonnaire de l'œuf de l'Oursin, *Paracentrotus lividus*. *C.R. Acad. Sci., Paris*, **248**, 1416.

LAMBKIN, S. & DESVIGNES, A. (1952). Influence de la streptomycine et de la pénicilline sur la phagocytose des bactéries. *C.R. Soc. biol., Paris*, **146**, 1923.

LANDSTEINER, K. & LEVINE, P. (1926). Group specific substances in spermatozoa. *J. Immunol.* **12**, 415.

LANNI, F. & LANNI, Y. T. (1953). Antigenic structure of bacteriophage. *Cold Spr. Harb. Symp. quant. Biol.* **18**, 159.

LANSBURY, J. & ROGERS, F. B. (1955). The hydralazine syndrome. *Bull. Rheum. Dis.* **5**, 85.

LAPORTE, R., HARDRE DE LOOZE, L. & SILLARD, R. (1955). Contribution à l'étude du complément. I. Inactivation spontanée dans les conditions de l'hémolyse. Action protectrice des ions calcium. *Ann. Inst. Pasteur*, **89**, 16.

LEDERBERG, J. & INO, T. (1956). Phase variation in *Salmonella*. *Genetics*, **41**, 743.

LEHNINGER, A. L. (1956). Physiology of mitochondria. In *Enzymes: Units of biological structure and function*, p. 127. Edited by O. H. Gaebler. New York: Academic Press.

LEONARD, C. G., HOUSEWRIGHT, R. D. & THORNE, C. B. (1958). Effects of some metallic ions on glutamyl polypeptide synthesis by *Bacillus subtilis*. *J. Bact.* **76**, 499.

LEOPOLD, A. C. & GUERNSEY, F. S. (1953). A theory of auxin action involving coenzyme A. *Proc. nat. Acad. Sci., Wash.* **39**, 1105.

LEOPOLD, A. C. & PRICE, C. A. (1956). The influence of growth substances upon sulphydryl compounds. In *The Chemistry and Mode of Action of Plant Growth Substances*, p. 271. Edited by R. L. Wain and F. Wightman. London: Butterworth.

LEWIN, R. A. (1953). Studies on the flagella of algae. II. Formation of flagella by *Chlamydomonas* in light and darkness. *Ann. N.Y. Acad. Sci.* **56**, 1091.

LIEB, M., WEIGLE, J. J. & KELLENBERGER, E. (1955). A study of hybrids between two strains of *Escherichia coli. J. Bact.* **69**, 468.

LINDBERG, O. & ERNSTER, L. (1954). Manganese, a co-factor of oxidative phosphorylation. *Nature, Lond.* **173**, 1038.

LINDBERG, R. B. (1951). The antigenic structure of *H. capsulatum*, particularly the yeast phase. *Microfilm Abstr.* **11**, 220.

LINDERSTRÖM-LANG, K. V. & SCHELLMAN, J. A. (1959). Protein structure and enzyme activity. In *The Enzymes*, p. 443. Edited by P. D. Boyer, H. Lardy and K. Myrbäck. Vol. I, 2nd ed. New York: Academic Press.

LINTON, R. W., SHRIVASTAVA, D. L. & SEAL, S. C. (1937). Studies on the specific polysaccharides of the vibrios. Part 1. The effect of the growth medium. *Indian J. med. Res.* **25**, 569.

LOMINSKI, I. & LENDRUM, A. C. (1942). The effect of surface-active agents on *B. proteus. J. Path. Bact.* **54**, 421.

LØVTRUP, S. (1959). Biochemical indices of embryonic differentiation. In *Biochemistry of Morphogenesis*. Edited by W. J. Nickerson. London: Pergamon Press.

LUNDBERG, A. & LAGET, P. (1949). L'influence du rapport calcium-potassium sur la thermosensibilité de la réponse propagée, le potentiel de membrane, et l'activité rythmique spontanée des racines rachidiennes de mammière. *Arch. Sci. Physiol.* **3**, 193.

LURIA, S. E. (1953). Host-induced modifications of viruses. *Cold Spr. Harb. Symp. quant. Biol.* **18**, 237.

LWOFF, A. (1957). Nutrition and metabolism in fields collateral to tissue culture. *J. nat. Cancer Inst.* **19**, 511.

MAAS, W. K. (1952). Pantothenate studies. III. Description of the extracted pantothenate-synthesizing enzyme of *Escherichia coli. J. biol. Chem.* **198**, 23.

MCILWAIN, H. (1957). *Chemotherapy and the Central Nervous System.* London: Churchill.

MCINTIRE, F. C., PETERSON, W. H. & RIKER, A. J. (1942). A polysaccharide produced by the crown gall organism. *J. biol. Chem.* **143**, 491.

MCINTOSH, J. & MCQUEEN, J. M. (1914). The immunity reaction of an inagglutinable strain of *B. typhosus. J. Hyg., Camb.* **13**, 409.

MACLEOD, R. A. & SNELL, E. E. (1948). The effect of related ions on the potassium requirement of lactic acid bacteria. *J. biol. Chem.* **176**, 39.

MALMSTRÖM, BO. G. (1955). The temperature dependence of the enolase reaction with different activating metal ions. *Biochim. biophys. Acta*, **18**, 285.

MANN, P. J. G., TENNENBAUM, M. & QUASTEL, J. H. (1939). Acetylcholine metabolism in the central nervous system. The effects of potassium and other cations on acetylcholine liberation. *Biochem. J.* **33**, 822.

MARGOLIN, P. (1956). An exception to mutual exclusion of the ciliary antigens in *Paramecium aurelia. Genetics*, **41**, 685.

MARGOLIN, P., LOEFER, J. B. & OWEN, R. D. (1959). Immobilizing antigens of *Tetrahymena pyriformis. J. Protozool.* **6**, 207.

DE MARS, R. I., LURIA, S. E., FISHER, H. & LEVINTHAL, C. (1953). The production of incomplete bacteriophage particles by the action of proflavine and the properties of the incomplete particles. *Ann. Inst. Pasteur*, **84**, 113.

MASCHMANN, E. (1943). Bakterien-Proteasen. *Ergebn. Enzymforsch.* **9**, 155.

MASRY, F. L. G. (1952). Production, extraction and purification of the haemagglutinin of *Haemophilus pertussis*. *J. gen. Microbiol.* **7**, 201.

MASSEY, V. (1953*a*). Studies on fumarase. 2. The effects of inorganic anions on fumarase activity. *Biochem. J.* **53**, 67.

MASSEY, V. (1953*b*). Studies on fumarase. 3. The effect of temperature. *Biochem. J.* **53**, 72.

MEYNELL, G. G. (1954). The antigenic structure of *Mycobacterium tuberculosis*, var. *hominis*. *J. Path. Bact.* **67**, 137.

MILES, A. A. (1939). The antigenic surface of smooth *Brucella abortus* and *melitensis*. *Brit. J. exp. Path.* **20**, 63.

MILTON, A. S. (1959). The relation between temperature and changes in ion concentration on ventricular fibrillation induced electrically. *Brit. J. Pharmacol.* **14**, 183.

MIRONESCO, T. G. (1899). Ueber einer besondere Art der Beeinflussung von Mikroorganismen durch die Temperatur. *Hyg. Rdsch.* **9**, 961.

MOMMAERTS, W. F. H. M. & SERAIDARIAN, K. (1947). A study of the adenosine triphosphatase activity of myosin and actomyosin. *J. gen. Physiol.* **30**, 401.

MORGAN, H. R. & BECKWITH, T. D. (1939). Mucoid dissociation in the colon-typhoid-salmonella group. *J. infect. Dis.* **65**, 113.

MORGAN, W. T. J. (1937). Studies in Immuno-chemistry. II. The isolation and properties of a specific antigenic substance from *B. dysenteriae* (Shiga). *Biochem. J.* **31**, 2003.

MORTON, R. K. (1955). The substrate specificity and inhibition of alkaline phosphatases of cow's milk and calf intestinal mucosa. *Biochem. J.* **61**, 232.

MOSKOVITZ, M. & CARB, S. (1957). Surface alteration and the agglutinability of red cells. *Nature, Lond.* **180**, 1049.

MOYLE, J. (1956). Some properties of purified *iso*citric enzyme. *Biochem. J.* **63**, 552.

MURRAY, J. (1952). Rh antigens of human and monkey blood. *J. Immunol.* **68**, 513.

MURRELL, W. G. (1955). *The Bacterial Endospore*. Thomas Lawrance Pawlett Lectures, University of Sydney.

MUSHIN, R. (1955). A study of antigenic structure of *Escherichia coli* $O_{111}B_4H_2$ possessing β antigen. *J. Hyg., Camb.* **53**, 297.

MUUS, J. (1953). Studies on salivary amylase with special reference to the interaction with chloride ions. *C.R. Lab. Carlsberg* (Sér. chim.), **28**, 317.

MYERS, D. K. (1953). Studies on cholinesterase. 9. Species variation in the specificity pattern of the pseudocholinesterases. *Biochem. J.* **55**, 67.

MYRBÄCK, K. (1926). Über Verbindungen einiger Enzyme mit aktivierenden Stoffen. II. *Z. physiol. Chem.* **159**, 1.

NACHMANSOHN, D. & JOHN, H. M. (1945). Studies on choline acetylase. I. Effect of amino acids on the dialysed enzyme. Inhibition by α-keto acids. *J. biol. Chem.* **158**, 157.

NAGINGTON, J. (1956). The sensitivity of *Salmonella typhi* to the bactericidal action of antibody. *Brit. J. exp. Path.* **37**, 397.

NEILL, J. M., SUGG, J. Y., HEHRE, E. J. & JAFFE, E. (1941). Influence of sucrose upon production of serologically reactive material by certain streptococci. *Proc. Soc. exp. Biol., N.Y.* **47**, 339.

NICOLLE, E. & TRENEL, M. (1902). Recherche sur le phénomène de l'agglutination variabilité de l'aptitude agglutinative et de la fonction agglutinogène. Leurs relations entre elles; leurs rapports avec la mobilité des microbes. *Ann. Inst. Pasteur*, **16**, 562.

NICOLLE, P. & JUDE, A. (1952*a*). Influence de la température d'incubation sur les propriétés Vi de la souche *Ballerup*. *C.R. Acad. Sci., Paris*, **234**, 1922.
NICOLLE, P. & JUDE, A. (1952*b*). Estimation quantitative de l'antigène Vi élaboré par quelques Entérobactériacées cultivées à différentes températures. *C.R. Acad. Sci., Paris*, **234**, 2313.
NICOLLE, P., JUDE, A. & DIVERNEAU, G. (1953). Antigènes entravant l'action de certains bactériophages. *Ann. Inst. Pasteur*, **84**, 27.
NORTHROP, J. H., KUNITZ, M. & HERRIOTT, R. M. (1948). *Crystalline Enzymes*, 2nd ed. New York: Columbia University Press.
NOVICK, A. & SZILARD, L. (1951). Virus strains of identical phenotype but different genotype. *Science*, **113**, 34.
NUNGESTER, W. J. (1929). Dissociation of *B. anthracis*. *J. infect. Dis.* **44**, 73.
NUNGESTER, W. J. & HALSEMA, G. VAN (1953). Reaction of certain phytoagglutinins with Flexner-Jobling carcinoma cells of the rat. *Proc. Soc. exp. Biol., N.Y.* **83**, 863.
OLITZKI, L., SHELUBSKY, M. & KOCH, P. K. (1946). A thermolabile substance of *Shigella dysenteriae* Shiga. *J. Hyg., Camb.* **44**, 271.
OLSON, J. A. & ANFINSEN, C. B. (1953). Kinetic and equilibrium studies on crystalline L-glutamic acid dehydrogenase. *J. biol. Chem.* **202**, 841.
ØRSKOV, F. (1956). Studies on *E. coli* K antigens. 1. On the occurrence of B antigens. *Acta path. microbiol. scand.* **39**, 147.
PEREIRA, F. B. & SOARES, M. (1936). Actions de plusieurs sels sur le système de l'aspartase. *C.R. Soc. Biol., Paris*, **121**, 255.
PERRY, S. V. (1951). The adenosine triphosphatase activity of myofibrils isolated from skeletal muscle. *Biochem. J.* **48**, 257.
PILLET, J., MERCIER, P. & ORTA, B. (1951). Influence du milieu de culture sur la production des agglutinogènes staphylococciques. *Ann. Inst. Pasteur*, **81**, 224.
POVITZKY, O. R. & WORTH, E. (1916). Agglutination in pertussis. Its characteristics and its comparative value in clinical diagnosis, and in determination of genus and species. *Arch. int. Med.* **17**, 279.
PRESTON, N. W. & MAITLAND, H. B. (1952). The influence of temperature on the motility of *Pasteurella pseudotuberculosis*. *J. gen. Microbiol.* **7**, 117.
PRIESTLEY, F. W. (1936). Some properties of the capsule of *Pasteurella septica*. *Brit. J. exp. Path.* **17**, 374.
QUADLING, C. (1958). The unilinear transmission of motility and its material basis in *Salmonella*. *J. gen. Microbiol.* **18**, 227.
RACE, R. R. & SANGER, R. (1958). *Blood Groups in Man*, 3rd ed. Oxford: Blackwell.
RAVEN, C. P. & ERKEL, VAN G. A. (1955). The influence of calcium on the effect of a heat shock in *Limnaea stagnalis*. *Exp. Cell Res.* (Suppl.), **3**, p. 294.
REID, J. (1958). Dinitrophenol and diabetes mellitus. *Brit. med. J.* ii, 724.
RENOUX, G. & MAHAFFEY, L. W. (1955). Sur l'existence probable de nouveaux antigènes des *Brucella*, avec un nouveau schéma proposé pour représenter la répartition de antigènes. *Ann. Inst. Pasteur*, **88**, 528.
RINGER, S. (1883). A further contribution regarding the influence of the different constituents of the blood on the contraction of the heart. *J. Physiol.* **4**, 29.
ROBERTSON, O. H. (1951). Factors influencing the state of dispersion of the dermal melanophores in rainbow trout. *Physiol. Zoöl.* **24**, 309.
ROSENTHAL, S. A. & COX, C. D. (1953). The somatic antigens of *Corynebacterium michiganense* and *Corynebacterium insidiosum*. *J. Bact.* **65**, 532.
RUPLEY, J. A., DREYER, W. J. & NEURATH, H. (1955). Structural changes in the activation of chymotrypsinogen. *Biochim. biophys. Acta*, **18**, 162.
SANADI, D. R. & LITTLEFIELD, J. W. (1953). Studies on α-ketoglutaric oxidase. III. Role of coenzyme A and diphosphopyridine nucleotide. *J. biol. Chem.* **201**, 103.

SAISON, R., GOODWIN, R. F. W. & COOMBS, R. R. A. (1955). The blood groups of the pig. *J. comp. Path.* **65**, 71.

SANDERS, M. & NATHAN, H. A. (1959). Protozoa as pharmacological tools: the antihistamines. *J. gen. Microbiol.* **21**, 264.

SCHECHTMAN, A. M. (1955). Ontogeny of the blood and related antigens and their significance for the theory of differentiation. In *Biological Specificity and Growth*, p. 3. Edited by E. G. Butler. Princeton: Princeton University Press.

SCHIFF, F. & NATHORFF, E. (1920). Untersuchungen zur Serologie des Fleckfiebers. *Z. ImmunForsch. I Orig.* **30**, 482.

SCHNEIDER, W. C. (1959). Mitochondrial metabolism. *Advanc. Enzymol.* **21**, 1.

SCHUBERT, O. (1952). Entstehung unbeweglicher Phasen bei begeisselten Bakterien. Einschlägige Beobachtungen bei einem neu isolierten Stamm. *Zbl. Bakt.* (1. *Abt. Orig.*), **158**, 540.

SCHUBERT, O. & SCHUBERT, R. (1953). Enstehung unbeweglicher Phasen bei begeisselten Bakterien. Beobachtung bei psychrophilen Bakterien. *Z. Hyg. InfektKr.* **136**, 639.

SCHÜTZE, H. (1932). Studies in *B. pestis* antigens. 1. The antigens and immunity reactions of *B. pestis*. *Brit. J. exp. Path.* **13**, 284.

SCHÜTZE, H. (1944). The agglutinability of *Bact. shigae*. *J. Path. Bact.* **56**, 250.

SEAL, S. C. (1951). Studies in *Pasteurella pestis*, *Pasteurella pseudotuberculosis* and their variants. Part II. Serology—agglutination, cross agglutination, absorption and cross absorption of agglutinins. *Ann. Biochem.* **11**, 143.

SHELUBSKY, M. & OLITZKI, L. (1948). The labile antigens of *Shigella dysenteriae Shiga*. *J. Hyg.*, *Camb.* **46**, 65.

SHIBAYAMA, G. (1905). Über die Agglutination des Pestbacillus. *Zbl. Bakt.* (1. *Abt. Orig.*), **38**, 482.

SKOOG, F. (1948). Relation between zinc and auxin in the growth of higher plants. *Amer. J. Bot.* **27**, 939.

SMITH, I. W. (1954). Flagellation and motility in *Aerobacter cloacae* and *Escherichia coli*. *Biochim. biophys. Acta*, **15**, 20.

SNEATH, P. H. A. (1956). Change from polar to peritrichous flagellation in *Chromobacterium* spp. *J. gen. Microbiol.* **15**, 99.

SONNEBORN, T. M. (1947). Developmental mechanisms in *Paramecium*. *Growth Symp.* **11**, 291.

SPECTOR, W. G. (1953). Electrolyte flux in isolated mitochondria. *Proc. roy. Soc.* B, **141**, 268.

SUSSMAN, M. (1958). A developmental analysis of cellular slime mold aggregation. In *The Chemical Basis of Development*, p. 264. Edited by W. D. McElroy and B. Glass. Baltimore: Johns Hopkins Press.

STERNE, M. (1937*a*). Variation in *Bacillus anthracis*. *Onderstepoort J. vet. Sci.* **8**, 271.

STERNE, M. (1937*b*). The effect of different carbon dioxide concentrations on the growth of virulent anthrax strains. *Onderstepoort J. vet. Sci.* **9**, 49.

STRATTON, F., RENTON, P. H. & HANCOCK, J. A. (1958). Red cell agglutinability affected by disease. *Nature, Lond.* **181**, 62.

STREISINGER, G. (1956). Phenotypic mixing of host range and serological specificities in bacteriophages T_2 and T_4. *Virology*, **2**, 388.

THAYER, P. S. & HOROWITZ, N. H. (1951). The L-amino acid oxidase of *Neurospora*. *J. biol. Chem.* **192**, 755.

THEORELL, H., NYGAARD, A. P. & BONNICHSEN, R. (1955). Studies on liver alcohol dehydrogenase. III. The influence of pH and some anions on the velocity constants. *Acta chem. scand.* **9**, 1148.

THIELE, E. H. (1950). Studies on the haemagglutinin of *Hemophilus pertussis*. *J. Immunol.* **65**, 627.

THORNE, C. B., GÓMEZ, C. G. & HOUSEWRIGHT, R. D. (1952). Synthesis of glutamic acid and glutamyl polypeptide by *Bacillus anthracis*. II. The effect of carbon dioxide on peptide production on solid media. *J. Bact.* **63**, 363.

THORNTON, F. L. (1935). The action of sodium, potassium, calcium and magnesium ions on the plasmagel of *Amoeba proteus* at different temperatures. *Physiol. Zoöl.* **8**, 246.

TOPLEY, W. W. C. & AYRTON, J. (1924). Further investigations into the biological characteristics of *B. enteritidis* (*Aertrycke*). *J. Hyg., Camb.* **23**, 198.

TORRIANI, A. M. & MONOD, J. (1949). Sur la réversibilité de la réaction catalysée par l'amylomaltase. *C.R. Acad. Sci., Paris*, **228**, 718.

TSUI, C. (1948). The role of zinc in auxin synthesis in the tomato plant. *Amer. J. Bot.* **35**, 172.

TSUYUKI, H. & MACLEOD, R. A. (1951). Ion antagonisms affecting glycolysis by bacterial suspensions. *J. biol. Chem.* **190**, 711.

TULASNE, R. (1951). Les formes L des bactéries. *Rev. Immunol.* **15**, 223.

TYLER, A. (1946). An auto-antibody concept of cell structure, growth and differentiation. *Growth*, **10** (*Suppl.* p. 7).

VALLEE, B. L., STEIN, E. A., SUMERWELL, W. N. & FISCHER, E. H. (1959). Metal content of α-amylases of various origins. *J. biol. Chem.* **234**, 2901.

VALLEE, B. L., WILLIAMS, R. J. P. & HOCH, F. L. (1959). The role of zinc in alcohol dehydrogenase. IV. The kinetics of the instantaneous inhibition of horse liver alcohol dehydrogenase by 1,10-phenanthroline. *J. biol. Chem.* **234**, 2621.

VELDSTRA, H. & ÅBERG, B. (1953). On auxin antagonists. Aryl and aryloxy-acetic acids with a 'bulky' α-substituent. *Biochim. biophys. Acta*, **12**, 593.

VENNES, J. W. & GERHARDT, P. (1959). Antigenic analysis of cell structures isolated from *Bacillus megaterium*. *J. Bact.* **77**, 581.

VESCIA, A. (1956). Ion-determined specificity of peptidase activity. *Biochim. biophys. Acta*, **19**, 174.

VAN WAGTENDONK, W. J. (1951). Antigenic transformations in *Paramecium aurelia* variety 4 stock 51 under the influence of trypsin and chymotrypsin. *Exp. Cell Res.* **2**, 615.

WARREN, G. H. & GRAY, J. (1954). The depolymerization of bacterial polysaccharides by hyaluronidase preparations. *J. Bact.* **67**, 167.

WEBB, E. C. & MORROW, P. F. W. (1959). The activation of an arylsulphatase from ox liver by chloride and other anions. *Biochem. J.* **73**, 7.

WEINER, W., LEWIS, H. B. M., MOORES, P., SANGER, R. & RACE, R. R. (1957). A gene, *y*, modifying the blood group antigen A. *Vox sanguinis*, **2**, 25.

WEITZEL, G., SCHNEIDER, F. & FRETZDORF, A. M. (1957). Zink-Komplexe von Histidyl-peptiden. *Z. phys. Chem.* **307**, 23.

WHITE, P. B. (1925). Serological studies with regard to the classification and behaviours of bacilli of the *Salmonella* group. In *Spec. Rep. Ser. med. Res. Coun., Lond.* no. 91.

WILKINSON, J. F., DUGUID, J. P. & EDMUNDS, P. N. (1954). The distribution of polysaccharide production in *Aerobacter* and *Escherichia* strains and its relation to antigenic character. *J. gen. Microbiol.* **11**, 59.

WILLMER, E. N. (1960). *Cytology and Evolution*. London: Academic Press.

WILSON, I. B. & BERGMANN, F. (1950). Studies on cholinesterase. VII. The active surface of acetylcholine esterase derived from effects of pH on inhibitors. *J. biol. Chem.* **185**, 479.

WOERDEMAN, M. W. (1955). Immunobiological approach to some problems of induction and differentiation. In *Biological Specificity and Growth*, p. 33. Edited by E. G. Butler. Princeton: Princeton University Press.

WOLFF, H. L. & DINGER, J. E. (1951). A new antigen of *Brucella abortus*. *J. Path. Bact.* **63**, 163.

WOLSTENHOLME, G. E. W. & O'CONNOR, C. M. (1959) (editors). Ciba Foundation Symposium on *Regulation of Cell Metabolism*. London: Churchill.

WOODBURY, D. M. (1955). Effect of diphenylhydantoin on electrolytes and radio-sodium turnover in brain and other tissues of normal, hyponatremic and postictal rats. *J. Pharmacol.* **115**, 74.

YOKOTA, K. (1925). Neue Untersuchungen zur Kenntnis der Bakteriengeisseln. *Zbl. Bakt.* (1. *Abt. Orig.*), **95**, 261.

YOKOYAMA, M., STACEY, S. M. & DUNSFORD, I. (1957). *Bx*—a new sub-group of the blood group B. *Vox sanguinis*, **2**, 348.

ZAMENHOF, J. (1946). Unstable strains of the colon bacillus. *J. Hered.* **37**, 273.

ZELITCH, I. (1955). The isolation and action of crystalline glyoxylic acid reductase from tobacco leaves. *J. biol. Chem.* **216**, 553.

THE CHEMICAL COMPOSITION OF MICRO-ORGANISMS AS A FUNCTION OF THEIR ENVIRONMENT

DENIS HERBERT

Microbiological Research Establishment, Porton, Wiltshire

There are few characteristics of micro-organisms which are so directly and so markedly affected by the environment as their chemical composition. So much is this the case that it is virtually meaningless to speak of the chemical composition of a micro-organism without at the same time specifying the environmental conditions that produced it. Such a statement as 'the ribonucleic acid content of *Bacillus cereus* is 16·2 %' is by itself as incomplete and misleading as the statement 'the boiling-point of water is 70°'. While the latter statement may be true (on the summit of Mount Everest, for example), it is incomplete and misleading because it says nothing of the effect of pressure on boiling-point. The former statement is likewise true but misleading, since the RNA* content of *B. cereus* can vary from about 3 to 30 %, depending on the environment in which the cells are grown.

The extent to which chemical composition is affected by the environment has come to be realized only fairly recently, and a great deal of early work on the chemical composition of micro-organisms (reviewed by Porter, 1946) is unfortunately invalidated through lack of adequate environmental control. No attempt will be made to review this early work, while even recent work is too extensive to be treated comprehensively in a single article. Attention will therefore be confined to a number of selected aspects of the subject, chosen because they illustrate some general principles or throw some light on problems of microbial growth or physiology. This aspect is emphasized, for there is little point in performing chemical analyses for their own sake; the important thing is what the results of the analyses tell us about microbiology in general.

The various factors comprising 'the environment' of micro-organisms may conveniently be grouped under the headings of: (1) physical or physico-chemical factors, and (2) chemical factors. So far as the former are concerned, there is little to relate; scarcely any studies have been made on the effects of pressure, temperature, etc., on the chemical

* Abbreviations used: RNA = ribonucleic acid; DNA = deoxyribonucleic acid; PHB = poly-β-hydroxybutyric acid; TPN = triphosphopyridine nucleotide.

composition of micro-organisms, but such data as exist indicate zero or negligible effects. As might be expected, the most important factor affecting the chemical composition of micro-organisms is the chemical composition of the environment. A significant fact here is that quantitative changes in the composition of the environment may be as important as qualitative ones. For example, in a culture medium containing only glucose, an ammonium salt and trace minerals, merely changing the ratio of the concentrations of glucose and ammonia may produce a more profound effect than replacing the ammonia by a mixture of amino acids.

The effects of such changes in the chemical environment on the chemical make-up of micro-organisms may be either qualitative (production in some environments of cell components completely absent in other environments), or quantitative (production of more or less of a cell component invariably present). The most striking examples of qualitative changes in cell composition are found in connexion with adaptive enzymes and certain types of antigen, but they will not be discussed here as they are the subjects of other contributions to this Symposium. This review therefore will be confined to the discussion of quantitative changes in major cell components, namely proteins, nucleic acids, polysaccharides (excluding antigens) and lipids.

PROTEINS AND NUCLEIC ACIDS

The nucleic acid content of bacteria can vary within very wide limits. In different bacterial species values for RNA content as low as 1·5 % and as high as 40 % have been recorded, while in a single strain a 16-fold variation in RNA content has been observed under different conditions of growth (Table 1). Changes in DNA content are less extreme, but still considerable.

The known association between nucleic acids and protein synthesis suggests that such changes would be intimately connected with the growth of the cell, and that environmental factors which affected growth would also affect the nucleic acid content. This is indeed the case, and in fact the effects of environment on the growth of the cell and on its chemical composition are here so intimately linked that it is virtually impossible to discuss one without the other. Some brief mention of the effect of environment on growth must therefore be made (without, it is hoped, trespassing too much on the subject-matter of another contribution to this Symposium).

When physico-chemical factors such as temperature and pH are favourable, the chemical composition of the environment determines not only the possibility of microbial growth but also its rate. The minimal requirements for growth vary with the synthetic ability of the organism; some, for example, can synthesize all the nitrogenous constituents of the cell from a single carbon compound and ammonia, while others are unable to synthesize one or more amino acids, which have to be present in the environment before the organism can grow. The rate of growth is affected by the extracellular concentrations of essential nutrients, the relationship between growth-rate and concentration being a type of 'saturation curve' (Monod, 1942) similar to the velocity against substrate–concentration curves found for enzyme action. Even when a cell is 'saturated' with all essential nutrients, it will grow still faster if additional (non-essential) nutrients are supplied, so that it no longer has to synthesize them. There are many such possibilities, so that cells may be made to grow at a number of different growth rates by supplying them with media of different chemical compositions. Alternatively, different growth rates may be induced by regulating the concentration of a single essential nutrient; the concentration may be held constant, in spite of utilization of the nutrient due to growth, by using a continuous culture apparatus of the 'chemostat' type (Monod, 1950). Both methods have been used to study the effect of growth rate on the chemical composition of bacteria. In addition, changes in chemical composition have been studied during the course of the so-called 'growth cycle' that occurs when bacteria are inoculated in the traditional manner into a vessel containing a limited volume of culture medium. This type of experiment, which will be referred to as 'growth in a closed system '(Herbert, 1960), is historically the earlier and will be discussed first.

Changes in chemical composition during growth in closed systems

When bacteria are inoculated into a limited volume of culture medium, the following fairly standard sequence of events occurs. (1) A period of lag, followed by (2) a period of accelerating growth; the occurrence and duration of these periods depends on the previous history of the inoculum. (3) A period of exponential growth at a constant growth rate (the so-called 'logarithmic phase'), when growth follows the equation:

$$\frac{1}{x}\frac{dx}{dt} = \frac{d(\ln x)}{dt} = \mu = \frac{\ln 2}{t_d} \tag{1}$$

(x being the dry weight of cells/ml., μ the exponential growth rate and t_d the doubling time). During this period extracellular nutrients are being rapidly used up. (4) A period of declining growth rate, whose onset (in an adequately aerated and buffered medium) is due to the exhaustion of some nutrient (not necessarily an essential one) and whose duration depends greatly on the complexity of the culture medium. (5) A stationary phase, during which cells remain viable but no longer grow, initiated by the exhaustion of an essential nutrient.

(The above sequence of events is usually called 'the growth cycle'—a term which, I suggest, should speedily be abandoned, since it conveys a quite misleading impression that this sequence is a necessary and inevitable feature of bacterial growth, whereas it is in reality a sequence forced upon the organisms by sequential environmental changes which are inevitable when growth occurs in a closed system. Unfortunately, bacteriologists have been inoculating flasks of culture media for so long now that they have come to regard this as part of the natural order of events, instead of as a convenient but highly artificial experimental procedure.)

Table 1. *Cell mass and RNA content in 'resting' and 'log phase' cells of a number of bacterial species (from Wade & Morgan,* 1961)

		Cell mass (picograms)†		RNA content (%)	
Organism	Medium*	Resting cells	Log phase cells	Resting cells	Log phase cells
Aerobacter aerogenes	CCY	0·11	0·40	4·4	26·6
Bacillus anthracis	TMB	—	—	1·5	24·0
Bacillus cereus	TMB	1·97	3·77	3·9	31·5
Chromobacterium prodigiosum	CCY	0·12	0·35	7·8	32·1
Chromobacterium violaceum	TMB	0·17	0·56	7·2	30·3
Clostridium welchii	TMB	0·91	2·19	32·2	42·2
Corynebacterium hofmannii	CCY	—	—	25·4	51·0
Salmonella typhi	CCY	0·19	0·34	10·5	35·9
Escherichia coli	CCY	0·12	0·41	15·5	37·0
Pasteurella pestis	TMB	0·13	0·15	5·9	20·1
Proteus vulgaris	TMB	0·18	0·36	12·6	35·0
Staphylococcus aureus	CCY	0·19	0·24	5·2	10·0

* CCY, casein-yeast extract medium; TMB, tryptic meat digest medium.
† One picogram = 10^{-12} g.

Bacteria grown in this way are most commonly studied during either phase (3) of the above sequence ('log phase cells'), or phase (5) ('stationary phase' or 'resting' cells). It has been known for some time that these two types of cell differ considerably both in size and in chemical composition. Log phase cells are large and have a high RNA content and low DNA content; resting cells (at least those produced by growth in the usual type of complex culture medium) are considerably smaller and have a much lower RNA content but a relatively higher DNA content. These generalizations are valid for a quite wide range of bacterial species, as illustrated by the data of Table 1.

More recently, changes in cell size, RNA content and DNA content have been followed for several different organisms throughout the growth of a culture. The type of result obtained is shown in Fig. 1, which is a composite (and slightly idealized) representation of the data of several workers (Hershey, 1938, 1939; Wade, 1952*a*, *b*; Mitchell &

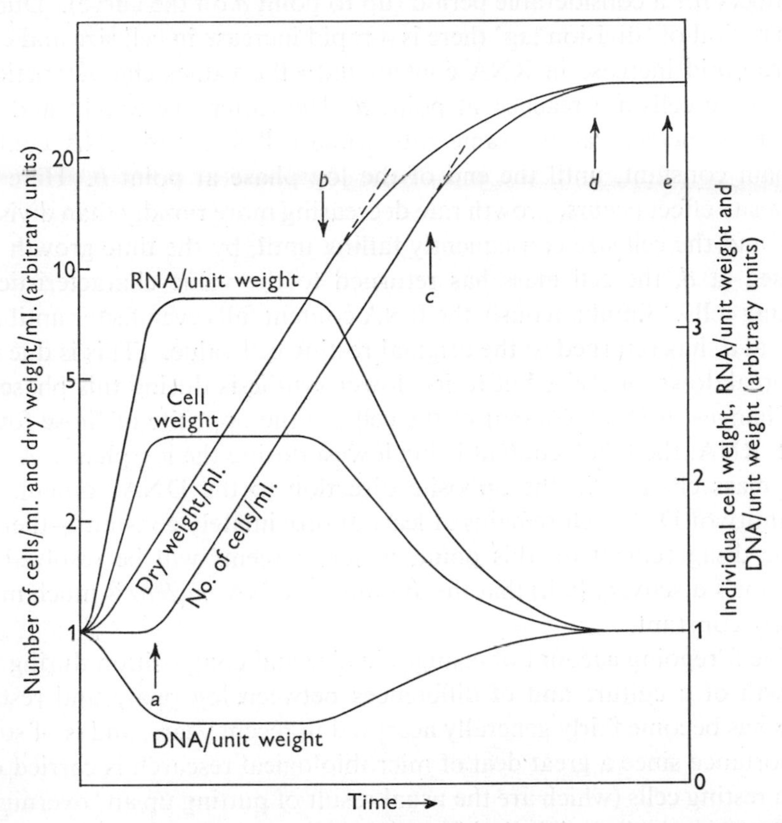

Fig. 1. Changes in cell size and chemical composition during growth of a bacterial culture. (Idealized curves after Wade and others.) The culture is inoculated at zero time with fairly young 'resting' cells. The increases in number of cells and dry weight/ml. are plotted on a logarithmic scale (left-hand ordinate). Values for individual cell weights (ratio of dry weight/ml. to total count/ml.), RNA content and DNA content are plotted on an arithmetic scale (right-hand ordinate). The units are arbitrarily chosen to make the initial values of all variables equal to 1·0.

Moyle, 1951; Gale & Folkes, 1953; Malmgren & Hedén, 1947). Bacterial dry weight/ml. and number of cells/ml. are plotted on a logarithmic scale, while RNA and DNA contents and cell mass (the ratio of dry weight to total count, i.e. the average dry weight of one cell) are plotted on arithmetic scales, the units being arbitrarily chosen to make the initial values of all variables equal to 1 (Maaløe, 1960). The culture is inoculated at zero time with resting cells taken from a previous culture at about the point marked by arrow *e*.

When such fairly 'young' resting cells are put into fresh medium, exponential growth (as measured by increase in dry weight) begins almost immediately, but there is no corresponding increase in cell numbers for a considerable period (up to point *a* on the curve). During this period of 'division lag' there is a rapid increase in cell size and even more rapid increase in RNA content until the values characteristic of log phase cells are reached at point *a*. Thereafter dry weight and cell numbers increase at the same rate while cell size and RNA content remain constant, until the end of the log phase at point *b*. Here the opposite effect occurs, growth rate decreasing more rapidly than division rate and the cell size consequently falling until, by the time growth has ceased at *d*, the cell mass has returned to the value characteristic of resting cells. Simultaneously the RNA content falls even faster until this too, at *d*, has returned to the original resting cell value. (This is due not to breakdown of RNA but to its slower synthesis during this phase.)

Changes in DNA content of the cell are the opposite of those found with RNA, the DNA content being lowest during the log phase. As the cell mass changes in the opposite direction to the DNA content, the quantity of DNA/cell remains at least approximately constant—there is some disagreement on this point, which it seems will be resolved by Maaløe's discovery (v.i.) that the amount of DNA/*nucleus* is much more nearly constant.

The foregoing account of changes in size and composition during the growth of a culture and of differences between log phase and resting cells has become fairly generally accepted in recent years, and is of some importance since a great deal of microbiological research is carried out with resting cells (which are the usual result of putting up an 'overnight' culture). As well as differing from log phase cells in size and chemical composition, resting cells are usually 'tougher' than log phase cells, being more resistant to physical disruption and osmotic shock and maintaining viability better on storage. However, work now in progress at M.R.E. (Fig. 2) suggests that all of the above applies only to growth in complex types of culture medium.

In Fig. 2, which is a previously unpublished experiment of Strange, Dark & Ness (1961), the growth of *Aerobacter aerogenes* is shown (*a*) in a complex (meat digest) medium, and (*b*) in a mannitol–ammonia–salts medium (with mannitol as limiting nutrient). The growth in complex medium is similar to the idealized curves of Fig. 1, the log phase being followed by an extended phase of deceleration before growth finally ceases. (As others have observed in such media (Monod, 1949), there are apparently two or perhaps three phases of exponential growth at successively decreasing rates, probably indicating successive exhaustion of non-essential but growth-accelerating nutrients.) Cells sampled in the log phase, deceleration phase and stationary phase show a progressive decline in RNA and increase in DNA content, as found by other workers. In the minimal medium, on the other hand, exponential growth is maintained almost to the end, ceasing abruptly and with a hardly detectable deceleration phase as the last traces of mannitol are exhausted. In this case the RNA and DNA contents of log phase cells and resting cells are virtually identical.

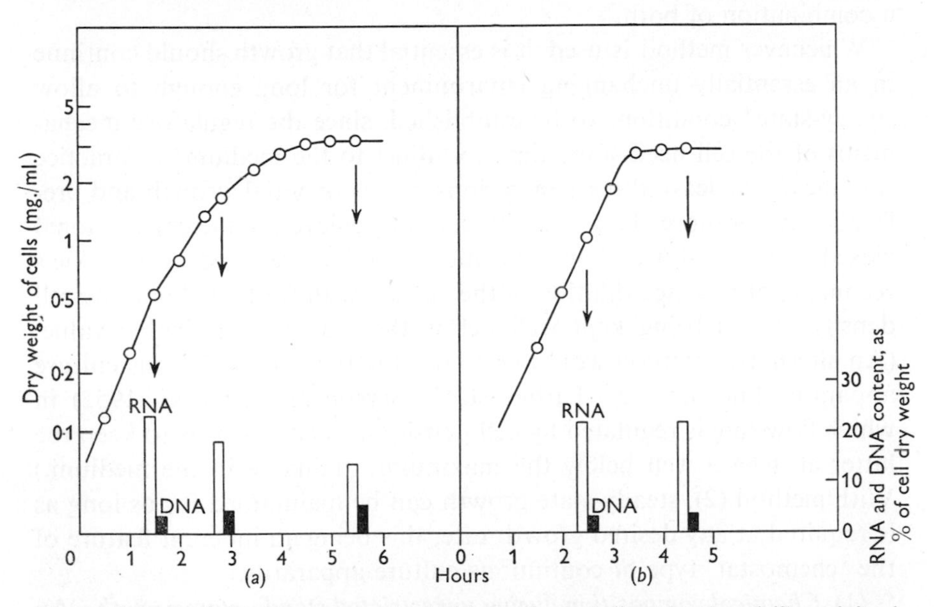

Fig. 2. Growth of *Aerobacter aerogenes* in (*a*) tryptic meat digest broth and (*b*) a minimal medium (NH_3—salts + 0·5 % mannitol as limiting factor) illustrating differences in chemical composition of 'resting' cells. From an unpublished experiment of Strange, Dark & Ness (1961); cultures grown in stirred fermenters with optimal aeration and continuous control of pH; samples for analysis taken at points indicated by arrows.

These results strongly suggest that the changes in RNA content at the end of growth in a complex medium are due to the extended period of growth at increasingly slow *rates*. This does not occur when growth terminates abruptly through exhaustion of a single essential nutrient, and the resting cells produced in such cases may be regarded as 'frozen' log phase cells. The effect of growth rate on the chemical composition of cells has been studied by both the methods described at the beginning of this section, with results described below.

Changes in chemical composition during steady-state growth at varying rates

As previously mentioned, the effect of growth rate on chemical composition has been studied by two methods: (1) unrestricted growth (i.e. all nutrients present in non-limiting concentrations) in media of different nutritional complexity, and (2) growth controlled by regulating the concentration of a single essential nutrient, using a flow-controlled type of continuous culture apparatus or 'chemostat'. Maaløe and co-workers at Copenhagen have used method (1), while at Porton we have used method (2); Magasanik, Magasanik & Neidhardt (1959) have used a combination of both.

Whichever method is used, it is essential that growth should continue in an essentially unchanging environment for long enough to allow steady-state* conditions to be established, since the regulatory mechanisms of the cell need some time to adjust to the medium; in practice this means at least three generations of exponential growth and preferably five or more. This can with care be achieved in the conventional closed-system culture, but very much to be preferred is Maaløe's technique of repeated dilution of the culture with fresh medium, the cell density always being kept well below the maximal attainable value. (An alternative method would be to use the type of continuous culture apparatus known as a 'Turbidostat' (Bryson & Szybalski, 1952) in which flow rate is regulated by cell density in such a way as to keep the latter at a level well below the maximum attainable in the medium.) With method (2), steady-state growth can be maintained for as long as is required at any desired growth rate, this being an inherent feature of the 'chemostat' type of continuous culture apparatus.

(1) *Chemical composition during unrestricted steady-state growth.* An extensive study of *Salmonella typhimurium* has been made by Schaechter, Maaløe & Kjeldgaard (1958), in which the organism was grown by the repeated dilution technique in twenty-two different growth media giving doubling times ranging from 22 to 97 min. The results were summarized by Maaløe (1960) in the graphs reproduced in Fig. 3, in which the following variables are plotted on a logarithmetic scale against the growth rate:† in (A), the dry weight/cell, weights of RNA and DNA/cell

* Campbell (1957) has introduced the expression 'balanced growth': 'growth is *balanced* over a time interval if, during that interval, every extensive property of the growing system increases by the same factor'. This is an admirably precise definition, but I cannot see that it means anything more than the well-established term 'steady-state growth'.

† The measure of growth rate used by Maaløe is the reciprocal of the doubling time i.e. $1/t_d$. This is $1/\ln 2$ or 1·44 times greater than the exponential growth rate μ of equation (1).

and number of nuclei/cell: in (B) the dry weight/nucleus, RNA/nucleus and DNA/nucleus. The ordinate scale units are chosen to make the values of all of these equal to 1·0 at zero growth rate (cf. Fig. 1). In Fig. 4 the data of Fig. 3 are replotted so as to show the RNA and DNA contents expressed, not as quantities/cell, but as percentages of the total dry weight.

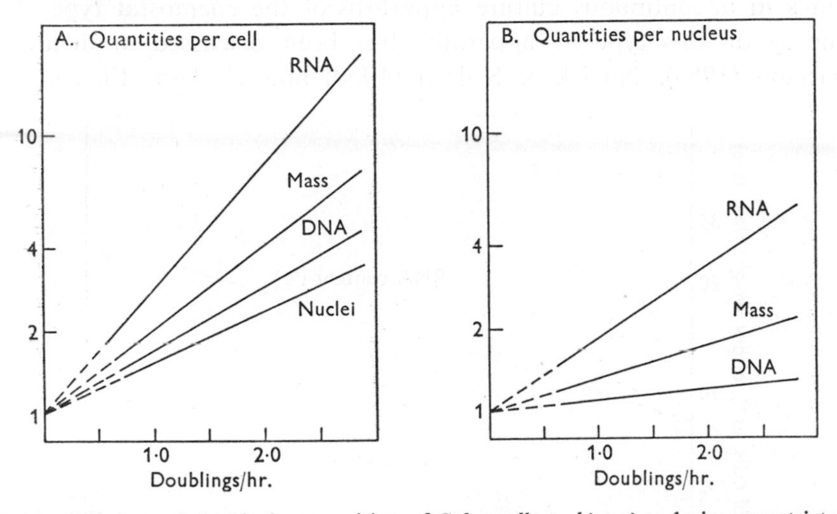

Fig. 3. Cell size and chemical composition of *Salmonella typhimurium* during unrestricted balanced growth (from Maaløe, 1960). The organism was grown in a series of culture media producing different rates of growth. The following values are plotted against growth rate: A, number of nuclei/cell, dry weight/cell, amount of RNA/cell and amount of DNA/cell; B, dry weight/nucleus, RNA/nucleus and DNA/nucleus. Dotted lines are extrapolations to zero growth rate from the regions covered by the experimental values. The units on the ordinate are as follows: 2·6 μg. DNA, 13 μg. RNA and 155 μg. dry weight/10^9 cells.

Figure 3A shows that all the variables plotted increase exponentially with growth rate, though at different rates. The amount of RNA/cell increases more rapidly than the cell mass as the growth rate is increased, while the DNA/cell increases less rapidly. In other words, the RNA *content* of the cell, expressed as a percentage of the dry weight, increases with increasing growth rate, while the percentage of DNA decreases, as shown in Fig. 4. In this organism the average number of nuclei/cell also increases exponentially with growth rate, at nearly the same rate as the amount of DNA/cell, so that, as shown in Fig. 3B, the weight of DNA/*nucleus* shows very little change with growth rate, although the RNA/nucleus increases considerably.

The dotted lines in Fig. 3 indicate extrapolations back to zero growth rate from the regions covered by the experimental values. It is an interesting fact that the zero growth rate values for cell size, average number of nuclei and chemical composition agree well with actual values obtained by the analysis of 'resting cells'.

(2) *Chemical composition during nutrient-limited steady-state growth.* Figures 5 and 6 show data obtained by Herbert *et al.* (1961) on cell mass and on RNA, DNA and protein contents of *Aerobacter aerogenes* and *Bacillus megaterium* grown over a wide range of growth rates in a continuous culture apparatus of the chemostat type. The theory of this type of apparatus has been discussed in detail by Monod (1950), Novick & Szilard (1950) and Herbert, Elsworth &

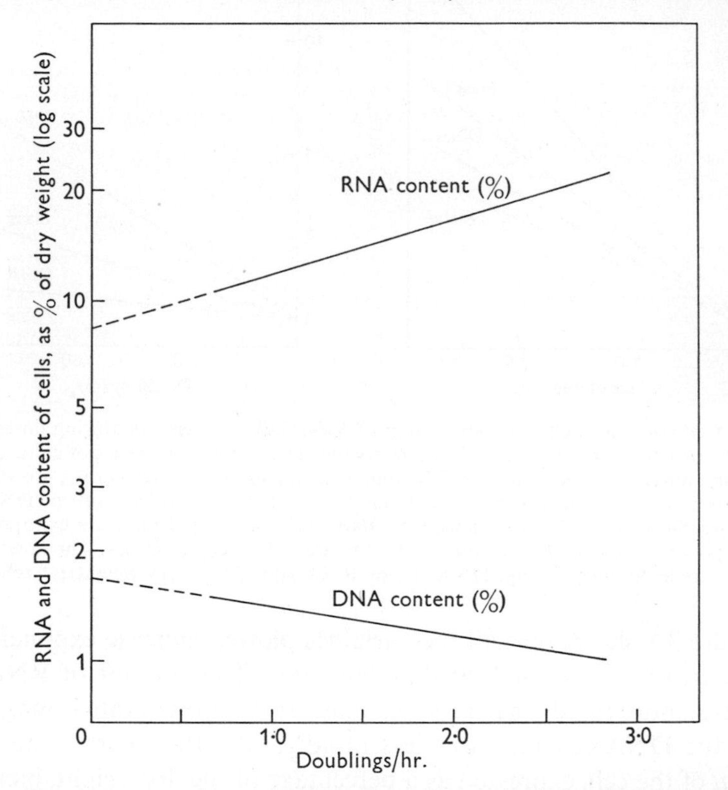

Fig. 4. Chemical composition of *Salmonella typhimurium* during balanced growth at different growth rates. The data of Maaløe *et al.* shown in Fig. 3 are plotted to show RNA and DNA contents as percentages of the dry weight (i.e. g. RNA or DNA/100 g. dry weight of cells).

Telling (1956). It will suffice here to say that if the dilution rate *D* (ratio of flow rate to culture volume) is held constant, the organisms grow up until the concentration of some essential nutrient is reduced to a level which makes the exponential growth rate μ equal to *D*. The system is now in a steady state, which is maintained indefinitely as long as *D* remains unaltered.

This means that, although medium of constant composition is fed to the growth vessel, the environment in which the cells are growing is different for each growth rate in respect of the concentration of limiting nutrient. In Fig. 5, for example, each experimental point corresponds

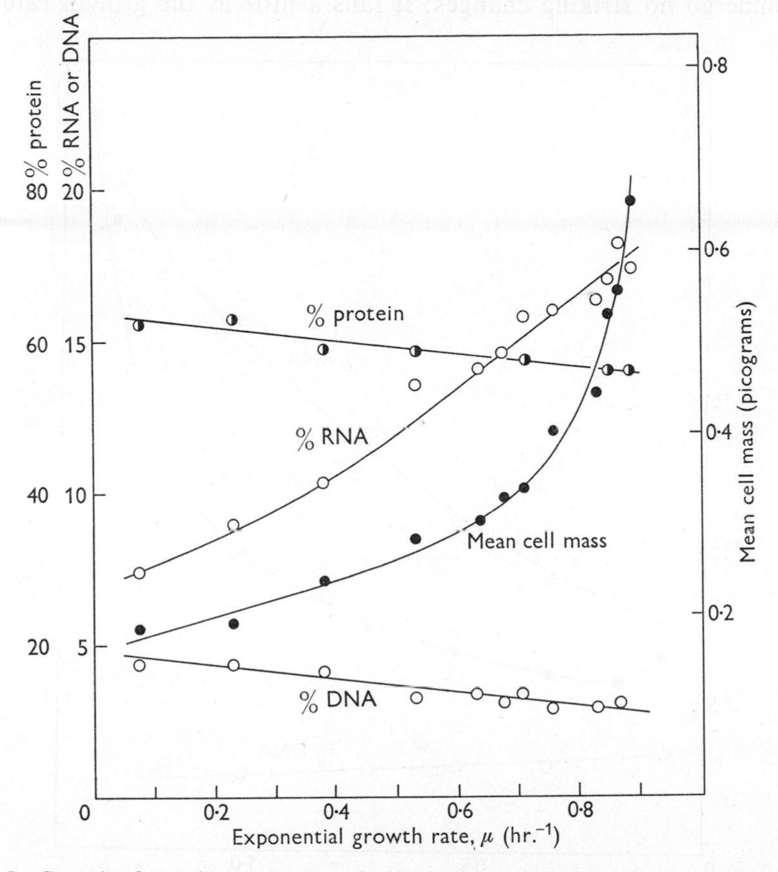

Fig. 5. Growth of *Aerobacter aerogenes* in continuous culture; protein and nucleic acid contents and mean cell mass as functions of the growth rate. The organism was grown in a continuous culture apparatus of the chemostat type at a number of different flow rates and the cells analysed after at least 2 days steady-state growth at each flow rate. Nucleic acid and protein contents expressed as percentage of cell dry weight; mean cell mass = dry weight/ml. divided by total count/ml. Culture medium: glycerol–NH_3–salts, with glycerol as limiting factor. (Data from Herbert *et al.* 1961.)

to the growth of the organism in a different concentration of glycerol, all other medium constituents being present in excess. The concentration of limiting nutrient can be made as low as desired simply by reducing the flow rate sufficiently; hence very much lower growth rates (doubling times up to 15 hr. or more) can be studied with this method than can be achieved by method (1).

The results of such experiments with the chemostat are broadly similar to those of Maaløe and his colleagues in respect of changes in cell size, RNA content and DNA content (the number of nuclei was not studied). The protein content of the cells was also measured, but is seen to undergo no striking changes; it falls a little as the growth rate in-

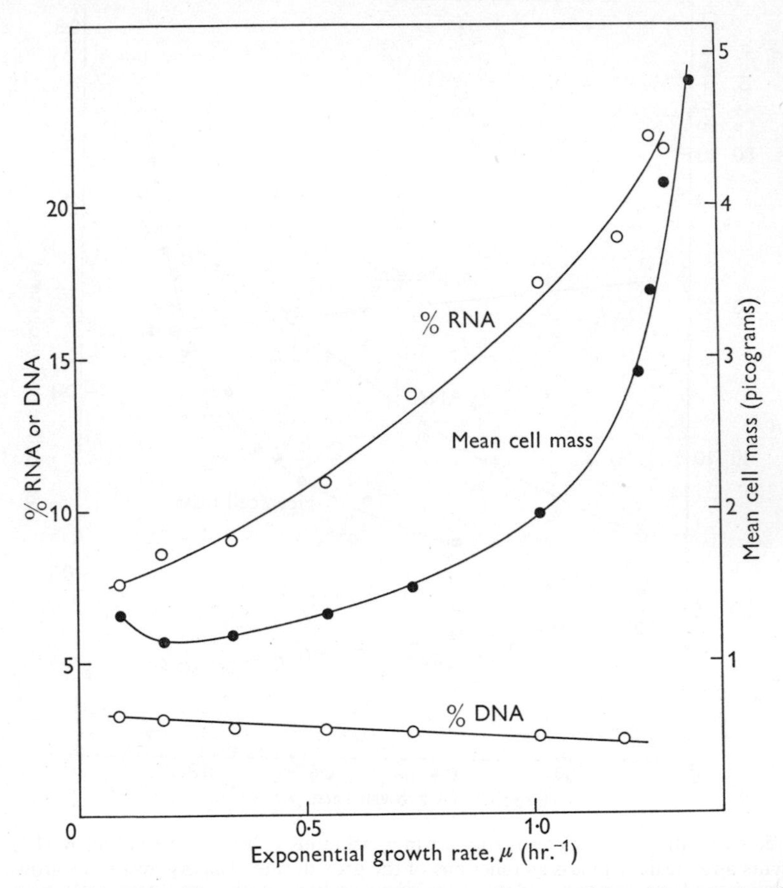

Fig. 6. Growth of *Bacillus megaterium* in continuous culture; nucleic acid content and cell mass as functions of growth rate. The organism was grown in a chemostat in a casein hydrolysate-mannitol medium and the cellular nucleic acid content and mean cell mass determined as in Fig. 5. (Herbert *et al.* 1961.)

creases, to an extent which can largely be accounted for by the increased proportion of RNA.

The data of Figs. 5 and 6 are plotted on an arithmetic scale; if logarithmic ordinates are used (cf. Figs. 3 and 4), the values for cell mass in both organisms, and for RNA content in the case of *Bacillus megaterium*, give approximately straight-line plots in the region of higher growth rates but fall off at lower growth rates; with *Aerobacter aerogenes* the change in RNA content with growth rate is closer to a linear than an exponential relationship. In this respect, therefore, there is not complete agreement between results obtained with the chemostat and by the unrestricted growth technique. In view of the different ranges of growth rate covered and the fact that species differences have already been observed in spite of the small number of organisms yet studied, this does not seem too serious; the general agreement in the over-all pattern of the results seems more important than the question of whether certain variables are exact exponential functions of growth rate, or only approximately so.

To the writer, at least, this general pattern of the results appears as one of the most striking examples of environmental control of chemical composition; it is remarkable that such profound changes in size, morphology and chemical make-up of the cell can be produced simply by altering the concentration of a single nutrient. Even more significant, however, is the fact that *the same pattern of changes seems to occur whatever the nature of the nutrient that is varied.* For example, the experiment of Fig. 5, with the nitrogen source in excess and glycerol as limiting nutrient, was repeated with a medium containing exactly the same chemical ingredients but with the carbon/nitrogen ratio altered so as to make the nitrogen source (NH_4^+ ion) the limiting nutrient. *The results were virtually identical*; not merely was the general pattern the same, but bacteria growing at the same rate in either experiment had the same size and chemical composition. Similarly, Schaechter *et al.* (1958) found that some of their growth media of quite different chemical composition gave identical growth rates during unrestricted growth; such media all produced cells of the same size and chemical composition.

Such results seem to show that the chemical composition of the cell is primarily dependent on the rate at which it is growing and is affected by the chemical composition of the growth medium only in so far as this affects the growth rate. The situation is not so simple as this, however, as was shown by the experiments of Schaechter *et al.* (1958) on the effect of temperature on chemical composition. On repeating the experiments of Fig. 3 at a lower temperature (25° instead of 37°), these workers found that for each different growth medium, cells of the same size and chemical composition were produced at both temperatures, although for all media the growth rate was halved at the lower temperature. Cell size and chemical composition would therefore seem to be determined by the pattern of metabolic activities imposed on the cell by the growth medium.

Chemical composition of cells during synchronous growth

All the data on cell size and chemical composition discussed so far were obtained from samples containing many billions of cells taken from unsynchronized cultures containing cells of all possible ages;* they therefore represent cells of some sort of average age and it is worth considering what this may be. The age distribution in a steadily growing culture is related to the distribution of generation times (Powell, 1956) and has the interesting property, not generally known, that the youngest organisms are present in the greatest numbers.

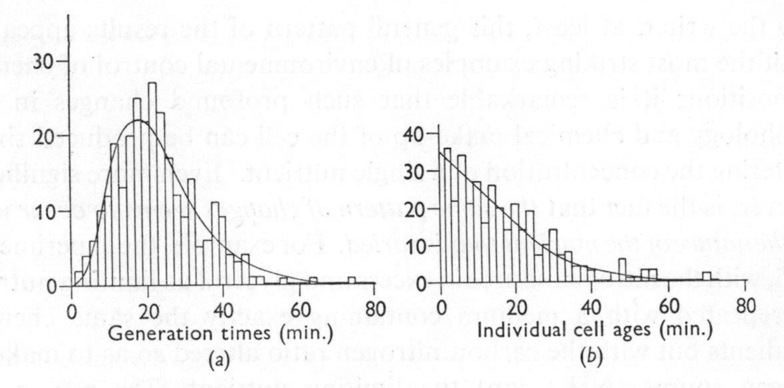

Fig. 7. Frequency distributions of (*a*) generation times, (*b*) individual cell ages, during unsynchronized growth of *Bacillus mycoides* (Powell, 1955, 1956).

Figure 7*a* shows the distribution of generation times in a culture of *Bacillus mycoides* during steady-state growth, from observations of Powell (1955); as is usually the case, the frequency distribution curve is unimodal and positively skew. Figure 7*b* shows the corresponding distribution of ages, i.e. the relative frequencies of occurrence of organisms of different ages in a sample taken from a culture at any instant. This has a J-shaped frequency distribution curve with a maximum at zero age. The mean generation time of this organism was 28·7 min., but the median age was only 11 min.; i.e. half of the organisms in a sample from such a culture would be less than 11 min. old. Data obtained from large samples are therefore representative of cells in rather early stages of their individual life cycles.

It would be of the greatest interest to know whether the chemical composition of the individual cell undergoes any systematic changes

* The 'age' of an organism (to be distinguished from the age of a culture) is the time that has elapsed since its inception through fission of its parent; a 'young' organism is one whose age is small compared with the mean generation time, whatever the value of this may be.

during the course of its life. A direct approach to this problem is beyond the range of the most ultra-micro analytical techniques known at present; it is true that the dry weights of individual bacteria have been determined with the interference microscope (Ross, 1957) and their total nucleic acid contents by ultraviolet microspectrophotometry (Malmgren & Hedén, 1947), but only the former could be used for repeated measurements on the same cell throughout its life and even these would not be very accurate. In a culture undergoing synchronous growth, however (that is, a culture in which all the cells are growing and dividing 'in step'), samples taken at intervals throughout a division cycle should give the information needed, provided that all the cells in the culture are undergoing normal steady-state growth. There is the further possibility of identifying changes in composition occurring at certain times in the cycle with the process of cell division.

So exciting are these possibilities that great interest was aroused by the discovery that growing cultures could be artificially synchronized by subjecting them to treatments such as temperature changes, illumination changes (in certain photosynthetic organisms), and starvation of certain nutrients (see review by Campbell, 1957). Interest was intensified when it was found that changes in chemical composition did in fact occur in the course of synchronous division cycles, discontinuous or stepwise synthesis of DNA, RNA and protein being reported by various workers (see Campbell, 1957). However, more recent evidence (e.g. Schaechter, Benzton & Maaløe, 1959) suggests that these effects are artefacts and that the treatments used to induce such 'forced synchrony' produce unbalanced growth; the changes in chemical composition observed are probably akin to those found by Kjeldgaard, Maaløe & Schaechter (1958) when cultures in steady-state growth are abruptly shifted from one medium to another. It seems doubtful, therefore, that such artificially synchronized cultures can be said to reproduce the normal division cycle, at any rate during the first one or two generations. (After several generations a condition of steady-state growth will be regained, but by then, owing to the normal scatter in individual generation times (Fig. 7), most of the synchrony will have disappeared.)

A better method of obtaining synchronous cultures appears to be the filtration technique of Maruyama & Yamagita (1956) which *selects* from a normally growing population those small cells which have just been formed by division. Abbo & Pardee (1960) introduced refinements of technique which avoided temperature shifts or medium starvation during the filtration process and their results (Fig. 8) appear to give a true picture of the normal division cycle.

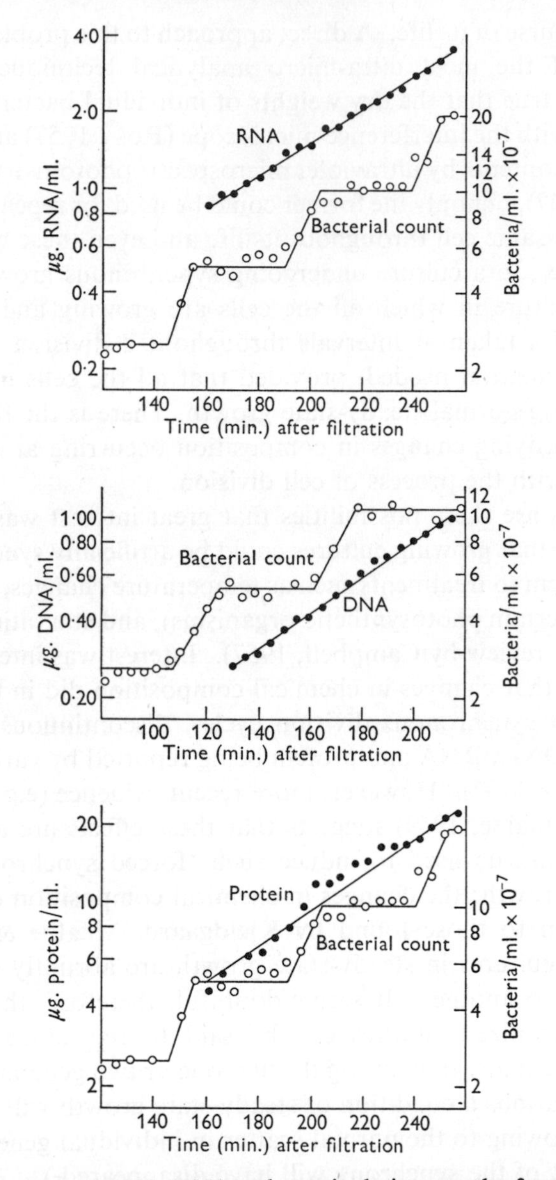

Fig. 8. Synthesis of RNA, DNA and protein during growth of synchronously dividing *Escherichia coli.* (From Abbo & Pardee, 1960.)

The results of Fig. 8 show that, contrary to what has been observed in artificially synchronized cultures, RNA, DNA and total protein all increase exponentially at the same constant rate over the whole division cycle; in other words, their *ratio* remains constant. So far as these components are concerned, therefore, we must conclude that the chemical constitution of the individual bacterial cell remains constant throughout its life cycle; undoubtedly this must be a fact of fundamental importance in bacterial physiology.

POLYSACCHARIDES

It is now time to turn to another class of major cell components, namely the polysaccharides. A great deal of work has been done in this field on the effect of environmental factors on capsular polysaccharides and the polysaccharide components of surface antigens; these are discussed elsewhere in this Symposium. The present contribution will be confined to the intracellular polysaccharides, of which by far the most important is glycogen.

The glycogen content of bacteria and yeasts (moulds appear to have been little studied) can vary very greatly; values ranging from 2 % to over 30 % have been recorded in the same organism under different growth conditions. Little progress was made in elucidating the factors necessary for glycogen storage so long as micro-organisms were grown in complex culture media of unknown constitution. The use of simple chemically defined media has in the last few years clarified the situation considerably.

Holme & Palmstierna (1956) grew *Escherichia coli* in a simple glucose-ammonium salt medium with either glucose or ammonia as limiting nutrient. When glucose was limiting, the glycogen content of the cells remained low throughout the growth of the culture; when ammonia was limiting, the glycogen content was also low during the early stages of growth but increased enormously during and shortly after the final stages (i.e. when the ammonia concentration was approaching zero). The results strongly suggested that glycogen deposition occurs when the ammonia concentration is very low; this was confirmed by Holme (1957) using the continuous culture technique, with results shown in Fig. 9.

In this experiment, the organisms were grown at a number of different flow rates in a chemostat type of apparatus, with ammonia as the growth-limiting nutrient. Under these conditions the ammonia is nearly all consumed so that its steady-state concentration is very low, increasingly so as the growth rate is decreased. As Fig. 9 shows, the glycogen content of the cells increases as the growth rate and ammonia concentration decrease, from *c.* 3 % of the dry weight at the highest growth rate tested to *c.* 23 % at the lowest. As the glycogen content of the cells increases, the protein content (as approximately measured by total N) decreases; this is not due to less protein being synthesized, but to its being 'diluted'

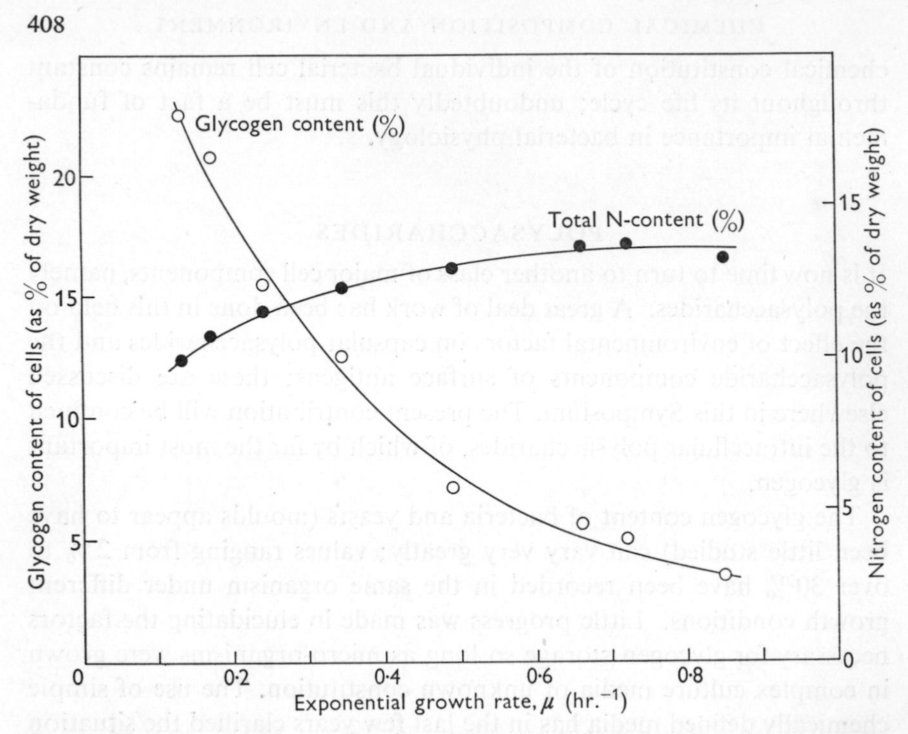

Fig. 9. Glycogen synthesis as a function of growth rate in *Escherichia coli* grown under conditions of nitrogen limitation. Plotted from data of Holme (1957, Table 2). The organism was grown in continuous culture at a number of different growth rates in a lactate–NH_3–salts medium with NH_3 as growth-limiting nutrient.

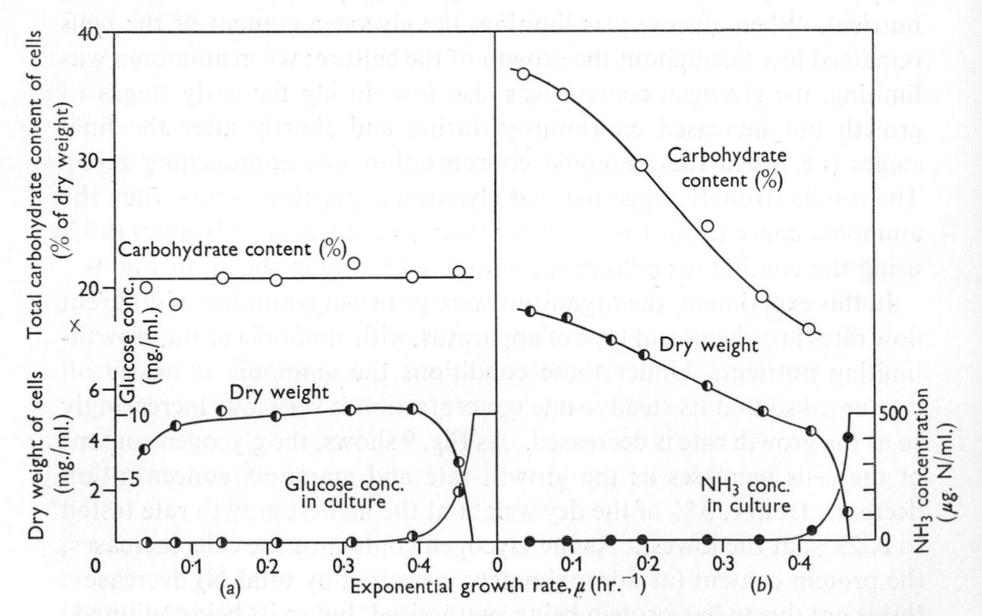

Fig. 10. Carbohydrate content of *Torula utilis* as a function of growth rate and limiting nutrient. (Unpublished data of Herbert & Tempest.) The organism was grown in continuous culture at a number of different growth rates in a glucose–NH_3–salts medium (*a*) with glucose as limiting nutrient, and (*b*) with NH_3 as limiting nutrient. Dry weight of cells in the culture and their total carbohydrate content (anthrone method), as well as steady-state levels of glucose and NH_3 in the culture, are plotted against growth rate.

with glycogen. Identical results were obtained when lactate replaced glucose as carbon source, the nitrogen content still being limiting.

Similar results to those of Holme have been obtained by the author and D. W. Tempest (unpublished) with a yeast (*Torula utilis*) and are shown in Fig. 10.

This figure shows the results of two continuous culture runs covering a wide range of flow rates (*a*) with glucose as limiting nutrient, and (*b*) with ammonia as limiting nutrient. With glucose limiting, the total carbohydrate content of the cells remains constant; while with ammonia limiting, the carbohydrate content increases with decreasing growth rate in the same manner as Holme observed with *Escherichia coli*. (It should be mentioned that the figures for 'total carbohydrate' were determined by the anthrone method and therefore include glucan and mannan as well as glycogen; separate analyses showed, however, that the changes were confined to the glycogen fraction.) The dry weight of cells obtained at each growth rate is also plotted and in Fig. 10*b* is seen to increase at low growth rates, the higher 'yield' being due to the higher glycogen content.

The results of Fig. 10*a* and *b* taken together suggest that high glycogen content is the result not of growth at a low rate but of growth in a low concentration of ammonia. The fact that synthesis of glycogen can occur at very low growth rates suggests that it will probably occur at zero growth rate, i.e. in resting cells; the results given in Fig. 11 show that this in fact occurs. In this experiment, washed yeast cells were shaken aerobically with glucose in a phosphate buffer, no nitrogen source being added so that growth, in the sense of protein or nucleic acid synthesis, could not occur. The cells doubled their dry weight in 2 hr., an increase which could entirely be accounted for in terms of glycogen synthesized (there was a barely significant increase in protein in the first half hour which we interpret as conversion to protein of the initial 'amino acid pool').

This experiment is by no means novel, numerous examples of 'carbon assimilation' in resting bacteria being known (Clifton, 1946), though the product of assimilation has by no means always been identified as glycogen. It has been included mainly to show the magnitude of the changes in chemical composition of the cell that can occur in such experiments; in this one, the carbohydrate content reached the very high level of 50 % of the dry weight of the cell, the protein content falling to *c*. 35 %. Such changes are sufficient to cause large changes in the refractive index of the cells, which have a characteristic appearance under the phase-contrast microscope.

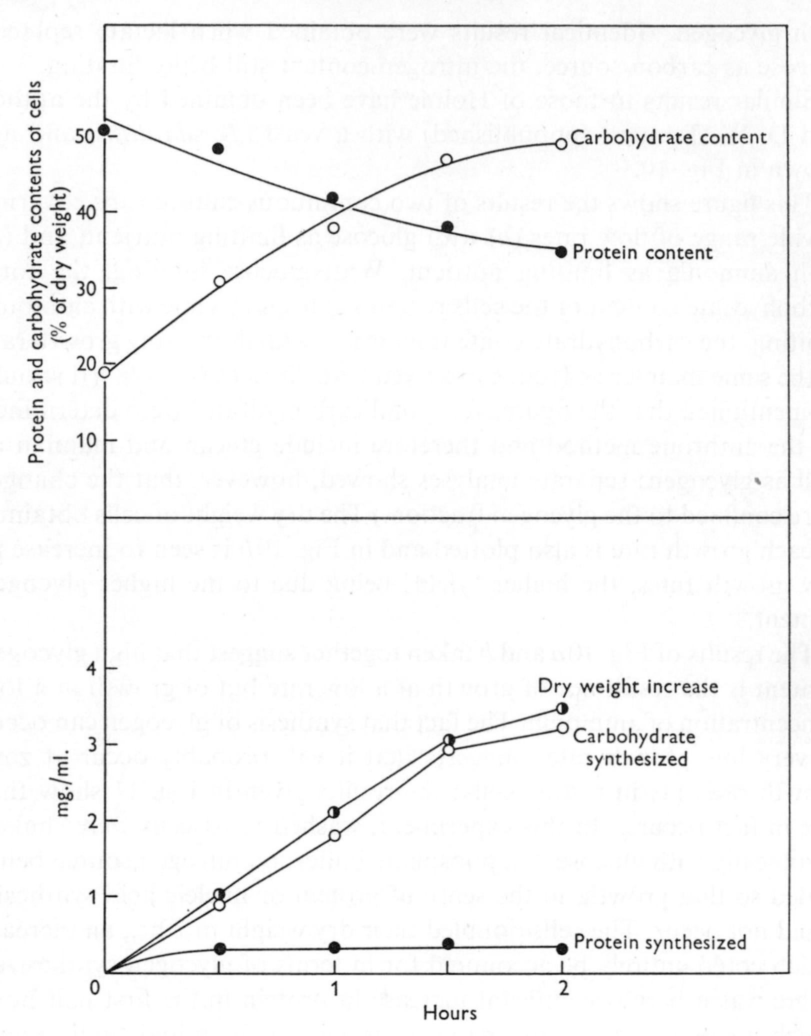

Fig. 11. Carbohydrate synthesis in *Torula utilis* in the absence of growth. Washed cells (4·3 mg./ml.) incubated aerobically in phosphate buffer (pH 5·5) containing 0·85 % glucose; dry weight, total carbohydrate and protein determined at intervals. (Unpublished data of Herbert & Tempest.)

LIPIDS

Work on bacterial lipids has virtually been confined to the *Mycobacteria* and has been mainly concerned with problems of lipid chemistry. A little more is known about yeasts and moulds, although most published work is confined to a few species of high fat content which have mainly been studied in the fermentation industry with a view to industrial fat production.

With such organisms as *Rhodotorula gracilis* and *Endomyces vernalis* there is a fair amount of evidence (reviewed by Kleinzeller, 1948) that high lipid content, like high polysaccharide content, is associated with growth in nitrogen-deficient media. This is certainly the case with the lipid poly-β-hydroxybutyric acid (PHB).

This interesting substance, a long-chain polyester of β-hydroxybutyric acid discovered in *Bacillus megaterium* by Lemoigne in 1927, was for many years regarded as something of a biochemical curiosity. Interest in it has revived with the discovery that it is in fact a fairly common constituent of bacteria, having in recent years been found in many *Bacillus* species, *Azotobacter*, a wide range of Gram-negative organisms and certain photosynthetic bacteria such as *Rhodospirillum rubrum* (Doudoroff & Stanier, 1959). It may form up to 45 % of the dry weight of the cell, and is a major component of the sudanophilic 'lipid granules' found in several of the above species, although it is not itself sudanophilic. Recent work indicates that PHB should be considered along with the triglyceride fats and glycogen as a major reserve storage material in many bacteria.

Macrae & Wilkinson (1958*b*) studied the formation of PHB in *Bacillus megaterium* growing in a glucose-NH_4Cl medium: (*a*) with glucose limiting, (*b*) with NH_4Cl limiting. Their results bear a striking resemblance to those of Holme & Palmstierna (1956) mentioned above on the formation of glycogen in *Escherichia coli*. In carbon-limited cultures the PHB content remained fairly uniform and low throughout the growth of the culture, while when nitrogen was limiting there was a rapid synthesis of PHB up to very high levels in the period immediately preceding and following the cessation of growth due to exhaustion of the nitrogen source.

Like glycogen again, PHB may be formed in large amounts by resting cells supplied with glucose (and certain other substances such as pyruvate, acetate and butyrate) in the absence of any nitrogen source (Macrae & Wilkinson, 1958*a*; Doudoroff & Stanier, 1959); the latter workers showed that PHB can also be formed by photo-assimilation of acetate and butyrate in *Rhodospirillum rubrum*.

To summarize, it would appear that PHB (and perhaps other lipids) resemble glycogen in that their formation in the cell is largely controlled by the nitrogen content of the environment. Continuous culture studies should assist in investigating this point further; it would also be interesting to know whether glycogen and PHB are strictly alternative reserve storage materials or whether they can both be formed in the same organism.

DISCUSSION

Although this contribution has been deliberately restricted to a few selected topics, sufficient evidence has been brought forward to illustrate how greatly the chemical composition of micro-organisms may vary in response to the chemical composition of the environment in which they are grown, or to which they are exposed after growth. Of the major cell constituents, the protein content varies least but even so may undergo variations of 50–100 %; the DNA content may vary by two- to three-fold while the RNA, polysaccharide and lipid contents may vary by ten-fold or more and may amount to anything between a few per cent and nearly half the dry weight of the cell. These data reinforce the statement made at the beginning of this review that it is quite meaningless to write down a *single* set of analytical figures and state that 'this is *the* chemical composition of micro-organism X'. The chemical composition of a micro-organism can within wide limits be what we choose to make it.

The major chemical components of the cell may be broadly divided into two main groups:

(i) storage materials (polysaccharides, lipids),
(ii) basal materials (nucleic acids, proteins).

The distinction is a crude one, but is based on the important fact that the cell can, at a pinch, do without the former altogether, while the latter are absolutely essential. There appears to be a considerable difference between the way in which the concentrations of the two types of component within the cell are affected by the environment.

Storage materials appear to be laid down within the cell whenever the external environment (*a*) contains the necessary small molecules (glucose, acetate, etc.) from which they can be synthesized, and (*b*) does *not* contain nitrogenous materials necessary for growth, or at any rate contains them only in suboptimal concentration. Synthesis of such storage materials therefore can and does take place independently of growth.

It seems highly probable that the synthesis or breakdown of such storage materials depends on the concentrations within the cell of relatively few small molecules such as phosphate, coenzymes, etc., alterations in the concentration of which can shift enzymic equilibria in the direction of synthesis or breakdown. This is made probable by the work of Holzer (1959) on the effect of small external concentrations of (NH_4^+) ions on the concentrations of phosphate, α-ketoglutarate, TPN, etc., *within* the yeast cell. The internal concentrations of all of these are very markedly affected by small traces of (NH_4^+) in the external environment, and it is obvious how the synthesis of glycogen by the enzyme phosphorylase, for example, could be affected by the internal concentration of phosphate.

Changes in the internal concentrations of the 'basal components' are affected by the external environment in a much more complex way, and one intimately bound up with cell growth; these components are, relatively speaking, inert in cells that are not growing. It is not proposed to discuss here the complex relationships between the synthesis of RNA, DNA and protein, first because to do so would inevitably re-tread much of the ground covered at the 1960 Symposium of the Society, and secondly because the writer strongly agrees with Pirie (1960) that this is a field in which speculation has far outrun facts. Only the following broad generalizations will be made.

(1) In the light of present knowledge, it seems reasonable to regard the microbial cell as *essentially* a complex autosynthetic system; unlike metazoan cells, which spend most of their lives in a non-growing state, a microbial cell is only behaving naturally when it is growing exponentially.

(2) When such a complex self-replicating system is in the process of steady-state exponential growth, its basal elements must be in a state of dynamic equilibrium with the *internal* concentrations of the numerous small molecules—amino acids, purines, pyrimidines, etc.—from which they are synthesized. Changing the *external* environment, as by placing a cell in a different medium, will inevitably cause a complex shift in the concentrations of all components of the internal environment, which in turn will react upon the rates of synthesis of the various large molecules, decreasing some rates and increasing others, thus altering the ratios of their steady-state concentrations; in other words, changing the chemical composition of the cell. Some such concept seems necessary to explain the complex inter-relationships between growth rate, chemical composition of the cell and chemical composition of the growth medium, discussed in the first section of this review.

(3) The distinction between cell growth (defined as increase in mass) and cell division has long been made. These processes, though distinct, must obviously be related. What might not have been expected, however, is the relationship disclosed by recent work between growth rate, cell size (intimately connected with the division process) and the chemical composition of the cell.

These generalizations are not particularly original and, even if approved of, may be thought too imprecise to be useful. They do suggest two conclusions, however: (i) that in future work in this field, more attention should be paid to the effect of the external environment upon the concentrations of components of the internal environment, such observations preferably being made upon cells during steady-state growth; (ii) more work needs to be done on the relations between the external environment, the composition of the cell *and the cell size*.

It is customary to conclude a review of this sort with an improving moral, and for good measure two will be provided, whose implications are not confined to the field of the review. The first is, that when studying effects of environment on micro-organisms it is a good idea to know precisely what the environment is—in other words, micro-organisms should be grown wherever possible in chemically defined media. Anyone who thinks this a truism would do well to read some of the hundreds of papers on the chemical composition of bacteria in the older literature, in which countless man-hours of painstaking analytical work are rendered virtually valueless by accompanying statements such as 'the organisms were grown in Difco Bacto-peptone horse serum broth'—or a similar decoction. It is certainly striking, on surveying the literature of this field, to note how almost all the results of real value were obtained by workers using chemically defined growth media. It may occasionally still be necessary to grow micro-organisms on decoctions of unknown composition, but how often is the reason not simple laziness?

The second moral is that useful deductions can seldom be drawn from growth experiments unless they are conducted in such a manner as to obtain *steady-state growth*. Again it is striking to notice, in the present field, how often the most valuable results have been obtained by workers who have fully realized this principle. To which may be added the rider that a method guaranteed to make it difficult to achieve steady-state growth is the usual one of placing some medium in a glass container, inoculating it with micro-organisms and allowing events to run their course.

ACKNOWLEDGEMENTS

The writer wishes to express his gratitude to his colleagues R. E. Strange, D. W. Tempest and H. E. Wade of the Microbiological Research Establishment, Porton, for allowing him to quote their unpublished results.

REFERENCES

ABBO, F. E. & PARDEE, A. B. (1960). Synthesis of macromolecules in synchronously dividing bacteria. *Biochim. biophys. Acta*, **39**, 478.

BRYSON, V. & SZYBALSKI, W. (1952). Microbial selection. *Science*, **116**, 45.

CAMPBELL, A. (1957). Synchronization of cell division. *Bact. Rev.* **21**, 263.

CLIFTON, C. E. (1946). Microbial assimilations. *Advanc. Enzymol.* **6**, 269.

DOUDOROFF, M. & STANIER, R. Y. (1959). Role of poly-β-hydroxybutyric acid in the assimilation of organic carbon by bacteria. *Nature, Lond.* **183**, 1440.

GALE, E. F. & FOLKES, J. P. (1953). Nucleic acid and protein synthesis in *Staphylococcus aureus*. *Biochem. J.* **53**, 483.

HERBERT, D. (1960). A theoretical analysis of continuous culture systems. In *Continuous Cultivation of Micro-organisms*. Soc. Chem. Ind. Monograph (in the Press).

HERBERT, D., ELSWORTH, R. & TELLING, R. C. (1956). The continuous culture of bacteria: a theoretical and experimental study. *J. gen. Microbiol.* **14**, 601.

HERBERT, D., SPURR, E., GOULD, G. W. & PHIPPS, P. J. (1961). In preparation.

HERSHEY, A. D. (1938). Factors limiting bacterial growth. II. Growth without lag in *Bacterium coli* cultures. *Proc. Soc. exp. Biol., N.Y.* **38**, 127.

HERSHEY, A. D. (1939). Factors limiting bacterial growth. IV. The age of the parent culture and the rate of growth of transplants of *Escherichia coli*. *J. Bact.* **37**, 285.

HOLME, T. (1957). Continuous culture studies on glycogen synthesis in *Escherichia coli* B. *Acta chem. scand.* **11**, 763.

HOLME, T. & PALMSTIERNA, H. (1956). Changes in glycogen and N-containing compounds in *E. coli* B during growth in deficient media. I. Nitrogen and carbon starvation. *Acta chem. scand.* **10**, 578.

HOLZER, H. (1959). Enzymic regulation of fermentation in yeast cells. In *Ciba Symp.* on *Regulation of Cell Metabolism*. London: J. and A. Churchill Ltd.

KJELDGAARD, N. O., MAALØE, O. & SCHAECHTER, M. (1958). The transition between different physiological states during balanced growth of *Salmonella typhimurium*. *J. gen. Microbiol.* **19**, 607.

KLEINZELLER, A. (1948). Synthesis of lipides. *Advanc. Enzymol.* **8**, 299.

LEMOIGNE, M. (1927). Études sur l'autolyse microbienne. Origine de l'acide β-oxybutyrique formé par autolyse. *Ann. Inst. Pasteur*, **41**, 148.

MAALØE, O. (1960). The nucleic acids and the control of bacterial growth. *Symp. Soc. gen. Microbiol.* **10**, 272.

MACRAE, R. M. & WILKINSON, J. F. (1958*a*). Poly-β-hydroxybutyrate metabolism in washed suspensions of *Bacillus cereus* and *Bacillus megaterium*. *J. gen. Microbiol.* **19**, 210.

MACRAE, R. M. & WILKINSON, J. F. (1958*b*). The influence of cultural conditions on poly-β-hydroxybutyrate synthesis in *Bacillus megaterium*. *Proc. Roy. phys. Soc. Edinb.* **27**, 73.

MAGASANIK, B., MAGASANIK, A. K. & NEIDHARDT, F. C. (1959). Regulation of growth and composition of the bacterial cell. In *Ciba Symp.* on *Regulation of Cell Metabolism*. London: J. and A. Churchill Ltd.

MALMGREN, B. & HEDÉN, C.-J. (1947). Studies on the nucleotide metabolism of bacteria. III. The nucleotide metabolism of the Gram-negative bacteria. *Acta path. microbiol. scand.* **24**, 448.

MARUYAMA, Y. & YANAGITA, T. (1956). Physical methods for obtaining synchronous culture of *Escherichia coli*. *J. Bact.* **71**, 542.

MITCHELL, P. & MOYLE, J. (1951). Relationships between cell growth, surface properties and nucleic acid production in normal and penicillin-treated *Micrococcus pyogenes*. *J. gen. Microbiol.* **5**, 421.

MONOD, J. (1942). *Recherches sur la croissance des cultures bacteriennes.* Paris: Hermann et Cie.

MONOD, J. (1949). The growth of bacterial cultures. *Annu. Rev. Microbiol.* **3**, 371.

MONOD, J. (1950). La technique de culture continue; théorie et applications. *Ann. Inst. Pasteur*, **79**, 390.

NOVICK, A. & SZILARD, L. (1950). Experiments with the Chemostat on spontaneous mutations of bacteria. *Proc. nat. Acad. Sci., Wash.* **36**, 708.

PIRIE, N. W. (1960). Biological replication considered in the general context of scientific illusion. *New Biology*, **31**, 117.

PORTER, J. R. (1946). *Bacterial Chemistry and Physiology.* New York: John Wiley and Sons.

POWELL, E. O. (1955). Some features of the generation times of individual bacteria. *Biometrika*, **42**, 16.

POWELL, E. O. (1956). Growth rate and generation time of bacteria, with special reference to continuous culture. *J. gen. Microbiol.* **15**, 492.

ROSS, K. F. A. (1957). The size of living bacteria. *Quart. J. micr. Sci.* **98**, 435.

SCHAECHTER, M., BENZTON, M. W. & MAALØE, O. (1959). Synthesis of desoxyribonucleic acid during the division cycle of bacteria. *Nature, Lond.* **183**, 1207.

SCHAECHTER, M., MAALØE, O. & KJELDGAARD, N. O. (1958). Dependency on medium and growth temperature of cell size and chemical composition during balanced growth of *Salmonella typhimurium. J. gen. Microbiol.* **19**, 592.

STRANGE, R. E., DARK, F. A. & NESS, A. G. (1961). Private communication.

WADE, H. E. (1952*a*). Observations on the growth phases of *Escherichia coli*, American type 'B'. *J. gen. Microbiol.* **7**, 18.

WADE, H. E. (1952*b*). Variation in the phosphorus content of *Escherichia coli* during cultivation. *J. gen. Microbiol.* **7**, 24.

WADE, H. E. & MORGAN, D. M. (1961). Private communication.